Resource Physiology of Conifers

Acquisition, Allocation, and Utilization

Edited by

William K. Smith
Department of Botany
University of Wyoming
Laramie, Wyoming

Thomas M. Hinckley
College of Forest Resources
University of Washington
Seattle, Washington

Academic Press

San Diego New York Boston London Sydney Tokyo Toronto

Cover photograph: Highly variable cone production in Engelmann spruce exemplifies the importance of seasonal resource acquisition and utilization in conifer tree species that experience the short growth seasons of high mountains and latitudes.

This book is printed on acid-free paper. ∞

Academic Press, Inc.
A Division of Harcourt Brace & Company
525 B Street, Suite 1900, San Diego, California 92101-4495

United Kingdom Edition published by
Academic Press Limited
24-28 Oval Road, London NW1 7DX

Library of Congress Cataloging-in-Publication Data

Resource physiology of conifers : acquisition, allocation, and utilization / edited by William K. Smith, Thomas M. Hinckley.
p. cm. -- (Physiological ecology)
Includes bibliographical references and index.
ISBN 0-12-652870-5
1. Conifers--Physiology. I. Smith, William K. (William Kirby), date. II. Hinckley, Thomas M. III. Series.
QK494.R47 1994
585' .2041--dc20 94-20834
CIP

PRINTED IN THE UNITED STATES OF AMERICA
94 95 96 97 98 99 QW 9 8 7 6 5 4 3 2 1

Contents

Contributors ix
Preface xi

Part I
Resource Acquisition

1. Photosynthetic Light Capture and Processing from Cell to Canopy
P. Stenberg, E. H. DeLucia, A. W. Schoettle, and H. Smolander
I. Introduction 3
II. Factors Influencing the Photosynthetic Response of Needles to Light 5
III. Interactions between Structure and Photosynthetic Light Response at Different Levels 18
IV. Concluding Remarks 31
References 32

2. Water and Nutrient Acquisition by Roots and Canopies
Ram Oren and David W. Sheriff
I. Introduction 39
II. Methodology 40
III. Fundamentals 41
IV. Uptake and Use of Water and Nutrients 58
References 67

3. Efficiency of Nutrient Acquisition by Fine Roots and Mycorrhizae
R. D. Yanai, T. J. Fahey, and S. L. Miller
I. Strategies of Tree Root Systems 75
II. A Model of Fine Root Efficiency 76
III. Model Limitations 82
IV. Root Longevity 85
V. Root Diameter 92

VI. Efficiency of Mycorrhizal Hyphae 94
VII. Summary 98
References 99

4. External and Internal Factors Regulating Photosynthesis
Robert O. Teskey, David W. Sheriff, David Y. Hollinger, and Richard B. Thomas

I. Introduction 105
II. Framework 106
III. Direct Effects 109
IV. Indirect Effects 117
V. Consequences for Growth and Survival 123
VI. Integration of Photosynthetic Rate over Time 124
VII. Scaling Photosynthetic CO_2 Exchange from Foliage to Canopies and Regions 126
References 131

Part II
Resource Allocation and Utilization

5. Resource Use Efficiency
D. W. Sheriff, H. A. Margolis, M. R. Kaufmann, and P. B. Reich

I. Introduction 143
II. Issues of Scaling 146
III. Resource Use Efficiencies at Various Spatial Scales 150
IV. Discussion 172
Notation of Symbols 173
References 174

6. Source–Sink–Storage Relationships of Conifers
R. J. Luxmoore, R. Oren, D. W. Sheriff, and R. B. Thomas

I. Introduction 179
II. Source–Sink Relationships for Five Conifer Species 180
III. Source–Sink–Storage Interactions 189
IV. Conifer Development and Source–Sink Relationships 197
V. Environmental Influences on Sources and Sinks 200
VI. General Discussion 205
References 208

7. Carbon Allocation and Accumulation in Conifers
Stith T. Gower, J. G. Isebrands, and David W. Sheriff

I. Introduction 217
II. Distribution of Conifer Forests 218

III. Controls on Carbon Assimilation 219
IV. Carbon Allocation Patterns in Conifers 222
V. Controls on Carbon Allocation 226
VI. Changes in Carbon Dynamics during Stand Development 238
VII. Multiple Resource Limitations 239
VIII. Future Research 242
References 244

8. Respiration from the Organ Level to the Stand

Douglas G. Sprugel, Michael G. Ryan, J. Renée Brooks, Kristiina A. Vogt, and Timothy A. Martin

I. Introduction 255
II. Foliar Respiration 261
III. Woody-Tissue Respiration (Stems, Branches, and Coarse Roots) 270
IV. Fine Root Respiration 277
V. Proxy Methods for Estimating Respiration without Tissue-Specific Measurements 283
VI. Conclusions 290
References 291

9. Water Transport Dynamics in Trees and Stands

S. G. Pallardy, J. Čermák, F. W. Ewers, M. R. Kaufmann, W. C. Parker, and J. S. Sperry

I. Water Absorption 301
II. Internal Transport and Storage 315
III. Water Stress Problems Unique to Conifers 327
IV. Transpiration: Trees to Stands 329
V. Physiological Response of Conifers to Water Deficits 349
VI. Conclusions 367
References 368

Index 391

Contributors

Numbers in parentheses indicate the pages on which the authors' contributions begin.

J. Renée Brooks[1] (255), College of Forest Resources, University of Washington, Seattle, Washington 98195

J. Čermák (301), Institute of Forest Ecology, Forestry Faculty-Agricultural University, 61300 Brno, Czech Republic

E. H. DeLucia (3), Department of Plant Biology, University of Illinois, Urbana, Illinois 61801

F. W. Ewers (301), Department of Botany and Plant Pathology, Michigan State University, East Lansing, Michigan 48824

T. J. Fahey (75), Department of Natural Resources, College of Agriculture and Life Sciences, Cornell University, Ithaca, New York 14853

Stith T. Gower (217), Department of Forestry, University of Wisconsin–Madison, Madison, Wisconsin 53706

David Y. Hollinger (105), Manaaki Whenua–Landcare Research, Christchurch, New Zealand

J. G. Isebrands (217), Forestry Sciences Laboratory, United States Department of Agriculture Forest Service, Rhinelander, Wisconsin 54501

M. R. Kaufmann (143, 301), Rocky Mountain Forest and Range Experiment Station, United States Department of Agriculture Forest Service, Fort Collins, Colorado 80526

R. J. Luxmoore (179), Oak Ridge National Laboratory, Environmental Sciences Division, Oak Ridge, Tennessee 37831

H. A. Margolis (143), Department of Forest Science, University of Laval, Sainte-Foy, Quebec, Canada G1K 7P4

Timothy A. Martin (255), College of Forest Resources, University of Washington, Seattle, Washington 98195

S. L. Miller (75), Department of Botany, University of Wyoming, Laramie, Wyoming 82071

Ram Oren (39, 179), School of the Environment, Duke University, Durham, North Carolina 27706

S. G. Pallardy (301), School of Natural Resources, University of Missouri, Columbia, Missouri 65211

[1]Present address: Department of Biology, University of Utah, Salt Lake City, Utah 84112.

W. C. Parker (301), Ontario Ministry of Natural Resources, Sault Sainte Marie, Ontario, Canada P6A 2M6

P. B. Reich (143), Department of Forest Resources, University of Minnesota, St. Paul, Minnesota 55108

Michael G. Ryan (255), Rocky Mountain Forest and Range Experiment Station, United States Department of Agriculture Forest Service, Fort Collins, Colorado 80526

A. W. Schoettle (3), Rocky Mountain Forest and Range Experiment Station, United States Department of Agriculture Forest Service, Fort Collins, Colorado 80526

David W. Sheriff (39, 105, 143, 179, 217), Plantation Forest Research Centre, Division of Forestry, CSIRO, Mount Gambier, South Australia 5290, Australia

H. Smolander (3), Suonenjoki Research Station, The Finnish Forest Research Institute, SF-77600 Suonenjoki, Finland

J. S. Sperry (301), Department of Biology, University of Utah, Salt Lake City, Utah 84112

Douglas G. Sprugel (255), College of Forest Resources, University of Washington, Seattle, Washington 98195

P. Stenberg (3), Department of Forest Ecology, University of Helsinki, SF-00014 Helsinki, Finland

Robert O. Teskey (105), Daniel B. Warnell School of Forest Resources, University of Georgia, Athens, Georgia 30602

Richard B. Thomas (105, 179), Department of Botany, Duke University, Durham, North Carolina 27706

Kristiina A. Vogt (255), Yale School of Forestry and Environmental Studies, New Haven, Connecticut 05611

R. D. Yanai (75), Boyce Thompson Institute for Plant Research, Cornell University, Ithaca, New York 14853

Preface

Coniferous forests may be the most economically important of our native plant communities. This importance stems primarily from the demand for building and paper products, but also from an increasing awareness of the value of preserving native forests for scientific, aesthetic, and recreational purposes. Both the field of physiological plant ecology and our view of coniferous forests have changed markedly over the past few decades. There have been dramatic discoveries in the ecological application of molecular genetics and the recognition that ecosystems are complex systems of the "global habitat." Our view of the forest has gone from largely commodity extraction and exploitation to an appreciation of the social and global roles that forests play. Terms such as carbon sequestration, sustainability, ecosystem management, and new forestry are frequently discussed by the media. Concomitant with the emergence of these new paradigms is the realization that a broader view is necessary. As stated by Allen and Hoekstra (1992) in their book entitled *Toward a Unified Ecosystem* (Columbia University Press, New York), "For any level of aggregation, it is necessary to look both to larger scales to understand the context and to smaller scales to understand mechanism; anything else would be incomplete." Physiological ecology emphasizes studies on the organism level and, thus, is directly linked to environmental adaptation and evolution. As a result, this research area provides a focal point from which one can scale upward to the ecosystem level or downward to the molecular level. The integration of the genetic mechanisms involved in an organism's response to its environment with the impact of changes in individual species on ecosystem dynamics spans the full breadth of the biological hierarchy. It is this biological "scaling" that is currently recognized as crucial to our ultimate understanding of such complex, contemporary issues as environmental change on a global scale.

The most general objective of this book is to provide a more synthetic view of the resource physiology of conifer trees with an emphasis on developing a perspective that can integrate across the biological hierarchy. This objective is in concert with more scientific goals of maintaining biological diversity and the sustainability of forest systems. Current management practices of our national and state agencies now reflect a more ecosystem-oriented approach. Regardless of the motivation, the preservation of coniferous forest ecosystems, in the face of important anthropogenic influences such as global climate change, is a major concern today. Without a basic understanding of the adaptive responses of individual conifer forest species, neither the molecular mechanisms of the response capability nor the impact at the ecosystem level can be evaluated. The response capabilities of an organism, as emphasized in physiological ecology, are the evolutionary mechanisms that establish a bridge between the molecular and ecological levels of the vast biological spectrum.

The ideas for the subject, content, and central theme of this book originated at a September 1991 meeting near Jackson Hole, Wyoming, at the University of Wyoming and the U.S. National Park Service Research Station. Our efforts at this workshop and since then have resulted in two companion volumes. This volume deals with the topics of resource acquisition, allocation, and utilization in conifers. The companion volume (*Ecophysiology of Coniferous Forests*) includes numerous other topics central to the field of conifer physiological ecology. We hope that the two books will provide a synthetic and contemporary view of the most recent and current research being undertaken in coniferous forests. We thank the staff at the Jackson Hole Research Station for their hospitality and for letting us share such a marvelous setting. We also thank the National Science Foundation, the U.S. Department of Energy, the U.S. Park Service, the U.S. Forest Service, and the Paper and Wood Products Industry for their support of the meeting and subsequent production of this book.

William K. Smith
Thomas M. Hinckley

I

Resource Acquisition

1

Photosynthetic Light Capture and Processing from Cell to Canopy

P. Stenberg, E. H. DeLucia, A. W. Schoettle, and H. Smolander

I. Introduction

Several aspects of needle and shoot structure, by influencing the photosynthetic response to irradiance, distinguish the function of evergreen coniferous trees from their broad-leafed counterparts. Some of these attributes, including needlelike foliage, the prolonged lifetime of the photosynthetic unit, and the complex geometrical arrangement of needles on shoots, require a different conceptual and methodological approach when addressing the photosynthetic light response of conifers. Moreover, these attributes influence the photosynthetic light response differently at various spatial and temporal scales. In this chapter we synthesize current understanding of the photosynthetic response to irradiance of conifers at different scales, and contrast functional and methodological characteristics of conifers and broad-leafed trees.

At the most elementary level, the photosynthetic organ of conifers, the needle, is functionally quite distinct from broad mesophytic leaves. Needles are optically nearly black (they transmit no visible light), inherently three-dimensional, and functionally symmetrical with regard to light harvesting and photosynthesis. The three-dimensional nature of conifer needles gives rise to the historical dilemma in ecophysiological research as to the appropriate denominator for expressing photosynthetic rates—one-sided, projected, or total leaf area? Sorting this issue out is central to any consideration of the evolutionary consequences of needle versus leaf structure.

Similarly, the extreme longevity of most conifer needles, often lasting from 4 to greater than 15 years, raises many special considerations for the influence of environmental factors on photosynthesis in conifers. Because of their longevity, needles that develop in high light levels may spend most of their lives ultimately operating in a shaded environment as the branch continues to elongate. Thus, individual needles must retain considerable capacity to acclimate to changing conditions during their prolonged life. Additionally, needles must withstand the characteristic environmental stresses of all four seasons, not just the spring and summer, which define the "growing season" for most broad-leafed species. The ability of conifer needles to tolerate or acclimate to the extreme cold of winter and resist photoinhibitory damage of photosynthesis, among other stresses, is discussed here as well as in Chapter 4 of this volume (see also Havranek and Tranquillini, 1994).

At the next level of spatial scale is the shoot, representing the aggregate arrangement of needles on a branch. The shoot is the basic element of light capture in conifers, in dramatic contrast to broad-leafed trees whereon the individual leaf provides the analogous function. When considering the photosynthetic response to irradiance, how does one measure the driving variable, light, for a complex shoot? Ultimately, the photosynthetic rate of the shoot is the integrated response of the individual needles, but light absorption of the shoot changes with the angle of incident light. Thus, to define the photosynthetic light response of a shoot a suitable scalar measure of light must be chosen. This measure must be stable in the sense that it reflects true differences in shoot geometry or in the physiological properties of the needles, not just changes in the angle of incident light. The mean irradiance, defined as the total intercepted irradiance divided by the total (all sides) leaf area, may be the appropriate driving variable for the shoot-level photosynthetic response to irradiance.

Characteristic differences in the mechanisms and measurement of light capture for photosynthesis of evergreen conifers and hardwoods are evident at all spatial scales, including the canopy and stand level. At the canopy level, for which statistical functions are typically used to describe light penetration, the clustering of needles on shoots and the nonuniform grouping of shoots in the canopy violate the assumption of random leaf display implicit in these statistical models. In some cases light attenuation along a shoot can be as great as it is through the canopy. We present some of the strong inroads that have been made into the understanding of light capture and photosynthesis for conifers, but recognize that more research in this area is needed to unravel the evolutionary significance of the conifer life form and to construct adequate predictive models of productivity.

II. Factors Influencing the Photosynthetic Response of Needles to Light

A. Interaction between Needle Structure and Light Absorption

The ability to harvest light for photosynthesis is determined by the interplay between the optical properties of leaves and their physiological–biochemical potential. A number of prominent structural features of the leaves of gymnosperms influence their optical properties, and interpretation of the photosynthetic response to incident photosynthetically active radiation (PAR; 400–700 nm) relies on an understanding of the interaction between leaf structure and function. With the exception of *Ginkgo*, *Larix*, and *Taxodium*, the leaves of most gymnosperms are needlelike and evergreen. Members of the Coniferales have been studied most intensively (Chamberlain, 1966), and the needles of species in this group are xeromorphic. This morphology is characterized by thick needles with low surface-area-to-volume ratios, a well-developed cuticular layer with abundant epicuticular waxes, a thick-walled epidermis often subtended by one or more layers of thick-walled hypodermal cells, and, in many cases, the presence of symmetrical photosynthetic mesophyll on ad- and abaxial surfaces (Esau, 1977). This suite of structural attributes has been alternatively interpreted as an adaptation to drought, nutrient deficiency, mechanical stress, or herbivory. In the context of light, these features influence the optical characteristics of conifer needles and therefore the ability to absorb light for photosynthesis.

Although much is known about the morphology and histology of conifer needles and reasonable efforts have been made to characterize the photosynthetic responses of many in this group, few attempts link needle structure and photosynthetic performance. In this section we explore the relationship between some relevant aspects of needle morphology and the ability to harvest light for photosynthesis. Our approach is to follow photons from their first interaction with the needle surface to their ultimate absorption by the pigment beds of the photosynthetic reaction centers. Where few data are available for conifer needles, liberal use is made of analogous properties of angiosperm leaves.

Light incident on a needle interacts first with a profuse covering of epicuticular waxes that is characteristic of needles of many coniferous species. These waxes may be tubular or platelike, and may be uniformly distributed or in tufts. The distribution, structure, and composition of epicuticular waxes are highly variable among species and may be altered by environmental conditions (Baker, 1974; Gunthardt and Wanner, 1982; Hunt and Baker, 1982; DeLucia and Berlyn, 1984). The stomatal antechamber of many coniferous species is completely occluded with anastomosing strands of tubular epicuticular waxes. In addition to affecting the diffusion of carbon dioxide and water vapor (Jeffree *et al.*, 1971), these waxes influence the spectral quality and quantity of PAR penetrating the leaf and reaching chromophores in the stomatal guard cells.

Spectra of epicuticular waxes dissolved in organic solvents show significant absorption below 400 nm (Bornman and Vogelmann, 1991), and these waxes may therefore provide screening against potentially damaging UV-B (280–320 nm) radiation (see Chapter 4). Absorption of PAR by epicuticular waxes is minimal; reflectance from 400 to 700 nm, however, can be substantial. For green needles of Douglas fir (*Pseudotsuga menziesii*), blue-green needles of Colorado spruce (*Picea pungens*), and blue-white needles of blue spruce (*P. pungens* var. *hoopsii*), reflectance was maximum in the PAR at 540 nm once the epicuticular waxes were removed (Clark and Lister, 1975a). Between 400 and 540 nm reflectance was significantly greater for blue spruce and Colorado spruce than for Douglas fir. These differences in reflectance alter the action spectrum for photosynthesis. The ratio of photosynthesis measured during irradiation with 455-nm light (blue) versus 619-nm light (red) was 0.60 for Douglas fir, and 0.21 and −0.28 for Colorado and blue spruce, respectively (Clark and Lister, 1975b). The negative ratio for blue spruce indicates that photosynthesis in blue light was below the light compensation point. Because of their structure, light reflected or transmitted through epicuticular waxes is probably highly diffuse, which may also alter penetration into the needle (Donahue, 1991).

Light transmitted or forward-scattered through the epicuticular waxes interacts with the needle cuticle and the epidermal–hypodermal layers. Relatively little is known about the influence of these layers on light penetration into the mesophytic leaves of angiosperms, and few examinations have been made of the optical characteristics of these layers in conifers. Measurements of light penetration through the cuticle and cutinized cell wall of several coniferous species indicate that these layers contain pigments with strong absorption below 320 nm but transmit light at longer wavelengths (Strack *et al.* 1988, 1989; DeLucia *et al.*, 1991a; Day *et al.*, 1992). The convex outer walls of epidermal cells of many angiosperms are effective lenses and may concentrate light within the leaf to as much as 20 times incident levels (Haberlandt, 1914; Bone *et al.*, 1985; Poulson and Vogelmann, 1990). It is postulated that light focusing by the epidermis may increase photosynthetic rates of localized groups of chloroplasts and may also direct chloroplast movements (Poulson and Vogelmann, 1990, and references therein). The needle epidermis of most coniferous species is, however, relatively planar, and it is unlikely that these cells produce a significant lens effect. Moreover, the light environment of the shoot may be dominated by isotropic or diffuse light, which is not focused by lenses.

Epidermal cells of conifer needles are thick walled with a greatly reduced cell lumen (Esau, 1977). These cells often are subtended by a structurally similar hypodermal layer (which may be one or more cell layers thick). The hypodermal layer is not continuous in all species. In *Abies balsamea*, for example, the hypodermis is most extensively developed in the midrib and in the sides of the needles in transverse section (DeLucia and Berlyn, 1984). The epidermal–hypodermal layers give conifer needles their characteristic rigidity but also carry a cost in terms of light penetration. Cell wall microfibrils scatter light intensely, and this scattering may contribute to nonchlorophyll absorption in the epidermis and to reflectance of PAR out of the leaf (McClendon, 1984). Direct measurements of the attenuation of 680-nm light with fiber-optic microprobes inserted in needles of Engelmann spruce (*Picea engelmannii*) and subalpine fir (*Abies lasiocarpa*) indicate that 20 to 40% of incident irradiance is attenuated in the epidermis–hypodermis (Fig. 1). These measurements were not adjusted for surface reflectance and propagation of forward-scattered light, and may therefore overestimate attenuation in the epidermal layer, which is nonetheless substantial.

As is typical of leaves of mesophytic angiosperms, the photosynthetic mesophyll in needles of some species of conifers is differentiated into palisade and spongy tissues. This differentiation occurs in *Abies*, *Sequoia*, *Taxus*, and *Torreya*, but in *Pinus*, *Picea*, and others the mesophyll is undifferentiated (Esau, 1977). The development of fiber-optic microprobes that can be inserted into leaves has permitted examination of the influence of structural attributes of the mesophyll on light propagation (Vogelmann *et al.*, 1991). The distribution of forward-transmitted light through spruce and fir needles is highly nonlinear (Fig. 1) and similar to the pattern of light attenuation observed in more mesophytic leaves. Although superficially resembling light attenuation predicted by the Beer–Lambert law, the distribution of pigments in the mesophyll is nonuniform and transmission is determined by the combined influence of absorbance, scattering, and the sieve effect. Thus, the Beer–Lambert law does not hold for light penetration through leaves.

As needles of spruce and fir develop, the depth of penetration of 680-nm light decreases (Fig. 1). This wavelength is near the absorption

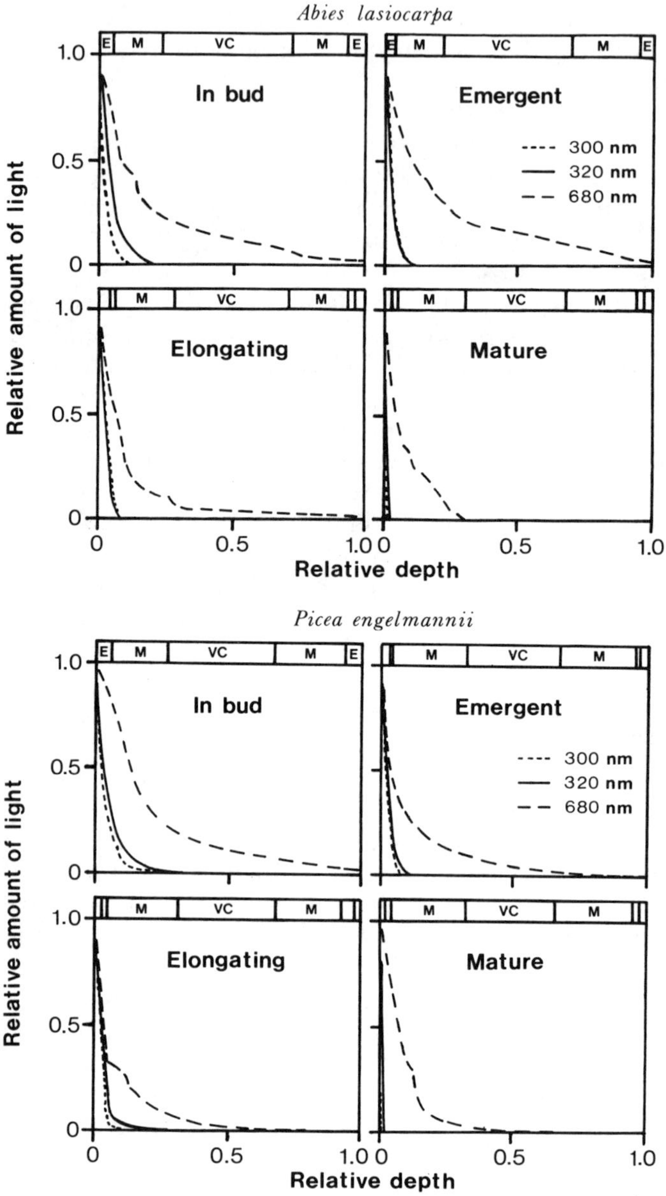

Figure 1 Penetration of UV and visible light into needles of *Abies lasiocarpa* and *Picea engelmannii.* E, Epidermis; M, mesophyll; VC, vascular cylinder.

peak for chlorophyll, and, by the time needles become mature, forward-propagating 680-nm light is almost completely absorbed before reaching the vascular cylinder in the middle of the needle. Scattered light, particularly at less strongly absorbed wavelengths, penetrates somewhat farther (Bornman and Vogelmann, 1991; Donahue, 1991). Spruce and fir needles are structurally asymmetrical in transverse section; however, the thickness and histology of mesophyll cells on the adaxial and abaxial sides of the vascular cylinder are similar and the profiles of light penetration are similar for both sides of the needle. As a result of the rapid attenuation of PAR through the mesophyll, the distal portion of the mesophyll relative to incident irradiance receives little light during unilateral irradiation, and cells in this portion of the mesophyll probably operate below the photosynthetic light compensation point. In contrast, average transmittance in the PAR of mesophytic angiosperm leaves is in the range of 5 to 10%, indicating significant irradiance reaches the distal spongy mesophyll. The steep gradient of light inside relatively thick conifer needles profoundly influences the characteristics of the photosynthetic response to irradiance.

B. Photosynthetic Light Response and Needle Structure

It is customary to express the rate of photosynthesis (P) per unit of leaf area (L) as a function of the irradiance (I) of PAR incident on the leaf surface. The photosynthetic response to incident irradiance is characterized by five regions that are influenced by different physiological attributes of leaves. (1) In the dark the rate of CO_2 evolution is influenced by the rate of mitochondrial respiration. (2) At low irradiances ($\leqslant 25\ \mu mol/m^2/sec$) there is an initial nonlinearity in the response, referred to as the Kok inflection. This inflection is caused by a combination of the effects of changing intercellular CO_2 on the measured rates of photosynthesis and a down-regulation of mitochondrial respiration in the light (Kirshbaum and Farquhar, 1987, and references therein). (3) Above the Kok inflection is a region in which photosynthesis increases linearly with irradiance. The slope of this linear region is the apparent quantum yield ("apparent" if light is measured as incident rather than as absorbed irradiance), and is influenced by leaf absorptance and the efficiency of the primary photochemical reactions and photosynthetic electron transport. (4) Above the linear region and below the light saturation is a convex region where absorptance and the internal distribution of light in the leaf influence the shape of the curve. (5) At high irradiance the rate of photosynthesis is saturated, and the rate at which saturation occurs is determined, in large part, by the biochemical capacity of the leaf and diffusional limitations to CO_2 flux. Leaf structure profoundly influences the degree of convexity of the light response.

The shape of the photosynthetic response to incident irradiance of isolated mesophyll cells (Terashima and Saeki, 1985) and of deep shade leaves of some angiosperms (DeLucia *et al.*, 1991c) resembles a Black-

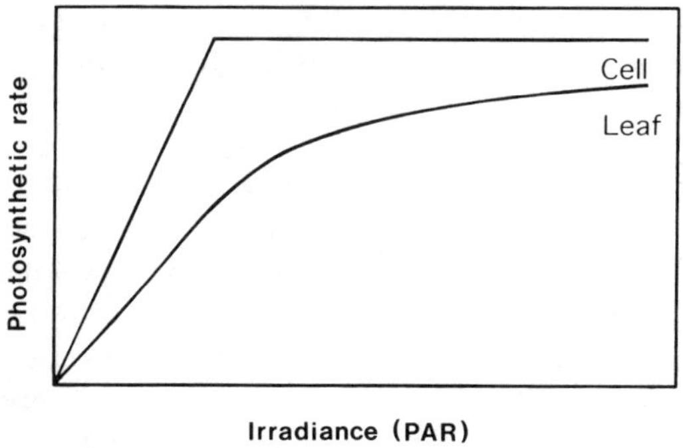

Figure 2 Response function of a leaf (lower curve) as a function of incident photon irradiance, assuming a Blackman response (upper curve) of the mesophyll cell. (From Lappi and Oker-Blom, 1992.)

man response. Here the rate of photosynthesis increases monotonically with increasing irradiance up to a sharp maximum, and then saturates (Fig. 2). For sun-type leaves and shoots (discussed in the next section), the initial linear region is short and is followed by a more gentle curve to the light-saturated photosynthetic rate. For individual needles this degree of curvature is strongly influenced by the light gradient inside leaves.

To quantify the convexity of the photosynthetic light response, several authors have used a mathematical model that includes an explicit term describing the degree of curvature (Prioul and Chartier, 1977; Terashima and Saeki, 1985; Leverenz, 1987; Zhang, 1989). In its general form this model (the nonrectangular hyperbola) is given by Eq. (1):

$$\Theta P^2 - (\alpha I + P_{max})P + \alpha I P_{max} = 0. \tag{1}$$

The sharpness (convexity) of the curve is described by Θ, which varies from 1 for a Blackman response to 0 for smooth rectangular hyperbola. The gross photosynthetic rate, apparent quantum yield, incident irradiance, and the maximum rate of gross photosynthesis at light saturation are designated as P, α, I, and P_{max}, respectively. The addition of a term for respiration converts assimilation rates to net rather than gross photosynthesis.

The effect of self-shading by more surficial chloroplasts on the photosynthetic response of a leaf can be illustrated as follows. Assume, for simplicity, that the leaf is sliced into "infinitely thin mesophyll layers" with area (L) equal to the (one-sided) leaf surface area. The irradiance incident on the layer may be written as cI, where I is the irradiance on the leaf surface and c denotes the effective proportion of I arriving at the layer. Let α be the initial slope and β the saturation point of a Blackman curve describing the photosynthetic response of such a layer. The rate of photosynthesis (p_l) of the layer, expressed as a function of I, is then given by Eq. (2):

$$p_l(I) = \begin{matrix} \alpha c I & \text{if} \quad I < \beta/c \\ \alpha\beta & \text{if} \quad I \geq \beta/c \end{matrix}; \tag{2}$$

i.e., it is a Blackman function with slope αc and saturating irradiance β/c.

The rate of photosynthesis (P) of the leaf equals the expected value of $p_l(I)$ with respect to the joint density function of α, β, and c, and is a concave function with initial slope equal to $E(\alpha c)$ and $P_{max} = E(\alpha\beta)$ (Fig. 2). The curve is linear for $I \leq (\beta/c)_{min}$, saturates at $I = (\beta/c)_{max}$, and has a nonlinear part between these values (Lappi and Oker-Blom 1992). Thus, the variation in c (the amount of absorbed PAR by chloroplasts) and differences in their photosynthetic characteristics (parameters α and β) decrease the curvature ("convexity") of leaf photosynthesis as a function of the irradiance incident on the leaf surface.

Mesophytic leaves of angiosperms may be structurally bifacial (dorsiventral) or unifacial. The adaxial epidermis is subtended by palisade mesophyll and the abaxial epidermis is subtended by spongy mesophyll in bifacial leaves. This asymmetry in leaf structure contributes to higher photosynthetic rates when bifacial leaves are irradiated on the adaxial surface than when irradiated on the abaxial surface (DeLucia *et al.*, 1991a). Unifacial leaves, in contrast, have palisade mesophyll under the adaxial and abaxial mesophyll, and the spongy layer may be reduced or absent. This type of structure is often observed in vertically oriented leaves of dicots and monocots, and results in similar photosynthetic rates during adaxial and abaxial irradiation. This symmetry in the photosynthetic response to light may be an adaptation enabling vertical leaves to maximize carbon gain under diffuse light conditions while minimizing water loss (Poulson and DeLucia, 1993). The steep light gradient in conifer needles causes them to function like unifacial leaves even though needle anatomy may be asymmetrical. The mesophyll of Engelmann spruce needles is arranged in symmetrical radial layers of palisade cells. In spite of gross asymmetry caused by a prominent midrib, the mesophyll of subalpine fir needles attenuates 680-nm light almost completely on each side of the leaf. Thus, the different leaf surfaces of spruce and fir rely on light incident on each surface and are functionally independent. This functional independence results in a low degree of convexity during unilateral irradiation and suggests that needles are adapted to highly isotropic (diffuse) light conditions.

Leverenz (1987) compared the photosynthetic response to incident irradiance of planar arrays of needles of several conifer species. With the exception of *Picea abies*, needles of *Picea mariana*, *Pinus sylvestris*, and

Picea sitchensis measured with unilateral irradiation incident on the adaxial surface had values of Θ from 0.44 to 0.58, indicating a gradual transition from linear to light-saturated regions of the photosynthetic light response. This gradual transition is caused by self-shading of chloroplasts within needles as indicated by the steep attenuation of PAR as it passes through the mesophyll. Adaptation of chloroplasts to their localized light environment within the mesophyll (Terashima and Inoue, 1985; Poulson and DeLucia, 1993) would minimize the effects of self-shading on convexity, but this does not seem to compensate fully for the steep light gradient observed in conifer needles. Bilateral irradiation or uniform irradiation in an integrating sphere increased Θ to >0.90 for these species (Leverenz, 1987). Thus, under unilateral irradiation many of the distal cells in conifer needles are operating below their light compensation point.

A number of the features that give conifer needles their characteristic toughness decrease the amount and alter the spectral quality of light penetrating to the photosynthetic mesophyll. Their thickness and the rapid attenuation of light in the mesophyll also reduce the efficiency with which conifer needles harvest unilateral light. However, the complex shoot structure and capacity to harvest light from many directions enhance the ability to function in a diffuse light environment. We remind the reader that many of the statements regarding the structure and function of conifer needles are deduced from the situation with mesophytic leaves. Estimates of the amount, quality, and directionality of light within conifer needles are largely lacking.

C. Structural and Functional Acclimation to Shade

In addition to the influence of structural features of conifer needles on the photosynthetic light response discussed above, the photosynthetic light response can also vary within and among conifer species. The variable light environment within a forest stand and individual crown can significantly influence the photosynthetic light response by altering the structure of the leaves as well as their biochemistry. This section will review the effects of the structural and functional acclimation responses to shade and their implications on photosynthetic light response.

Conifer species are adapted for growth under a specific range of light conditions. Species vary in their capacity to acclimate to different irradiances within their tolerance range. These light ranges dictate the conditions under which the species can survive and affect their position along the successional series. Those species that can live only in high-light environments (shade-intolerant species) tend to be pioneer species, and those that can tolerate shade (shade-tolerant species) will tend to be late successional species. The photosynthetic light response of the leaves is not the only factor that contributes to the shade tolerance of a species. Other factors, such as carbon and nutrient allocation patterns and drought tolerance, also affect the shade tolerance of the species, yet are beyond the scope of this chapter. This discussion will be limited to the photosynthetic light response aspects of shade acclimation at the leaf scale.

Structurally, leaves developed in high light levels (sun leaves) have additional layers of mesophyll (Turrel, 1936; Nobel *et al.*, 1975), causing the needles to be more robust and to have a lower specific leaf area (SLA; cm^2/g) than shade leaves. The SLA of a leaf is inversely related to the daily sunlight under which the leaf developed within the canopy (Del Rio and Berg, 1979; Björkman, 1981; Lewandowska and Jarvis, 1977; Gutschick and Wiegel, 1988; Hager and Sterba, 1985). Because sun leaves have a greater number of cell layers compared to shade leaves, the convexity of the photosynthetic light response curve is less, as discussed in previous sections. In addition, the light-saturated rate of net photosynthesis, on a projected leaf area basis, is inversely related to specific leaf area (Oren *et al.*, 1986).

Because SLA varies within the crown [increases with depth from 121 to 293 cm^2/g for *Pinus radiata* (Rook *et al.*, 1987)], it is especially important to specify the basis (dry weight or leaf surface area) on which biochemical processes are reported and compared (Smith *et al.*, 1991). Chlorophyll concentration on an area basis is higher in sun (upper crown) than in shade (lower crown) leaves (Tucker and Emmingham, 1977; Lewandowska and Jarvis, 1977), yet is the same in leaves throughout the crown when expressed on a leaf dry weight basis in *P. radiata* (Wood, 1973). The ratio of chlorophyll *a* to *b* decreases with depth in the crown, suggesting that production of light-harvesting pigments is greater in shade than in sun leaves.

Physiologically, leaves are said to be shade acclimated if shade leaves fix more carbon than sun leaves under low irradiance (Björkman, 1981). This is often achieved by a lower respiration rate in shade leaves compared to sun leaves, resulting in a shift in the linear portion of the net photosynthetic light response curve toward lower light levels (and a lower light compensation point) rather than a change in the slope (quantum yield) of that portion of the relationship (Ehleringer and Björkman, 1977). In addition, the PAR required to saturate net photosynthesis and the light-saturated rate of photosynthesis are commonly greater for leaves developed under high irradiance compared to those grown under low irradiance (Boardman, 1977).

The characteristics of sun and shade leaves discussed above are a result of leaves developing in different light environments, however, mature plant tissues can also experience different light environments over time. For example, leaves become increasingly shaded as they age due to natural crown development (Schoettle and Smith, 1991) or to competitors. The foliar biomass of a conifer can be the accumulation of 1.5 to up to 40 years of leaf production, therefore the capacity for altering the photosynthetic light response of mature leaves over time would provide an efficient means of keeping old leaves productive in their progressively shaded environment.

Physiological changes in mature nonconifer leaves with shading and age have been attributed to shade acclimation and resource relocation within the crown (Field, 1981; Hirose and Werger, 1987). In conifers, the light-saturated photosynthesis (*Pinus contorta*; Schoettle, 1990) and respiration (*Abies amabilis*; Brooks *et al.*, 1991; *P. contorta*; A. W. Schoettle, unpublished data, 1989) decrease with leaf age, as would be expected in response to shading. Although the photosynthetic efficiency of C_3 plants has been shown to be relatively constant among species (Ehleringer and Björkman, 1977) and light environments, no studies have been done to test the effect of shading of mature conifer leaves on quantum yield. Chlorophyll concentration on an area basis was unchanged with leaf age in *P. radiata* (Wood, 1973), yet would have been predicted to increase if shade acclimation was occurring. There is some indication that the chlorophyll *a*:*b* ratio does shift in mature leaves with shading in *A. amabilis*, suggesting some degree of shade acclimation with age (Brooks, 1993).

D. Environmental Factors Affecting the Photosynthetic Light Response

Photosynthesis can be affected by environmental factors other than light at the capacity or performance level. The photosynthetic capacity of the leaf is a function of the amount of biochemical machinery present in the leaf; photosynthetic performance is the expression of that capacity within the current environmental constraints. For example, it has been known for many years that water stress causes stomatal closure, thereby reducing CO_2 availability to the chloroplasts and reducing the rate of photosynthesis (performance). At the same water potentials that cause stomatal closure, the photosynthetic capacity can also be affected, i.e., the quantum yield of photosystem II is reduced as is the light-saturated rate of photosynthesis (Björkman and Powles, 1984).

The photosynthetic capacity of a leaf is largely influenced by membrane characteristics (photochemical reactions) and enzyme activity (carbon fixation cycle), and both of these are sensitive to temperature. As a result, most of the research on photosynthetic light response in relation to environmental factors has focused on temperature effects. In this section we focus on the effects of temperature on four of the five phases of the photosynthetic light response relationship described previously: dark respiration, linear low-light response (quantum efficiency), the transition from the linear portion to the light-saturated rate (convexity), and the light-saturated rate of photosynthesis.

Most conifer leaves are long-lived, such that they persist through all four seasons of the year in the temperate zone. As a result, they experience temperature variation both within and among seasons. The effects of temperature on the photosynthetic light response of a leaf are nonuniform throughout the year. In this section the effects of temperature during the growing and dormant periods of the year are discussed.

The optimal temperature for net photosynthesis can vary with species, ecotype, site, and time of the year (Larcher, 1983). The temperature conditions under which the population has adapted may or may not have a profound effect on the temperature optimum for net photosynthesis (Fig. 3). Normal variation in the air temperature within a growing season does not affect the photosynthetic capacity of leaves to a large extent, but due to the effect of temperature on enzyme activity, photosynthetic performance can vary considerably. Respiration is especially sensitive to temperature, with a temperature coefficient Q_{10} of approximately 2 (res-

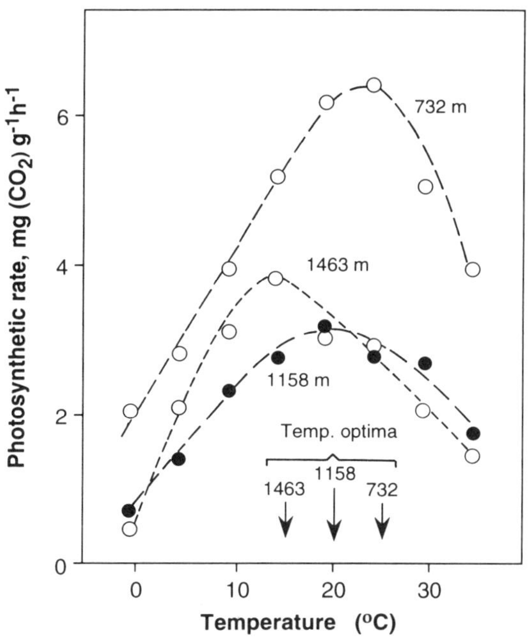

Figure 3 Temperature response curves for CO_2 uptake in *Abies balsamea* from populations originating at elevations of 1463, 1158, and 732 m but grown under uniform conditions. (From Fryer and Ledig, 1972.)

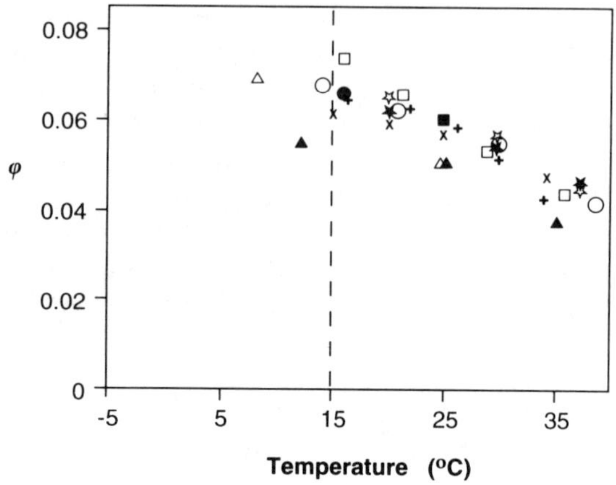

Figure 4 The relationship between quantum yield (ϕ) and leaf temperature in a number of C_3 plants, including some conifers. (From Leverenz and Öquist, 1987.)

piration doubles when the temperature rises by 10 degrees), whereas the Q_{10} values of the electron transport and enzyme reactions of photosynthesis are 1 and 2–3, respectively (Fitter and Hay, 1987). As a result, the light-saturated rate of net photosynthesis for most conifer species during the growing season is greatest at approximately 20°C and is depressed below 10–15°C and above 30°C due to suppression of Calvin cycle enzymes and to increased respiration, respectively.

The quantum yield, the slope of the low-light portion of the response curve, of C_3 species during the growing season increases as temperature decreases from 39 to 13°C (Ehleringer and Björkman, 1977) (Fig. 4). This is also observed in conifers under controlled conditions, e.g., *Picea sitchensis* (Leverenz and Jarvis, 1979) and *Pinus sylvestris* (Öquist and Strand, 1986). At least two mechanisms may be involved in the changing pattern of quantum yield with temperature: a direct response of ribulose 1,5-bisphosphate carboxylase oxygenase (Rubisco) activity to temperature (Ehleringer and Björkman, 1977) and/or carbohydrate accumulation at greater temperatures may suppress photosynthesis (Azcon-Bieto, 1983). The convexity of the photosynthetic light curve is a function of coincidence of the light saturation of all the chloroplasts within the leaf (Leverenz, 1987), and because this is not related to air temperature, temperature would not be expected to affect convexity.

A freeze during the growing season reduces the light-saturated rate of net photosynthesis and causes stomatal closure. The reduction in photosynthesis is caused by the CO_2 limitations imposed by stomatal closure as well as an irreversible reduction in apparent quantum yield and photosynthetic efficiency, as shown by DeLucia and Smith (1987) for *Picea engelmannii*. The reduction in net photosynthesis caused by freezing temperatures increased during the growing season and was proportional to the initial photosynthetic rate, therefore the processes necessary to attain high photosynthetic capacities also increased the sensitivity to freezing temperatures (DeLucia and Smith, 1987).

Winter hardening of conifers is associated with many cellular changes and results in the development of resistance to freezing temperatures (Levitt, 1980) during the transition from the growing to the dormant season. Extreme air temperatures during the dormant period can significantly affect the photosynthetic capacity of leaf tissues, such that even under favorable conditions they cannot perform well. The temperature dependence of the low-light response of photosynthesis is depressed for leaves collected in the winter compared to those collected in the summer, as shown by Leverenz and Öquist (1987) for *P. sylvestris* (Scotch pine), yet the quantum yield during the winter season is consistently lower than during the growing season (Fig. 5). The seasonality of quantum yield may be due to membrane permeability changes associated with frost hardening (Griffith *et al.*, 1982) such that protons leak across the thylakoids, resulting in a reduction in photophosphorylation (Leverenz and Öquist, 1987). Chlorophyll content is also reduced, yet the chlorophyll $a:b$ ratio is unchanged. It is suggested that the hardening process alters the chemical interactions within and among chlorophyll–protein complexes (Öquist and Strand, 1986) and may help reduce maintenance costs.

Freezing temperatures increase the sensitivity of conifer photosynthesis to photoinhibition, even in hardened tissues and under low irradiance levels (Strand and Öquist, 1985; Öquist and Huner, 1991). Photoinhibition is detected by a reduction in quantum yield and, in many cases, the light-saturated rate of net photosynthesis (Powles, 1984). The depression of photosynthesis after exposure to freezing temperatures is

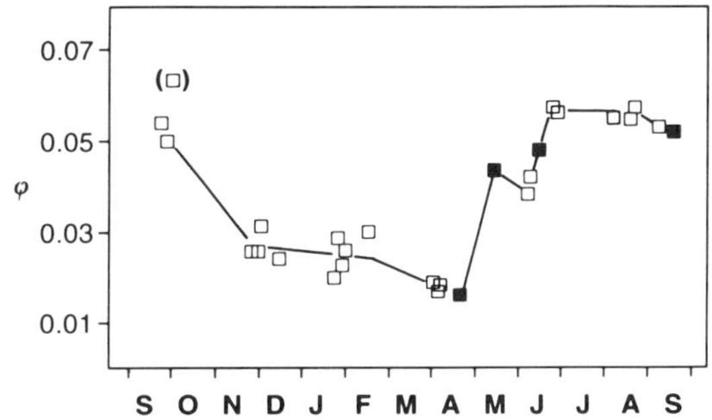

Figure 5 The annual variation in quantum yield (ϕ) (measured at 25°C) for *Pinus sylvestris*. (From Leverenz and Öquist, 1987.)

caused by the combination of temperature-suppressed Calvin cycle enzymes and/or photophosphorylation followed by photoinhibition of photosystem II (Strand and Öquist, 1988). With warming air above freezing, the photoinhibition of photosystem II recovers more quickly than the temperature-induced enzyme limitation to photosynthesis (Ottander and Öquist, 1991). Photoinhibition following exposure to low air temperatures may protect the photosynthetic machinery by dissipating excess absorbed energy from photosystem II (Strand and Öquist, 1985; Krauss, 1988; Ottander and Öquist, 1991).

Low soil temperatures can suppress conifer photosynthesis throughout the winter and well into the growing season at high elevations (DeLucia and Smith, 1987; Day *et al.*, 1989; DeLucia *et al.*, 1991b). Low soil temperatures reduce stomatal conductance (Running and Reid, 1980; Teskey *et al.*, 1984; DeLucia *et al.*, 1991b) and the light-saturated rate of net photosynthesis (DeLucia, 1986; Day *et al.*, 1991). The severe reduction in net photosynthesis on soil chilling to 1°C in *P. sylvestris* (DeLucia *et al.*, 1991b) was not a result of photoinhibition, and may be a result of carbohydrate-induced down-regulation of the photosynthetic enzymes.

III. Interactions between Structure and Photosynthetic Light Response at Different Levels

In this section we explore the interactions between structure, light environment, and photosynthetic response of an aggregated unit (shoot, crown, canopy). As discussed in the previous section, structure affects the photosynthetic response at the needle level. However, as we move up from needle to shoot, crown, and canopy, the relation between structure and photosynthetic light response becomes increasingly complex. An understanding of this relation is necessary not only to be able to scale up but also in "scaling down," i.e., to be able to separate the effects of physiological and structural factors on a measured response function of an aggregated unit. Specific problems involved in the scaling process that will be addressed in this section are (1) how the shoot and canopy structures influence the distribution of irradiance on the needle surface area, and (2) how the the photosynthetic response of an aggregated unit changes as a result of the spatial and temporal variation in irradiance.

A. Scaling from Needle to Shoot

The commonly accepted basic element for experimental observations of the photosynthesis of broad-leafed trees is the individual leaf. For coniferous trees the analogous element would be the individual needle, but in experimental studies this causes methodological problems, particularly in species with short and densely packed needles (e.g., *Picea, Larix*). Therefore, the next hierarchical element—the annual shoot—is used as the basic element in most experimental studies (e.g., Richter, 1974; Linder, 1979, 1981). Because of the complex arrangement of needles to form shoots and the accompanying changes in the shoot light environment, there are qualitative and quantitative differences in the light response of individual needles and shoots. To interpret measurements at shoot level we need to be able to separate the effect of shoot structure from environmental and physiological factors.

In contrast to most broad leaves, the thickness and efficient PAR absorption of conifer needles cause them to act functionally as three-dimensional, optically "black" surfaces with respect to photosynthesis. Because the PAR incident on one side of a needle is negligibly transmitted to the other side, needles should be treated as unifacial and the assimilating leaf area should be defined based on a total surface area rather than on a basis of a one-sided area. Assuming a particular photosynthetic response function (P_n) of a needle surface area element, the photosynthetic response of a shoot can be expressed as the integrated response of these elements, and the rate of shoot photosynthesis (P_s) is obtained [Eq. (3)] (Oker-Blom, 1985),

$$P_s = \frac{1}{A_n} \int_{A_n} P_n[I(r)] \, dr = \int_0^{I_{max}} P_n(I) \, dF_n(I), \tag{3}$$

where A_n denotes the total needle surface area of the shoot and F_n is the distribution function of PAR irradiance (I) over A_n.

In a given radiation environment, the distribution of irradiance over the needle surface of a shoot depends on shoot structure, which then is reflected in the photosynthetic radiation response of the shoot. Because the distribution of radiation over the needle surface cannot be represented by a scalar measure, it is not clear how we should quantify the driving variable "irradiance" of the response function of the shoot. For example, if (as is commonly done) the irradiance on a (horizontal) plane surface beside the shoot is used as the driving variable, the response function varies with direction of the shoot relative to the direction of radiation because it changes both the amount of intercepted PAR and its distribution over the needle surface (Fig. 6).

Smolander *et al.* (1987) showed that the main part of the variation in photosynthesis of Scotch pine shoots in direct light could be eliminated by expressing the rate of shoot photosynthesis, P_s, as a function of the mean irradiance on the needle surface area, $\bar{I}$ (the intercepted photon

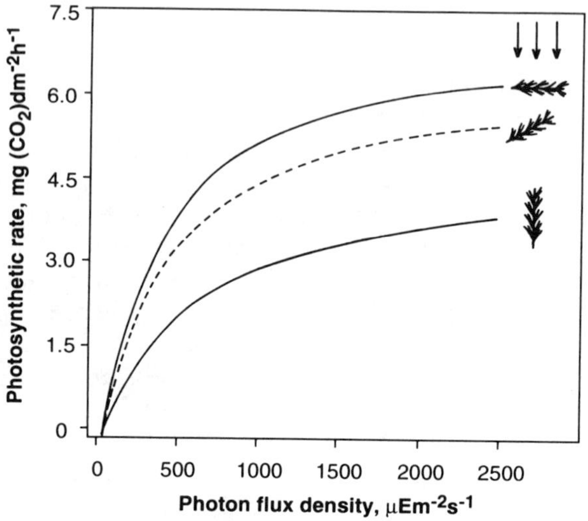

Figure 6 Photosynthetic rate of a Scotch pine shoot, inclined at three different angles (0°, 45°, and 90°) to the radiation beam, expressed as a function of incident photon flux density. (From Smolander *et al.*, 1987.)

flux divided by the total needle surface area of the shoot). In direct radiation the intercepted PAR is proportional to the *shoot silhouette area* (shadow area on a plane normal to the light beam), and the mean irradiance on the shoot's needle surface area can be expressed as the silhouette area to total needle area ratio (STAR) (Carter and Smith, 1985; Oker-Blom and Smolander, 1988) of the shoot multiplied by the incident photon irradiance. The same approach can be used for diffuse radiation, in which case the intercepted PAR is proportional to the average shoot silhouette area with respect to the directional distribution of radiation.

In conclusion, use of the mean irradiance as the driving variable for shoot photosynthesis minimizes the variation in the response function that is caused by shoot structure. It is easily recognized that by this approach the initial slopes of the shoot and needle surface response functions are equal, i.e., that $\lim(P_s(\bar{I})/\bar{I}) = \lim(P_n(I)/I)$ when $\bar{I}, I \rightarrow 0$ (see Oker-Blom, 1985). If there was no variation in the irradiance over the needle surface, the photosynthetic response function of the shoot would be similar to that of a needle surface area element, $P_s(\bar{I}) = P_n(\bar{I})$. However, because P_s represents a mean of concave functions, variation in irradiance decreases the curvature of P_s as a function of $\bar{I}$ and for this reason different response functions are obtained, e.g., in direct and diffuse radiation. In a diffuse light environment the irradiance is more evenly distributed over the needle surface area and, thus, at the same

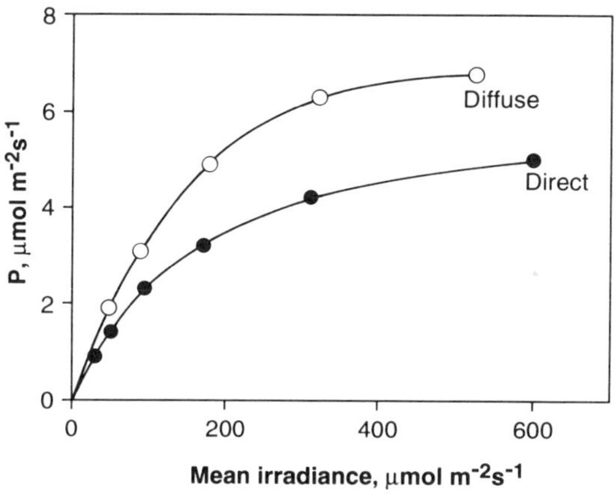

Figure 7 Rates of photosynthesis (P) of a Scotch pine shoot in diffuse and direct radiation, expressed as a function of the mean irradiance on the needle surface. (From Oker-Blom *et al.*, 1992.)

value of $\bar{I}$ (implying equal interceptance) the rate of photosynthesis is higher (Fig. 7, Oker-Blom *et al.*, 1992).

1. Methodological Remarks In conifers we have the same kind of problem as in cacti of defining the proper assimilating leaf area. By analogy to broad leaves, for which the assimilating leaf area commonly is defined on a one-sided area basis, the projected area of horizontally lying needles is often used as a measure of needle area in expressing the rate of photosynthesis of conifers. However, the analogy suffers for several reasons. In contrast to the case with planar leaves, the ratio of the total and projected areas of a needle is not constant, but varies within and between conifer species, being close to π for many pine species (Johnson, 1984; Lang, 1991). More importantly, the projected area of a horizontally lying needle is not a well-defined measure because needles are curved, axially asymmetrical, and often twisted. In addition, due to the thickness and efficient PAR absorption of conifer needles, the "upper" and "lower" sides of a needle function rather independently of each other. Thus, in contrast to thin, planar leaves, conifer needles should not be treated as one-sided but rather as "all-sided," and needle photosynthesis should be expressed on a total surface area basis.

Shoot structure, involving mutual shading and different inclinations of needles in the shoot, affects the *apparent* assimilating leaf area and causes ambiguity in the interpretation of photosynthetic responses measured at shoot level. This was noticed by Kramer and Clark as early as 1947, in a study in which they compared rates of photosynthesis of individual needles and entire seedlings of loblolly pine (*Pinus taeda* L.). At low light intensities, seedlings had a significantly lower rate of photosynthesis compared to individual needles, leading the authors to conclude that "it is believed that the relatively low rate of entire pine seedlings in the shade results principally from mutual shading of the needles by another." Later Zelawski *et al.* (1973) in their classical study demonstrated experimentally that the photosynthetic capacity of conifers was comparable to that measured for broad-leafed trees when mutual shading was eliminated using a diffuse radiation field.

Smith *et al.* (1991) proposed that "photosynthesis measured on basis of silhouette area of leaves attached to the shoot in its natural orientation to the sun may provide a better indication of photosynthetic capacity." This approach results in a more stable response function to incident photon irradiance (see Leverenz and Jarvis, 1979; Carter and Smith, 1985). However, the silhouette area is a relevant measure for light capture but not for gas exchange. A more correct approach, therefore, is to express gas exchange (e.g., rates of photosynthesis or transpiration) on the basis of total needle surface area (or in some cases volume, weight, or chlorophyll), and take into account the silhouette area (light capture) in defining the light environment of a shoot. As shown above, there exists no single scalar measure for the light environment of the geometrically complex shoot. By treating the shoot as a collection of independently acting needle surface area elements [Eq. (3)], we can describe the light environment of the shoot using the distribution of irradiance on its needle surface. The mean irradiance seems to be the most important factor causing variation in the photosynthetic response of a shoot. Using the STAR multiplied by incident photon irradiance as a measure of the light environment of a shoot and expressing the rate of photosynthesis on a total needle surface area basis, we get the mean rate of photosynthesis as a function of mean irradiance. If a more detailed description is needed, e.g., in the case of diffuse versus direct radiation, higher moments (e.g., variance and skewness) of the irradiance distribution on the photosynthetic units considered may be needed. Because there is no handy method to measure the irradiance distribution on the needle surface, simulations based on a geometrical model of shoot structure seem to be the most realistic method to produce estimates of these irradiance distributions (Oker-Blom *et al.*, 1992).

The measurement method must carefully match the question being posed. There is no elegant method, but also no reason, to study the physiology of photosynthesis at the shoot level. In the future, physiological studies should be done at the needle level when possible, and shoot level measurements should be used (1) to study implications of physiological processes at the shoot level and (2) for monitoring shoot carbon balance over longer periods.

B. Relations between Canopy Structure and Light Environment

In progressing from the leaf (shoot) to the canopy we need information on the effects of canopy structure on the amount of, and the spatial and temporal distributions of, intercepted radiation on the leaves. The amount of absorbed PAR by a canopy is determined by incident (above-canopy) radiation and canopy structure, involving the amount and dispersion of leaves, as well as their geometrical and optical properties. Because of the high absorption of PAR by conifer needles, the scattered component of radiation is often ignored, and the irradiance of PAR within a coniferous canopy is modeled as a function of shading leaf area.

In canopy-radiation models, a statistical approach is commonly used to describe the spatial distribution of foliage in the canopy. Numerous canopy-radiation models have been developed since the classical model of Monsi and Saeki (1953), which essentially is Beer's law applied to plant canopies. Application of Beer's law to plant canopies is based on the assumption that foliage elements (e.g., leaves or shoots) are located independently of each other. The mean relative direct solar irradiance of PAR at a given point (level) in the canopy, defined as the probability of a gap (p_0) through the canopy in the sun's direction, can then be expressed as

$$p_0 = \exp(-G\rho d), \tag{4}$$

where d is the length of the path of the solar beam within the canopy, ρ is the mean leaf area density along this path, and G is the mean projection of unit foliage area (Nilson, 1971). The mean relative diffuse irradiance is correspondingly obtained by averaging the gap probabilities in different directions of the upper hemisphere with respect to the distribution of sky radiance.

If the leaf area density (ρ) does not vary in space (leaves are uniformly distributed in the canopy), the leaf area density integrated along the solar path (ρd) is equal to $L/\cos\theta_s$, where L is the (downward cumulative) leaf area index and θ_s is the solar zenith angle. In many canopies of agricultural crops the assumption of randomly dispersed leaves has been shown to hold approximately true. In coniferous canopies, however, this assumption is proposed not to be valid because of the clustering of needles on shoots and the grouping of shoots into whorls and crowns (Norman and Jarvis, 1975; Whitehead *et al.*, 1990; Nilson, 1992). Thus, the development of radiation models applicable to conifers has concentrated on finding proper ways to describe the clumped spatial distribution of needles.

Grouping into crowns can be considered by letting the leaf area density (ρ) in Eq. (4) be zero outside the crowns, in which case d represents the total length of the path of the solar beam through the crowns. The foliage area density within a crown is commonly derived by dividing the crown envelope into cells, determining the foliage area in each of them, and fitting a statistical distribution to the data. A variety of statistical distributions, e.g., the beta distribution (*P. sitchensis*; Wang *et al.*, 1990) and the normal distribution (*Tsuga heterophylla*; Kershaw, 1993), have been used. However, the leaf area density in coniferous crowns can be very discontinuous (e.g., *P. radiata*; Whitehead *et al.*, 1990) and the small-scale variation in leaf area density caused by grouping of needles into shoots might not be reflected in these distributions.

An alternative approach, therefore, has been to consider the annual shoot as a discrete entity in modeling the spatial distribution of leaf area in a crown (*P. sylvestris*; Stenberg *et al.*, 1993). If shoots, instead of needles, are assumed to be independently located, Eq. (4) is applicable when the parameter G is defined as the ratio of projected shoot area to needle area. If the needle area density (ρ) is expressed on a total needle surface area basis, G should then be replaced by a needle area weighted average of the silhouette needle area to total needle area ratio (STAR) of the shoots (Stenberg *et al.*, 1994). Grouping into shoots and crowns increases light penetration (gap probability) at a fixed L (Fig. 8).

1. Spatial and Temporal Variation of Irradiance in a Canopy The spatial variation of irradiance in the canopy at a given moment may be

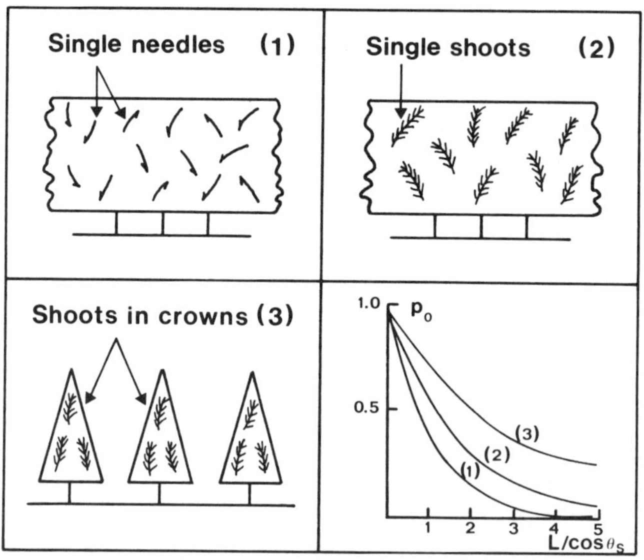

Figure 8 The effect of grouping of needles into shoots (2) and crowns (3) on light penetration, as compared to a nongrouped needle area distribution (1).

separated as resulting from two components: differences in the amount of shading leaf area at different locations in the canopy, and a "stochastic" component, which mainly depends on the occurrence of a gap in the sun's direction. The first component creates gradients of mean irradiance in the canopy, and is accounted for in Eq. (4) by variation in the leaf area density integrated along the path of the solar beam (ρd). At a fixed value of the mean irradiance (gap probability), however, there is variation in the actual fraction of visible sun, creating areas of sunflecks, full shadows, and penumbras (half shadows). In addition, when considering the distribution of irradiance on the leaf surface, a third component of variation results from different leaf orientations.

Assuming a uniform brightness of the sun, the direct solar irradiance $I_s(r)$ at a point (r) on a leaf surface inclined at an angle (α) to the sun can be written as $I_{s0}\sin(\alpha)f_s(r)$, where $f_s(r)$ denotes the fraction of visible sun and I_{s0} is the solar irradiance on a plane perpendicular to the sun. The mean value of $f_s(r)$ equals the probability of a gap (p_0) [Eq. (4)]. By definition, we have $0 \leq f_s \leq 1$. For $f_s = 0$ the sun is totally obscured and for $f_s = 1$ the sun is fully visible as seen from r. If the sun is partially blocked, $0 < f_s < 1$, the point r lies in a penumbra.

The statistical behavior of $f_s(r)$ can be studied by expressing it as a mean of random variables, i.e., as

$$f_s(r) = \frac{1}{S}\int_S \delta(r, \omega)\, d\omega, \tag{5}$$

where $\delta(r, \omega)$ is defined such that it is 1 or 0 depending on whether there is a free line of sight (a gap) through the canopy from r in the direction ω, and S denotes the solid angle subtended by the sun.

It can be shown that the variance of $f_s(r)$ [and $I_s(r)$] decreases with the ratio between projected leaf width and depth in the canopy (Stenberg, 1994).The fact that the sun has an angular diameter of approximately 0.5° (0.0094 radians) implies that a needle of 1 mm in width situated further away than ~10.6 cm from the target point cannot obscure the whole solar disk but creates a penumbra (Miller and Norman, 1971). In short canopies with fairly large leaves, the penumbra occurs rarely and has only a small effect on the irradiance distribution and rates of photosynthesis in clear sky conditions (Denholm, 1981). In contrast, the usually deep crown and small dimensions of needles, characteristic for conifers, are very efficient in creating penumbra. Simulations for Scotch pine (Fig. 9) (Stenberg, 1994) showed that the shading from a shoot is completely penumbral at a distance of 1 m to the target point. In contrast, mutual shading of needles on the same shoot, occurring between needles situated close (<<10 cm) together, usually resulted in a total blocking of the sun.

The statistical behavior (scale of variation) of diffuse radiation is quite

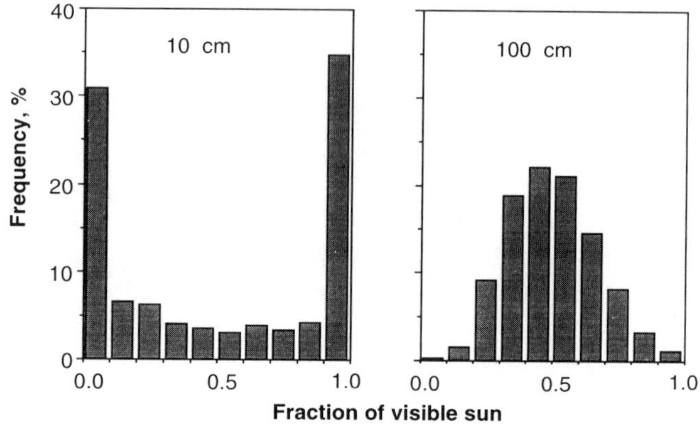

Figure 9 Simulated frequency distribution of the fraction of visible sun (relative direct solar irradiance) at a point shaded by a shoot situated at a distance of 10 and 100 cm away. (From Stenberg, 1994.)

different from that of direct radiation because it is formed as an integral over the solid angle of the hemisphere, which exceeds the solid angle of the sun by a factor of almost 10^5. Because the degree of statistical dependency (covariance) between the occurrence of a gap in different directions decreases rapidly with angular distance between these directions, the diffuse irradiance may be considered as an average of nearly independent "random" variables and, therefore, has a small variance (Gutschick, 1984).

In conclusion, the clustering of needles on shoots, as well as the grouping of shoots at other levels (whorl, crown) of the structural hierarchy, lead to a nonuniform leaf area density in the canopy and, subsequently, reduce light interception per unit of leaf area (see Fig. 8). More important, yet, is the effect of grouping into shoots on the spatial distribution of radiation. The gradient of radiation within a shoot may be as large as within the whole canopy and, thus, the irradiance at a point on the needle surface is to a high degree determined by shoot structure and the relative position of the point in the shoot. In addition, the statistical behavior of within-shoot shading is quite different from that of between-shoot shading because it occurs between needles situated close together.

Temporal variation in irradiance at a specific point in the canopy is caused by changes in incoming radiation (changing solar angle, cloud movement) and by foliage movement. The passage of the sun across the sky gives rise to a deterministic annual and diurnal variation of incoming radiation, which can be calculated using sun and earth geometry (e.g., Gates, 1980). The power's spectra of temporal variation caused by clouds depend on cloud types; broken clouds cause fluctuations with periods of minutes in incoming radiation.

Within a canopy the movement of sunflecks and shaded area due to changing solar angle causes a slow fluctuation of irradiance. The movement of foliage is the main source of rapid temporal variation in irradiance within a canopy. Experimental data on the temporal statistical variability of irradiance within a canopy are scarce. Norman and Tanner (1969) have published the spectral density functions for soybean and maize canopies. Most of the power was contained at low frequencies (<1 Hz). The distribution of power at higher frequencies was connected to the size of leaves. As far as we know, the same kind of power spectra have not been published for coniferous canopies. The measurement of those canopies is possible only using sensors with a diameter smaller than half the characteristic leaf width, because spatial integration caused by the area of the sensor gives biased power spectra.

Smolander (1984) studied the temporal microvariation in a Scotch pine stand by calculating standard deviations for periods from 20 to 200 sec, and analyzing serial autocorrelation functions. For 20-sec periods, the standard deviation was over 15 W m^{-2} in 25% of the cases, and for 200-sec periods it was 80 W m^{-2}. These estimates of temporal variation are, however, underestimates because the diameter of sensors used in the study was 6 mm. The conclusion was that the temporal microvariation should be measured by integrating it over the period for which photosynthesis is measured. Invariance was not found in the stochastic properties of temporal variation of irradiance with autocorrelation analysis, because autocorrelation functions varied widely.

Gross (1982) summarizes the information on the effects of irradiance fluctuations on photosynthesis as follows: (1) irradiance fluctuations of high frequencies (a few seconds or less) are averaged by the leaf, i.e., the photosynthetic rate achieved corresponds to the mean irradiance level, and (2) low-frequency irradiance fluctuations (about few minutes or longer) cause averaging of photosynthetic rates and not of irradiance. Hence it appears that most high-frequency fluctuations due to wind do not seem to affect photosynthesis appreciably. However, the low-frequency fluctuations due to the movement of the sun and intermittent clouds should be accounted for in the computation of canopy photosynthetic rates.

2. Methodological Remarks Distributions of irradiance in a canopy can be estimated by a theoretical model, based on a description of canopy structure, or by radiation measurements. These methods are by no means mutually exclusive; on the contrary, measurements are always needed to test the models and, on the other hand, the theoretical background for the relation between structure and radiation must be known to interpret the measurements and derive the appropriate distribution function.

It is generally recognized that, due to the large spatial and temporal variation of radiation in a canopy, a very large number of sensors is needed to get reliable estimates of the distributions of irradiance over time and space. This causes considerable practical problems in radiation measurements. However, less attention has been paid to the fact that, for the purpose of estimating photosynthesis, a problem far more serious than accuracy is the validity of radiation measurements (Anderson, 1965). The estimation requires the irradiance distribution on the photosynthetically active elements (e.g., leaves); however, radiation is measured with artificial surfaces that differ from these elements with respect to size, structure, arrangement, and directional distribution. If, as commonly done, irradiance is measured on horizontal planes at different depths in the canopy, we obtain an estimate of the mean irradiance on leaves in different layers of the canopy, but the statistical distribution of irradiance on the leaf surface area does not coincide with the measured one. In addition, measurements always involve averaging over space and so, for instance, penumbras cannot be adequately measured with standard sensors having a much larger diameter than the characteristic needle width.

C. Photosynthetic Light Response of a Canopy

1. Integration of the Photosynthetic Light Response over Time and Space

The mean rate of canopy photosynthesis during a given time period can formally be expressed as an explicit integral over time and space of a lower level (cell, needle, shoot) photosynthetic response function. As a full characterization of all variables in the response function that vary in time or space is usually not available, and because the variation in irradiance constitutes the most severe source of aggregation error, the rate of photosynthesis is expressed as a "fixed-response" function of irradiance, and integration is made over temporal and spatial distributions of irradiance. Assuming an invariant photosynthetic response (P_n) of a needle surface area element to irradiance (I), the mean rate of canopy photosynthesis ($\bar{P}_c$) during the time period (T) can then formally be expressed as

$$\bar{P}_c = \int_0^{I_{max}} P_n(I)\, dF_{n,T}(I), \tag{6}$$

where $F_{n,T}$ is the distribution function of I at the canopy needle surface during the given time period. Equation (6) generally describes scaling of the photosynthetic response to irradiance from a lower level of the structural hierarchy to a higher. Formally, the integration can be made over all chloroplasts in the canopy; however, the chloroplast is not an operational unit in modeling canopy photosynthesis, and, instead, a needle (needle surface area element) or an entire shoot is commonly used as the basic element. [Note that the choice of basic element affects the definition of the driving variable (I) and, consequently, the shape of the element response curve and the distribution function (F).]

The assumption that the response function (P_n) does not change spatially (between needles) or temporally, allowing the integration of a fixed response function in Eq. (6), is rarely true because (as discussed in Section II) various structural, physiological, and environmental factors cause variation in the photosynthetic response. However, the mean response can be calculated conditional on given assumptions on these factors, i.e., "subaggregates" with fixed (parameters of the) response function, and variation in the response function over time and space, can be accounted for by partitioning into fixed-response aggregates and averaging. For example, the rate of photosynthesis as a function of irradiance is commonly described as a nonrectangular hyperbola with parameters (initial slope, convexity, and maximum value) that depend on different structural, physiological, and environmental variables (Section II). The mean rate of photosynthesis over the distribution of irradiance can be estimated assuming fixed values of the parameters of the response function, and the variation in response function (parameters) between the components in the aggregation can be incorporated by taking a weighted mean of different fixed-response estimates.

2. Intercepted PAR and Dry Matter Production The shape of the curve describing the relation between the rate of canopy photosynthesis [Eq. (6)] and the mean irradiance (the amount of intercepted radiation per unit of leaf area and time) depends on the distribution function (F), which is induced by incident radiation and canopy structure. If there was no variation in irradiance (I) on the needle surface, $P_c(I)$ would be similar to the response function of a needle surface area element, $P_n(I)$. Temporal and spatial variation in irradiance, however, decreases the curvature of the canopy response function, making it apparently more linear ("flat").

We proceed to analyze the relation between intercepted radiation and photosynthetic production. (Note that the same curve describes the mean rate of canopy photosynthesis as a function of the mean irradiance.) The intercepted radiation is obtained as the integral of irradiance over leaf area and time. Consequently, an increase in intercepted PAR follows from an increase in (1) incoming irradiance, (2) effective leaf area, and/or (3) time, a fact that should be carefully considered in the interpretation of growth in terms of intercepted radiation (Demetriades-Shah *et al.*, 1992). When interception is increased by one of these factors separately (assuming the others to remain unchanged), the dependence of photosynthetic production on intercepted PAR is concave for (1), convex for (2) (see Oker-Blom *et al.*, 1989), and linear for (3). Thus, the relation between photosynthetic production and intercepted radiation depends on which factor(s) caused the increase in interception. As pointed out by Demetriades-Shah *et al.* (1992), "there will always be a good correlation between interception and anything under illumination which increases in size, even when illumination is not a limiting factor." For example, if the accumulated biomass production by a crop is plotted against accumulated intercepted PAR, the mere fact that both increase in time will inevitably produce good correlation and, in addition, the relationship will be approximately linear unless there are drastic changes in growth rate during the time period considered. The slope of the regression line gives an estimate of the mean light use efficiency in the particular case; however, the close correlation is of no consequence.

More relevant is to compare dry matter production for different crops or stands during a fixed time period. In a study by Linder (1985) the relationship between annual aboveground biomass production (y) and estimated intercepted PAR (x) in forest stands (including the coniferous species *Pinus radiata*, *Picea abies*, *Pinus sylvestris*, *Pinus nigra*, and *Picea sitchensis*) at different locations (Australia, New Zealand, the United Kingdom, and Sweden), was well approximated by a regression line of the type $y = ax + b$ with slope a equal to 1.7 g/MJ. The slope represents the increase in biomass production per unit of intercepted PAR; however, it is not equal to the ratio of produced biomass to intercepted PAR (y/x, the "light use efficiency"), unless the line goes through the origin ($b = 0$). In the data of Linder (1985) the intercept b was negative (implying negative dry matter production when light interception is zero), which is physiologically sound because of respiration. As a result, however, the light use efficiency (defined as y/x) varied between 0.27 and 1.6 g/MJ in the individual stands.

The ratio of dry matter production to intercepted PAR depends on (case 1) the efficiency of conversion of intercepted PAR to photosynthates, and (case 2) losses from maintenance respiration and the conversion of photosynthates into biomass (growth respiration). Various environmental factors may limit the photosynthetic response to light (Section II) and, in addition, the concavity of the photosynthetic light response function implies that whenever there is (spatial or temporal) variation in irradiance the ratio (case 1) of photosynthesis to intercepted PAR is always smaller than the initial slope (apparent quantum yield) of the light response curve of an individual leaf (needle). Generally speaking, the ratio (case 1) is smaller the larger is the fraction of leaf area operating at light levels above saturation. The conversion efficiency would then be smaller, e.g., in clear sky conditions and at low latitudes, and larger in canopies with high leaf area index, as shown also in a simulation study for Scotch pine (Oker-Blom *et al.*, 1989). However, despite a large variation in modeled stand structure and incoming radiation, the simulations did not produce a large variation in the conversion efficiency (case 1), which averaged 1.7 g/MJ. If this is in accordance with reality, differences in dry matter production at a fixed amount of intercepted PAR are caused mainly by environmental stress and respiratory losses. Moreover, a linear relationship between dry matter production and intercepted PAR during a specified time period, as found by Linder (1985), would follow only if also the respiratory loss (case 2) was constant (in this case independent of stand structure and geographical location).

In conclusion, there is empirical evidence that the light use efficiency, defined as the ratio of dry matter production to intercepted PAR ($a = y/x$), varies dramatically in (coniferous) forests. It is possible (and physiologically more sound), however, that a relationship of the type $y = ax + b$ could apply more generally, and that the slope a in that case might be less variable. This kind of relationship would be useful for estimating potential productivity under a well-defined set of conditions; however, it does not allow us to predict growth solely based on intercepted radiation.

IV. Concluding Remarks

We have addressed the unique structural features of conifers, as they relate to photosynthetic production, at different levels of organization (from needle to canopy). Because most of the protocols in ecophysiological research have been developed for broad leaves, many concepts and measures must be defined for conifers so that they are consistent with the structural properties of needles and shoots. Consistency is needed also in comparing the photosynthetic performance of conifers and broad leaves, wherein it is important to distinguish the effect of structural factors on light capture from differences in the photosynthetic response at a fixed interception.

Needles differ from broad leaves both with respect to inner structure and external shape, which includes a continuum from nearly flat to cylindrical. The area of flat leaves is for obvious reasons measured as the plan area or one-sided area. For nonflat three-dimensional objects such as conifer needles, total surface area is the natural measure (Lang, 1993). The meaning of the one-sided area of needles is not clear, but consistency requires that it be defined as half the total needle surface area, as concluded by Chen and Black (1992). By that approach, for example, the mean ratio of silhouette area (light-intercepting area) and one-sided area (the light extinction coefficient) is 0.5 for both flat and nonflat leaves (Lang, 1991). Because there exists an exact proportionality between total and half of the total surface area, these can be used in a parallel fashion without causing confusion.

Characteristic structural factors of conifers that affect their ability to harvest light are a deep canopy combined with a small needle size, which create an important penumbra effect, and the clustering of needles on shoots, which creates a discontinuous distribution of needle area. These factors imply that, at a fixed leaf area index, the intercepted PAR would be smaller in coniferous than in broad-leafed canopies, but the vertical gradient of light in conifers is less steep and light reaching the lower canopy is all penumbral (diffuse). In addition, the gradient of light within a coniferous shoot may be as large as that within the whole canopy. Conifers can maintain a higher leaf area index, and this may be accomplished by a more even distribution of light between shoots at different locations in the canopy and also because shade shoots have a structure that effectively intercepts light (Leverenz and Hinckley, 1990).

It has been widely recognized that broad leaves in general have higher maximum photosynthetic rates than do needles, and yet conifers are at least equally productive on a stand basis (Sprugel, 1989, and references therein). Possible reasons for this might be a longer growing season combined with a larger leaf area and/or that conifers disperse light more evenly (the light-saturated leaf area is smaller) (Sprugel, 1989; Stenberg *et al.*, 1994). However, most comparisons of the photosynthetic capacity of broad leaves and cônifers have been done in direct light, and the difference between intercepting and assimilating area has not been recognized. Considerable confusion has been caused by comparing the rate of photosynthesis of single broad leaves to the rate of photosynthesis of intact coniferous shoots, wherein only a small portion of the needle area is fully irradiated.

References

Anderson, M. C. (1965). Some problems of simple characterization of the light climate in plant communities. *In* "Light as an Ecological Factor" (R. Bainbridge, G. C. Evans, and O. Rackham, eds.), pp. 77–90. Blackwell Scientific Publications, Oxford.

Azcon-Bieto, J. (1983). Inhibition of photosynthesis by carbohydrates in wheat leaves. *Plant Physiol.* 73:681–686.

Baker, E. A. (1974). The influence of environment on leaf wax development in *Brassica oleracea* var. gemmifera. *New Phytol.* 73:955–966.

Björkman, O. (1981). Responses to different quantum flux densities. *In* "Physiological Plant Ecology I: Responses to the Physical Environment. Encyclopedia of Plant Physiology" (O. L. Lange, P. S. Nobel, C. B. Osmond, and H. Ziegler, eds.), New Series Vol. 12A, pp. 57–107. Springer-Verlag, New York.

Björkman, O. and Powles, S. B. (1984). Inhibition of photosynthetic reactions under water stress: Interaction with light level. *Planta* 161:490–504.

Boardman, N. K. (1977). Comparative photosynthesis of sun and shade plants. *Annu. Rev. Plant Physiol.* 28:355–377.

Bone, R. A., Lee, D. W., and Norman, J. M. (1985). Epidermal cells functioning as lenses in leaves of tropical rainforest shade plants. *Appl. Opt.* 24:1408–1412.

Bornman, J. F., and Vogelmann, T. C. (1991). Effect of UV-B radiation on leaf optical properties measured with fibre optics. *J. Exp. Bot.* 42:547–554.

Brooks, J. R. (1993). Light acclimation in *Abies amabilis* canopies. Ph.D. Dissertation. University of Washington, Seattle.

Brooks, J. R., Hinckley, T. M., Ford, E. D., and Sprugel, D. C. (1991). Foliage dark respiration in *Abies amabilis* (Dougl.) Forbes: Variation within the canopy. *Tree Physiol.* 9: 325–338.

Carter, G. A., and Smith, W. K. (1985). Influence of shoot structure on light interception and photosynthesis in conifers. *Plant Physiol.* 79:1038–1043.

Chamberlain, C. J. (1966). Gymnosperms. *In* "Structure and Evolution." Dover Publications, Inc. New York.

Chen, J. M., and Black, T. A. (1992). Defining leaf area index for non-flat leaves. *Plant, Cell Environ.* 15:421–429.

Clark, J. B., and Lister, G. R. (1975a). Photosynthetic action spectra of trees. I. Comparative photosynthetic action spectra of one deciduous and four coniferous tree species as related to photorespiration and pigment complements. *Plant Physiol.* 55:401–406.

Clark, J. B., and Lister, G. R. (1975b). Photosynthetic action spectra of trees. II. The relationship of cuticle structure to the visible and ultraviolet spectral properties of needles from four coniferous species. *Plant Physiol.* 55:407–413.

Day, T. A., DeLucia, E. H., and Smith, W. K. (1989). Influence of cold soil and snowcover on photosynthesis and leaf conductance in two Rocky Mt. conifers. *Oecologia* 80: 546–552.

Day, T. A., Heckathorn, S. A., and DeLucia, E. H. (1991). Limitation of photosynthesis in *Pinus taeda* L. (loblolly pine) at low temperatures. *Plant Physiol.* 96:1246–1254.

Day, T. A., Vogelmann, T. C., and DeLucia, E. H. (1992). Are some plant life forms more effective than others in screening out ultraviolet-B radiation? *Oecologia* 92:513–519.

Del Rio, E., and Berg, A. (1979). Specific leaf area of Douglas-fir reproduction as affected by light and needle age. *For. Sci.* 25:183–186.

DeLucia, E. H. (1986). Effect of low root temperature on net photosynthesis, stomatal conductance, and carbohydrate concentration in Engelmann spruce (*Picea engelmannii* Parry ex. Engelm.) seedlings. *Tree Physiol.* 2:143–154.

DeLucia, E. H., and Berlyn, G. P. (1984). The effect of increasing elevation on leaf cuticle thickness and cuticular transpiration in balsam fir. *Can. J. Bot.* 62:2423–2431.

DeLucia, E. H., and Smith, W. K. (1987). Air and soil temperature limitations on photosynthesis in Engelmann spruce during summer. *Can. J. For. Res.* 17:527–533.

DeLucia, E. H., Day, T. A., and Vogelmann, T. C. (1991a). Ultraviolet-B radiation and the Rocky Mountain environment: Measurement of incident light and penetration into foliage. Current Topics in Plant Biochemistry and Physiology, University of Missouri, Columbia 10:32–48.

DeLucia, E. H., Day, T. A., and Öquist, G. (1991b). The potential for photoinhibition of *Pinus sylvestris* L. seedlings exposed to high light and low soil temperature. *J. Exp. Bot.* 42:611–617.

DeLucia, E. H., Shenoi, S. L., Naidu, S. L., and Day, T. A. (1991c). Photosynthetic symmetry of sun and shade leaves of different orientation. *Oecologia* 87:51–57.

Demetriades-Shah, T. H., Fuchs, M., Kanemasu, E. T., and Flitcroft, I. (1992). A note of caution concerning the relationship between cumulated intercepted solar radiation and crop growth. *Agric. For. Meteorol.* 58:193–207.

Denholm, J. V. (1981). The influence of penumbra on canopy photosynthesis. II. Canopy of horizontal circular leaves. *Agric. Meteorol.* 25:167–194.

Donahue, R. A. (1991). Interaction of ambient light directionality with leaf anatomy and effects on photosynthesis of sun and shade leaves of *Thermopsis montana*. Dissertation, University of Wyoming, Laramie.

Ehleringer, J., and Björkman, O. (1977). Quantum yields for CO_2 uptake in C_3 and C_4 plants. Dependence on temperature, CO_2 and O_2 concentration. *Plant Physiol.* 59: 86–90.

Esau, K. (1977). "Anatomy of Seed Plants," 2nd Ed., Wiley, New York.

Field, C. (1981). Leaf age effects on the carbon gain of individual leaves in relation to microsite. *In* "Components of Productivity of Mediterranean Climate Regions—Basic and Applied Aspects" (M. S. Margaris and H. A. Mooney, eds.), pp. 41–50. Dr. W. Junk Publ., The Hague.

Fitter, A. H., and Hay, R. K. M. (1987). "Environmental Physiology of Plants." Academic Press, San Diego.

Fryer, J. H., and Ledig, F. T. (1972). Microevolution of photosynthetic temperature optimum in relation to the elevational complex gradient. *Can. J. Bot.* 50:1231–1235.

Gates, D. M. (1980). "Biophysical Ecology." Springer-Verlag, New York.

Griffith, M., Brown, G. N., and Huner, N. P. A. (1982). Structural changes in thylakoid proteins during cold acclimation and freezing of winter rye (*Secale cereale* L. cv. Puma). *Plant Physiol.* 70:418–423.

Gross, L. J. (1982). Photosynthetic dynamics in varying light environments: A model and its application to whole leaf carbon gain. *Ecology* 63:84–93.

Gunthardt, M. S., and Wanner, H. (1982). Die menge des cuticularen Wachses auf Nadel von *Pinus cembra* L. and *Picea abies* (L.) Karsten in Abhangigkeit von Nadelalter und Standort. *Flora (Jena)* 172:125–137.

Gutschick, V. P. (1984). Statistical penetration of diffuse light into vegetative canopies: Effect on photosynthetic rate and utility for canopy measurement. *Agric. Meteorol.* 30: 327–341.

Gutschick, V. P., and Wiegel, F. W. (1988). Optimizing the canopy photosynthetic rate by patterns of investment in specific leaf mass. *Am. Nat.* 132:67–86.

Haberlandt, G. (1914). "Physiological Plant Anatomy," 4th Ed., pp. 613–630. Macmillan. London.

Hager, H., and Sterba, H. (1985). Specific leaf area and needle weight of Norway spruce (*Picea abies*) in stands of different densities. *Can. J. For. Res.* 15:389–392.

Havranek, W. M., and Tranquillini, W. (1994). Physiological processes during winter dormancy and their ecological significances. *In* "Ecophysiology of Coniferous Forests" (W. K. Smith and T. M. Hinckley, eds.), pp. 95–124. Academic Press, San Diego.

Hirose, T., and Werger, M. J. A. (1987). Maximizing daily canopy photosynthesis with respect to the leaf nitrogen allocation pattern in the canopy. *Oecologia* 72:520–526.

Hunt, G. M., and Baker, E. A. (1982). Developmental and environmental variations in plant epicuticular waxes: Some effects on penetration of naphthylacetic acid. *In* "The Plant Cuticle" (D. F. Cutler, K. L. Alvin and C. E. Price, eds.), pp. 279–292. Academic Press, New York.

Jeffree, C. E., Johnson, R. P. C., and Jarvis, P. G. (1971). Epicuticular waxes in the stomatal antechamber of Sitka spruce and its effect on the diffusion of water vapour and carbon dioxide. *Planta* 98: 1–10.

Johnson, I. D. (1984). A rapid technique for estimating total surface area of pine needles. *For. Sci.* 30:913–921.

Kershaw, J. A. (1993). Crown structure and stem form development in young stands of western hemlock. Ph.D. Dissertation, College of Forest Resources, University of Washington, Seattle.

Kirschbaum, M. U. F., and Farquhar, G. D. (1987). Investigation of the CO_2 dependence of quantum yield and respiration in *Eucalyptus pauciflora*. *Plant Physiol.* 83: 1032–1036.

Kramer, P. I., and Clark, W. S. (1947). A comparison of photosynthesis in individual pine needles and entire seedlings at various light intensities. *Plant Physiol.* 22:51–57.

Krauss, G. H. (1988). Photoinhibition of photosynthesis. An evaluation of damaging and protective mechanisms. *Physiol. Plant.* 74:566–574.

Lang, A. R. G. (1991). Application of some of Cauchy's theorems to estimation of surface areas of leaves, needles and branches of plants, and light transmittance. *Agric. For. Meteorol.* 55: 191–212.

Lang, A. R. G. (1993). Cauchy's theorems and estimation of surface areas of leaves, needles and branches. *In* "Crop Structure and Light Microclimate. Characterization and Applications" (C. Varlet-Grancher, R. Bonhomme, and H. Sinoquet, eds.), pp. 175–182. INRA, Paris.

Lappi, J., and Oker-Blom, P. (1992). Characterizing photosynthetic radiation response or other output function as a mean of element responses. *J. Theor. Biol.* 154:371–389.

Larcher, W. (1983). "Physiological Plant Ecology," 2nd Ed., Translated by M. A. Biederman-Thorson, Springer-Verlag, New York.

Leverenz, J. W. (1987). Chlorophyll content and the light response curve of shade-adapted conifer needles. *Physiol. Plant.* 71:20–29.

Leverenz, J. W., and Hinckley, T. M. (1990). Shoot structure, leaf area index and productivity of evergreen conifer stands. *Tree Physiol.* 6: 135–149.

Leverenz, J. W., and Jarvis, P. G. (1979). Photosynthesis in Sitka spruce. VIII. The effects of light flux density and direction on the rate of net photosynthesis and the stomatal conductance of needles. *J. Appl. Ecol.* 16:919–932.

Leverenz, J. W., and Öquist, G. (1987). Quantum yields of photosynthesis at temperatures between −2°C and 35°C in a cold-tolerant C_3 plant (*Pinus sylvestris*) during the course of one year. *Plant, Cell Environ.* 10:287–295.

Levitt, J. (1980). "Responses of Plants to Environmental Stresses. Volume 1. Chilling, Freezing, and High Temperature Stresses," 2nd Ed. Academic Press, New York.

Lewandowska, M., and Jarvis, P. G. (1977). Changes in chlorophyll and carotenoid content, specific leaf area and dry weight fraction in Sitka spruce (*Picea sitchensis*), in response to shading and season. *New Phytol.* 79:247–256.

Linder, S. (1979). Photosynthesis and respiration in conifers: A classified reference list 1891–1977. *Stud. For. Suec.* 149: 1–71.

Linder, S. (1981). Photosynthesis and respiration in conifers: A classified reference list. Supplement 1. *Stud. For. Suec.* 161: 1–32.

Linder, S. (1985). Potential and actual production in Australian forest stands. *In* "Research for Forest Management" (J. J. Landsberg and W. Parsons, eds.), pp. 11–34. CSIRO, Australia.

McClendon, J. (1984). The micro-optics of leaves. I. Patterns of reflection from the epidermis. *Am. J. Bot.* 71: 1391–1397.

Miller, E. E., and Norman, J. M. (1971). A sunfleck theory for plant canopies. I. Lengths of sunlit segments along a transect. *Agron. J.* 63:735–738.

Monsi, M., and Saeki, T. (1953). Über den Lichtfactor in den Pflanzengesellschaften und seine bedeutung für die Stoff-production. *Jpn. J. Bot.* 14:22–52.

Nilson, T. (1971). A theoretical analysis of the frequency of gaps in plant stands. *Agric. Meteorol.* 8:25–38.

Nilson, T. (1992). Radiative transfer in nonhomogeneous plant canopies. *In* "Advances in Bioclimatology 1," (G. Stanhill, ed.), pp. 59–88. Springer-Verlag, Berlin and Heidelberg.

Noble, P. S., Zaragosa, L. J., and Smith, W. K. (1975). Relationship between mesophyll surface area photosynthetic rate, and illumination level during development of leaves of *Plactrantnus parviflorus* Henekel. *Plant Physiol.* 55:1067–1070.

Norman, J. M., and Jarvis, P. G. (1975). Photosynthesis in Sitka Spruce (*Picea sitchensis* (Bong.) Carr.). V. Radiation penetration theory and a test case. *J. Appl. Ecol.* 12: 839–877.

Norman, J. M., and Tanner, C. B. (1969). Transient light measurements in plant canopies. *Agron. J.* 61:847–849.

Oker-Blom, P. (1985). Photosynthesis of a Scots pine shoot: Simulation of the irradiance distribution and photosynthesis of a shoot in different radiation fields. *Agric. For. Meteorol.* 34:32–40.

Oker-Blom, P., and Smolander, H. (1988). The ratio of shoot silhouette area to total needle area in Scots pine. *For. Sci.* 34:894–906.

Oker-Blom, P., Pukkala, T., and Kuuluvainen, T. (1989). Relationship between radiation interception and photosynthesis in forest canopies: Effect of stand structure and latitude. *Ecol. Modell.* 49:73–87.

Oker-Blom, P., Lahti, T., and Smolander, H. (1992). Photosynthesis of a Scots pine shoot: A comparison of two models of shoot photosynthesis in direct and diffuse radiation fields. *Tree Physiol.* 10:111–125.

Öquist, G., and Huner, N. P. A. (1991). Effects of cold acclimation on the susceptibility of photosynthesis to photoinhibition in Scots pine and in winter and spring cereals; a flourescence analysis. *Funct. Ecol.* 5:91–100.

Öquist, G., and Strand, M. (1986). Effects of frost hardening on photosynthetic quantum yield, chlorophyll organization, and energy distribution between the two photosystems in Scots pine. *Can. J. Bot.* 64:748–753.

Oren, R., Schulze, E.-D., Matyssek, R., and Zimmermann, R. (1986). Estimating photosynthetic rate and annual carbon gain in conifers from specific leaf weight and leaf biomass. *Oecologia* 70:187–193.

Ottander, C., and Öquist, G. (1991). Recovery of photosynthesis in winter-stressed Scots pine. *Plant, Cell Environ.* 14:345–349.

Poulson, M. E., and DeLucia, E. H. (1993). Photosynthetic and structural acclimation to light direction in vertical leaves of *Silphium terebinthinaceum*. *Oecologia* 95:393–400.

Poulson, M. E., and Vogelmann, T. C. (1990). Epidermal focussing and effects upon photosynthetic light-harvesting in leaves of *Oxalis*. *Plant, Cell Environ.* 13:803–811.

Powles, S. B. (1984). Photoinhibition of photosynthesis induced by visible light. *Ann. Rev. Plant Physiol.* 35:15–44.

Prioul, J. L., and Chartier, P. (1977). Partitioning of transfer and carboxylation components of intracellular resistance to photosynthetic CO_2 fixation: A critical analysis of the methods used. *Ann. Bot.* 41:789–800.

Richter, H. (1974). Erhöhte Saugspannungswerte und morphologische Veränderungen durch transversale Einschnitte in einem *Taxus* Stamm. *Flora (Jena)* 163:291–309.

Rook, D. A., Bollmann, M. P., and Hong, S. O. (1987). Foliage development within the crowns of *Pinus radiata* trees at two spacings. *N.Z. J. For. Sci.* 17:297.

Running, S. W., and Reid, C. P. (1980). Soil temperature influences on root resistance of *Pinus contorta* seedlings. *Plant Physiol.* 65:635–640.

Schoettle, A. W. (1990). Importance of shoot structure to sunlight interception and photosynthetic carbon gain in *Pinus contorta* crowns. Ph.D. Dissertation, University of Wyoming, Laramie.

Schoettle, A. W., and Smith, W. K. (1991). Interrelation between shoot characteristics and solar irradiance in the crown of *Pinus contorta* ssp. *latifolia*. *Tree Physiol.* 9:245–254.

Smith, W. K., Schoettle, A. W., and Cui, M. (1991). Importance of the method of leaf area measurement to the interpretation of gas exchange of complex shoots. *Tree Physiol.* 8: 121–127.

Smolander, H. (1984). Measurement of fluctuating irradiance in field studies of photosynthesis. *Acta For. Fenn.* 187: 1–56.

Smolander, H., Oker-Blom, P., Ross, J., Kellomäki, S., and Lahti, T. (1987). Photosynthesis of a Scots pine shoot: Test of a shoot photosynthesis model in a direct radiation field. *Agric. For. Meteorol.* 39:67–80

Sprugel, D. G. (1989). The relationship of evergreenness, crown architecture, and leaf size. *Am. Nat.* 133:465–479.

Stenberg, P. (1994). Penumbra in within-shoot and between-shoot shading in conifers and its significance for photosynthesis. *Ecol. Modell.* in press.

Stenberg, P., Smolander, H., and Kellomäki, S. (1993). Description of crown structure for light interception models: Angular and spatial distribution of shoots in young Scots pine. *Stud. For. Suec.* 191:43–50.

Stenberg, P., Kuuluvainen, T., Kellomäki, S., Grace, J., Gholz, H., and Jokela, E. (1994). Crown structure and light interception in relation to productivity. *In* "A Comparative Analysis of Pine Forest Productivity. (H. Gholz, S. Linder, and R. McMurtrie, eds.), Ecological Bulletin, Stockholm, in press.

Strack, D., Heilemann, J., Mimken, M., and Wray, V. (1988). Cell wall-conjugated phenolics from Coniferae leaves. *Phytochemistry* 27:3517–3521.

Strack, D., Heilemann, J., Wray, V., and Dirks, H. (1989). Structures and accumulation patterns of soluble and insoluble phenolics from Norway spruce needles. *Phytochemistry* 28:2071–2078.

Strand, M., and Öquist, G. (1985). Inhibition of photosynthesis by freezing temperatures and high light levels in cold-acclimated seedlings of Scots pine (*Pinus sylvestris*). I. Effects on the light-limited and light-saturated rates of CO_2 assimilation. *Physiol. Plant.* 64: 425–430.

Strand, M., and Öquist, G. (1988). Effects of frost hardening, dehardening and freezing stress on *in vivo* chlorophyll fluorescence of seedlings of Scots pine (*Pinus sylvestris* L.). *Plant, Cell Environ.* 11:231–238.

Terashima, I., and Inoue, Y. (1985). Palisade tissue chloroplasts and spongy tissue chloroplasts in spinach: Biochemical and ultrastructural differences. *Plant Cell Physiol.* 26: 63–75.

Terashima, I., and Saeki, T. (1985). A new model for leaf photosynthesis incorporating the gradients of light environment and of photosynthetic properties of chloroplasts within a leaf. *Ann. Bot.* 56:489–499.

Teskey, R. O., Grier, C. C., and Hinckley, T. M. (1984). Change in photosynthesis and water relations with age and season in *Abies amabilis*. *Can. J. For. Res.* 14:77–84.

Tucker, G. F., and Emmingham, W. H. (1977). Morphological changes in leaves of residual western hemlock after clear and shelterwood cutting. *For. Sci.* 23:195–203.

Turrel, F. M. (1936). The area of the internal exposed surface of dicotyledon leaves. *Am. J. Bot.* 23:255–264.

Vogelmann, T. C., Martin, G., Chen, G. and Buttry, D. (1991). Fibre optic microprobes and measurement of the light environment within plant tissues. *Adv. Bot. Res.* 18: 231–270.

Wang, Y. P., Jarvis, P. G., and Benson, M. L. (1990). Two-dimensional needle area density distribution within the crowns of *Pinus radiata* trees. *For. Ecol. Manage.* 32:217–237.

Whitehead, D., Grace, J. C., and Godfrey, M. J. S. (1990). Architectural distribution of foliage in individual *Pinus radiata* D. Don crowns and the effects of clumping on radiation interception. *Tree Physiol.* 7:135–155.

Wood, G. B. (1973). Spatial variation in leaf chlorophyll within the crown of a radiata pine sapling. *Aust. For. Res.* 6:5–14.

Zelawski, W., Szaniawski, R., Dybczyriski, W., and Piechurowski, A. (1973). Photosynthetic capacity of conifers in diffuse light of high illuminance. *Photosynthetica* 7:351–357.

Zhang, J. (1989). Modelling irradiance response of leaf photosynthetic rate in wheat and bamboo. *Photosynthetica* 22:526–534.

2

Water and Nutrient Acquisition by Roots and Canopies

Ram Oren and David W. Sheriff

I. Introduction

Water and nutrient supply rates, as well as internal (plant) and external (soil) deficits, can have major effects on physiological activity and growth. Effects of water or nutrient deficits on growth can be demonstrated separately, but they often interact, as shown by Allen *et al.* (1990) for several *Pinus* species, and by Turner (1982) for *Pinus radiata*. Moist soil and wet canopy surfaces facilitate nutrient uptake through roots and foliage, respectively. Water uptake is affected by the number and distribution of roots in relation to the distribution of soil moisture, and by the wetness and hydraulic permeability of foliage. Nutrient uptake is similarly affected by tissue characteristics and nutrient concentration, but also depends on the moisture regime in the bulk soil and in the vicinity of absorbing surfaces.

In this chapter, we discuss generalities based on results from observational studies of unmanipulated plants and of stands. We also consider information from experimental manipulation of nutrient and water availability. A more thorough treatment of the effects of mycorrhizae and anthropogenic pollution on water and nutrient acquisition is given in Chapter 3 (this volume) and in Matyssek *et al.* (1994), respectively.

II. Methodology

A. Foliar Acquisition

Much of the knowledge about uptake of water and nutrients through foliage was acquired via research on three very different topics: (1) uptake of nutrients and dissolved CO_2 by aquatic plants; (2) uptake of foliar applied fertilizers, herbicides, plant growth regulators, and systematic insecticides; and (3) uptake of water and nutrients from wet canopy surfaces, particularly in relation to anthropogenically enriched, or polluted, air, cloud, and fog water. In this chapter we report on processes involved in foliar acquisition of water and nutrients by terrestrial plants, and provide examples from research on conifers.

B. Acquisition by Roots

Various methods have been employed to investigate the uptake of nutrients by whole plants. Assessing relationships between nutrient availability and uptake is most often accomplished using the following methods:

1. Measurement of uptake rate of an element per unit of root (area, length, or weight). Uptake by the plant is calculated by using a measured or estimated quantity for the root system.
2. Determining nutrient pools in the whole plant at different times using measured concentrations of elements in samples of plant organs and total dry weights of each organ.
3. Determining nutrient pools and fluxes in soil by sampling, and calculating uptake as the amount of nutrient that cannot be accounted for.

In most cases, a theoretical or empirical relationship is evaluated with regard to supply of a certain ion (using either its input to the soil or an index of availability in the soil) and its uptake. Sometimes the effect of supply of one element on uptake of others is also assessed. This can often lead to problems in interpretation because, for example, increased supply of one element may result in an increase in growth rate. Greater growth will usually increase absolute rates of uptake of all nutrients, but can reduce concentrations in plant tissues of those not in abundant supply.

Other problems in interpretation can be caused by experimental conditions not being representative of the range of temporal and spatial variations found in the field. There may also be errors in estimation of components of biomass, especially those belowground, which can result in unreliable estimates of rates of uptake. This problem increases with plant size, and because measurement errors are smaller in proportion to fluxes, they may decrease as the period between successive measurements increases.

When interpretation is based on measurements of nutrient concentrations in plant tissues, use of several labeled elements in a single study may overcome some difficulties in estimating both uptake and interactions between the supply and uptake of different elements. However, estimation of the belowground component is still difficult. This can be overcome by considering accumulation and fluxes in only one or more aboveground organs, but in this case nutrient storage and changes in nutrient concentrations of belowground tissues are ignored. It is, nevertheless, a practical compromise, best supported where possible (e.g., in monocultures) by uptake determined from fluxes in the soil.

The operationally simplest technique in the field that does not necessitate destructive sampling of plants is to determine nutrient fluxes and pools in the soil (e.g., Smethurst and Nambiar, 1989). Such techniques are subject to several possible sources of error (e.g., Smethurst and Nambiar, 1989), but often provide a desirable combination of a good index of uptake *in situ* and minimal disturbance to the system.

III. Fundamentals

A. Foliar Acquisition

Thorough reviews of the topic have been published by Kannan (1986), Raven (1988), Riederer (1989), and Schönherr and Riederer (1989). Here we provide a short overview of fundamental mechanisms that determine rates of water and nutrient uptake by conifer shoots and canopies.

The canopy of conifers may take up water and nutrients by mass flow (Katz *et al.*, 1989a; Riederer, 1989) and by diffusion, but will lose these resources only by diffusion. Diffusion of elements and water in a gaseous form occurs mostly through stomata, whereas diffusion in an aqueous phase occurs only through the cuticle (Riederer, 1989). Under certain circumstances, mass flow of water and nutrients may contribute temporarily significant amounts of water and, to a lesser degree, nutrients. However, it is diffusion of water and nutrients that dominate the rates of nutrient and water uptake or loss.

Depending on the species, either stomatous or astomatous surfaces may be more permeable to aqueous-phase diffusion (Kannan, 1969). It may be expected that stomatous surfaces would be more permeable to this flux because cuticles over cells in the substomatal cavity are thinner (Kannan, 1986). However, due to the geometry and the hydrophobic nature of stomatal surfaces, naturally occurring aqueous solutions cannot penetrate stomatal pores (Schönherr and Bukovac, 1972). Thus, an aqueous continuum does not develop from the surface of a leaf to its interior, so a flow of solution through the pores is not possible (Riederer, 1989). Entry of aqueous solutions is, therefore, only through the cuticular wax and cuticle on a leaf's external surface. These present the main, almost exclusive, barrier to transport of solutes into leaves (Schönherr, 1976; Schönherr and Bukovac, 1970).

In leaves (and probably in expanding twigs), cuticular permeability to water and nutrients is high during periods of expansion, and declines rapidly thereafter. Completion of the cuticle (Lange and Schulze, 1966) and of an epicuticular wax layer (Schönherr and Riederer, 1989) after full expansion may be the cause for this increased resistance. However, damage to leaf cuticles, and rearrangement of cuticular wax caused by wind, can subsequently increase the permeability of leaf surfaces (Grace, 1988; Grace and Russell, 1977). This damage often increases with the age of mature leaves, and this will cause permeability to increase with leaf age after maturity. Leaf cuticles have a permeability coefficient of $\sim 10^{-6}$ m g^{-1}, although this varies with humidity: water sorption by cuticles of *Abies alba* increased with atmospheric humidity (Chamel *et al.*, 1992). Permeability to water and nutrients of the cuticle is high relative to that of epicuticular wax, which is ~300 to 500 times less permeable to water compared to the cuticle (Schönherr, 1976). The permeability of a cuticle may depend on the presence in it of a large pectin fraction of high shrink–swell properties at dehydration–hydration states (Kannan, 1986). Cuticle permeability to ions and organic molecules changes depending on the characteristics (pH, chemical composition, concentration) of the penetrating solute, the solution at the end of the pathway, and the charge characteristics of both the cuticle and the specific ion or molecule (McFarlane and Berry, 1974; Swietlik and Faust, 1984; Riederer, 1989).

Movement through a cuticle is mostly by diffusion, but movement may be facilitated for nitrogenous compounds (Yamada *et al.*, 1965). Facilitated movement for organic nitrogenous compounds is indicated by a cuticular penetration pattern different from a simple first-order reaction, which describes the diffusive penetration of organic solutes and inorganic cations and anions through cuticular membranes. This diffusive movement is inversely related to the size and valency of the ion or organic molecule. Facilitated movement of organic nitrogenous compounds may be due to a change in the structure of the cuticular membrane, resulting in enhanced rates of penetration. A change in membrane structure is indicated by an enhanced penetration of inorganic cations and ions [e.g., calcium (Ca^{2+}) and phosphate] in the presence of urea.

In general, organic compounds penetrate the cuticle faster than do nonorganic ions and cations (Yamada *et al.*, 1965). Both nonelectrolytes (e.g., water and certain organic solutes) and electrolytes (mostly inorganic solutes) move across cuticular membranes at a rate determined by the concentration of the substance, its mobility in the cuticle, and the driving force acting on it (Riederer, 1989). Unlike a nonelectrolyte, movement of which is affected only by its own concentration difference, movement of one electrolyte is affected by the movement of all others. It is difficult to predict the rate or direction of movement of a given electrolyte in multiion systems.

A cuticle contains dissociable groups, which result in it having a cation exchange capacity ranging from 250 to 500 meq/kg, and an order of magnitude lower anion exchange capacity (Schönherr, 1982). Under most conditions the cuticle is weakly acidic, with calcium (Ca^{2+}) on most of its negative exchange sites. When weak electrolyte solutions from the cell wall come into contact with the inner cuticle, or those from acidic precipitation come into contact with the outer cuticle, cations in solution may exchange with those at negative sites. However, under such conditions anion uptake by the cuticle cannot occur. Only when solution concentration exceeds the concentration of fixed charge in the cuticle will it operate as a simple system in relation to the uptake of both cations and anions (Riederer, 1989). Empirical results demonstrate that nutrient uptake and loss, both as electrolytes and nonelectrolytes, by conifer canopies play an important role in plant nutrition under certain conditions (Tukey and Tukey, 1962; Bowden *et al.* 1989; Carleton and Kavanagh, 1990). For example, in *Picea abies* stands in northeastern Bavaria, canopy uptake of nitrogen (N) and loss of potassium (K) can account for 15 and 30% of annual plant uptake, respectively (Horn *et al.*, 1989).

Once ions have traversed the cuticle and reached cells, they are absorbed by processes that are mediated by metabolic energy (Kannan, 1986). Ions subsequently may move through either apoplastic or symplastic pathways to vascular tissues, and undergo further distribution within the leaf and plant. Ion movement into vascular tissues is passive, and its rate for each element depends much on the affinity of cells sited between the cuticle and nearby vascular tissues. Calcium and iron (Fe), for which cells have large affinity, reach vascular tissues at relatively low rates. Movement of ions in vascular tissues can be acropetal in xylem or, more often, basipetal in phloem. Most elements applied to foliage exhibit some degree of long-distance transport (out of the "treated" leaf). In general, N, phosphorus (P), and K are freely mobile, whereas Ca and magnesium (Mg), and most of the microelements, are partially mobile. Magnesium is partially mobile in plant tissues, but Ca, once sequestered in the middle lamella of cell walls, is quite immobile.

Another pathway for the movement of water and solutes into the xylem of twigs is via the bark and xylem ray cells (Katz *et al.*, 1989a). Movement depends on the creation of an aqueous continuum after bark is

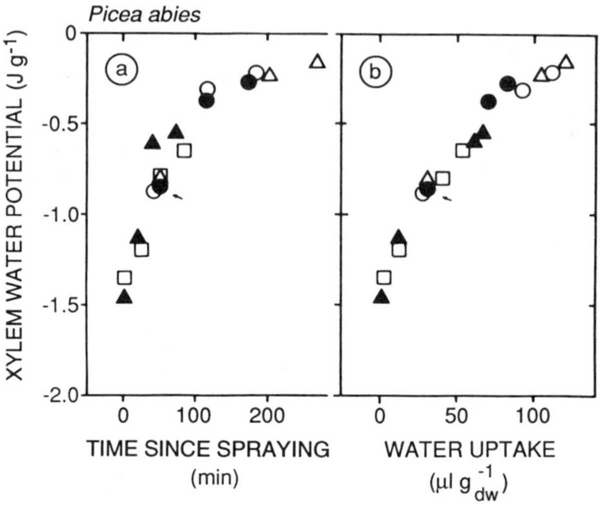

Figure 1 Increase in xylem water potential of *Picea abies* twigs in relation to (a) the time from initial spraying of the twig surface and (b) the cumulative amount of water absorbed. Reprinted from Katz, C., Oren, R., Schulze, E.-D., and Milburn, J. A. (1989). Uptake of water and solutes through twigs of *Picea abies* (L.) Karst. *Trees* 3:33–37. Copyright (1989) by Springer-Verlag.

saturated by precipitation. Movement of nutrients is primarily by diffusion, but both water and nutrients may move by mass flow. Diffusion may occur in either direction. However, because bark is the first part of the pathway to dry between precipitation events, mass flow out of trees is rare. Mass-flow uptake of water from the surface of bark depends on the driving force: flow into *P. abies* seedlings increases with decreasing plant water potential (Fig. 1).

Under mesic conditions, uptake of water by wet canopies is not important to the water budget of plants, particularly in light of the large decrease in demand for water due to lowered transpiration from wet foliage (Werk *et al.*, 1988). This absorbed water may, however, be important to recovery of a favorable plant water status as plant water potential increase. In regions and seasons with dry soil, uptake of water by the canopy may play an important role in the recovery of plant hydration when there are heavy fog, dew, or light rain events (e.g., Stone, 1965), or sometimes when the atmosphere is merely near saturation with water vapor (e.g., Slatyer, 1956).

The concentration of most nutrients, except for N and sulfur (S) in some regions, is typically low in precipitation water. Moreover, between rain events, most nutrients do not accumulate as dry deposition on canopy surfaces in quantities large enough to enrich canopy surface rainwater, again with the exception of N and S (Katz *et al.*, 1989b; Klemm, 1989). Thus, although uptake of Mg from twig surfaces, both by mass

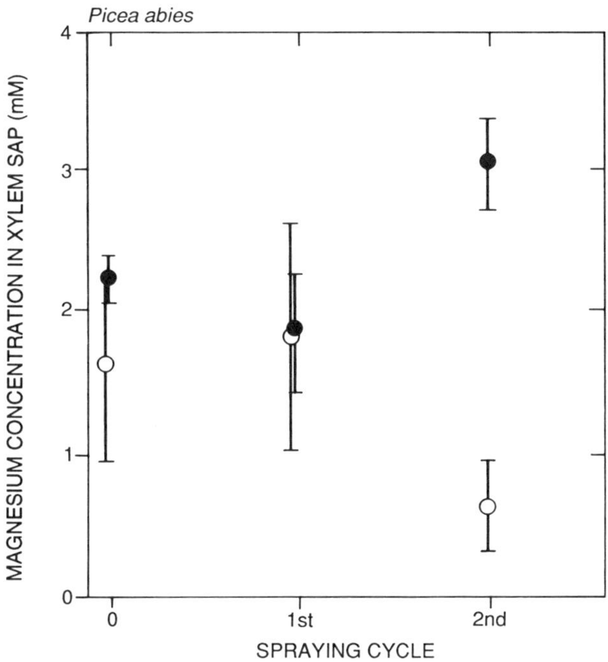

Figure 2 Magnesium concentration in the xylem sap of twigs that have a water potential of − 1.0 MPa, following repeated spraying with 2.0 m*M* magnesium chloride solution (●) or distilled water (○). Reprinted from Katz, C., Oren, R., Schulze, E.-D., and Milburn, J. A. (1989). Uptake of water and solutes through twigs of *Picea abies* (L.) Karst. *Trees* 3:33–37. Copyright (1989) by Springer-Verlag.

flow and diffusion, is of limited importance to a plant's Mg budget (Fig. 2), an uptake pathway can be seen to exist. This pathway may be used for uptake of N and S.

Nutrient uptake through wet canopies may reduce the demand for and uptake of these resources from soil. From 7 to 15% of the annual demand for N by *P. abies* stands was obtained through canopy uptake (Horn *et al.*, 1989): it is difficult to evaluate whether plants could have simply absorbed more N from the soil had pollution not bathed their canopies with so much N. Nevertheless, canopy absorption of elements is an efficient pathway for nutrient uptake. This is indicated by the faster uptake of Mg by *P. abies* stands after foliar dressing with Mg than after Mg was applied to the soil (Kaupenjohann *et al.*, 1989).

B. Acquisition by Roots

The importance of roots to trees can be seen from the proportion of assimilated carbon they invest annually belowground, estimated as 24 to 66% for forest crops generally (Cannell, 1985), and as 24 to 80% for *P. radiata* (Sheriff and Rook, 1990). Different root types have been identified (e.g., Kozlowski, 1971). Here we are interested in roots that are major sites of water and nutrient absorption.

Acquisition of soil resources depends on the amount (surface area) of root, on the proportions of various root types and their distribution, and on their rate of growth. These will be influenced by soil conditions, especially the availability of resources in the soil (see later), by the soil bulk density (e.g., Gliñski and Lipiec, 1990 for *P. radiata*), which is negatively associated with fine root mass, by the species, and by plant age. Generally, fine (surface) roots, mostly in the top ~300 to 400 mm of soil (e.g., Nambiar, 1990; Roberts, 1976), are the principal sites of absorption of nutrients and water when surface soil is moist. Deep (tap or sinker) roots will become increasingly important for water uptake as the surface soil dries. When all parts of a root system were healthy, fine roots of *Pinus taeda* were found to absorb more water per unit root weight and per unit root area compared to tap or lateral roots. However, tap roots absorbed more water per root length than did either fine roots or laterals (MacFall *et al.*, 1991). Woody roots can be of major importance for water uptake. In *Pinus elliottii* seedlings they comprise as much as 95% of total root length, and removal of nonwoody roots does not change water absorption in this species (Van Rees and Comerford, 1990). In contrast, zones of maximum water uptake in *P. abies* are the nonwoody apical zone and woody zones in which new lateral roots are formed (Häussling *et al.*, 1988).

Factors affecting the acquisition of water and nutrients interact, but involve different processes. Here we first examine acquisition of water, and then acquisition of nutrients by roots. Detailed discussions of various aspects of nutrient and water uptake can be found elsewhere (e.g., Barber *et al.*, 1984; Loughman *et al.*, 1989; Gliñski and Lipiec, 1990; Kozinka, 1992; Kolek and Holobrada, 1992).

1. Water There are three ways that roots can access new water in soil: (1) if the soil is wet, new input from, for example, rain will drain past roots and resupply areas where water has been removed; (2) water will move by flow or diffusion from wetter, more distant areas to drier areas of soil nearer roots; (3) axial growth can take (new) absorbing areas of a root into wetter parts of the soil. Rates of root growth and the volume of soil explored are reduced by high soil bulk density (Squire *et al.*, 1978) and by high soil strength (Sands *et al.*, 1979).

For any soil water status, plant water deficits depend on hydraulic conductances of the soil and roots. The region between root xylem and soil mass is normally the site of the smallest hydraulic conductance, and so of the largest gradient in water potential in the liquid phase of the soil, plant, atmosphere continuum. Conductances in this part of the pathway are influenced by soil conductivity, root conductivity, the pathway from the surrounding soil mass to the root zone (soil conductance), and the pathway between a root and nearby soil (rhizosphere conductance).

The hydraulic conductance of soils depends on soil pore and particle sizes, and so varies with soil type and, for a given soil type, with bulk density. Conductance declines more rapidly in coarse soils than in fine-textured soils as they dry. Coarse-textured soils also have a lower water-holding capacity and undergo a more rapid decrease in water potential with drying. Thus, soil and plant water deficits are expected earlier in coarse soils than in fine-textured soils (e.g., Campbell, 1985).

Several factors influence the surface area of a root in contact with surrounding soil, and therefore affect conductances in the rhizosphere. The hydraulic conductance of the radial path of roots of most plants is less than that of wet soil (e.g., Passioura, 1988; Sands and Theodorou, 1978). Nevertheless, the water uptake of 15- to 18-year-old *Abies amabilis* trees was unaffected even when one-third of their root system was severed (Teskey *et al.*, 1985). Leaf water potential and conductance declined only after one-half of the root system was severed. However, when soil water availability is not optimal, the small root length density of fine roots of conifers [500–53,000 m/m^3 vs. 50,000–500,000 m/m^3 for herbaceous plants (Nambiar, 1981)], especially of conifer seedlings [600–1,800 m/m^3 for young *P. radiata* vs. 320,000–340,000 m/m^3 for competing weeds (Nambiar, 1990)], will often result in the path between roots and the soil mass having a small conductance because of the small absorbing surface area. Thus, the water potential in a conifer will be lower than that of herbaceous species when the rates of transpiration are similar. The relationship between water supply and use of a stand will, therefore, depend on stand structure and composition.

Contact between a root surface and the immediately surrounding soil varies, and so affects the conductance of the pathway between roots and soil mass. Fine roots can grow easily in large spaces or cracks or in old root channels because of the very small physical resistance to root growth. However, here roots will be in poor contact with surrounding soil. Clumping of roots in these spaces will further reduce contact between roots and soil mass (Passioura, 1991). Shrinkage of roots or soil during periods of water deficit will often reduce contact between soil and roots for periods (e.g., Huck *et al.*, 1970; Atkinson and Wilson, 1979), drastically lowering the conductance of this part of the pathway.

Despite their low rooting density, deep roots play a major role in supply of water to trees in some environments. Water can be withdrawn from several meters to several tens of meters deep in a profile, depending on the species of conifer (Stone and Kalisz, 1991) and soil profile. This supply is important because it increases the effective store of available soil water, a source that may not be available to competing understory species, especially herbaceous species. Water use by the understory can be appreciable; for example, Roberts *et al.* (1984) showed that up to 52% of water transpired from a *Pinus sylvestris* forest can originate from the *Pteridium aquilinum* understory. There are few data on the relative amounts of water supplied by surface and deep roots of conifers. However, because deep roots have a lower rooting density, and the transpiration stream pathway is much longer, this route will have a lower hydraulic conductance than that through surface roots, and so is likely to provide a significant proportion of transpirational loss only when surface soils are dry. Effects of dry surface soil and deep roots on internal water deficits may sometimes be most clearly seen in the first few years after establishment of a plantation. In the summer following planting, Sands and Nambiar (1984) found *P. radiata* and weeds competed for water in the surface soil, whereas by their second summer the trees removed water to at least a 2-m depth, where roots of weeds could not reach. Extraction of water from depth can be important for survival during drought or as an additional supply during growth. Following establishment of a *P. radiata* plantation, depletion of water in the summer at a 1.5-m depth was first measured by Mitchell and Correll (1987) 33 months after establishment. By 48 months, withdrawal occurred to 2.7 m. Access to deeper soil allowed the young plantation to use, at least for a period, considerably more water than was supplied by current rainfall.

Effects of root density and deep rooting habit can be seen from the results of a simple simulation (Fig. 3). As the soil around its roots dries,

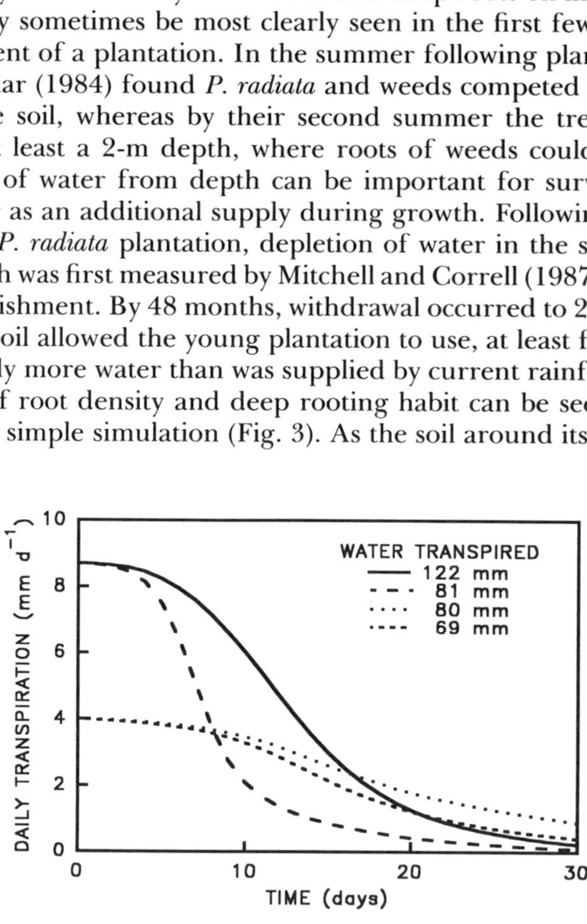

Figure 3 Simulation of daily transpiration determined over the period of a drying cycle for plants with shallow and deep roots at a high root density (—), shallow and deep roots at a low root density (· · ·), shallow roots only at a high root density (- - -), and shallow roots only at a low root density (– – –). Values were calculated by the technique of Campbell (1985) and are relative only. A comparison of total amounts of water transpired during the period of the simulation is given on the figure.

the rate of transpiration of a plant falls because of a reduction in stomatal conductance. This is caused by a lowering of "static" soil and plant water potentials, which is additional to the drop in water potential due to water flow through hydraulic conductances in the plant and soil. Hydraulic conductances in the soil are usually negatively related to water deficit. A low density of deep roots increases the size of the water resource available to a plant as compared with the same amount of root confined to surface soils. This means that transpiration does not decline as rapidly in deep-rooted as in shallow-rooted plants. Small hydraulic conductances resulting from a low rooting density cause water potentials within a plant to decline sooner than is observed with a high rooting density. If all other plant characteristics are the same, this results in an earlier reduction in transpiration in plants with a low root density, and so conserves water (Fig. 3). The same simulation indicates that over the period of a day, transpiration is most reduced (by low water potentials) in the middle of a day, and that this reduction increases as water deficits increase (Fig. 4). In addition, many species will exhibit a stomatal response to humidity (not taken into account in this simulation), which reduces stomatal conductance in drier air, and which therefore usually has most effect around the middle of the day. Generally, there is a positive association between transpiration efficiency (TE) (see Chapter 5, this volume) and the reduction in transpiration (and carbon assimilation) in the middle of the day (e.g., Cowan and Farquhar, 1977). Thus, in a water-limited environment it is of advantage to a plant to increase the

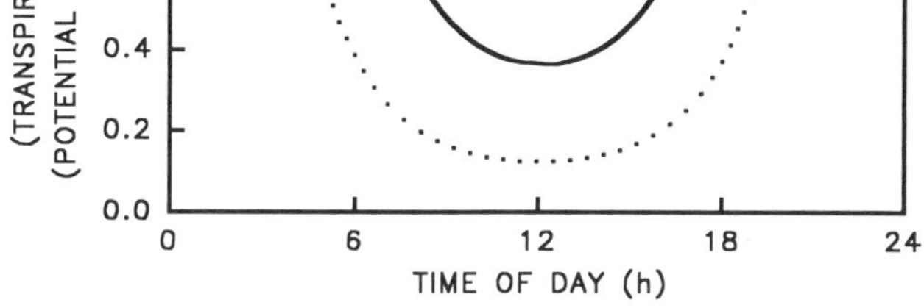

Figure 4 Simulation of diurnal rates of transpiration as a proportion of potential rates of transpiration at four times during a drying cycle: (- - -) beginning, (—) after 10 days, (– – –) after 20 days, and (· · ·) after 30 days. Values were calculated by the technique of Campbell (1985) and are relative only. Simulated potential rates of transpiration were greatest in the middle of the day, and followed a sinusoidal pattern.

possibility of midday stomatal closure when growing alone or in a community of plants with similar characteristics. This can be achieved by having a low rooting density and by apportioning some of those roots to deep exploration. In a mixed community the optimum tactics and strategies of a plant will depend on those used by neighboring species: there is no advantage to an individual to save water for others at its own expense.

If water is readily available, a higher growth rate is achieved by a dense root system, which maximizes uptake. Because plant water status is positively associated with hydraulic conductance, large conductances allow high rates of physiological activity and growth when the soil is moist. At sites where significant water deficits occur in the soil for periods, large hydraulic conductances in the root, soil, and rhizosphere could be disadvantageous in a monoculture or stand of species with similar characteristics, because it will lead to more rapid soil drying, and therefore to an earlier onset of critical water shortages. This will reduce both long-term growth of a stand and its TE (e.g., Passioura, 1972, 1983).

Hydraulic conductances of roots and soils change with space and time. Environmental factors can have large effects. At moderate temperatures low hydraulic conductances of roots and soil can lower plant water potentials and restrict rates of transpiration (e.g., Kaufmann, 1977; Day *et al.*, 1991). Low soil temperatures increase root resistance and the viscosity of water, so causing lower water potentials in transpiring plants. Low temperatures also restrict root growth, which in the longer term can cause water deficits within plants by restricting development of the water supply network, perhaps most noticeably in the first year after establishment of some plantations (e.g., Nambiar *et al.*, 1979). Anoxia can lower the hydraulic conductance of roots (e.g., Everard and Drew, 1987), a potential problem in waterlogged soils. The hydraulic conductance of roots is also lowered by nutrient deficiencies (below).

The hydraulic conductance of a root is a dynamic property, affected by plant–water and soil–water relations. Root conductance is not constant: the drop in water potential across a root is often not directly proportional to the rate of water flow through it (Passioura, 1988). There may be hysteresis in this relationship, and it can vary with time of day and rooting medium (Passioura, 1984; Passioura and Munns, 1984). Small soil water deficits may reduce absolute rates of growth, but they often increase root growth as a proportion of plant growth (Kramer and Kozlowski, 1979). More severe deficits have several effects. As soil dries its conductance declines (e.g., Campbell, 1985), root mortality increases (e.g., Coutts, 1982; Santantonio and Hermann, 1985; Santantonio, 1989), root growth declines (Leshem, 1965), and root conductivity decreases at least in part because of suberization (Sands *et al.*, 1982). Root conductivity that has become lower in dry soil may remain low for several days after rewetting (e.g., Passioura, 1988). Water uptake in dry soil may be aided by mycorrhizae (R. Sands unpublished, cited by Sands and Mulligan, 1990), which not only greatly increase contact with the soil (Lamhamedi *et al.*, 1992), but may also increase the hydraulic conductivity of the root. However, the effects of mycorrhizae on uptake are not always simply predictable: the hydraulic conductivity of roots of *Pseudotsuga menziesii* seedlings increased when they were inoculated with one ectomycorrhiza-forming fungal species, whereas two other fungal species had no effect (Coleman *et al.*, 1990). This difference may result from differences in hyphal thickness, or in the quantity of the extramatrical phase, or both (e.g., Lamhamedi *et al.*, 1992), or could be a secondary effect resulting from the removal of P deficiency.

*2. **Nutrients*** Much of the above discussion applies to acquisition of nutrients and to water in the same way, but additional factors are also important. Nutrients are absorbed into roots as ions dissolved in water, and may be absorbed against a concentration gradient with expenditure of energy. Acquisition of mobile ions involves their movement in soil solution with the mass flow of water taken into roots, or diffusion in soil solution to regions of depletion near root surfaces. Often both occur (e.g., Nye and Tinker, 1977). Immobile ions are acquired mainly by roots extending into regions of new supply. Root elongation may be enhanced by nutrient deficiencies (e.g., Anuradha and Narayanan, 1991).

Roots regulate nutrient uptake by controlling intake or efflux, or both, of ions. Sometimes plant tissue concentrations are maintained constant independent of external concentrations (e.g., Letey *et al.*, 1982). At least some roots modify their rhizosphere by secretions that make nutrients more available, although this may be to the detriment of the plant's nitrogen balance (e.g., Griffiths and Robinson, 1992). Secreted polysaccharides cohere soil particles, improving both porosity and contact between root and soil (e.g., Vermeer and McCully, 1982; McCully and Canny, 1988; Letey *et al.*, 1982), and provide a substrate for microorganisms, enhancing microbial activity in the rhizosphere. Enzymes (e.g., Hedley *et al.*, 1982a,b), or protons (e.g., Gardner *et al.*, 1981), most often secreted in response to a nutrient deficiency, enhance the uptake of nutrients, especially those that are relatively immobile. There is little information about the role of deep roots in the absorption of nutrients; it is generally accepted that most absorption is usually through fine surface roots except possibly with very dry or nutrient-deficient topsoils (Comerford *et al.*, 1984).

Both nonwoody and woody roots participate in nutrient uptake. Total K uptake by roots of *P. elliottii* seedlings was not changed when the nonwoody roots were removed (Van Rees and Comerford, 1990). However, in *P. abies* roots the highest rate of K uptake occurred in the root elongation zone, especially if lateral roots were developed. Rates of Mg and Ca uptake were highest at the apical zone, but Ca, like water, was taken up at a high rate in zones of formation of new laterals (Häussling *et al.*, 1988).

Uptake of phosphorus, an immobile nutrient, by roots depends on active processes. The rate of P uptake per root surface area, by *Pinus contorta*, *Picea sitchensis* (Dighton and Harrison, 1983), and *Pinus taeda* (Pennell *et al.*, 1990), from P-deficient sites was higher than that of roots from similar sites that were fertilized with P. This uptake was reduced by 50% when the temperature was reduced from 25 to 7°C, and uptake stopped when root metabolism was inhibited. P uptake is often enhanced with successful mycorrhizal formation, particularly in soils low in P. Ectomycorrhizas increase P uptake by *Pseudotsuga menziesii* (Coleman *et al.*, 1990) and *P. contorta* (Carleton and Read, 1991). Thus, factors that inhibit mycorrhizal formation are likely to reduce P uptake. For example, the formation of mycorrhizae is inversely related to the Al:Ca ratio in the mineral soil of *P. abies* stands (Meyer *et al.*, 1988). Nitrate (NO_3^-) has been found to inhibit the formation of mycorrhizal associations with roots of *P. radiata* seedlings under both acidic and alkaline soil conditions; the effect under alkaline conditions was exacerbated by a reduction in growth of the fungi due to high pH (Theodorou and Bowen, 1969).

Nitrogen may be present in soil in any of several available, mobile forms, of which NH_4^+ and NO_3^- are the only forms taken up by plants in significant amounts. In forest soils, NH_4^+ is often more available than NO_3^-, but its uptake may not be proportionally higher due to the greater mobility of NO_3^-. Some species of conifers (e.g., *P. abies*, *P. elliottii*, *P. menziesii*, and *Larix kaempferi*) take up NH_4^+ "in preference" to NO_3^-, but in most forest stands uptake of both N forms may be similarly available, or uptake of NO_3^- may be greater if it is the most available form (Arnold and van Diest, 1991; Buchmann *et al.*, 1992; Marschner *et al.*, 1991; Olsthoorn *et al.*, 1991; van den Driessche, 1972). The concentrations of NH_4^+ were higher than that of NO_3^- in solutions collected below the organic layer in both moderately polluted and relatively unpolluted *P. abies* stands. However, the opposite was true for solutions collected from all mineral soil horizons of both stands (Schneider *et al.*, 1989). Uptake of both NH_4^+ and NO_3^- is sensitive to root temperature. The uptake by roots of *P. abies* was significantly lower at 10°C (NH_4^+) or 15°C (NO_3^-) relative to uptake at 20°C.

There is often a positive association between a plant's shoot:root ratio and the nutritional quality of a site. For example, a large NH_4^+ addition caused seedlings of *P. menziesii* to have reduced root growth and root length per unit of biomass while shoot growth was enhanced (Olsthoon *et al.*, 1991). At nutritionally poor sites there is often a local proliferation of roots in nutritionally "rich" microenvironments; for example, N-deficient *P. menziesii* seedlings grew a disproportionate amount of roots in zones with high N (Friend *et al.*, 1990). This, and greater partitioning of biomass to roots at nutritionally poor sites, can considerably increase the carbon cost of a plant's root system, although roots in nutritionally poor microenvironments are more efficient users of respired carbon (e.g., Clarkson, 1985; Smucker, 1984; St. John *et al.*, 1983).

The spatial pattern of roots may be affected by, and will influence, their environment, and this will have a potentially large effect on nutrient uptake. Roots in *P. elliottii* stands were found to be randomly distributed and unaffected both by the understorey and by a plant's microsite (Escamilla *et al.*, 1991). However, in the vicinity of roots, zones depleted of nutrients were larger in infertile compared with fertile soils, causing more intense competition between neighboring roots. Because competing vegetation often has a higher root density, the efficiency of nutrient uptake by conifers is often lowered proportionally more by low soil fertility than is uptake by average roots in a stand.

3. Effects of Water on Nutrient Uptake from Soil The mass flow of mobile ions to roots is proportional to the rate of transpiration. The rate at which ions can diffuse in soil solution to the surface of a root is proportional to the number of water-filled soil pores: as soil dries its pores drain of water in sequence according to their size. Water deficits reduce both the number of water-filled pores and the rate of transpiration (through stomatal closure), and so reduce rates of both diffusion and mass flow to roots. These effects are, to a limited extent, offset by an increase in concentration of the soil solution as the soil dries (Nye and Tinker, 1977). Lower soil moisture causes a reduction in permeability of roots to nutrients as well as to water. Nutrient acquisition will, therefore, be constrained by slower rates both of nutrient transport and of nutrient uptake. Moreover, lower soil moisture also reduces the mineralization rate and the availability of N, thus lowering the uptake of N even in fertilized plots (e.g., Raison *et al.*, 1990). In the case of P, the concentration in solution is determined by the buffering capacity of the soil, and this in turn is a function of soil water content. As the soil dries, the concentration of P increases, shifting the equilibrium more toward P sorption, rather than desorption. Overall, soil water deficits have a smaller effect on uptake of immobile than of mobile ions.

4. Effects of Nutrients on Water Uptake from Soil A positive effect of nutrition on the water status of conifers has been reported in a number of studies, for example, by Hillerdal-Hagstromer *et al.* (1982) for *P. sylvestris* and by Brix and Mitchell (1986) for *P. menziesii*, but the mechanism has not been completely identified. It is not always possible to be certain of the cause of this nutritional effect because better nutrition may improve stomatal control of water loss, or may increase the hydraulic conductance of roots, or may do both. These both raise plant water potential, but potentially have opposite effects on water use: better stomatal control will reduce use, while greater hydraulic conductance may increase it. The effect on stomata is well-known, and there is accumulating evidence that improved nutrition increases the hydraulic conductance of conifer roots. Myers (1988) found that predawn water potentials of *P. radiata* foliage in an irrigation/fertilizer experiment were higher in fertilized than in unfertilized trees of each irrigation treatment, an effect that was independent of soil water content (Linder *et al.*, 1987). Predawn water potential measurements on unirrigated *P. radiata* with different fertilizer treatments (D. N. Fife, E. K. S. Nambiar and Woods, unpublished cited in Nambiar, 1990; D. W. Sheriff and E. K. S. Nambiar, unpublished, 1990) support these observations. Plant water potentials measured predawn are normally the highest found during a day, at a time when rates of transpiration are very low. These results indicate that greater nutrient supply or higher nutrient concentrations in plant tissues, or both, increase the ability of roots to absorb water. There is evidence for this from nonconiferous species for which deficiencies of N or P reduce the hydraulic conductance of roots severalfold (e.g., Radin and Eidenbock, 1986; Radin, 1990; Salim, 1991). Radin and Eidenbock (1984) found this effect can be great enough to lower the turgor pressure of cells sufficiently to limit cell expansion growth. However, this is an extreme case. It is generally found that water deficits limit nutrient uptake much more than nutrient deficits limit water uptake.

A greater biomass partitioning aboveground at higher nutrient supply may result in a lowering of a plant's water potential because of disparity between the plant's ability to take up water and its transpirational demand. Local proliferation of roots at nutritionally better microsites in soils poor in nutrients lowers the ability of plants to absorb water for a particular carbon investment in root biomass, and makes them more vulnerable to drought. However, a lower demand for water by the smaller shoots of N-deficient plants will at least somewhat offset this reduced capacity.

Excess of some salts in the soil, most notably NaCl, will also reduce the hydraulic conductance of roots (e.g., Salim, 1991), adding to a plant's problems of extracting water from a soil of low water potential.

5. Effect of Availability of Nutrients in the Soil on the Uptake of Others The availability of some elements in soil can affect the ability of a plant to take up others, both by changing the availability of other elements in the soil, and by interactions during the uptake process. Interactions between nutrients result from biological and chemical effects. Because of the large requirement by plants for N and P, much of the information on the topic is about interactions between these elements.

Effects of availability of one element on the availability of others vary with the soil, and with the elements in question. Availability may be altered by a change in pH, which can, for example, alter the balance between soluble and insoluble forms of mineral salts, or by addition of an ion that changes the solubility of another ion species, usually one with opposite charge. Similarly, death of decomposer organisms, caused by a change in soil chemistry, can result in a short-term increase in availability of at least some mineral elements. Alternatively, a change in soil chemistry may increase the activity of decomposer organisms, and so increase mineralization. Longer term changes in the activity of decomposers that are of consequence to nutrient availability can be expected only from a very drastic perturbation of the stand or soil, which produces conditions that prevent organisms from stabilizing their environment. Any one or combination of these effects may influence the availability of nutrients, although sometimes it is not possible to determine causes of such changes. For example, Maftoun and Pritchett (1970) could not find a clear cause for N fertilizer reducing the availability of P in coastal plain *P. elliottii* soils. The reduction in P availability may have been the result of a rapid depletion of available P from the inherently low P spodosols as growth was stimulated by N addition.

In addition to modifying availability in the soil, one nutrient can affect processes involved in the uptake of others. Ion transport across membranes involves binding of an ion to a carrier on the surface of the membrane. Competition between different ions for these sites is greatest between ions with similar physicochemical properties, and causes suppression of uptake of all competing ions. However, some ions have greater affinity than others for binding sites, and uptake of those with lowest affinity will be most suppressed. For example, the highly hydrated Mg^{2+} ion has low affinity for binding sites, and its uptake can be greatly suppressed by Ca^{2+} and K^+: Schimansky (1981) found a 45% reduction in Mg^{2+} uptake by Ca^{2+} and a 92% reduction by a combination of Ca^{2+} and K^+. Alternatively, uptake of one element can increase that of another by concurrent uptake, i.e., by causing movement across a membrane of an ion of charge opposite of that being taken up actively.

Effects of soil pH on nutrient uptake are caused by competition between H^+ and other cations, especially at low pH, and between OH^-

and other anions, especially at high pH. These effects could result from competition for membrane binding sites, or from production of electropotentials across membranes due to an accumulation of H^+ or OH^- in cells at low and high pH. When uptake of anions or cations is suppressed by pH, uptake of ions with the opposite charge is enhanced.

A strong effect on chemical processes in the rhizosphere and, thus, on availability of nutrients to plants is exerted by the combined effect of soil pH and the prevailing form of N. Although the concentration of NO_3^- in solution may be greater than NH_4^+ in some forest soils, the stocks of NH_4^+ on exchange sites are large, and NH_4^+ is well buffered. Hence uptake of NH_4^+ may still exceed that of NO_3^-. A study using ^{15}N application of both N forms to a 12-year-old *P. abies* stand revealed similar rates of NO_3^- and NH_4^+ uptake (Buchmann *et al.*, 1992). Uptake of NH_4^+ is associated with a net H^+ production, and uptake of NO_3^- with a net H^+ consumption. Large consumption of H^+ causes alkalinization of the rhizosphere relative to that of the bulk soil. This increases or decreases the availability of other nutrients, depending on the pH and soil chemical properties (Marschner *et al.*, 1991). Such changes are likely to alter the species composition and abundance of microorganisms in the rhizosphere, further altering nutrient availability.

At a more holistic level, uptake of one element may affect that of others by its effects on growth, on partitioning of carbon to roots, on root permeability, morphology and distribution, and on mycorrhizal development. Thus, for example, if plant growth increases with N uptake, uptake of other elements will also increase, provided they are sufficiently available, although their concentration in plant tissues may decline. Increasing the supply to *P. abies* and *P. sylvestris* seedlings of any of the macroelements from limiting to sufficient increased growth and the uptake of all measured elements (Ingestad, 1959, 1979). However, although an increase in Mg supply only slightly affected N uptake, a greater N supply resulted in a large increase in Mg uptake (A. Berger, unpublished, 1991). Thus, when growth is increased by increasing the supply of a limiting resource, the uptake of some other elements may be enhanced proportionally less, reducing their concentration in tissues and causing localized deficiencies (Ingestad and Ågren, 1988; Oren and Schulze, 1989). In contrast, Hunter and Hunter (1991) found that addition of P to *P. radiata* plantations had little effect on foliar N concentrations, which were marginal to adequate. Effect of P addition on N uptake varied, being small at two sites, where increased P caused aboveground dry weights to be ~1.1 to 1.3 times greater than the control, and 41% greater at a site where aboveground weights of P-treated trees were ~1.5 times greater than the control. It is generally found that the absolute uptake of each element increases with growth, and any factor that increases growth is likely to enhance nutrient uptake.

An increase in N uptake resulting from a sudden increase in the concentration of soil N can produce a significant reduction in either root length/weight, or in mycorrhizae (Meyer, 1985). A lower uptake of relatively soil-immobile elements (e.g., P) may thus be an indirect effect of an increased N concentration.

At any site we can expect to find direct effects of a change in uptake of one element on uptake of others. An indirect effect through alteration of growth and roots is most likely to occur only after a heavy application of imbalanced fertilizer. Except for such fertilizer application, the overall plant response to increased uptake of one element is likely to have no effect, or to increase the uptake of other elements.

6. Effects of Demand on Uptake of Nutrients and Water Uptake of nutrients and water depends not only on the availability of these resources, but also on the requirement (or "demand") by a plant for them. During periods when growth is greatest, e.g., when there is a flush of shoot extension growth, there is a large demand by the new tissues for nutrients. Except in young seedlings this can, in the short-term, be partly met by retranslocation of nutrients, especially from adjacent green, nonsenescent, older foliage (e.g., Sheriff *et al.*, 1986; Oren and Schulze, 1989). This internal redistribution of nutrients does not reduce the requirement for them from the external environment, but lengthens the period of uptake; retranslocation from senescent foliage does reduce demand on the external environment. Following a period of active growth, tissues that lost nutrients are replenished (e.g., Sheriff *et al.*, 1986; Oren *et al.*, 1988). The buffering capacity that retranslocation provides means that growth, and the accompanying demand for and uptake of nutrients, are more rapidly affected by lack of soil water than by lack of soil nutrients.

Sometimes growth is slow, or ceases because of factors other than water deficit [e.g., phenology or environmental factors such as light (Ingestad, 1987) or temperature (Berger, 1989)]. Under these conditions uptake of nutrients may be limited more by low demand for nutrients than by the availability of nutrients, as determined, for example, by the amount of nutrient at absorbing surfaces and by the supply of water. Ingestad (1987) showed that an increase in the supply of light caused the demand for all nutrients to increase proportionally. Similarly, studies have demonstrated that phenology affects the demand for and uptake of nutrients. Absorption of P by roots of *P. taeda* growing on soils poor in P was apparent only during the growing season (Pennell *et al.*, 1990);

Mg uptake was high only during periods of flush growth, and was greatly reduced during the intervening, nonflush periods (Kelly and Barber, 1991).

On long temporal scales (seasons to years), water uptake will strongly reflect the amount and distribution of roots, their hydraulic properties and mycorrhizal associations, and the leaf area of plants and canopies. These are influenced by the species, and by the environment in which plants are growing.

The effect of soil water and nutrient status on partitioning of assimilates aboveground and belowground will not be discussed in detail in this chapter. In general, the ratio of carbon allocated belowground to that allocated aboveground decreases with increasing supply of resources in the soil. However, the absolute amount allocated to root does not decrease, or may even increase with increase in soil resources. It is the comparatively large increase in the absolute amount allocated aboveground that drives down the root–shoot allocation. Thus, when *P. sylvestris* stands were provided with water, fertilizer, or with a combination of both resources, carbon allocated to the shoot increased relative to that allocated to fine roots by ~0.5, 2.4, and 3.8 times, respectively (Axelsson, 1981; Waring and Schlesinger, 1985). This was only as a result of increased shoot growth; there was no change in fine root production (Linder and Axelsson, 1982).

IV. Uptake and Use of Water and Nutrients

A. Principles

Water and nutrients are both gained and lost from plants of all ages. They have different residence times, which vary (usually increase) with plant size. There is a change in patterns of acquisition, allocation, and internal recycling as plants grow.

The amount of water stored and available for transpiration in seedlings is small relative to that in saplings and trees. Water is stored in tissues that can undergo elastic changes in volume as their water content changes (cambium, fine roots, immature tracheids, leaves, parenchyma, and phloem), and in sapwood, where changes in water storage are by cavitation. Some of this volume will not vary until the occurrence of a catastrophic event, such as death of an organ or plant. This portion is important to maintenance of integrity and vital function, but not to use of water in quantity by a plant. This, and some of the volume changes due to growth involve plastic changes in plant water content. Other parts of the total volume are more readily exchangeable and will contribute to water transpired each day and during drying cycles, this being replaced at night and on rewetting. The importance of this exchangeable water depends on its quantity in relation to the volume of water transpired over a period of time. Although there is variation between species, generally in any environment the importance of this exchangeable water depends on its amount per leaf area. This water is not all equally exchangeable. A particular decline in water potential of a tissue will, over a period of time sufficient for a new equilibrium to be established, cause the tissue to release a particular amount of water. A further decline in tissue water potential will cause release of more water. A similar sequence occurs in reverse during rehydration, often with hysteresis. The amount exchanged will depend on aspects of the state of the tissue, and cannot be expected to remain the same for any tissue over long periods of time. This exchangeable water constitutes the hydraulic capacitance [(quantity of water exchanged)/(change in water potential)] of a plant and its tissues. Hydraulic resistances and hydraulic capacitances in the transpiration stream together result in a change in flow of xylem sap occurring later than the causative change in transpiration or root conductance, with greater delays furthest from the source of the change (e.g., Jarvis, 1975). The magnitude of a perturbation declines with "distance" from its source: here distance is defined by hydraulic properties such that it is greater when the combination of hydraulic capacitance and resistance is larger. Because of larger hydraulic capacitance and resistance between water in the bulk soil and the foliage, bigger trees often show greater lags in flow of xylem sap and uptake behind transpiration than do small trees.

Residence times of nutrients are usually much longer than those of water, so we often view their use during residence and their retranslocation differently. Small seedlings retain in their tissues all of the nutrients they absorb. As they grow larger some tissues, especially older leaves and fine roots, die and some nutrients are lost in this biomass. Internal cycling of nutrients lessens the amount plants need to absorb from their environment. The capacity for recycling increases with plant size, and the capacity to fulfil demand by recycling is often negatively related to the rate of growth. Recycling of nutrients within conifers was recently reviewed by Nambiar and Fife (1991).

Forest canopies may act as a source or sink for N and S, depending on the availability of these resources to the plant and their concentration in the atmosphere or precipitation water. Data from different species and stands indicate a range of scenarios. The uptake of N from cloud water by *Picea rubens* is small relative to demand, but accumulation in the stem suggests an alternative mechanism for long-term N retention by spruce stands (Bowden *et al.*, 1989). In contrast, the uptake of N and S from precipitation by *P. abies* needles and twigs may provide a significant portion of the demand for these nutrients if their concentration is high (Katz *et al.*, 1989b; Klemm, 1989). Gaseous emission of S by *P. abies* branches is related to stomatal conductance, but losses are small and decrease under limited availability of soil S (Rennenberg *et al.*, 1990). In contrast, NO_2 is taken up and reduced by *P. abies* crowns in relation to their stomatal conductance and the atmospheric concentration of NO_2 (Thoene *et al.*, 1991). However, uptake of gaseous oxides of N by *Pinus strobus* branches is small relative to demand (Vose and Swank, 1990), and *P. sylvestris* forests in Sweden appear to act as a negligible sink for atmospheric NO_x (Johansson, 1987). Forest canopies may act as a source of gaseous ammonia if the atmospheric concentration of ammonia is low, and as a sink if the concentration is high (Langford and Fehsenfeld, 1992). The compensation point is likely to vary depending on the species composition, age, and structure of a forest, as well as on environmental variables that affect stomatal conductance and biochemical reactions. One example of this is the greater uptake of N and smaller loss of Ca and Mg by younger than by older stands of *Picea mariana* (Carleton and Kavanagh, 1990), perhaps reflecting a greater demand for these nutrients by the more vigorously growing younger stands. In general, nutrients other than N and S are leached from plant and forest canopies by precipitation (Tukey and Tukey, 1962; Horn *et al.*, 1989).

The high frequency of root grafts in conifers indicates that viewing trees as discrete individuals when they are in a stand may be simplistic (Eis, 1972). Thus, care must be taken in collection and interpretation of data on the acquisition and allocation of resources. Root grafts (e.g., Eis, 1972; Kozlowski and Cooley, 1961) increase in number during development of dense stands, from none early in the life of a tree. They have potential to supply additional nutrients or water from some plants to other members of the same species, possibly favoring those individuals that become dominant (Eis, 1972). There is inadequate information to indicate how large this supply is, but it seems that individuals may not **greatly** increase their access to soil resources by this route (e.g., Bormann, 1966). Mycelial strands that join roots of different plants have a similar role in interplant transfer of resources (e.g., Brownlee *et al.*, 1983).

B. By Seedlings

In seedlings, instantaneous rates of water uptake and transpiration are often similar. For example, in *P. abies* seedlings with adequate water supply, the rate of water uptake was found to be highly correlated with

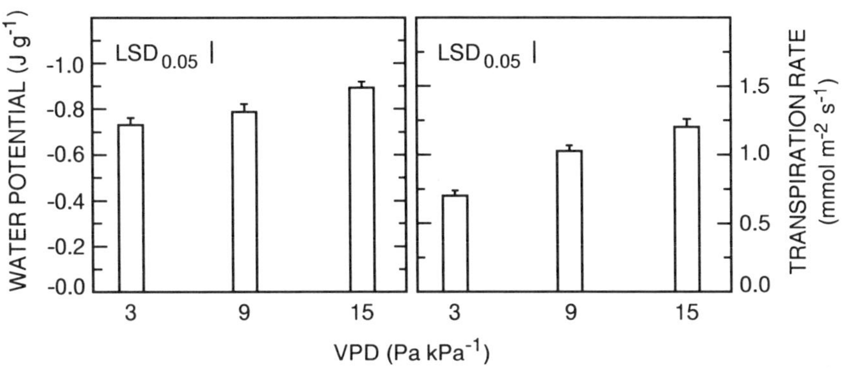

Figure 5 Water uptake and xylem water potential of seedlings after 9 hours of transpiration under different levels of vapor pressure deficit. After Berger (1989).

the vapor pressure deficit of the air after stomata opened at the onset of the light period, although plant water potential declined (Fig. 5) (Berger, 1989). A reduction in water content of a peat–vermiculite rooting medium to 25% of holding capacity did not affect the water potential of a plant or its water uptake (Fig. 6) (Berger, 1989). Only when the water content was reduced to less than 20% of capacity did both water potential and water uptake of plants decline rapidly with a relatively small decrease in water content of the root medium.

In the above studies, K, Ca, and Mg concentrations in xylem sap did not decrease over mild or severe drought cycles, but N concentration declined even before water uptake was reduced. This indicates that as the soil dried N uptake was reduced because N availability decreased, but the availability of other elements was not so rapidly affected. This is similar to findings for young stands of *P. abies* (Osonubi *et al.*, 1988). Alternatively, the ability of roots to take up N may be impaired in dry soil (Runge, 1983). Under the more severe drought conditions Gross and Pham Nguyen (1987) found seedling growth was reduced. A reduction in growth would result in a lower demand for elements, which could be met by a lower uptake rate (Nye and Tinker, 1977; Clarkson, 1985).

Mycorrhizal development has long proved essential for development of hardy and vigorous seedlings (see Chapter 3, this volume), a result of several possible effects. Ectomycorrhizal seedlings of *Pinus pinaster* have enhanced transport in the phloem of reduced N supplied as NO_3^-, but there is no effect on absorption (Scheromm *et al.*, 1990). Seedlings of *Larix laricina* inoculated with ectomycorrhizal fungus grew at a rate similar to that of uninoculated seedlings when nutrient supply was high (Chakravarty and Chatarpaul, 1990). When the supply of nutrient was

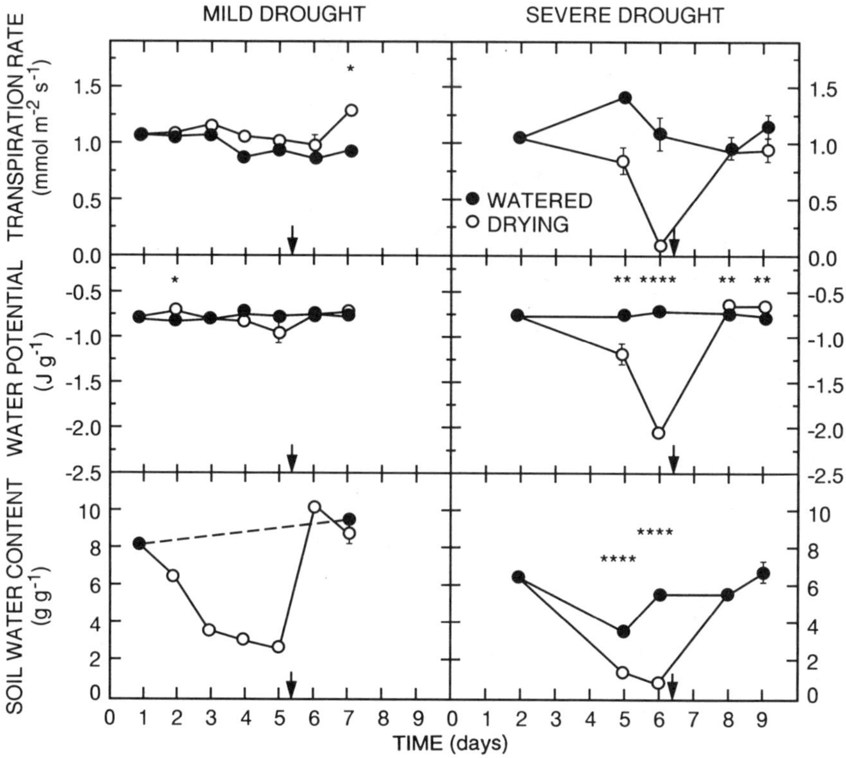

Figure 6 Daily water uptake and xylem water potential after 9 hours of transpiration for seedlings subjected to a vapor pressure deficit of 9 Pa/kPa while experiencing a drying and rewetting cycle. After Berger (1989).

low, inoculated seedlings grew faster and had higher concentrations of all macronutrients compared to uninoculated seedlings. The root length and biomass of inoculated seedlings were greater regardless of fertility, indicating that mycorrhizal seedlings are better equipped to face periods of low nutrient and water supply. Mycorrhizal development is enhanced when soil N and P are low: in *Pinus caribaea* seedlings mycorrhizal development was found to be reduced by high soil NO_3^- and high total N (Richards and Wilson, 1963). Mycorrhizal development was also enhanced by high availability of soluble carbohydrates relative to N in roots. This represents an "ideal symbiosis" for seedlings: their fungal partner is fully developed under conditions of low nutrient availability, when it can be of greatest help in sequestering nutrients. It is more poorly developed and so does not cost much to maintain when its services are not necessary.

C. In Mature Stands

Water storage of sapwood of large conifers can be as much as 300 tons ha^{-1}, the quantity transpired in 5–10 days (Waring and Running, 1978). Supply of this water to transpirational loss of *P. menziesii* was found by Waring and Running (1978) to be nearly continuous during a summer drought, with periods of partial recharge from summer rains. Over short periods 30–50% of water transpired by *P. sylvestris* trees can be supplied from their sapwood, and water available from the sapwood is equivalent to this proportion of daily transpiration (Waring *et al.*, 1979). Thus, large trees have considerable capacity to buffer against a short-term increase in transpirational demand and against periods of soil water deficit. They are able to rehydrate tissues when transpiration is small, during rain and at night, and to maintain some carbon assimilation during periods of soil water deficit. This buffering capacity enables large trees to function with a transpiration stream of considerably smaller hydraulic conductance than would otherwise be necessary. The trees, thus, do not need to invest as much carbon in roots that can readily absorb water or in the greater sapwood area (i.e., greater hydraulic conductance) they would need without the reservoir.

Nutrients required by foliage of large trees must be originally taken up from the external environment. However, once the foliage of a tree has acquired an appreciable nutrient capital it is able to recycle some of this internally (see above). The proportion recycled depends on several factors: the nutrient in question, the species, stand age, and various site factors, including availability of nutrient in the soil. For example, D. N. Fife and E. K. S. Nambiar (unpublished data, 1988–1990) found that 68–81% of the N and 35–69% of the P in foliage of 7-year-old *P. radiata* trees, as measured at their maximum levels, had moved out of the needles prior to abscission. This retranslocation was highest in trees with greatest N supply: increasing N supply stimulated growth, thus enhancing the demand placed on retranslocation. Turner and Lambert (1986) estimated that 50–60% of the annual P requirement was provided by retranslocation in *P. radiata*. However, Son and Gower (1991) estimated that only 27 and 30% of the annual N and P requirements of *Pinus resinosa*, *Pinus strobus*, and *P. abies* foliage were met by retranslocation, whereas 81 and 72%, respectively, were met in *Larix decidua*, a deciduous conifer on the same site. Internal cycling of nutrients will supply more of the N and P, and possibly other nutrients required by new foliage when trees have established a constant leaf area index, than is possible earlier in their life. Internal cycling means that not as much carbon need be invested in fine roots, in a high absorption capacity, or in specialized storage tissues as would be needed if this cycling did not occur.

Although it is important because annual loss of N and P from a plant through fine root turnover is almost twice the annual loss in abscinded needles (Nambiar and Fife, 1991), there is not much information on this topic. Nambiar (1987) concluded there is little retranslocation of N or P from fine roots of *P. radiata* prior to their death. There was also no clear evidence of nutrient retranslocation from old to young roots of *P. radiata* in a root regeneration experiment (Nambiar and Fife, 1991), although such retranslocations was found by van den Driessche (1985) in *P. menziesii.*

Root elongation and distribution in mature conifer stands depends on soil chemical conditions and on the accompanying vegetation. Fertilization in *P. sylvestris* stands reduced the root tips to one-third of unfertilized stands, perhaps due to a reduction in mycorrhizal aggregates (Ahlström *et al.*, 1988). High aluminum content in the soil solution reduced root elongation in *P. abies* stands (Meyer *et al.*, 1988). Moreover, roots of *P. sylvestris* and *P. abies* become more sensitive to soil drought with increasing soil acidity, which means that they require a longer time to recover and resume root elongation after drought (Bartsch, 1987). With increasing soil acidification, either by continuous acidic deposition or through forest management practices, a reduction in root length extension into deep soil horizons and reduced mycorrhizal development (Meyer *et al.*, 1988) are likely to result in lower capacity for water and nutrient uptake.

D. Limitations to Growth

The rate of water or nutrient uptake by roots or canopy may limit growth. To assess this it is necessary to evaluate the potential of each species to acquire and use these resources in relation to their availability, to atmospheric demand for water, and to effects of competing vegetation. As discussed earlier, understories can contribute a major proportion of the total transpired water. This can be seen, in more detail than previously discussed, from studies of a *P. abies* stand near Leningrad that had been observed before clear felling, and subsequently for 90 years of stand regrowth (Fig. 7). The vegetation established following removal of the forest used less water than the preceding forest, and within 8 years the regenerating stand used more water than the understory vegetation. At no time during the establishment phase of the new forest did water use by all vegetation exceed precipitation. When the stand reached 30 years of age, understory vegetation was shaded out by the trees and did not play a further role in the water cycle.

In contrast, water uptake by *Pinus ponderosa* increased as the canopy leaf area developed to $\sim 1.5\ m^2/m^2$, at which point most of the available

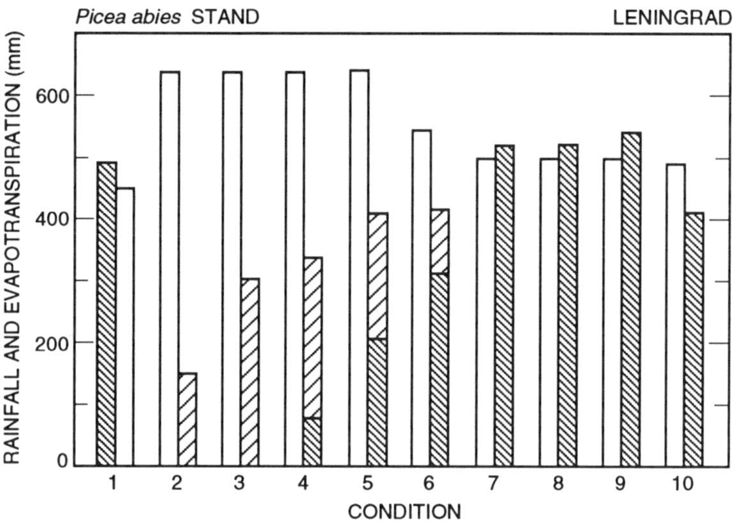

Figure 7 Water balance of a stand of spruce near Leningrad (observations taken over many years) for (1) a 100-year-old stand (crown closure 0.8); (2) an area of clear felling, with *Vaccinium myrtillus*; (3) an area of earlier clear felling, with *Polytrichum commune* and the onset of waterlogging; (4) an even older deforestation, with *Sphagnum* cover and tree saplings; (5) a young plantation, 5–8 years old with spruce and pioneer plants; (6) a young stand of spruce, 12–15 years old (crown closure 0.9–1.0); (7) a stand 25–30 years old; (8) a stand 50–60 years old; (9) a stand 70–80 years old, ready for felling; (10) a stand 85–90 years old. Data are presented for rainfall (mm) ( ) and evapotranspiration (mm) (\\\\\\) from the tree stand and (/ / /) from other plants. Reprinted from Walter, H., and Breckle, S.-W. (1986). "Ecological Systems of the Geosphere 3: Temperate and Polar Zonobiomes of Northern Eurasia," Figure 1.3.3. Copyright (1986) by Springer-Verlag.

water was used during the growing season (Fig. 8). Further increase in pine canopy leaf area to $4m^2/m^2$ must have been accompanied by an increasing efficiency of water use by the larger canopy, because growth continued to increase with leaf area. Where not eradicated, the understory vegetation consumed as much water as did a conifer canopy with a leaf area of $\sim 1\ m^2/m^2$, resulting in a significant reduction in both conifer stem growth and leaf area development. The impact of the understory was particularly severe in this stand because hardly any of the relatively low (600 mm) precipitation fell during the 6-month dry season, during which the ambient vapor pressure deficit is high. The effect of the understory was increased by limitation of the rooting zone to 2 m, and restriction of root elongation to a fraction of a meter per year by the abrasive pumice soil.

Unlike *P. abies* in the former example, *P. ponderosa* does not develop a canopy leaf area high enough to shade out the understory, which will

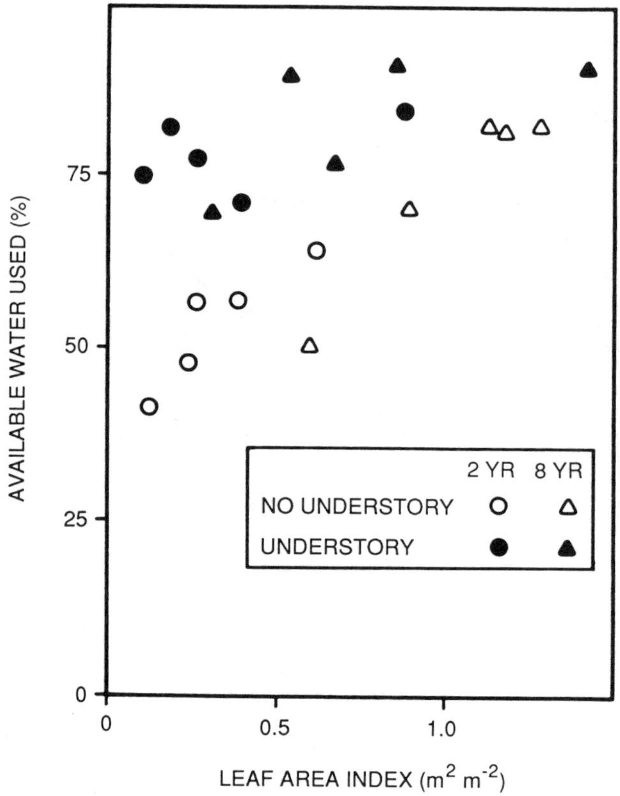

Figure 8 Relationship between pine leaf area index and the percentage of available soil water used to the end of the growing season by thinned stands of ponderosa pine. Data were collected 2 and 8 years after thinning from stands of five tree densities, with and without understory vegetation. Reprinted from Oren *et al.* (1987), with permission.

continue to compete with the pine for water and nutrients. This is one of the most severely limiting conditions for water and nutrient uptake: water availability is low, soil exploration by roots is highly restricted, light summer rains are not available for a direct canopy uptake of water, and the understory vegetation is an efficient competitor. In this situation, controlling the understory may be the single most effective way for forest managers to enhance growth. However, had the soil permitted better root development, had there been a greater or a more evenly distributed precipitation, or had the conifer been a species that develops high canopy leaf area (e.g., as in the first example), controlling the understory may result in only a marginal increase in growth. Clearly, to predict successfully the effect of any manipulation on stand growth, the anticipated increase in water and nutrient availability cannot be regarded in isolation.

Acknowledgments

We thank Dr. Daniel D. Richter, School of the Environment, Duke University, Durham, North Carolina; Dr. J. Clive Carlyle, Plantation Forest Research Centre, Division of Forestry, CSIRO, Mount Gambier, South Australia; and Dr. Philip Polglase and Mr. Peter Snowdon, Division of Forestry, CSIRO, Canberra, ACT, Australia, for helpful comments on the manuscript.

References

Ahlström, K., Persson, H., and Börjesson, I. (1988). Fertilization in a mature Scots pine (*Pinus sylvestris* L.) stand—Effects on fine roots. *Plant Soil* 106:179–190.

Allen, H. L., Dougherty, P. M., and Campbell, R. G. (1990). Manipulation of water and nutrients—Practice and opportunity in southern U.S. pine forests. *For. Ecol. Manage.* 30:437–453.

Anuradha, M., and Narayanan, A. (1991). Promotion of root elongation by phosphorus deficiency. *Plant Soil* 136:273–275.

Arnold, G., and van Diest, A. (1991). Nitrogen supply, tree growth and soil acidification. *Fert. Res.* 27:29–38.

Atkinson, D., and Wilson, S. A. (1979). The root soil interface and its significance for fruit tree roots of different ages. *In* "The Soil Root Interface" (J. L. Harley and R. S. Russell, eds.), pp. 259–271. Academic Press, London.

Axelsson, B. (1981). Site differences and yield differences in biological production or in redistribution of carbon within trees. *Res. Rep. No. 9 Dept. Ecol. Environ. Res. Swed. Univ. Agric. Sci., Uppsala.*

Barber, S. A., Bouldin, D. R., Kral, D. M., and Hawkins, S. L. (1984). "Roots, Nutrient and Water Influx, and Plant Growth." Soil Science Society of America, Madison, Wisconsin.

Bartsch, N. (1987). Responses of root systems of young *Pinus sylvestris* and *Picea abies* plants to water deficits and soil acidity. *Can. J. For. Res.* 17:805–812.

Berger, A. (1989). Nährstofftransport im xylemsaft der Fichte (*Picea abies* (L.) Karst.) in Beziehung zur Tageszeit, Transpiration, zum Bodenwassergehat und zum Nährstoffkonzentration in der Bodenlösung. Diplom work, University of Bayreuth, Bayreuth.

Bormann, F. H. (1966). The structure, function, and ecological significance of root grafts in *Pinus strobus* L. *Ecol. Monogr.* 36:1–26.

Bowden, R. D., Geballe, G. T., and Bowden, W. B. (1989). Foliar uptake of ^{15}N from simulated cloud water by red spruce (*Picea rubens*) seedlings. *Can. J. For. Res.* 19:382–386.

Brix, H., and Mitchell, A. K. (1986). Thinning and nitrogen effects on soil and tree water stress in a Douglas-fir stand. *Can. J. For. Res.* 2:467–478.

Brownlee, C., Duddridge, J. A., Malibari, A., and Read, D. J. (1983). The structure and function of mycelial systems of ectomycorrhizal roots with special reference to their role in forming inter-plant connections and providing pathways for assimilate and water transport. *Plant Soil* 71:433–443.

Buchmann, N., Oren, R., Gebauer, G., Dietrich, P., and Schulze, E.-D. (1992). The use of stable isotopes in ecosystem research—First results of a field study with ^{15}N. *Die Isotopenpraxis* 28:51–59.

Campbell, G. S. (1985). "Soil Physics with Basic." Elsevier, Amsterdam.

Cannell, M. G. R. (1985). Dry matter partitioning in tree crops. *In* "Attributes of Trees and Crop Plants" (M. G. R. Cannell and J. E. Jackson, eds.), pp. 160–193. Institute of Terrestrial Ecology, Huntington, England.

Carleton, T. J., and Kavanagh, T. (1990). Influence of stand age and spatial location on throughfall chemistry beneath black spruce. *Can. J. For. Res.* 20:1917–1925.

Carleton, T. J., and Read, D. J. (1991). Ectomycorrhizas and nutrient transfer in conifer—Feather moss ecosystems. *Can. J. Bot.* 69:778–785.

Chakravarty, P., and Chatarpaul, L. (1990). Effect of fertilization on seedling growth, ectomycorrhizal symbiosis, and nutrient uptake in *Larix laricina*. *Can. J. For. Res.* 20: 245–248.

Chamel, A., Escoubes, M., Baudrand, G., and Girard, G. (1992). Determination of water sorption by cuticles isolated from fir tree needles. *Trees* 6:109–114.

Clarkson, D. T. (1985). Factors affecting mineral nutrient acquisition by plants. *Annu. Rev. Plant Physiol.* 36:77–115.

Coleman, M. D., Bledsoe, C. S., and Smit, B. A. (1990). Root hydraulic conductivity and xylem sap levels of zeatin riboside and abscisic acid in ectomycorrhizal Douglas-fir seedlings. *Tree Phytol.* 115:275–284.

Comerford, N. B., Kidder, G., and Mollitor, A. V. (1984). Importance of subsoil fertility to forest and non-forest plant nutrition. *In* "Forest Soils and Treatment Impacts. Proceedings of the Sixth North American Forest Soils Conference" (E. L. Stone, ed.), pp. 381–404. University of Tennessee, Knoxville.

Coutts, M. P. (1982). Water relations of Sitka spruce seedlings after root damage. *Picea sitchensis. Ann Bot.* 49(5):661–668.

Cowan, I. R., and Farquhar, G. D. (1977). Stomatal function in relation to leaf metabolism and environment. *Symp. Soc. Exp. Biol.* 31:471–505.

Day, T. A., Heckathorn, S. A., and DeLucia, E. H. (1991). Limitations of photosynthesis in *Pinus taeda* L. (Loblolly pine) at low soil temperatures. *Plant Physiol.* 96:1246–1254.

Dighton, J., and Harrison, A. F. (1983). Phosphorus nutrition of lodgepole pine and Sitka spruce stands as indicated by a root bioassay. *Forestry* 56:33–43.

Eis, S. (1972). Root grafts and their silvicultural implications. *Can. J. For. Sci.* 2:111–120.

Escamilla, J. A., Comerford, N. B., and Neary, D. G. (1991). Spatial pattern of slash pine roots and its effect on nutrient uptake. *Soil Sci. Soc. Am. J.* 55:1716–1722.

Everard, J. D., and Drew, M. C. (1987). Mechanisms of inhibition of water movement in anaerobically treated roots of *Zea mays* L. *J. Exp. Bot.* 38:1154–1165.

Friend, A. L., Eide, M. R., and Hinckley, T. (1990). Nitrogen stress alters root proliferation in Douglas-fir seedlings. *Can. J. For. Res.* 20:1524–1529.

Gardner, W. K., Parbery, D. G., and Barber, D. A. (1981). Proteoid root morphology and function in *Lupinus albus*. *Plant Soil* 60:143–148.

Gliński, J., and Lipiec, J. (1990). "Soil Physical Conditions and Plant Roots." CRC Press, Boca Raton, Florida.

Grace, J. (1988). Plant response to wind. *Agric. Ecosyst. Environ.* 22/23:71–88.

Grace, J., and Russell, G. (1977). The effect of wind on grasses. III. Influence of continuous drought or wind on anatomy and water relations in *Festuca arundinacea* Schreb. *J. Exp. Bot.* 28:268–278.

Griffiths, B., and Robinson, D. (1992). Root-induced nitrogen mineralization: A nitrogen balance model. *Plant Soil* 139:253–263.

Gross, K., and Pham Nguyen, T. (1987). Einfluss von langfristigem konstanten Wassermangelstress auf die Netto-Photosynthese und das Wachstum junger Fichten (*Picea abies* (L.) Karst) und Douglasien (*Pseudotsuga menziesii* (Mirb.) Franco) im Freiland. *Forstwiss. Centralbl.* 106:7–26.

Häussling, M., Jorns, C. A., Lehmbecker, G., Hecht-Buchholz, Ch., and Marschner, H. (1988). Ion and water uptake in relation to root development of Norway spruce (*Picea abies* (L.) Karst.). *J. Plant Physiol.* 133:486–491.

Hedley, M. J., Nye, P. H., and White, R. E. (1982a). Plant induced changes in the rhizosphere of rape (*Brassica napus* var. Emerald) seedlings. II. Origin of the pH change. *New Phytol.* 91:31–44.

Hedley, M. J., White, R. E., and Nye, P. H. (1982b). Plant induced changes in the rhizosphere of rape (*Brassica napus* var. Emerald) seedlings. III. Changes in L value, soil phosphate fractions and phosphatase activity. *New Phytol.* 91:45–56.

Hillerdal-Hagstromer, K., Mattson-Djos, E., and Hellkkvist, J. (1982). Field studies of water relations and photosynthesis in Scots pine. II. Influence of irrigation and fertilization on needle water potential of young pine trees. *Physiol. Plant.* 54:295–301.

Horn, R., Schulze, E.-D., and Hantschel, R. (1989). Nutrient balance and element cycling in healthy and declining Norway spruce stands. *In* "Forest Decline and Air Pollution: A Study of Spruce (*Picea abies*) on Acid Soils, Ecological Studies 77" (E.-D. Schulze, O. L. Lange, and R. Oren, eds.), pp. 444–455. Springer-Verlag, Berlin.

Huck, M. G., Klepper, B., and Taylor, H. M. (1970). Diurnal variations in root diameter. *Plant Physiol.* 45:529–530.

Hunter, I. R., and Hunter, J. A. C. (1991). Apparent phosphorus uptake and change in nitrogen content of *Pinus radiata* growing on soils of different phosphorus retention, treated with superphosphate and A-grade rack phosphate. *N. Z. J. For. Sci.* 21:50–61.

Ingestad, T. (1959). Studies on the nutrition of forest tree seedlings. II. Mineral nutrition of spruce. *Physiol. Plant.* 12:568–593.

Ingestad, T. (1979). Mineral nutrient requirements of *Pinus sylvestris* and *Picea abies* seedlings. *Physiol. Plant.* 45:373–380.

Ingestad, T. (1987). New concepts on soil fertility and plant nutrition as illustrated by research on forest trees and stands. *Geoderma* 40:237–252.

Ingestad, T., and Ågren, G. I. (1988). Nutrient uptake and allocation at steady-state nutrition. *Physiol. Plant.* 72:450–459.

Jarvis, P. G. (1975). Water transfer in plants. *In* "Heat and Mass Transfer in the Biosphere. Part I. Transfer Processes in the Plant Environment" (D. A. deVries, and N. H. Afgan, eds.), pp. 369–394. Scripta Books (Wiley), New York.

Johansson, C. (1987). Pine forest: A negligible sink for atmospheric NO_x in rural Sweden. *Tellus* 39B:426–438.

Kannan, S. (1969). Penetration of iron and some organic substances through isolated cuticular membranes. *Plant Physiol.* 44:517–521.

Kannan, S. (1986). Foliar absorption and transport of inorganic nutrients. *Crit. Rev. Plant Sci.* 4:341–375.

Katz, C., Oren, R., Schulze, E.-D., and Milburn, J. A. (1989a). Uptake of water and solutes through twigs of *Picea abies* (L.) Karst. *Trees* 3:33–37.

Katz, C., Oren, R., Schulze, E.-D., and Steudle, E. (1989b). Uptake of water and solutes through bark and needle surfaces. *In* "Forest Decline and Air Pollution: A Study of Spruce (*Picea abies*) on Acid Soils, Ecological Studies 77" (E.-D. Schulze, O. L. Lange, and R. Oren, eds.), pp. 218–221. Springer-Verlag, Berlin.

Kaufmann, M. R. (1977). Soil temperature and drying cycle effects on water relations of *Pinus radiata*. *Can. J. Bot.* 55:2413–2418.

Kaupenjohann, M., Zech, W., Hantschel, R., Horn, R., and Schneider, B. U. (1989). Effects of fertilization. *In* "Forest Decline and Air Pollution: A Study of Spruce (*Picea abies*) on Acid Soils, Ecological Studies 77" (E.-D. Schulze, O. L. Lange, and R. Oren, eds.), pp. 418–424, Springer-Verlag, Berlin.

Kelly, J. M., and Barber, S. A. (1991). Magnesium uptake kinetics in loblolly pine seedlings. *Plant Soil* 134:227–232.

Klemm, O. (1989). Leaching and uptake of ions through above-ground Norway spruce parts. *In* "Forest Decline and Air Pollution: A Study of Spruce (*Picea abies*) on Acid Soils, Ecological Studies 77" (E.-D. Schulze, O. L. Lange, and R. Oren, eds.), pp. 210–237. Springer-Verlag. Berlin.

Kolek, J., and Holobrada, M. (1992). Ion uptake and transport. *Dev. Plant Soil Sci.* 46: 204–285.

Kozinka, V. (1992). Uptake and transport of water. *Dev. Plant Soil Sci.* 46: 129–203.

Kozlowski, T. T. (1971). "Growth and Development of Trees Volume 1." Academic Press, New York.

Kozlowski, T. T., and Cooley, J. H. (1961). Natural root grafting in northern Wisconsin. *J. For.* 59: 105–107.

Kramer, P. J., and Kozlowski, T. T. (1979). "Physiology of Woody Plants." Academic Press, New York.

Lamhamedi, M. S., Bernier, P. Y., and Fortin, J. A. (1992). Hydraulic conductance and soil water potential at the soil-root interface of *Pinus pinaster* seedlings inoculated with different dikaryons of *Pisolithus* sp. *Tree Physiol.* 10: 231–244.

Lange, O. L., and Schulze, E.-D. (1966). Untersuchungen über die Dickenentwicklung der kutikularen Zellwandschichten bei der Fichtennadel. *Forstwiss. Centralbl.* 85: 27–38.

Langford, A. O., and Fehsenfeld, F. C. (1992). Natural vegetation as a source or sink for atmospheric ammonia: A case study. *Science* 255: 581–583.

Leshem, B. (1965). The annual activity of intermediary roots of the Aleppo pine. *For. Sci.* 11: 291–298.

Letey, J., Jarrell, W. M., and Valora, N. (1982). Nitrogen and water uptake patterns and growth of plants at various minimum solution nitrate concentrations. *J. Plant Nutr.* 5: 73–89.

Linder, S., and Axelsson, B. (1982). Changes in carbon uptake and allocation patterns as a result of irrigation and fertilization in a young *Pinus sylvestris* stand. *In* "Carbon Uptake and Allocation: Keys to Management of Subalpine Ecosystems. IUFRO Workshop." (R. H. Waring, ed.), pp. 38–44. Forest Research Laboratory, Oregon State Univ., Corvallis.

Linder, S., Benson, M. L., Myers, B. J., and Raison, R. J. (1987). Canopy dynamics and growth of *Pinus radiata*. I. Effects of irrigation and fertilization during a drought. *Can. J. For. Res.* 17: 1157–1165.

Loughman, B. C., Gasparikova, O., and Kolek, J. (1989). "Structural and Functional Aspects of Transport in Roots." Kluwer Academic Publ., Dordrecht, The Netherlands.

MacFall, J. S., Johnson G. A., and Kramer, P. J. (1991). Comparative water uptake by roots of different ages in seedlings of loblolly pine (*Pinus taeda* L.). *New Phytol.* 119: 551–560.

Maftoun, M., and Pritchett, W. L. (1970). Effects of added nitrogen on the availability of phosphorus to slash pine on two lower coastal plain soils. *Soil Soc. Am. Proc.* 34: 685–690.

Marschner, H., Häussling, M., and George, E. (1991). Ammonium and nitrate uptake rates and rhizosphere pH in non-mycorrhizal roots of Norway spruce [*Picea abies* (L.) Karst]. *Trees* 5: 14–21.

Matyssek, R., Reich, P., Oren, R., and Winner, W. (1994). Response mechanisms of conifers to air pollutants. *In* "Ecophysiology of Coniferous Forests" (W. K. Smith and T. M. Hinckley, eds.), pp. 255–308. Academic Press, San Diego.

McCully, M. E., and Canny, M. J. (1988). Pathways and processes of water and nutrient movement in roots. *Plant Soil* 111: 159–170.

McFarlane, J. C., and Berry, W. L. (1974). Cation penetration through isolated leaf cuticles. *Plant Physiol.* 53: 723–727.

Meyer, F. H. (1985). Einfluss des Stickstoff-Faktors auf den Mykorrihizabesatz von Fichtensämlingen im Humus einer Waldschadenfläche. *Allg Forstz* 40: 208–219.

Meyer, J., Schneider, B. U., Werk, K. S., Oren, R., and Schulze, E.-D. (1988). Performance of two *Picea abies* (L.) Karst. stands at different stages of decline. V. Root tips and ectomycorrhiza development and their relation to above-ground and soil nutrients. *Oecologia* 77:7–13.

Mitchell, B. A., and Correll, R. L. (1987). The soil water regime in a young radiata pine plantation in southeastern Australia. *New For.* 4:273–289.

Myers, B. J. (1988). Water stress integral—A link between short-term stress and long-term growth. *Tree Physiol.* 4:315–323.

Nambiar, E. K. S. (1981) Ecological and physiological aspects of development of roots: From nursery to forest. *In* "Proceedings of the Australian Forest Nutrition Workshop Productivity in Perpetuity," pp. 117–129. CSIRO, Canberra.

Nambiar, E. K. S. (1987). Do nutrients retranslocate from fine roots? *Can. J. For. Sci.* 17:913–918.

Nambiar, E. K. S. (1990). Interplay between nutrients, water, root growth and productivity in young plantations. *For. Ecol. Manage.* 30:213–232.

Nambiar, E. K. S., and Fife, D. N. (1991). Nutrient retranslocation in temperate conifers. *Tree Physiol.* 9:185–207.

Nambiar, K. D. S., Bowen, G. D., and Sands, R. (1979). Root regeneration and plant water status of *Pinus radiata D. Don* seedlings transplanted to different soil temperatures. *J. Exp. Bot.* 30(119):1119–1131.

Nye, P. H., and Tinker, P. B. (1977). "Solute Movement in the Soil–Root System." Blackwell, Oxford.

Olsthoorn, A. F. M., Keltjens, W. G., Van Baren, B., and Hopman, M. C. G. (1991). Influence of ammonium on fine root development and rhizosphere pH of Douglas-fir seedlings in sand. *Plant Soil* 133:75–81.

Oren, R., and Schulze, E.-D. (1989). Nutritional disharmony and forest decline: A conceptual model. *In* "Forest Decline and Air Pollution: A Study of Spruce (*Picea abies*) on Acid Soils, Ecological Studies 77" (E.-D. Schulze, O. L. Lange, and R. Oren, eds.), pp. 428–443. Springer-Verlag, Berlin.

Oren, R., Waring, R. H., Stafford, S. G., and Barrett, J. W. (1987). Twenty-four years of Ponderosa pine growth in relation to canopy leaf area and understorey competition. *For. Sci.* 33:538–547.

Oren, R., Schulze, E.-D., Werk, K. S., and Meyer, J. (1988). Performance of two *Picea abies* (L.) Karst. at different stages of decline. VII. Nutrient relations and growth. *Oecologia* 77:163–173.

Osonubi, O., Oren, R., Werk, K. S., Schulze, E.-D., and Heilmeier, H. (1988). Performance of two *Picea abies* (L.) Karst. at different stages of decline. IV. Xylem sap concentration of magnesium, calcium, potassium and nitrogen. *Oecologia* 77:1–6.

Passioura, J. B. (1972). The effect of root geometry on the yield of wheat growing on stored water. *Aust. J. Agric. Res.* 23:745–752.

Passioura, J. B. (1983). Roots and drought resistance. *Agric. Water Manage.* 7:265–280.

Passioura, J. B. (1984). Hydraulic resistance of plants. I. Constant or variable? *Aust. J. Plant Physiol.* 11:333–339.

Passioura, J. B. (1988). Water transport in and to roots. *Ann. Rev. Plant Physiol. Plant Mol. Biol.* 39:245–265.

Passioura, J. B. (1991). Soil structure and plant growth. *Aust. J. Soil Res.* 29:717–728.

Passioura, J. B., and Munns, R. (1984). Hydraulic resistance of plants. II. Effects of rooting medium, and time of day, in barley and lupin. *Aust. J. Plant Physiol.* 11:341–350.

Pennell, K. D., Allen, H. L., and Jackson, W. A. (1990). Phosphorus uptake capacity of 14-year-old loblolly pine as indicated by a ^{32}P root bioassay. *For. Sci.* 36:358–366.

Radin, J. W. (1990). Responses of transpiration and hydraulic conductance to root temperature in nitrogen- and phosphorus-deficient cotton seedlings. *Plant Physiol.* 92: 855–857.

Radin, J. W., and Eidenbock, M. P. (1984). Hydraulic conductance as a factor limiting leaf expansion of phosporus-deficient cotton plants. *Plant Physiol.* 75:372–377.

Radin, J. W., and Eidenbock, M. P. (1986). Carbon accumulation during photosynthesis in leaves of nitrogen- and phosphorus-stressed cotton. *Plant Physiol.* 82:869–871.

Raison, R. J., Khanna, P. K., Connell, M. J., and Falkiner, R. A. (1990). Effects of water availability and fertilization on nitrogen cycling in a stand of *Pinus radiata*. *For. Ecol. Manage.* 30:31–43.

Raven, J. A. (1988). Acquisition of nitrogen by the shoots of land plants: Its occurrence and implications for acid–base regulation. *New Phytol.* 109:1–20.

Rennenberg, H., Huber, B., Schröder, P., Stahl, K., Haunold, W., Georgii, H. W., Slovik, S., and Pfanz, H. (1990). Emission of volatile compounds from spruce trees. *Plant Physiol.* 92:569–564.

Richards, B. N., and Wilson, G. L. (1963). Nutrient supply and mycorrhiza development in Caribbean pine. *For. Sci.* 9:405–412.

Riederer, M. (1989). The cuticles of conifers: Structure, composition and transport properties. *In* "Forest Decline and Air Pollution: A Study of Spruce (*Picea abies*) on Acid Soils. Ecological Studies 77" (E.-D. Schulze, O. L. Lange, and R. Oren, eds.), pp. 157–192. Springer-Verlag, Berlin.

Roberts, J. (1976). A study of root distribution and growth in a *Pinus sylvestris* L. (Scots pine) plantation in East Anglia. *Plant Soil* 44:607–621.

Roberts, J., Wallace, J. S., and Pitman, R. M. (1984). Factors affecting stomatal conductance of bracken below a forest canopy. *J. Appl. Ecol.* 21:643–655.

Runge, M. (1983). Physiology and ecology of nitrogen nutrition. *In* "Physiological Plant Ecology III. Responses to the Chemical Environment," pp. 163–200, Springer-Verlag, New York.

Salim, M. (1991). Change in water conducting properties of plant roots by nutrition and salt stress. *J. Agron. Crop Sci.* 166:285–287.

Sands, R., and Mulligan, D. R. (1990). Water and nutrient dynamics and tree growth. *For. Ecol. Manage.* 30:91–111.

Sands, R., and Nambiar, E. K. S. (1984). Water relations of *Pinus radiata* in competition with weeds. *Can. J. For. Res.* 14:233–237.

Sands, R., and Theodorou, C. (1978). Water uptake by mycorrhizal roots of radiata pine seedlings. *Aust. J. Plant Physiol.* 5:301–309.

Sands, R., Greacen, E. L., and Gerard, C. J. (1979). Compaction of sandy soils in radiata pine forests. I. A penetrometer study. *Aust. J. Soil Res.* 17:101–113.

Sands, R., Fiscus, E. L., and Reid, C. P. P. (1982). Hydraulic properties of pine and bean roots with varying degrees of suberization, vascular differentiation and mycorrhizal infection. *Aust. J. Plant Physiol.* 9:559–569.

Santantonio, D. (1989). Dry-matter partitioning and fine-root production in forests—New approaches to a difficult problem. *In* "Biomass Production by Fast-Growing Trees," pp. 57–72. Kluwer Academic Publ., Dordrecht, The Netherlands.

Santantonio, D., and Hermann, R. K. (1985). Standing crop, production, and turnover of fine roots on dry, moderate and wet sites of mature Douglas fir in western Oregon. *Ann. Sci. For.* 42:113–142.

Scheromm, P., Plassard, C., and Salsac, L. (1990). Nitrate nutrition of maritime pine (*Pinus pinaster* Soland in AIT.) ectomycorrhizal with *Hebeloma cylindrosporum* Romagn. *New Phytol.* 114:93–98.

Schimansky, C. (1981). Der Einfluss eininger Versuchsparameter auf das Fluxverhalten von ^{28}Mg bei Gerstenkeimpflanzen in Hydrokulturversuchen. *Landwirtsch. Forsch.* 34: 154–165.

Schneider, B. U., Meyer, J., Schulze, E.-D., and Zech, W. (1989). Root and mycorrhizal development in healthy and declining Norway spruce stands. *In* "Forest Decline and Air Pollution: A Study of Spruce (*Picea abies*) on Acid Soils. Ecological Studies 77" (E.-D. Schulze, O. L. Lange, and R. Oren, eds.), pp. 428–443. Springer-Verlag, Berlin.

Schönherr, J. (1976). Water permeability of isolated cuticular membranes: The effect of cuticular waxes on diffusion of water. *Planta* 131: 159–164.

Schönherr, J. (1982). Resistances of plant surfaces to water loss: Transport properties of cutin, suberin and associated lipids. *In* "Physiological Plant Ecology II, Encyclopedia of Plant Physiology" (O. L. Lange, P. S. Nobel, C. B. Osmond, and H. Ziegler, eds.), Vol. 12B, pp. 153–179. Springer-Verlag, Berlin.

Schönherr, J., and Bukovac, J. (1970). Preferential polar pathways in the cuticle and their relationship to ectodesmata. *Planta* 92: 189–201.

Schönherr, J., and Bukovac, J. (1972). Penetration of stomata by liquids. Dependence on surface tension, wettability, and stomatal morphology. *Plant Physiol.* 49: 813–819.

Schönherr, J., and Riederer, M. (1989). Foliar penetration and accumulation of organic chemical in plant cuticles. *Rev. Environ. Contam. Toxicol.* 108: 1–70.

Sheriff, D. W., and Rook, D. A. (1990). Wood density and aboveground growth in high and low wood density clones of *Pinus radiata* D. Don. *Aust. J. Plant Physiol.* 17: 615–628.

Sheriff, D. W., Nambiar, E. K. S., and Fife, D. N. (1986). Relationships between nutrient status, carbon assimilation and water use efficiency in *Pinus radiata* (D. Don) needles. *Tree Physiol.* 2: 73–88.

Slatyer, R. O. (1956). Absorption of water from atmospheres of different humidity and its transport through plants. *Aust. J. Biol. Sci.* 9: 552–558.

Smethurst, P. J., and Nambiar, E. K. S. (1989). An appraisal of the *in situ* soil-core technique for measuring nitrogen uptake by a young *Pinus radiata* plantation. *Soil Biol. Biochem.* 21: 939–942.

Smucker, A. J. M. (1984). Carbon utilization and losses by plant root systems. *In* "Roots, Nutrient and Water Influx, and Plant Growth" (S. A. Barber and D. R. Bouldin, eds.), pp. 27–48. Soil Science Society of America, Crop Science Society of America, American Society of Agronomy, Madison, Wisconsin.

Son, Y., and Gower, S. T. (1991). Aboveground nitrogen and phosphorus use by five plantation-grown trees with different leaf longevities. *Biogeochemistry* 14: 167–191.

Squire, R. O., Marks, G. C., and Craig, F. G. (1978). Root development in a *Pinus radiata* D. Don plantation in relation to site index, fertilizing and soil bulk density. *Aust. For. Res.* 8: 103–14.

St. John, T. V., Coleman, D. C., and Reid, C. P. P. (1983). Growth and spatial distribution of nutrient-absorbing organs: Selective exploitation of soil heterogeneity. *Plant Soil* 71: 487–493.

Stone, E. C. (1965). The role of dew in pine survival in soils below the wilting point and its measurement. *UNESCO Arid Zone Res.* 25: 421–427.

Stone, E. L., and Kalisz, P. J. (1991). On the maximum extent of tree roots. *For. Ecol. Manage.* 46: 59–102.

Swietlik, D., and Faust, M. (1984). Foliar nutrition of fruit crops. *Hortic. Rev.* 6: 287–355.

Teskey, R. O., Grier, C. C., and Hinckley, T. M. (1985). Relation between root system size and water inflow capacity of *Abies amabilis* growing in a subalpine forest. *Can. J. For. Res.* 15: 669–672.

Theodorou, C., and Bowen, G. D. (1969). The influence of pH and nitrate on mycorrhizal association of *Pinus radiata* D. Don. *Aust. J. Bot.* 17: 59–67.

Thoene, B., Schröder, P., Papen, H., Egger, A., and Rennenberg, H. (1991). Absorption of atmospheric NO_2 by spruce (*Picea abies* L. Karst.) trees. I. NO_2 influx and its correlation with nitrate reduction. *New Phytol.* 117:575–586.

Tukey, J. R., and Tukey, S. R. (1962). The loss of organic and inorganic materials by leaching from leaves and other aboveground plant parts. *In* "Radioisotopes in Soil–Plant Nutrition Studies." International Atomic Energy Agency, Vienna.

Turner, J. (1982). Long-term superphosphate trial in regeneration of *Pinus radiata* at Belanglo state forest, N.S.W. *Aust. For. Res.* 12:1–9.

Turner, J., and Lambert, M. J. (1986). Nutrition and nutritional relationships of *Pinus radiata. Annu. Rev. Ecol. Sys.* 17:325–350.

van den Driessche, R. (1972). Different effects of nitrate and ammonium forms of nitrogen on growth and photosynthesis of slash pine seedlings. *Aust. For.* 36:125–137.

van den Driessche, R. (1985). Late-season fertilization, mineral nutrient reserves, and retranslocation in planted Douglas-fir (*Pseudotsuga menziesii* (Mirb.) Franco) seedlings. *For. Sci.* 31:485–496.

Van Rees, K. C. J., and Comerford, N. B. (1990). The role of woody roots of slash pine seedlings in water and potassium absorption. *Can. J. For. Res.* 20:1183–1191.

Vermeer J., and McCully, M. E. (1982). The rhizosphere in *Zea mays* cultivar Senecachief; new insight into its structure and development. *Planta* 156:45–61.

Vose, J. M., and Swank, W. T. (1990). Preliminary estimates of foliar absorption of ^{15}N-labeled nitric acid vapor (HNO_3) by mature eastern white pine (*Pinus strobus*). *Can. J. For. Res.* 20:857–860.

Walter, H., and Breckle, S.-W. (1986). "Ecological Systems of the Geosphere 3: Temperate and Polar Zonobiomes of Northern Eurasia." Springer-Verlag, Berlin.

Waring, R. H., and Running, S. W. (1978). Sapwood water storage: Its contribution to transpiration and effect upon water conductance through the stems of old-growth Douglas-fir. *Plant, Cell Environ.* 1:131–140.

Waring, R. H., and Schlesinger, W. H. (1985). "Forest Ecosystems Concepts and Management." Academic Press, Orlando, Florida.

Waring, R. H., Whitehead, D., and Jarvis, P. G. (1979). The contribution of stored water to transpiration in Scots pine. *Plant Cell Environ.* 2:309–317.

Werk, K. S., Oren, R., Schulze, E.-D., Zimmermann, R., and Meyer, J. (1988). Performance of two *Picea abies* (L.) Karst. at different stages of decline. III. Canopy transpiration of green trees. *Oecologia* 76:519–524.

Yamada, Y., Wittwer, S. H., and Bukovac, M. J. (1965). Penetration of organic compounds through isolated cuticular membranes with special reference to ^{14}C urea. *Plant Physiol.* 40:170–175.

3

Efficiency of Nutrient Acquisition by Fine Roots and Mycorrhizae

R. D. Yanai, T. J. Fahey, and S. L. Miller

I. Strategies of Tree Root Systems

As successful dominant plants in many dry and infertile environments, coniferous trees presumably exhibit structural and functional adaptations in the root system that provide a competitive edge when soil resources are in short supply. With this as a precept we advance the general notion that certain properties of the root systems of conifers have evolved to optimize soil resource acquisition under different environmental regimes. Formulating an expression for root system optimality is made difficult by the complex life history of perennials and the multiple functions performed by roots. Moreover, quantitative analysis of the optimal structure or function of root systems has been constrained by the difficulty of root observations and measurements and by the spatial and temporal variation in the soil environment. Nevertheless, recent workers have invoked arguments concerning root foraging strategies to explain variation among root systems of different types of plants or of plants in different environments. For example, root longevity should vary with soil fertility, according to Grime *et al.* (1991):

> Plant strategy theory predicts that mineral nutrition on fertile soils will involve "active foraging" which consists of patch exploitation by morphologically dynamic root systems, in which the life span of individual fine roots is short. On infertile soils it is predicted that root systems will be less dynamic but by remaining functional throughout the year will be capable of intercepting brief mineralization pulses even when these coincide with harsh environmental conditions.

Harper *et al.* (1991) discuss environments that favor strategies of fine-scale branching versus extension growth:

> Water and some nutrients (especially nitrates) can move relatively freely in the wetter soils. Very close proximity of the root surface to the location of the resource is then relatively unimportant and branching at a fine scale will bring little reward. Under these conditions extension growth will tend to be favored over fine-scale branching. . . . If we were to design an ideal root system for growth in a phosphate-limited environment, or for one in which the water status was regularly close to the wilting point, it would have to have a fine-scale and intimate branching pattern and probably have to sacrifice extensive elongation (or develop mycorrhizal association).

Tiffney and Niklas (1985) suggest that rooting density should respond to the patchiness of the environment:

> Root-bearing plants, in contrast to rhizomatous plants, thereby possess both the ability to explore for resources (linear growth) and the ability to exploit resources efficiently when encountered (clumped growth).

In this chapter, we present a quantitative treatment of the strategies of nutrient uptake by tree root systems building on a biophysical model of the soil–root system. The model is used as a tool for synthesizing existing information on conifer fine root dynamics and the soil environment. Recognizing inherent limitations of the information base and the current model formulation, we also consider in qualitative terms some likely implications of crucial root system characteristics such as root diameter, longevity, and mycorrhizal association.

Our principal objectives are to illustrate an approach for improving basic understanding of root system structure and function and to suggest directions for future research on this complex topic. Ultimately, we hope that this approach will be useful both for explaining natural ecological patterns and for guiding forest management.

II. A Model of Fine Root Efficiency

A. Calculating Carbon Expenditure and Nutrient Uptake

A quantitative model of the carbon expenditure and nutrient acquisition of fine root systems makes it possible to explore some of the above-stated claims about the foraging strategies of roots. Specifically, we want to test which of the many variations in root properties across species and environments could be explained as adaptations to optimize the efficiency of nutrient acquisition by roots. Some properties of roots will not be explained by the nutrient-uptake function of roots, such as those required to anchor and support the plant. An analysis parallel to the study of the efficiency of nutrient acquisition could be undertaken to help identify adaptations of roots that improve the efficiency of water acquisition. The present analysis is appropriate to situations in which the availability of nutrients limits plant growth more than does lack of water or support.

An underlying assumption of the present analysis is that plants are energy limited and that there should be an advantage to acquiring nutrients efficiently. Therefore, we define the efficiency of nutrient acquisition by roots as the amount of carbon expended per unit of nutrient taken up, averaged over the lifetime of the root. In simplest terms, then,

$$E = UPTAKE/COST,$$

where E is the efficiency of nutrient acquisition by roots (grams nutrient/gram C expended), *UPTAKE* is nutrient gain (grams nutrient/gram fine root/day), and *COST* is carbon cost (grams C/gram fine root/day), averaged over the lifetime of the root. We do not subtract the nutrient required to construct the root nor do we consider nutrient or carbon resorption on senescence. Current evidence suggests that the amount of retranslocation is small (Nambiar, 1987).

The calculation of carbon cost, *COST*, includes carbon contained in the root and carbon lost in growth respiration and maintenance respiration. These are averaged over the life span of the root to obtain the cost per unit root per day.

$$COST = (C_{\mathrm{root}} + Y_{\mathrm{g}} + LY_{\mathrm{m}})/L,$$

where C_{root} is root C content (grams C/gram root), Y_{g} is growth respiration (grams C/gram root), Y_{m} is maintenance respiration (grams C/gram root/day), and L is root longevity (days). This equation could be modified to include other carbon costs such as exudation and mycorrhizae.

The amount of nutrient taken up, *UPTAKE*, depends on properties of the soil as well as those of the root. *UPTAKE* is calculated using the steady-state solution to equations of solute movement to the root surface by diffusion and mass flow and solute uptake at the root surface (Nye and Spiers, 1964). This method uses the average concentration in the soil to calculate the concentration at the root surface (Baldwin *et al.*, 1973; Nye and Tinker, 1977). The rate of uptake is limited at high concentrations by saturation kinetics, rather than being a linear function of concentration (Yanai, 1994). The assumption of a steady-state condition (the rate of solute uptake at the root surface equals the rate of delivery to the root by diffusion and mass flow) means that a single rate of nutrient uptake can be used for a root in a given soil environment. Specifically,

$$UPTAKE = \mathrm{SRL}(2\pi r_0 \alpha C_0\, \Delta t),$$

where SRL is specific root length (centimeters/gram root), r_0 is the radius of the root (centimeters), α is the root absorbing power (centimeters/

sec), C_0 is the concentration of substance at the root surface (moles/cm^3); and Δt is the model timestep (seconds); α is calculated from Michaelis–Menton uptake parameters:

$$\alpha = I_{\max}/(k_m + C_0),$$

where $I_{\max}$ is the maximum rate of uptake (moles/cm^2/sec), and k_m is the concentration at the root surface at half of $I_{\max}$ (moles/cm^3). C_0 can be calculated from C_{av}, the average solution concentration, from the following proportion:

$$\frac{C_0}{C_{av}} = v_0\left[\alpha + (v_0 - \alpha)\left(\frac{2}{2-\gamma}\right)\frac{(r_x/r_0)^{2-\gamma} - 1}{(r_x/r_0)^2 - 1}\right]^{-1},$$

where v_0 is the inward radial velocity of water at the root surface (centimeters/sec), r_x is the average radial distance to the next root's zone of influence (centimeters), and $\gamma = r_0 v_0/Db$ (dimensionless), where D is the effective diffusion coefficient (cm^2/sec) and b is the buffer capacity of the soil, or the ratio between exchangeable and dissolved nutrient (dimensionless). These equations are presented by Baldwin *et al.* (1973) and Nye and Tinker (1977) and are derived by Yanai (1994).

The steady-state rate of nutrient uptake provides an estimate of nutrient availability that is appropriate to linking specified environmental conditions to specified root properties. It does not simulate the depletion of the nutrient resource by root uptake. In this model, there is no feedback between exploitation of the soil and the availability of nutrients. Differences in soil fertility are represented by different values of the average concentrations of nutrients in the soil (C_{av}); each value is assumed to apply for the lifetime of the root under consideration. Similarly, roots are represented by a single set of characteristics; we do not consider the distibution of root characteristics in a population of roots of differing morphology, age, and uptake characteristics. Spatial heterogeneity in root density and soil properties is also not considered in this analysis. To be relevant to whole-plant foraging strategies, root properties should be chosen to describe the average root active in nutrient uptake. As a result of this simplification, optimal values of root properties defined by this model apply to the average value rather than to a distribution of values.

B. Parameter Values for Roots

Many of the plant and soil parameters used in our calculations are difficult to measure or estimate. We have used a combination of values appropriate to phosphorus uptake but from different plants and soil types. As a result, the magnitude of individual estimates of carbon costs or nutrient uptake can be highly uncertain. Fortunately, because demonstration of optimal behavior depends on varying those parameter val-

Table I Parameter Values Used in Calculating E

Parameter	Values used in base case calculation	Range of values used in sensitivity analysis	Values for mycorrhizal hyphae
Carbon costs			
C_{root} (g C/g dry mass)	0.48	—	0.48
Y_g (g C/g root)	0.51	0.25 to 4	0.51
Y_m (g C/g root/day)	0.042	—	0.042
L (months)	12	4–40	0.25–3
Uptake, root properties			
r_0 (cm)	0.025	0.01–0.05	0.0003
r_x (cm)	0.4	—	0.044
SRL (m/g root)	20	—	80,000
v_0 (cm/sec)	5.66×10^{-7}	1.1 E-6–3.0×10^{-6}	1.3×10^{-7}
I_{max} (pmol/cm²/sec)	0.268	0.29–0.066	0.268
k_m (pmol/cm³)	16	—	16
Uptake, soil properties			
C_{av} (μmol/liter)	190	50–190	190
D (cm²/sec)	2.43×10^{-8}	—	2.43×10^{-8}
b (dimensionless)	5.84	—	5.84

ues and observing the effect on nutrient acquisition effeciency, E, it is the directional change in E rather than its quantitative value that is most important. It is possible, however, for the values of one parameter to affect the sensitivity of calculated uptake to another parameter (Yanai, 1994), such as the value of C_{av} determining whether D or I_{max} limits uptake.

The parameters required by the model include those associated with carbon costs of root growth and maintenance, as well as root and soil properties affecting nutrient uptake (Table I). Because uptake is usually calculated on the basis of root dry mass, the carbon concentration of fine roots (C_{root}) is a required parameter, and the standard value of 0.48 g C/g dry mass was used in these calculations (Fahey *et al.*, 1988).

The growth respiration coefficient (Y_g) is difficult to estimate directly (Veen, 1981) and a variety of assumptions about the energetic costs of biochemical synthesis are usually employed to calculate Y_g for roots (Lambers *et al.*, 1983; Johnson, 1990). Considerable uncertainty exists in Y_g values because of limited information on the importance of the alternative respiration pathway (Szaniawski, 1981). The value of Y_g should vary with the biochemistry of root tissues because of differences in the cost of synthesizing different compounds. For example, the cost of lignin synthesis is relatively high because decarboxylation reactions are involved and because part of the substrate is catabolized (Lambers *et al.*, 1983). For our calculations, we used a base case value of Y_g of 0.51 g C/g root, based on measurements of Scotch pine seedlings by Szaniawski

(1981); in other scenarios we varied this value with assumed root chemistry as suggested by Lambers *et al.* (1983) and Johnson (1990).

Maintenance respiration (Y_m) of roots has been estimated for herbaceous (e.g., van der Werf *et al.*, 1988) and woody species (Ledig *et al.*, 1976; Szaniawski, 1981). Szaniawski (1981) estimated Y_m for Scotch pine seedlings growing in solution culture by determining the relationship between total root respiration and root growth rate and extrapolating to zero growth rate. We used this estimate (0.042 g dry mass/g root/day) in the base case scenario, although it should be noted that this value was for seedlings and may be higher than that for roots of older trees.

Schoettle and Fahey (1994) summarized estimates of fine root longevity for *Pinus* species, all of which were calculated from the ratio of mean fine root biomass to annual production. These estimates ranged from a few months to several years; for our base case calculations we employed a value of 12 months.

Diameter distributions of fine roots vary markedly among woody species (Eissenstat, 1992; Fahey, 1992) and between soil horizons (Fahey and Hughes, 1994); few values have been published for coniferous forest trees (Eissenstat, 1992). We used a root radius (r_0) of 0.025 cm (diameter of 0.5 mm), which was less than that measured by Kelly *et al.* (1992) for loblolly pine seedlings, but more than that characteristic of mature northern hardwoods (0.015 to 0.020 cm) (Fahey and Hughes, 1994).

Specific root length, the length per gram of root, is not commonly reported. Fahey and Hughes (1994) measured a SRL of 26.1 m/g for roots less than 1 mm in diameter in mineral soil of mixed northern hardwoods. Roots (<0.5 mm) of red spruce saplings had a SRL of 13 m/g in the mineral soil and 10 m/g in the forest floor (unpublished data). We assumed a SRL of 20 m/g, which is intermediate between these two estimates. A value less than 26 m/g is consistent with the smaller r_0 of our base case (0.025 cm) compared to the length-weighted r_0 of the northern hardwoods of Fahey and Hughes (1994). This value is also in line with an estimate of 21 m/g for slash pine roots <1 mm reported by Eissenstat and Van Rees (1994).

The interroot radial distance (r_x) required in the nutrient uptake calculation is generally calculated from root length density, assuming a uniform distribution of root length in soil (i.e., evenly spaced roots). Kelly *et al.* (1992) report an r_x of 2.0 cm. A much lower value of 0.4 cm was calculated from data of Fahey (1994) for mineral soil roots in a northern hardwood forest, where fine roots had an average SRL of 26.1 m/g and fine root biomass was 220 g/m^2 in the upper 0.3 m of soil. We used this value in our calculations.

The values of v_0, I_{max}, k_m, and C_{av} are those reported by Kelly and Barber (1991) and Kelly *et al.* (1992) for uptake of P by loblolly pine seedlings growing in pots in modified A horizon soil collected from a fine loamy silicious, mesic Typic Hapludult (Lilly series). The values of D and b differ from those published by Kelly *et al.* (1992) because they have been adjusted to include the solution-phase contribution as recommended by Van Rees *et al.* (1990) (J. M. Kelly, personal communication, 1992).

C. Parameter Values for Hyphae

The parameter values required to calculate the efficiency of mycorrhizal hyphae in nutrient uptake (Table I) were estimated as follows. The C content per gram and the rates of growth respiration and maintenance respiration were unchanged from the values used for fine roots, because the few data available give similar values (P. Rygiewicz, personal communication, 1993; Marshall and Perry, 1987). The carbon required for the formation of fruiting bodies and rhizomorphs not active in nutrient uptake is not included in the model calculation. We used a hyphal diameter of 6 μm, based on a variety of ectomycorrhizal fungi observed in symbiosis and in culture. The length per unit mass was 80,000 m/g, based on a tissue density of 0.44 g (dry weight)/cm^3 (Paul and Clark, 1989). The rate of water uptake, v_0, was calculated from a transpiration rate of 3.3 liters/m^2/day (on a stand area basis) (Knight *et al.*, 1981) and a hyphal mass of 2000 kg/ha, assuming all water uptake was through hyphae. The interroot distance, r_x, was calculated by assuming that the hyphae are evenly distributed in the top 10 cm of soil. We used the I_{max} and k_m for P uptake by roots, because the few measurements of I_{max} and k_m for P uptake by hyphae suggest that the mechanisms and rates of P uptake are similar (Thomson *et al.*, 1990).

These parameter values were selected to characterize ectomycorrhizal fungi. Endomycorrhizae might be expected to be more efficient at carbon and nutrient exchange because of their penetration directly into root cortical cells. Most conifer species in the Northern Hemisphere are largely ectomycorrhizal, but many are also endomycorrhizal (Harley and Smith, 1983). These include genera such as *Cupressus*, *Thuja*, and *Metasequoia*. The genus *Juniperus* has been found to be both ecto- and endomycorrhizal. Gymnosperms from the Southern Hemisphere, including *Podocarpus*, are largely endomycorrhizal.

Little physiological research has been accomplished with endomycorrhizal conifers. Physiological work on endomycorrhizal herbaceous and crop plants suggests that some physiological parameters may be similar to those of ectomycorrhizal plants, whereas others may be vastly different. For example, the range of inflow rates of P into endomycorrhizal white clover roots (1.6 to 46 $\times$ 10^{-15} mol/cm root/sec) (Smith, 1982) was similar to that measured for ectomycorrhizal *Salix* roots (1.4 to 32 $\times$

10^{-15} mol/cm root/sec) (Jones *et al.*, 1991). On the other hand, George *et al.* (1992) found that the hyphae of the endomycorrhizal fungus *Glomus mosseae* were not able to transport substantial amounts of water to the host plant, whereas Duddridge *et al.* (1980) suggested that the rhizomorphs of many ectomycorrhizal fungi have a transport capacity sufficient to permit significant rates of water transport to host trees.

III. Model Limitations

Before the results of our root efficiency calculations can be correctly understood, it is essential that the current limitations of our modeling approach be fully appreciated. Our effort to develop a quantitative theory of tree root system function was stimulated by the frustration of attempts to evaluate objectively arguments about root foraging strategy, such as those quoted in Section I. Caldwell (1979) had suggested that explanations of patterns of root growth and distribution must await such quantification, and our experience with this quantitative approach has revealed some promising avenues for future development. However, several difficulties appear to limit the explanatory power of any quantitative theory of tree root system function. These problems arise from our use of a variant of resource optimization theory (e.g., Bloom *et al.*, 1985) under the assumption that energy availability ultimately limits tree fitness. We discuss the limitations of this approach to quantifying tree root system function in three categories: (1) the complexity of tree life history, (2) optimal versus maximal function of roots, and (3) constraints in model formulation and parameterization. Some of these limitations are specific to our particular treatment whereas others may limit the development of any unifying theory of tree root function.

A. Life History Complexity

Natural selection probably has acted most strongly in that set of environments or stages of plant development in which survival is most tenuous. Thus, our conception of E being maximized for the mature sporophyte may be erroneous. Studies of tree populations indicate that very strong selective filters operate at several stages in life history: seed set, survival, and germination; seedling survival and establishment; and growth to sexual maturity (Harper, 1977). Trees have achieved an elaborate integration of life history characteristics that maximizes the chances of completion of the life cycle in suitable environments, and the design of the tree root system must have been selected toward this goal. In resource allocation terms, this goal has been likened to strategies of minimizing risks (Bloom *et al.*, 1985).

If root system traits that maximize survival at some more critical stage result in constraints on achieving optimal function at another stage, then apparently suboptimal root system strategies could be retained. For example, Fitter (1989) has suggested that mycorrhizal symbioses may be retained only to achieve maximum nutrient delivery at crucial times, such as the seedling stage and during fruit production. Similarly, intolerant tree species usually become established under conditions of generally high resource availability but potentially high risk from environmental extremes; they undergo severe resource limitations during growth to maturity, and then they persist for decades under conditions of low soil resource availability but high light availability. The optimal root properties in each stage might differ, but the earlier stages must at least partly define the root properties of later stages. In contrast, tolerant species often establish under severe limitation of both energy and soil resources, but must respond to release when both sets of resources are briefly supplied in surplus. Hence, it seems likely that some stages in the tree life cycle operate at suboptimal root efficiency. Perhaps most importantly, for a dominant overstory tree, the assumption of primary limitation by energy availability must be questioned. The availability of carbon should affect the optimal exchange rate between carbon and soil resources, as discussed below.

B. Optimal versus Maximal Root Efficiency

Root foraging strategy can be accurately predicted by an analysis of root efficiency only if root systems have been selected entirely for maximizing nutrient uptake under conditions of energy limitation. However, we know that selection has favored other aspects of root function and that energy is not always limiting to tree fitness. As a result, the maximum value of E may not be optimal for the plant. Whole-plant allocation strategy should be adjusted so that growth limitation is most nearly equal for all resources (Bloom *et al.*, 1985), a situation that may be achieved at less than maximal root efficiency. For example, if nutrients are in short supply while carbon is available in excess, the optimal E for plant growth will be reached at higher nutrient uptake and carbon expenditure than the maximum E. Clearly, the estimation of E is necessary but not sufficient for an economic analysis of allocation strategies. A more sophisticated approach to whole-plant or whole-stand modeling would be required to identify the optimal E for plant growth; our approach can only show the values of root parameters optimal in a given environment for maximizing E.

Departures of root system traits from theoretical predictions of maximum efficiency may provide insights into the trade-offs that are necessary to optimize total root function. For example, despite the undoubted advantage in nutrient uptake efficiency of reducing root diameter, coniferous trees retain relatively coarse root systems (Eissenstat and Van

Rees, 1994). This may be because of limits on design associated with their multiple functions, such as storage, anchorage, and transport.

Another reason why root system strategies might deviate markedly from maximal efficiency is that persistence in the face of temporally varying stresses is more important to tree survival than is maximal growth (Gutschick, 1981). This is true not just as the tree passes through different life stages, as discussed above, but also on a much shorter time scale. Bowen (1985) pointed out that selection for survival under seasonal or less frequent environmental stresses may result in carbon partitioning between roots and shoots that differs from the apparent optimum. In the face of possible climatic change, such an assertion takes on practical importance for plantation forestry.

Because most tree species occupy a wide range of habitats and exhibit only a limited degree of genetic variability and phenotypic plasticity in root system traits, it seems likely that most natural stands operate below both the maximal and the optimal root efficiency. It will be a continuing challenge to practical forestry to further develop the basis for selection of root system traits that improve root uptake efficiency while recognizing that optimal effeciency may differ from maximal efficiency.

C. Constraints of Current Model Formulation

The steady-state approach of our model allows for a straightforward analysis of the costs and benefits of roots with different properties in different environments. Only one calculation is required to characterize the lifetime average rate of nutrient uptake, assuming that properties of both roots and soil are constant for the life of the root or that average values may be used to represent them. Further, because each calculation is made for an individual root, variation in properties of the root system and the soil are ignored. This approach obviously suffers certain limitations; neither root systems nor soils are homogeneous or static.

Most forest soils are patchy; soil resource availability is extremely heterogeneous. Root system traits might be selected as much for efficiently exploiting patchy soil resources as for static exploitation of a homogeneous soil, but optimization of E under such conditions could be complex. For example, species from fertile sites appear to be better at exploiting patchiness because they exhibit higher root growth rates (Grime *et al.*, 1991; Eissenstat and Van Rees, 1994), whereas intraspecific differences in root proliferation within patches may be more closely tied to internal carbon allocation strategies [e.g., for *Pseudotsuga menziesii* (Mirb.) Franco seedlings] (Friend *et al.*, 1990).

Another limitation of our model formulation concerns the use of a single fertility level (nutrient concentration) to describe the soil environment. Although root depletion zones are taken into account, they are assumed to be at steady state; the ability of roots to draw down the nutrient content of soil over time is not considered. Reynolds (1975) observed nonsynchronous growth of the feeder roots of *P. menziesii* even when bulk soil conditions seemed to be invariant, and he attributed this pattern to the need for relocation of sites of uptake owing to local nutrient depletion. Reynolds applied the term "tactical organization of the root system" to this nonsynchronous growth, suggesting the probable importance of root tip relocation in defining the maximal or optimal E.

Future exploration of the trade-offs between carbon expenditure and nutrient uptake by roots should take into account both spatial and temporal variation in root systems and the soil environment (Yanai and Eissenstat, 1994). To simulate the local depletion of soil by root uptake would require the model to be implemented over time. The optimal deployment of roots in a patchy environment could also be accommodated with only slight modification to the current model. To address the question of what root strategies optimize plant growth, survival, or reproductive success, as opposed to simply maximizing root E, would require incorporating these notions of carbon costs and nutrient gain by roots with a whole-plant approach to resource values and exchanges.

IV. Root Longevity

A. Model Predictions

1. Root Longevity Our first exploration is of the relationship between root longevity and soil fertility. Grime *et al.* (1991), quoted in Section I, suggested that roots should be long-lived in infertile soil and short-lived in fertile soil. We varied root longevity at different levels of soil fertility to find how optimal longevity (defined as the longevity that gives maximum efficiency) depended on soil fertility.

First, for the case in which nutrient uptake is constant with root age, and hence constant with root longevity (Fig. 1a), average lifetime carbon cost declines with root longevity (Fig. 1b) because the construction cost is spread out over a longer period. The maintenance cost per day is assumed to be constant; at very long life spans, the cost per day approaches the maintenance cost. The effect of carbon cost, therefore, is that long-lived roots are more efficient than short-lived roots (Fig. 1c). Why, then, do plants construct roots that are ephemeral as well as roots that are long-lived? Why does root longevity vary both among species and among roots in different soil environments, even within a single plant (Shoettle and Fahey, 1994)? We will consider how construction cost, nutrient uptake, and water uptake might contribute to differences in the efficiencies of roots of different longevity.

2. Root Longevity and Carbon Cost The construction cost of long-lived roots might be higher than that of short-lived roots, if long-lived roots

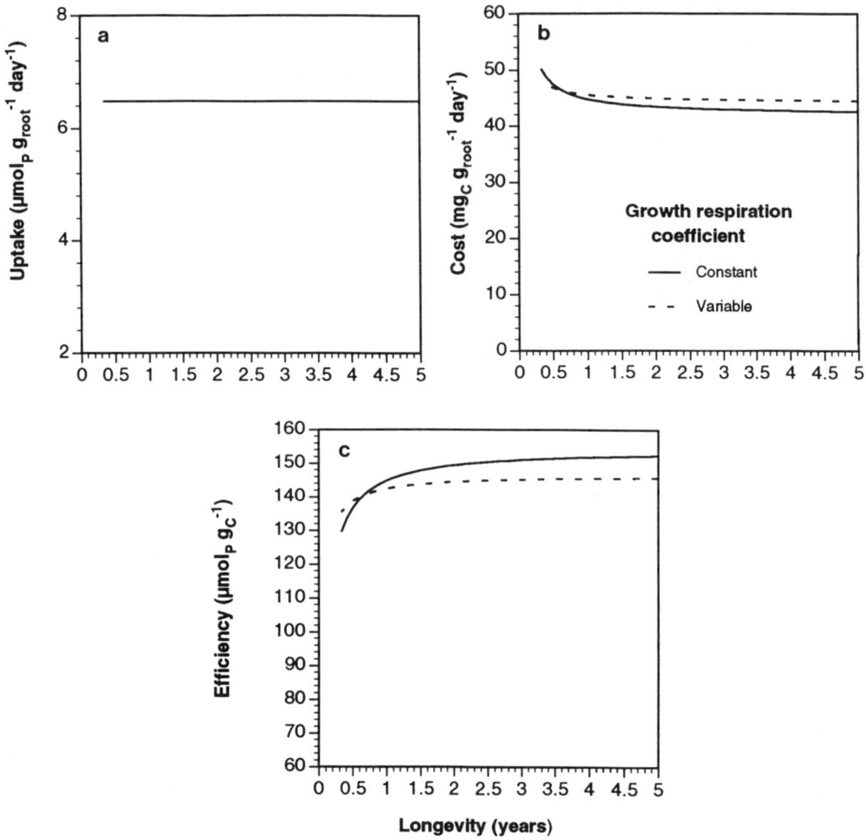

Figure 1 Average lifetime uptake (a), cost (b), and efficiency (c) as a function of root longevity for constant and for varying values of growth respiration. Variation in the growth respiration coefficient was linear with root longevity, ranging from 0.25 g C/g root at a longevity of 4 months to 3.7 g C/g root grown at a longevity of 40 months.

contain higher concentrations of more energetically costly biochemicals, such as lignin and secondary chemicals [Chapin (1989), however, found no basis for this assumption in leaves]. For example, whereas the conversion efficiency of sucrose to cellulose is essentially 1.0, the conversion efficiency of sucrose to lignin is 0.47 (Lambers *et al.*, 1983). When we varied the growth respiration of roots as a function of longevity (Fig. 1b), we found that even a very steep relationship of growth respiration to longevity did not make cost increase at high longevity, because the tendency of increasing longevity to reduce cost was so strong. Variation in construction cost, therefore, fails to indicate an optimal longevity. Further, an optimal longevity derived by this means would not be sensitive to soil fertility.

3. Root Longevity and Nutrient Uptake Another possibility to consider in explaining the advantage of root turnover is that short-lived roots are more effective at nutrient uptake than are older roots. Although it seems likely that uptake rates vary with root age, largely as a result of progressive suberization and mycorrhiza formation (e.g., Queen, 1967; Chung and Kramer, 1975), the magnitude and shape of this relationship are unknown. To explore this mechanism, we assumed a linear relationship between longevity and I_{max}, the saturation constant for uptake at the root surface. The resultant relationship between uptake and longevity is nearly linear at high soil fertility, with uptake highest in the most ephemeral roots (Fig. 2a). In an infertile soil, there is less advantage to roots of

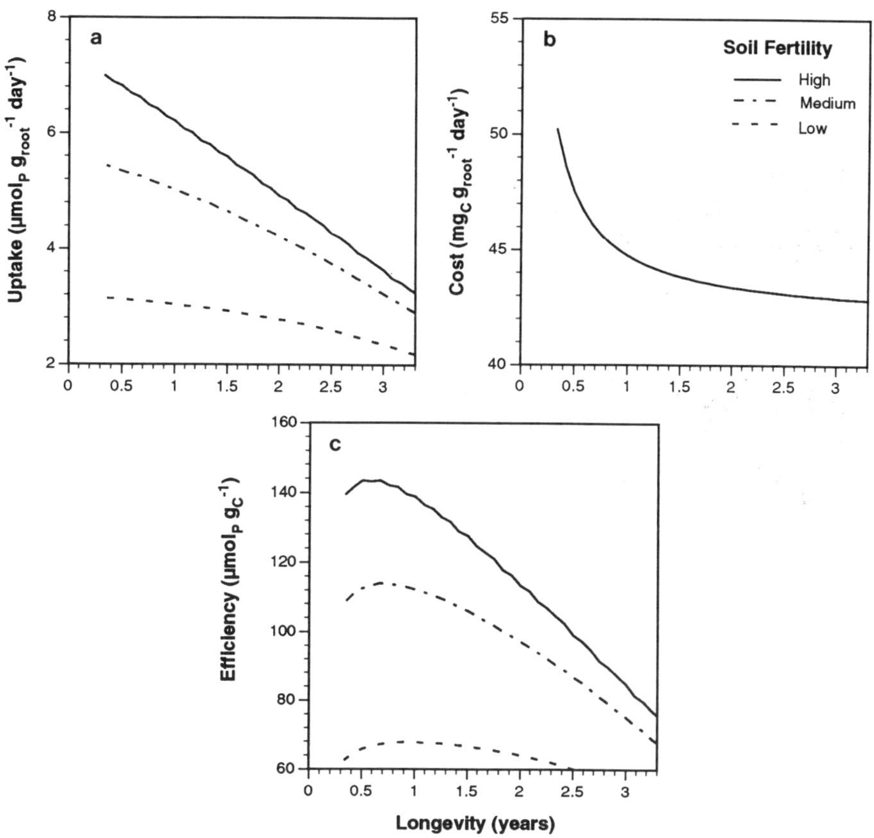

Figure 2 Average lifetime uptake (a), cost (b), and efficiency (c) as a function of root longevity, with nutrient uptake varying with root longevity. Maximum P influx rate (I_{max}) declined linearly from 0.29 pmol/cm^2/sec at a longevity of 4 months to 0.13 pmol/cm^2/sec at a longevity of 40 months. Soil of high, medium, and low fertility had P concentrations of 190, 100, and 50 μmol P/liter.

very high uptake capacity, because modeled nutrient uptake is limited by the rate of solute movement to the root surface, not by the kinetics of uptake by the root. As a result, short-lived roots in infertile soil absorb less nutrient than the same roots in more fertile soil. Long-lived roots, in contrast, are limited by the kinetics of nutrient uptake in our example, and uptake by long-lived roots is similarly low in fertile and infertile soil. This result follows from our assumption that an average I_{max} characterizes the root throughout its life. If the calculation were made with I_{max} declining as the root aged, the advantage of fertile soil to young roots would contribute to the average uptake by long-lived roots.

The effect of carbon cost (Fig. 2b) is to make ephemeral roots inefficient; the effect of low nutrient uptake rates at high longevity tends to make long-lived roots inefficient as well (Fig. 2c). The optimal longevity depends on the combination of values of the other parameters in the model. In our example, the optimal longevity in fertile soil is about 6 months, with the optimal longevity in infertile soil about 1 year. Of course, roots in infertile soil are less efficient than roots in more fertile soil, in terms of carbon expended to obtain a unit of nutrient.

4. Root Longevity and Water Uptake Water uptake rates affect solute uptake, especially when uptake is limited by the rate of delivery of solute to the root surface (Yanai, 1994). Changes in the rate of water uptake with root longevity might provide an alternate explanation of optimal root longevity. To explore this possibility, we assumed a linear relationship between longevity and the velocity of water uptake at the root surface (v_0). There is some justification for this assumption; for example, Sands *et al.* (1982) observed that water uptake rates were about half as high for suberized as for nonsuberized roots of loblolly pine. The effect of declining water uptake rates with root age was to reduce nutrient uptake at high longevity (Fig. 3a). In infertile soil, nutrient uptake increases almost linearly with increasing v_0. In more fertile soil, the advantage of short-lived roots with high v_0 is diminished; nutrient uptake is limited by I_{max}, the saturation constant for nutrient uptake.

Assuming a constant growth respiration coefficient and constant maintenance respiration costs (Fig. 3b), the effect on efficiency of making long-lived roots more resistant to water uptake was similar to that of making them slow at nutrient uptake. More ephemeral roots were inefficient because of their high carbon cost, whereas more long-lived roots were inefficient because of their low nutrient uptake. However, the effect of soil fertility on optimal longevity in the case of varying v_0 was the opposite of the case of varying I_{max}. Optimal root longevity for maximizing the efficiency of nutrient uptake in the fertile soil was quite broad,

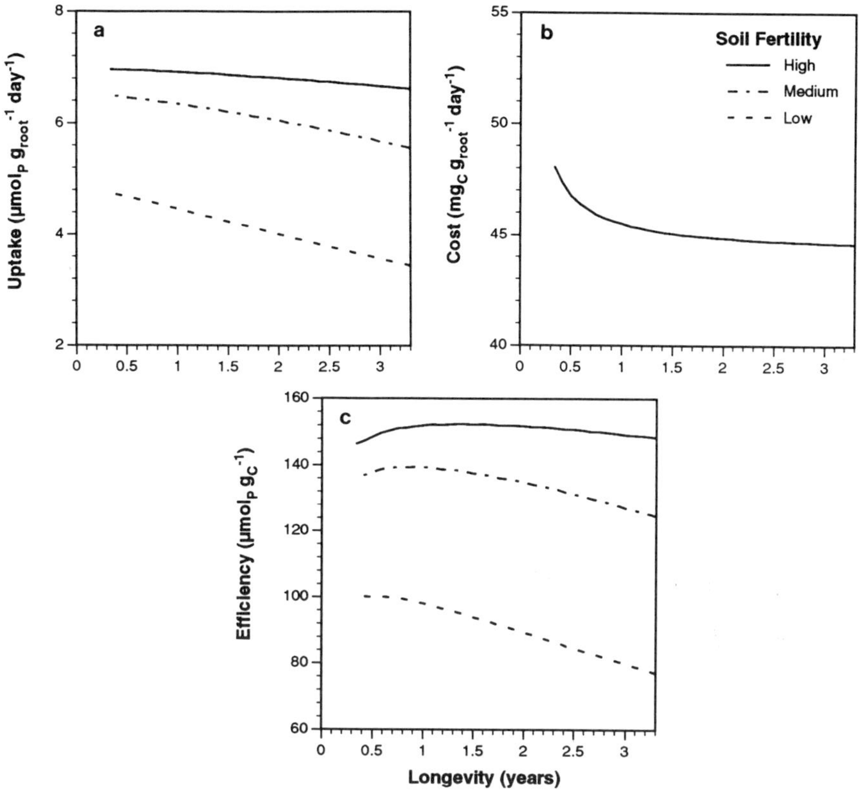

Figure 3 Average lifetime uptake (a), cost (b), and efficiency (c) as a function of root longevity, with water uptake varying with root longevity. The radial velocity of water uptake at the root surface (v_0) was assumed to vary linearly with longevity from 3×10^{-5} cm/sec at 4 months to 51.1×10^{-5} cm/sec at 40 months. Soil of high, medium, and low fertility had P concentrations of 190, 100, and 50 μmol P/liter.

between 1 and 2.5 years. Optimal longevity decreased with decreasing soil fertility, because of the higher advantage given by high v_0 at low fertility.

B. Discussion

Previous theoretical arguments (Grime *et al.*, 1991), as well as limited empirical evidence (Nadelhoffer *et al.*, 1985, Schoettle and Fahey, 1994), suggest that tree root longevity is higher on infertile soils than on fertile soils. Similarly, our model suggests that if I_{max} declines with increasing longevity, the maximum efficiency of root uptake is exhibited at values of longevity that increase with decreasing fertility. This pattern results from model predictions that nutrient uptake in infertile soil is limited by solute movement to the root surface more than by root uptake kinetics; thus, roots in infertile soil have little advantage in expressing high uptake capacity and short-lived roots should be at an energetic disadvantage there.

A variety of other patterns of variation in tree root longevity have been proposed on the basis of field surveys or experimental manipulations. Fine root longevity may be related to the form of N taken up by trees, lower longevity being associated with soils exhibiting high nitrification rates (Nadelhoffer *et al.*, 1985). Soil temperature may influence longevity either through effects on maintenance costs (Hendrick and Pregitzer, 1993; Marshall and Waring, 1985) or on rates of decline in root uptake activity with root age (Eissenstat and Van Rees, 1994). In patchy soil environments, root longevity may be higher in fertile than in infertile microsites (Pregitzer *et al.*, 1993; Fahey and Hughes, 1994), contrary to model predictions for uniform soil. Finally, thinning of conifers on nutrient-rich sites reduces root longevity (Santantonio and Santantonio, 1987). Each of these patterns of root longevity is discussed below.

The extra energy cost of NO_3 uptake is associated mostly with the high cost of NO_3 reduction; the theoretical Y_g value of protein synthesis from NH_4 is 0.84 whereas for NO_3 it is 0.58 (Johnson, 1990). Although this difference depends on the proportion of NO_3 reduction occurring in foliage, current evidence suggests that for most coniferous forest trees this proportion is probably very low (Andrews, 1986). We incorporated the influence of the differing energy costs of NH_4 versus NO_3 uptake on the longevity associated with maximum root efficiency, with inconclusive results. Not surprisingly, efficiency declined markedly as the proportion of NO_3 absorbed increased from 0 to 1. However, the longevity at maximum efficiency did not change appreciably across this gradient. The longevity at maximum root efficiency was not sensitive to the form of N absorbed because the cost differences were not stongly dependent on longevity. Nadelhoffer *et al.* (1985) observed root longevity differences between nitrifying and nonnitrifying soils that may be explained in part by root uptake costs, but it is uncertain whether these differences would tend to maximize uptake efficiency. Many questions about nitrogen economy and root function that could influence such considerations remain unanswered: What are the differences in costs associated with ion leakage (Johnson, 1990)? What proportion of nitrate reduction and amino acid synthesis actually occurs in the roots (Gutschick, 1981)? What is the energy cost of protein turnover, especially for nitrate reductase (Amthor, 1984)?

Root longevity could be sensitive to soil temperature because of the lower maintenance respiration costs at lower temperatures (Ryan, 1991).

Hendrick and Pregitzer (1993) observed a slight increase in root longevity of sugar maple-dominated forests along a gradient of decreasing soil temperature, in agreement with some of the experimental results of Marshall and Waring (1985). Model calculations show a shift in optimal longevity to shorter life spans with increasing temperature, assuming I_{max} decreases with longevity (R. D. Yanai and D. M. Eissenstat, unpublished data, 1994). Again, the question of whether root longevity is actually optimized across a soil temperature gradient awaits further research, as does the important practical question of whether such variation reflects phenotypic plasticity or ecoclinal, genetic variation as has been observed for a variety of other temperature-regulated traits of trees (Muona, 1990).

The role of patchiness in soil fertility in regulating root longevity adds further complexity to the optimization of root longevity. Although fine root longevity appears to be lower in fertile than in infertile soils, two recent studies in northern hardwood forests have documented higher longevity of roots in fertilized microsites than in adjacent unfertilized areas (Pregitzer *et al.*, 1993; Fahey and Hughes, 1994). Both studies also showed about a sixfold increase in root growth within fertilized microsites. These results further suggest extending our modeling approach to incorporate spatial variation in soil resource availability and the depletion of soil by root uptake. The root proliferation and higher longevity observed in fertile microsites probably could be explained as a whole-root system strategy to maximize uptake efficiency, and might result mechanistically from the absence of depletion zones or from the relatively high overall nutrient transport along mother root axes that penetrate fertile microsites (Fahey and Hughes, 1994).

Changes in canopy structure could influence root longevity. For example, observed effects of overstory thinning on root longevity might be connected with respiratory costs: decreased longevity of *Pinus radiata* fine roots on fertile sites in New Zealand (Santantonio and Santantonio, 1987) might be associated with increased soil temperatures. However, many other influences could be postulated, including changes in carbon allocation.

Finally, it should be noted that the mechanisms of root mortality are poorly understood (Schoettle and Fahey, 1994), and the importance of internal cues for root sloughing compared with external causes, such as drought, anoxia, and herbivory, has not been resolved. For example, recent evidence suggests the possible importance of soil fauna in root mortality (Atkinson, 1992). If external effects that are largely outside the control of the tree play a major role in root turnover, then the notions of optimality developed here must be modified to reflect these effects.

V. Root Diameter

A. Model Predictions

We were unsuccessful in our attempts to establish an optimal diameter of roots for maximum nutrient uptake efficiency. The model indicates that uptake per gram of root increases indefinitely as roots get finer (Fig. 4a). This is because root length and root surface area per unit mass of root increase as diameter decreases. If the carbon cost per unit mass of root were constant, the optimal root for nutrient absorption would be infinitely fine (Fig. 4c). Other factors, such as structural constraints, presumably limit the diameter of fine roots. This illustration supports the long-held view that the efficiency of mycorrhizal hyphae is due to their having smaller diameters than are possible for roots.

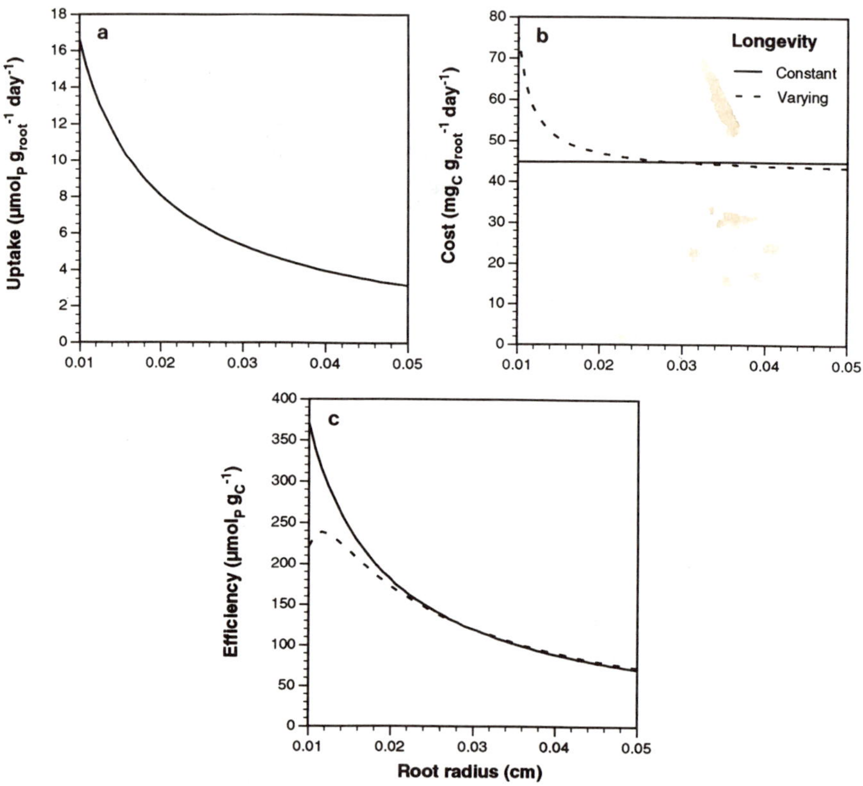

Figure 4 Average lifetime uptake (a), cost (b), and efficiency (c) as a function of root radius (r_0), assuming either constant longevity or longevity varying with root diameter, from 1 month at $r_0 = 0.01$ cm to 24 months at $r_0 = 0.05$ cm.

We explored whether a higher carbon cost associated with finer roots could increase optimal root diameter. A shorter life span associated with finer roots would produce such a carbon cost. Assuming a relationship between diameter and longevity creates a curve of C cost with diameter (Fig. 4b) that looks like that of C cost with longevity (Fig. 1b). The resulting curve does have a maximum efficiency and optimal diameter (Fig. 4c); the value of this optimal diameter is a reflection of the relationship we assumed with root longevity, not a quantitatively supported prediction.

B. Discussion

The results of our simulations indicated that, unless root longevity is strongly tied to root diameter, average lifetime root efficiency increases continuously with decreasing diameter. Most likely, other root functions or anatomical constraints limit the minimum diameter of roots. For example, larger diameter roots may be able to withstand greater soil impedence before buckling (Whitely *et al.*, 1982). The root cortex may serve as a short-term reservoir for nutrient storage. Glass and Siddiqi (1984) observed that under K limitation, shoot K remained constant while root concentration declined markedly, indicating that K stored in the root cortex had been transported to maintain shoot K concentrations. In pines, only diarch and tetrarch roots (i.e., those of relatively large diameter) appear to be morphogenetically capable of becoming mother roots (Wilcox, 1968), and hence of long-distance, exploratory growth, illustrating a trade-off necessitated by existing root system design. The mycorrhizal symbiosis provides absorbing surfaces of much smaller diameters than roots, with consequent advantages, as we have illustrated. In addition, root hairs might be considered a plant adaptation in the direction of finer absorbing organs. The effect of root hairs on nutrient uptake can be simulated by increasing the effective radius of the root by the average length of the root hairs; it would be misleading to consider them independently of the roots because root hairs, to a greater extent than hyphae, operate within the depletion zone of roots. Alternatively, the principal role of root hairs might be to prevent the development of root–soil gaps and the resulting resistance to solute transport to the root surface (Caldwell, 1979), rather than to enhance the absorbing area of roots. To the degree that the formation of root–soil gaps increases with root age, this factor is represented in our analysis of root uptake rate decreasing with longevity, and would contribute to the benefits of rapid root replacement. If large-diameter roots are more likely to suffer root–soil gaps, then the advantages of small-diameter roots in nutrient uptake would include the benefit of more continuous nutrient uptake.

VI. Efficiency of Mycorrhizal Hyphae

A. Model Predictions

Mycorrhizal associations are common in conifers (Miller, 1982) and are known to be beneficial in cases of nutrient scarcity, particularly of N and P. Absorption rates of N and P, reported on the basis of root mass or root length, are increased by mycorrhizal infection of plants at low fertility (e.g., Smith, 1982). A partial explanation for this increased nutrient uptake ability can be found by comparing the uptake and efficiency of roots and mycorrhizal hyphae. If the C cost per unit mass of absorbing tissue (root or hypha) and the P uptake kinetics are assumed to be the same for roots and hyphae (Table I), then the efficiency of hyphae is orders of magnitude greater than the efficiency of roots (Table II), due to the increase in length of the absorbing organ per unit mass, as predicted from the previous exploration of variation in root diameter. Differences in the uptake rates of roots and hyphae may also be important, although few measurements for extramatrical hyphae and ectomycorrhizal roots are available and none appears appropriate to our modeling approach (Miller and Allen, 1992; Finlay, 1992). Mycorrhizae may also increase P uptake by increasing P in solution (Harley and Smith, 1983; Bowen, 1973); this effect is not included in the model. The efficiency of the mycorrhizal association, calculated using the cost of roots plus hyphae and assuming that uptake is achieved through the hyphae (Table II), is intermediate between that of the hyphae alone

Table II Uptake, Cost, and Efficiency of Roots and Hyphae[a]

Fertility (μmol P/liter)	Root	Hypha	Root + hyphae
	Uptake (μmol P/day)		
190	6.7	323	165
100	6.3	302	154
50	5.4	264	134
	Cost (g C/day)		
All	0.045	0.075	0.060
	Efficiency (μmol P/g C)		
190	150	4310	2760
100	140	4010	2570
50	122	3520	2250

[a] The combined effect of roots and hyphae assumes equal amounts of each. Hyphae were assumed to turn over monthly, roots annually. All other parameter values are shown in Table I.

(which does not include the support of the root system) and the roots alone (which includes neither the cost nor the benefit of association with the fungus). A more complete evaluation would include the effects of mycorrhizal association on root longevity and respiration; these refinements, however, are unlikely to change the result that, from the point of view of the efficiency of carbon expenditure for nutrient gain, it would seem advantageous to expend carbon for mycorrhizal development rather than for roots wherever possible.

Although conifer species are universally mycorrhizal (Meyer, 1973), ectomycorrhizae generally are more fully developed on infertile than on fertile soils or where nutrients become available in seasonal flushes (Högberg, 1986; Harley and Smith, 1983). We tested the hypothesis that differences in mycorrhizal development in these different environments could be explained by differential efficiency. We found, however, that hyphae are much more efficient than roots at P uptake regardless of the soil concentrations of P (Table II), which is not consistent with experimental results (Rousseau and Reid, 1990). The lack of mycorrhizal development on fertile soils, therefore, is most likely due to other factors (discussed below).

Turnover rates of ectomycorrhizal hyphae remain largely unquantified in the field. Although Fogel and Hunt (1983) were able to calculate the overall monthly changes in the standing crop of soil and litter hyphae for a coniferous forest site in Oregon, it was not possible to follow the fate of any given cohort of hyphae. Because of the rapid growth rate and relatively small size of fungal hyphae compared to roots, turnover rates are undoubtedly much more rapid for fungal hyphae than for roots. We explored whether the short life span of mycorrhizal hyphae could be explained by consideration of the efficiency of uptake. As in the case of roots (Fig. 1), if uptake is assumed to be constant with hyphal longevity, then efficiency only increases with longevity, due to the amortization of construction costs. We assumed a rapid decrease of I_{max} with hyphal age, resulting in a decrease of uptake with longevity (Fig. 5). The resulting longevity to maximize E is much less than for roots because of the rate at which we assumed uptake to decline with hyphal age. Optimal longevity did not vary, in this case, with soil fertility.

B. Discussion

The extraordinary efficiency of mycorrhizal hyphae compared to roots should make it advantageous, in terms of C conservation, for any plant to use hyphae rather than roots for nutrient uptake. We need look no further to explain why coniferous forests have obligate relationships with ectomycorrhizal fungi; the mystery, perhaps, is why some plants have less mycorrhizal development than others. Specifically, why is it that

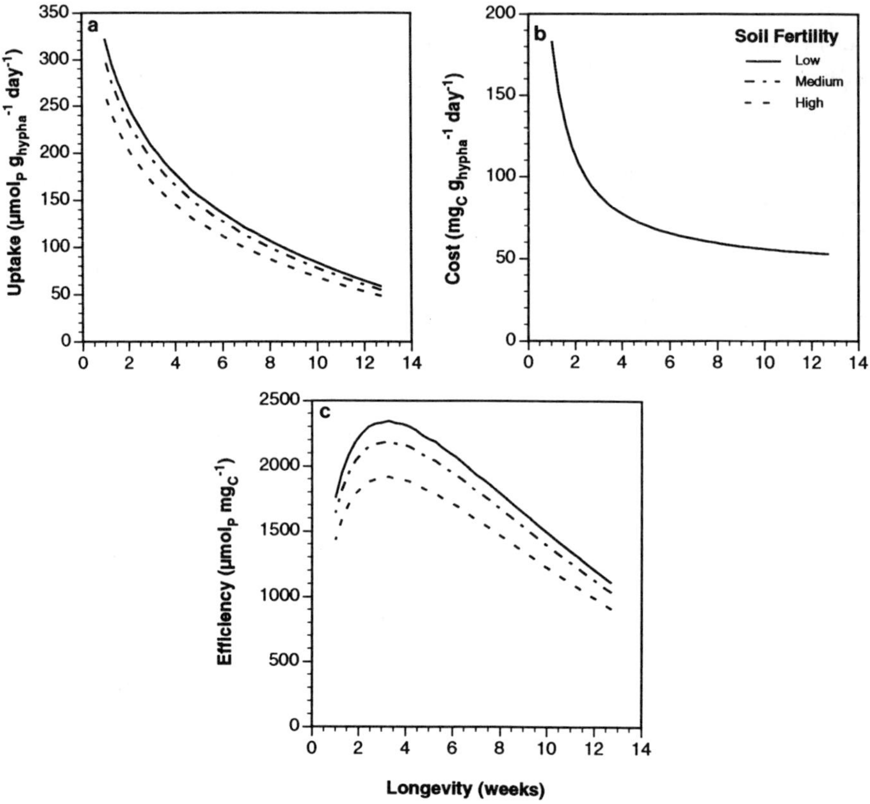

Figure 5 Average lifetime uptake (a), cost (b), and efficiency (c) of mycorrhizal hyphae at three levels of soil fertility, assuming I_{max} declined from 0.27 pmol/cm^2/sec at a longevity of 1 week to 0.05 pmol/cm^2/sec at a longevity of 13 weeks. Soil of high, medium, and low fertility had P concentrations of 190, 100, and 50 μmol P/liter.

trees are more likely to develop ectomycorrhizae in poor soils? There are possible causes of variation other than differences in carbon and nutrient availability.

Hemlock (*Tsuga*) species thrive in wet and nutrient-rich soils, often surviving for a year or more before developing mycorrizae (Kropp, 1982). Fewer species of fungi may inhabit these soils (*Elaphomyces granulatus* being one notable exception). Anaerobiosis, high nitrogen availability, and acid soil may contribute to the scarcity of ectomycorrhizae in these conditions. To the degree that nutrient-rich soils tend to develop in more favorable environments—with higher organic matter contents, higher water-holding capacity, and lower pH—soil characteristics other than nutrient supply may contribute to the variation in fungal availability across habitats (Harley and Lewis, 1969). Nutrient supply, however, also appears to have a direct effect: laboratory studies have shown that high levels of N fertilization decreased ectomycorrhizal formation (Hacskaylo and Snow, 1959).

Variations in the relative importance of nutrient acquisition and carbon conservation may also help explain variation in mycorrhizal formation across habitats. First, plants with little or no mycorrhizal development might be unlikely to survive in an infertile soil but might be at much less of a disadvantage in nutrient-rich soil. In other words, the higher value of nutrients in the nutrient-poor condition increases the value of the benefit obtained for carbon investment belowground. The value of carbon, too, varies with the plant environment. In the nutrient-poor condition, where growth is limited by nutrient availability, carbon is available for investment in mycorrhizal fungi with few competing sinks elsewhere in the plant. In contrast, where nutrients are plentiful, plant growth may be limited by carbon availability, and the expense of allocating carbon belowground is greater. In short, the relative value of C and nutrient in the accounting system of the whole plant will vary with environmental conditions, affecting the optimal efficiency of carbon–nutrient exchange, as discussed in Section III,B.

One limitation to nutrient supply to plants via mycorrhizae is the transport of nutrient from fungal tissues to root tissues. If the rate of nutrient absorption by the root is still limited by its own uptake kinetics (Smith and Smith, 1990; Cairney and Smith, 1992), then increases of orders of magnitude, as shown in Table II, are unlikely to be realized. The rate of transport along the hyphae may also limit the transfer of nutrients to the plant. Neither of these factors has been considered in this analysis.

Another factor that has generally been ignored but that may be very important is the effect of ectomycorrhizal fungi on the availability of soil resources. For example, ectomycorrhizal fungi have now been shown to take up N from complex organic molecules (Abuzinadah and Read, 1986a,b; Abuzinadah *et al.*, 1986; Dighton *et al.*, 1987; Finlay *et al.*, 1992); much of this N becomes directly available to the plant. Because there is far more organic N than mineralized N in soils from coniferous forests, our conception of N availability to plants (that only mineralized N is taken up) may require substantial revision.

The supply of carbon to mycorrhizae through this saprotrophic activity has also been previously unrecognized. Finlay *et al.* (1988, 1989) found that ^{15}N-labeled organic N is taken up primarily as amino acids by ectomycorrhizal fungal mycelia. These amino acids can either be transported to the mycorrhizas or stripped of their N and the carbon skeletons utilized by the fungus for a variety of physiological processes (Finlay, 1992). Finlay (1992) further suggests that amino acids derived from the incorporation of ammonium into carbon skeletons from fungal sugars are released to the plant root, providing a route for the reverse translocation of carbon from fungus to host.

In the present context, it may be helpful to view the energy required to acquire N from organic sources by ectomycorrhizal fungi as a belowground ecological pyramid. Because ectomycorrhizal fungi are able to cycle N (originating as discarded material from the host primary producer) directly back to the host (also the primary consumer), the classical N mineralization cycle is circumvented. More N is therefore available to the host with much less immobilization and much less total energy loss than if the same organic N were completely mineralized before it became available again.

The study of plant ecophysiology and the study of mycorrhizal fungi have long been independent, to the detriment of both. The importance of mycorrhizae in carbon budgets (the biomass of extramatrical hyphae can be considerable) and nutrient uptake is often ignored. Because the uptake parameters I_{max} and k_m might be different for fungi than for roots, calculating root uptake based on uptake parameters measured for roots alone could be misleading. More research in the area of mycorrhizal contributions to tree function is clearly warranted.

VII. Summary

It is difficult to assess claims about the adaptive advantages of root foraging strategies without a conceptual model specific enough to allow quantitative prediction and testing. Application of a solute uptake model in combination with a calculation of carbon costs provides a means of assessing the efficiency of carbon expenditures in procuring nutrients from soil. Although many of the parameters required to calculate root efficiency are poorly known at present, such that costs and benefits cannot be accurately quantified, the model can be used to test hypotheses for internal consistency and for correspondence to observed patterns of root growth in different soil environments.

We analyzed the optimal values of root properties, such as longevity, diameter, and mycorrhizal association, that maximized the efficiency of carbon exchange for nutrient uptake in different environments. Optimal longevity was found to decrease with increased soil fertility if the kinetics of nutrient uptake were assumed to decline with increased root longevity. Optimal diameter was found to be smaller than observed in roots, suggesting that other constraints on root structure or function limit their minimum diameter. Mycorrhizal hyphae were found to be more efficient than roots regardless of soil fertility.

The steady-state approach to calculating carbon costs and nutrient gain enabled combinations of root and soil properties to be very simply evaluated. However, this approach ignored spatial heterogeneity and temporal variation in root and soil properties, such as aging of roots and patchiness of soil fertility. Furthermore, finding the values of root parameters that maximize root *E* may not predict the optimal root deployment for the plant, which depends on the relative value of carbon and nutrients in the whole plant. Estimation of the rate of exchange of carbon and nutrients in roots is a necessary step toward an economic analysis of allocation strategies; it also reveals areas of ignorance and helps to identify future research needs.

Acknowledgments

We are grateful to Dave Eissenstat and Tom Hinckley for review of the manuscript and to Gail Blake and Carol Zollweg for technical support. Financial support was provided to TJF by National Science Foundation Grant BSR-87002331, to SLM by National Science Foundation Grant BSR-8805983 and United States Department of Agriculture Grant 91-37101-6759, and to RDY by the Boyce Thompson Institute for Plant Research, the Electric Power Research Institute under research contract number RP2799-1, and the USDA Forest Service, Southeastern Forest Experiment Station (Southern Global Change Program), through contract TV-8644OV with the Tennessee Valley Authority. The Southern Global Change Program, a coordinated effort of the Southeastern and Southern Forest Experiment Stations, is one component of the Forest Service's Global Change Program. This paper has not been subjected to Forest Service policy review and should not be construed to represent the policies of the agency.

References

Abuzinadah, R. A., and Read, D. J. (1986a). The role of proteins in the nitrogen nutrition of ectomycorrhizal plants. I. Utilization of peptides and proteins by ectomycorrhizal fungi. *New Phytol.* 103:481–493.

Abuzinadah, R. A., and Read, D. J. (1986b). Utilization of protein by ectomycorrhizal fungi and by their host plants. *In* "Physiological and Genetical Aspects of Mycorrhizae" (V. Gianinazzi-Pearson and S. Gianinazzi, eds.), pp. 395–401. Institut National de la Recherche Agronomique, Paris.

Abuzinadah, R. A., Finlay, R. D., and Read, D. J. (1986). The role of proteins in the nitrogen nutrition of ectomycorrhizal plants. II. Utilization of proteins by mycorrhizal plants of *Pinus contorta*. *New Phytol.* 103:495–506.

Amthor, J. S. (1984). The role of maintenance respiration in plant growth. *Plant, Cell Environ.* 7:561–569.

Andrews, M. (1986). The partitioning of nitrate assimilation between root and shoot of higher plants. *Plant, Cell Environ.* 9:511–519.

Atkinson, D. (1992). How long is the life span of a root? *Trends Ecol. Evol.* 7:173–174.

Baldwin, J. P., Nye, P. H., and Tinker, P. B. (1973). Uptake of solutes by multiple root systems from soil. III. A model for calculating the solute uptake by a randomly dispersed root system developing in a finite volume of soil. *Plant Soil* 38:621–635.

Bloom, A. J., Chapin, III, F. S., and Mooney, H. A. (1985). Resource limitation in plants: An economic analogy. *Annu. Rev. Ecol. System.* 16:363–392.

Bowen, G. D. (1973). Mineral nutrition of ectomycorrhizae. *In* "Ectomycorrhizae: Their ecology and physiology." (G. C. Marks and T. T. Kozlowski, eds.), pp. 151–205. Academic Press, New York.

Bowen, G. D. (1985). Roots as a component of tree productivity. *In* "Attributes of Trees as Crop Plants" (M. G. R. Cannell and J. E. Jackson, eds.), pp. 303–315. Institute of Terrestrial Ecology, Edinburgh.

Cairney, J. W. G., and Smith, S. E. (1992) Effect of monovalent cations on efflux of phosphate from the ectomycorrhizal fungus *Pisolithus tinctorius*. *In* "Mycorrhizas in Ecosystems" (D. J. Read, D. H. Lewis, A. H. Fitter, and I. J. Alexander, eds.), pp. 352–355. C.A.B. International, Cambridge Univ. Press, Cambridge, England.

Caldwell, M. M. (1979). Root structure: The considerable cost of belowground function. *In* "Topics in Plant Population Biology" (O. T. Solbrig, S. Jain, G. B. Johnson, and P. H. Raven, eds.), pp. 408–427. Columbia Univ. Press, New York.

Chapin, III, F. S. (1989). The cost of tundra plant structures: evaluation of concepts and curriencies. *Am. Nat.* 133:1–19.

Chung, H. H., and Kramer, P. J. (1975). Absorption of water and ^{32}P through suberized and unsuberized roots of lolbolly pine. *Can J. For. Res.* 10:348–356.

Dighton, J., Thomas, E. D., and Latter, P. M. (1987). Interactions between tree roots, mycorrhizas, a saprotrophic fungus and the decomposition of organic substrates in a microcosm. *Biol. Fertil. Soils* 4:145–150.

Duddridge, J. A., Malibari, A., and Read, D. J. (1980). Structure and function of mycorrhizal rhizomorphs with special reference to their role in water transport. *Nature (London)* 287:834–836.

Eissenstat, D. M. (1992). Costs and benefits of constructing roots of small diameter. *J. Plant Nutr.* 15:763–782.

Eissenstat, D. M., and Van Rees, K. C. J. (1994). The growth and function of pine roots. *Ecol. Bull.* 43 (in press).

Fahey, T. J. (1992). Mycorrhizal and forest ecosystems. *Mycorrhiza* 1:83–89.

Fahey, T. J., and Hughes, J. W. (1994). Fine root dynamics in a northern hardwood forest ecosystem, Hubbard Brook Experimental Forest, N.H. *J. Ecol.* in press.

Fahey, T. J., Hughes, J. W., Mou Pu, and Arthur, M. A. (1988). Root decomposition and nutrient flux following whole-tree harvest of northern hardwood forest. *For. Sci.* 34: 744–768.

Finlay, R. D. (1992). Uptake and translocation of nutrients by ectomycorrhizal fungal mycelia. *In* "Mycorrhizas in Ecosystems" (D. J. Read, D. H. Lewis, A. H. Fitter, and I. J. Alexander, eds.), pp. 91–97. C.A.B International, Cambridge Univ. Press, Cambridge, England.

Finlay, R. D., Ek, H., Odham, G., and Söderström, B. (1988). Mycelial uptake, translocation and assimilation of nitrogen from ^{15}N-labelled ammonium by *Pinus sylvestris* plants infected with four different ectomycorrhizal fungi. *New Phytol.* 110:59–66.

Finlay, R. D., Ek, H., Odham, G., and Söderström, B. (1989). Uptake, translocation and assimilation of nitrogen from ^{15}N-labelled ammonium and nitrate sources by intact ectomycorrhizal systems of *Fagus sylvatica* infected with *Paxillus involutus*. *New Phytol.* 113:47–55.

Finlay, R. D., Frostegård, Å., and Sonnerfeldt, A.-M. (1992). Utilization of organic and inorganic nitrogen sources by ectomycorrhizal fungi in pure culture and in symbiosis with *Pinus contorta* Dougl. ex Loud. *New Phytol.* 120:105–115.

Fitter, A. H. (1989). The role and ecological significance of vesicular–arbuscular mycorrhizas in temperate ecosystems. *Agric. Ecosyst. Environ.* 29:137–151.

Fogel, R., and Hunt, G. (1983). Contribution of mycorrhizae and soil fungi to nutrient cycling in a Douglas-fir ecosystem. *Can. J. For. Res.* 13:219–232.

Friend, A. L., Eide, M. R., and Hinckley, T. M. (1990). Nitrogen stress alters root proliferation in Douglas-fir seedlings. *Can. J. For. Res.* 20:1524–1529.

George, H., Haussler, K., Kothari, S. K., Li, X.-L., and Marschner, H. (1992). Contribution of mycorrhizal hyphae to nutrient and water uptake of plants. *In* "Mycorrhizas in Ecosystems" (D. J. Read, D. H. Lewis, A. H. Fitter, and I. J. Alexander, eds.), pp. 42–47. C.A.B. International, Cambridge Univ. Press, Cambridge, England.

Glass, A. D. M., and Siddiqi, M. Y. (1984). The control of nutrient uptake rates in relation to the inorganic composition of plants. *In* "Advances in Plant Nutrition, Volume 1" (P. B. Tinker and A. Lauchli, eds.), pp. 103–147. Praeger, New York.

Grime, J. P., Campbell, B. D., Mackey, J. M. L., and Crick, J. C. (1991). Root plasticity, nitrogen capture and competitive ability. *In* "Plant Root Growth: An Ecological Perspective" (D. Atkinson, ed.), pp. 381–397. Special Publication series of the British Ecological Society No. 10. Blackwell, Oxford.

Gutschick, V. P. (1981). Evolved strategies in nitrogen acquisition by plants. *Am. Nat.* 118:607–637.

Hacskaylo, E., and Snow, A. G., Jr. (1959). Relation of soil nutrients and light to prevalence of mycorrhizae on pine seedlings. Northeast Forest Experiment Station, Station Paper 125.

Harley, J. L., and Lewis, D. H. (1969). The physiology of ectotrophic mycorrhizas. *Adv. Microb. Physiol.* 3:53–81.

Harley, J. L., and Smith, D. E. (1983). "Mycorrhizal Symbiosis." Academic Press, London.

Harper, J. L. (1977). "Population Biology of Plants." Academic Press, New York.

Harper, J. L., Jones, M., and Sackville Hamilton, N. R. (1991). The evolution of roots and the problems of analysing their behavior. *In* "Plant Root Growth: An Ecological Perspective" (D. Atkinson, ed.), pp. 3–22. Special Publication series of the British Ecological Society No. 10. Blackwell, Oxford.

Hendrick, R. L., and Pregitzer, K. S. (1993). Patterns of fine root mortality in two sugar maple forests. *Nature (London)* 361:59–61.

Högberg, P. (1986). Soil nutrient availability, root symbioses and tree species composition in tropical Africa: A review. *J. Trop. Ecol.* 2:359–372.

Johnson, I. R. (1990). Plant respiration in relation to growth, maintenance, ion uptake and nitrogen assimilation. *Plant, Cell Environ.* 13:319–328.

Jones, M. D., Durall, D. M., and Tinker, P. B. (1991). Fluxes of carbon and phosphorus between symbionts in willow ectomycorrhizas and their changes with time. *New Phytol.* 119:99–106.

Kelly, J. M., and Barber, S. A. (1991). Magnesium uptake kinetics in loblolly pine seedlings. *Plant Soil* 134:227–232.

Kelly, J. M., Barber, S. A., and Edwards, G. S. (1992). Modeling magnesium, phosphorus, and potassium uptake by loblolly pine seedlings using a Barber–Cushman approach. *Plant Soil* 135:209–218.

Knight, D. H., Fahey, T. J., Running, S. W., Harrison, A. T., and Wallace, L. L. (1981). Transpiration from 100-yr-old lodgepole pine forests estimated with whole-tree potometers. *Ecology* 62:717–726.

Kropp, B. R. (1982). Formation of mycorrhizae on nonmycorrhizal Western hemlock outplanted on rotten wood and mineral soil. *For. Sci.* 28:706–710.

Lambers, H., Szaniawski, R. K., and de Visser, R. (1983). Respiration for growth, maintenance and ion uptake. An evaluation of concepts, methods, values and their significance. *Physiol. Plant.* 58:556–563.

Ledig, F. T., Drew, A. R., and Clark, J. G. (1976). Maintenance and constructive respiration, photosynthesis, and net assimilation rate in seedlings of pitch pine (*Pinus rigida* Mill.). *Ann. Bot.* 40:289–300.

Marshall, J. D., and Perry, D. A. (1987). Basal and maintenance respiration of mycorrhizal and nonmycorrhizal root systems of conifers. *Can. J. For. Res.* 17:872–877.

Marshall, J. D., and Waring, R. H. (1985). Predicting fine root production and turnover by monitoring root starch and soil temperature. *Can J. For. Res.* 15:791–800.

Meyer, F. H. (1973). Distribution of ectomycorrhizae in native and man-made forests. *In* "Ectomycorrhizae—Their Ecology and Physiology" (G. C. Marks and T. T. Kozlowski, eds.), pp. 79–105. Academic Press, New York.

Miller, O. K. (1982). Taxonomy of ecto- and ectendomycorrhizal fungi. *In* "Methods and Principles of Mycorrhizal Research" (N. C. Schenck, ed.), pp. 91–101. American Phytopathological Society, St. Paul, Minnesota.

Miller, S. L., and Allen, E. B. (1992). Mycorrhizae, nutrient translocation and interaction between plants. *In* "Mycorrhizal Functioning—An Integrative Plant-Fungal Process (M. F. Allen, ed.), pp. 301–332. Chapman & Hall, New York.

Muona, O. (1990). Population genetics in forest tree improvement. *In* "Plant Population Genetics, Breeding, and Genetic Resources" (A. H. D. Brown, M. T. Clegg, A. L. Kahler, and B. S. Weir, eds.), pp. 282–298. Sinauer, Sunderland, Massachusetts.

Nadelhoffer, K. J., Aber, J. D., and Melillo, J. M. (1985). Fine roots, net primary production, and soil nitrogen availability: A new hypothesis. *Ecology* 66:1377–1390.

Nambiar, E. K. S. (1987). Do nutrients retranslocate from fine roots? *Can. J. For. Res.* 17:913–918.

Nye, P. H., and Spiers, J. A. (1964). Simultaneous diffusion and mass flow to plant roots. *8th Int. Congr. Soil Sci.* 3:535–541.

Nye, P. H., and Tinker, P. B. (1977). "Solute Movement in the Soil–Root System." Blackwell, Oxford.

Paul, E. A., and F. E. Clark. (1989). "Soil Microbiology and Biochemistry." Academic Press, San Diego.

Pregitzer, K. S., Hendrick, R. L., and Fogel, R. (1993). The demography of fine roots in response to patches of water and nitrogen. *New Phytol.* 125:575–580.

Queen, W. H. (1967). Radial movement of water and ^{32}P through suberized and unsuberized roots of grape. Ph.D. Dissertation. Duke University, Durham, North Carolina.

Reynolds, E. R. C. (1975). Tree rootlets and their distribution. *In* "The Development and Function of Roots" (J. G. Torrey and D. T. Clarkson, eds.), pp. 163–178. Academic Press, New York.

Rousseau, J. V. D., and Reid, C. P. P. (1990). Effects of phosphorus and ectomycorrhizas on the carbon balance of loblolly pine seedlings. *For. Sci.* 36:101–112.

Ryan, M. G. (1991). Effects of climate change on plant respiration. *Ecol. Appl.* 1:157–167.

Sands, R., Fiscus, E. L., and Reid, C. P. P. (1982). Hydraulic properties of pine and bean roots with varying degrees of suberization, vascular differentiation and mycorrhizal infection. *Aust. J. Plant Physiol.* 9:559–569.

Santantonio, D., and Santantonio, E. (1987). Effects of thinning on production and mortality of fine roots in a *Pinus radiata* plantation on a fertile site in New Zealand. *Can. J. For. Res.* 17:919–928.

Schoettle, A. W., and Fahey, T. J. (1994). Foliage and fine root longevity in pines. *Ecol. Bull.* 43 (in press).

Smith, S. E. (1982). Inflow of phosphate into mycorrhizal and non-mycorrhizal plants of *Trifolium subterraneum* at different levels of soil phosphate. *New Phytol.* 90:293–303.

Smith, W. E., and Smith, F. A. (1990) Structure and function of the interfaces in biotrophic symbioses as they relate to nutrient transport. *New Phytol.* 114:1–38.

Szaniawski, R. K. (1981). Growth and maintenance respiration of shoot and roots in Scots pine seedlings. *Z. Pflanzenphysiol.* 101:391–398.

Thomson, B. D., Clarkson, D. T., and Brain, P. (1990). Kinetics of phosphorus uptake by the germ-tubes of the vesicular-arbuscular mycorrhizal fungus, *Gigaspora margarita*. *New Phytol.* 116:647–653.

Tiffney, B. H., and Niklas, K. J. (1985). Clonal growth in land plants: A paleobotanical perspective. In "Population Biology and Evolution of Clonal Organisms" (J. B. C. Jackson, L. W. Buss, and R. E. Cook, eds.), pp. 35–66. Yale Univ. Press, New Haven, Connecticut.

Veen, B. W. (1981). Relation between root respiration and root activity. *Plant Soil* 63: 73–76.

van der Werf, A., Kooijman, A., Welschen, R., and Lambers, H. (1988). Respiratory energy costs for the maintenance of biomass, for growth and for ion uptake in roots of *Carex diandra* and *Carex acutiformis*. *Physiol. Plant.* 72:483–491.

Van Rees, K. C. J., Comerford, N. B., and Rao, P. S. C. (1990). Defining soil buffer power: Implications for ion diffusion and nutrient uptake modeling. *Soil Sci. Soc. Am. J.* 54:1505–1507.

Whitely, G. M., Hewitt, J. S., and Dexter, A. R. (1982). The buckling of plant roots. *Physiol. Plant.* 54:333–342.

Wilcox, H. (1968). Morphological studies of the roots of red pine, *Pinus resinosa*. I. Growth characteristics and patterns of branching. *Am. J. Bot.* 55:247–254.

Yanai, R. D. (1994). A steady-state model of nutrient uptake improved to account for newly-grown roots. *Soil Sci. Soc. Am. J.* 58 (in press).

4

External and Internal Factors Regulating Photosynthesis

Robert O. Teskey, David W. Sheriff,
David Y. Hollinger, and Richard B. Thomas

I. Introduction

Photosynthesis is simultaneously regulated by many factors. Environmental factors, consisting primarily of light, water, heat, carbon dioxide, and other gases, can have direct effects on photosynthesis when they alter rates of chemical processes in the photosynthetic pathway. Environmental factors, along with nutrients and sink demands for carbohydrates, may also have indirect effects on photosynthesis. Indirect effects are the result of changes in nonphotosynthetic processes that, in turn, alter the rate of photosynthesis. A final category of regulators, photosynthetic framework, consists of substances that compose the physiologically active compounds and structures that form the basis of the light and dark reactions responsible for photosynthesis. The framework determines the photosynthetic capacity of a plant or organ, i.e., the rate of photosynthesis achievable when direct and indirect effects are nonlimiting. In this chapter we have divided the discussion of factors that regulate photosynthesis in conifers into these three categories, framework, direct effects, and indirect effects, because of an expanding appreciation in the field of environmental physiology that external and internal factors can simultaneously regulate photosynthesis by both direct and indirect means. We offer this outline as a logical way of presenting and discussing these issues.

II. Framework

The main components of the photosynthetic framework are the physical structure and arrangement of photosynthesizing organs, and the nutrients involved in the photosynthetic pathway. Different aspects of physical structure are important at each level of structural organization: chloroplasts within mesophyll cells, architecture of mesophyll tissue, and canopy structure. These determine the efficiency of light interception and absorption for a given organization of the photosynthetic pathway.

Within a cell, chloroplasts are usually moved by cytoplasmic streaming and are able to rotate to make "best use" of available light, so the arrangement of chloroplasts in a cell will not often be a factor limiting photosynthesis. However, arrangement of the photosynthetic apparatus within a leaf can significantly affect its rate of photosynthesis. Because the amount of light declines with distance from the illuminated surface, leaf photosynthesis will be maximized if electron transport and ribulose bisphosphate carboxylase/oxygenase (Rubisco) capacities are greater nearest this surface (e.g., Farquhar, 1989). There is an analogous situation in canopies, where photosynthetic capacity is often greatest in the upper part of the canopy and declines with irradiance toward the lower crown (see Chapter 1, this volume). Photosynthetic capacity is also affected by the structural characteristics of individual leaves, and an excellent correlation was found between specific leaf mass (dry mass per unit of projected area) and seasonal mean photosynthesis (Oren *et al.*, 1986). Because of their effects on light interception, plant density and crown shape can affect rates of photosynthesis both per plant and per unit ground area (e.g., Rook *et al.*, 1985). Generally, under conditions that do not severely limit rates of photosynthesis, a greater plant density results in photosynthetic production being lower per tree and higher per ground area. A horizontal broadening of a plant's canopy increases its photosynthetic gain, whereas making the canopy deeper has little effect (Rook *et al.*, 1985).

A gradient in CO_2 concentration exists within air spaces of a photosynthesizing leaf, with $[CO_2]$ highest near stomata. To maximize photosynthetic capacity of a leaf, the diffusive conductance of mesophyll air spaces must vary in relation to the photosynthetic capacities of each cell vis-à-vis its light environment (e.g., Farquhar, 1989). Other arrangements will cause photosynthesis to be reduced, and light to be used less efficiently. This factor will be of greatest importance when leaves are thick and stomatal densities small.

A. Nutrients

The physicochemical "structure" of the photosynthetic system is determined by a number of compounds that contain mineral elements (e.g., Gerwick, 1982). Shortage within the foliage of any of these below an optimum, or appreciable allocation to nonphotosynthetic compounds at any foliar nutrient concentration, will reduce rates of photosynthesis per unit area or weight of foliage under conditions that are not otherwise limiting. The effect of nutrient status on photosynthetic capacity has been clearly demonstrated in experiments that prevent the retranslocation of nutrients out of older foliage into the current year's growing foliage by removing the terminal buds of individual branches (Lange *et al.*, 1987). When new foliage was prevented from growing, both the nutrient content and photosynthetic capacity of 1-year-old foliage were higher than in foliage on branches where the terminal bud was left intact. Modifications of nutritional status produced by acidic mist have been shown to change many aspects of photosynthesis, including carboxylation efficiency, quantum yield, light- and CO_2-saturated rates, light and CO_2 compensation point, chlorophyll content, and percentage stomatal limitation (Eamus and Fowler, 1990) (Fig. 1). However, the effects of individual nutrients on these processes cannot be distinguished.

The effect of an individual nutrient on photosynthetic rate is most evident for nitrogen (Fig. 2) (Evans, 1989; Smolander and Oker-Blom, 1989; Thompson and Wheeler, 1992; Reich and Schoettle, 1988) and phosphorus (Black, 1988; Rousseau and Reid, 1990; Sheriff *et al.*, 1986). An effect of increased concentrations of foliar nutrients other than nitrogen and phosphorus on photosynthesis has only occasionally been demonstrated in conifers (e.g., Linder and Rook, 1984). The effect of magnesium deficiency on photosynthesis has been demonstrated (Zimmermann *et al.*, 1988), although magnesium appeared to affect the rate of photosynthesis only when foliar levels were very low [<5 μmol/g (dry wt)]. However, nutritional effects may be more easily observed when other constraints on photosynthesis are reduced, as was shown by the greater effect of phosphorus on photosynthesis of *Pinus radiata* at elevated compared to ambient [CO_2] (Conroy *et al.*, 1990).

B. Carbon Balance

The measurement of photosynthesis that is most often made, and which is of most interest in relation to carbon gain and hence increase in plant biomass, is that of net foliar photosynthesis. When using this as a measure of carbon accumulation in relation to nutrition, allowance

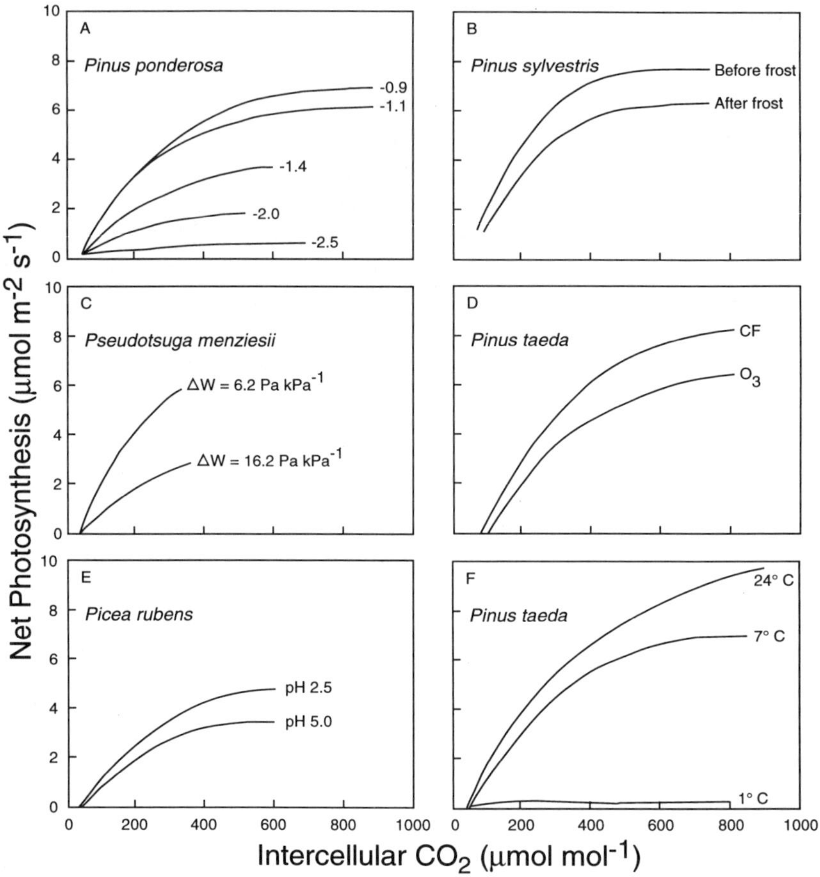

Figure 1 Effects of direct and indirect influences on the rate of net photosynthesis at various intercellular CO_2 concentrations. (A) Water stress: soil water potentials (MPa) shown next to each curve (Data from DeLucia and Heckathorn, 1989). (B) Effect of frost (Data from Strand and Öquist, 1985). (C) High and low vapor pressure deficits (ΔW) (Data from Grieu *et al.*, 1988). (D) Comparison of photosynthesis under charcoal-filtered and twice-ambient ozone treatments after 2 years of treatment (Data from Sasek and Richardson, 1989). (E) Indirect effect of acid mist on photosynthesis (Data from Eamus and Fowler, 1990). (F) Indirect effect of soil temperature; soil temperature at the time of treatment shown next to each curve (Data from Day *et al.*, 1991).

must be made for the positive association between nutrient concentration and the rates of respiration of tissues (e.g., Linder and Rook, 1984). For example, the ratio of net photosynthesis to foliar respiration has been linked to tree vigor on stressful sites (McLaughlin *et al.*, 1990). These responses are to be expected because photosynthetic capacity and

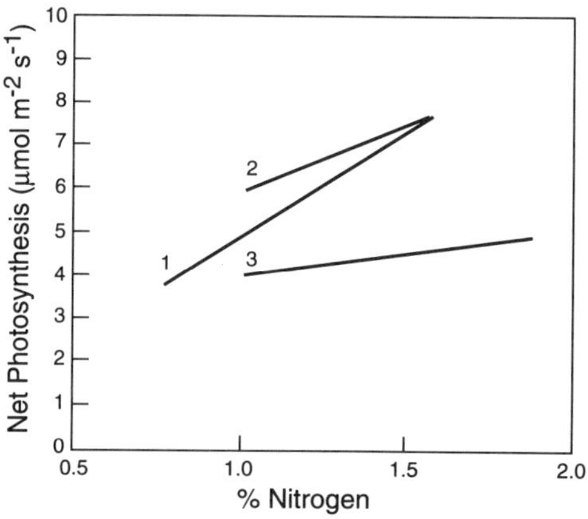

Figure 2 Net photosynthesis as a function of the percentage of nitrogen in the leaves. (1) *Pinus sylvestris* (Data from Smolander and Oker-Blom, 1989); (2) *Pseudostuga menziesii* (Data from Brix, 1981); and (3) *Pinus radiata* (Data from Thompson and Wheeler, 1993).

rate of dark respiration are positively correlated in both coniferous and broad-leafed tree species (Ceulemans and Saugier, 1991).

III. Direct Effects

A. Light

The textbook biphasal response of leaf photosynthesis to irradiance is well known: as the amount of light increases, irradiance and photosynthesis are first linearly related, then the increase in photosynthesis declines at each successive increase in light until it saturates (e.g., Long and Hällgren, 1985). This relationship is affected by the level of irradiance, usually in a leaf's environment, such that for shade leaves the linear portion of the relationship occurs over a smaller range of (low) irradiances, and photosynthesis saturates at a lower irradiance than in exposed foliage. Usually shaded leaves are relatively more efficient at light capture (Kull and Koppel, 1987), and subsequent parts of the photochemical pathway have a smaller capacity than exists in exposed leaves: shade leaves have proportionately more of their nitrogen in thylakoids than in ribulose bisphosphate carboxylase compared with sun leaves (Yushkov and Zav'yalova, 1989). Chlorophyll content and the convexity of the light response curve are well correlated in conifers, supporting the hypothesis that gradients in light within the leaf alter the shape of the light response curve (Leverenz, 1987). Numerous other factors also alter the shape of the curve by affecting the maximum rate of photosynthesis or

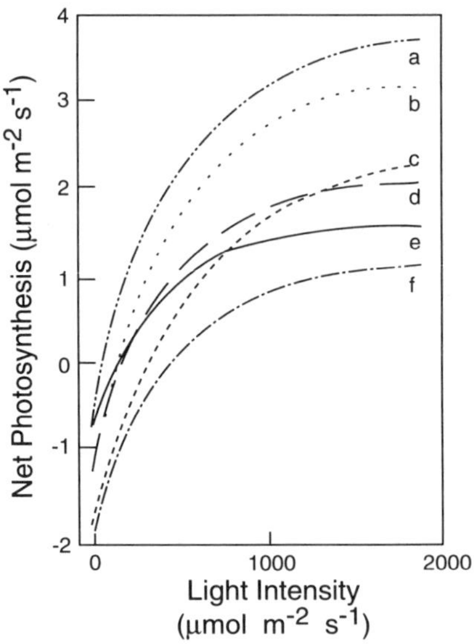

Figure 3 Variation in the response of net photosynthesis to light for 2-year-old *Pinus contorta* seedlings. Various factors affected the response: (a and e) plants were greenhouse-grown; (b and c) roots were heated; (d) after planting in the field; (f) before planting (Data from van den Driessche, 1987).

the rate of respiration. These include the environment [e.g., temperature and CO_2 concentration (Leverenz, 1988)], foliar nutrient contents (Smolander and Oker-Blom, 1989), air pollutants (Eamus and Fowler, 1990), and leaf geometry (Jordan and Smith, 1993). An illustration of the variation that is possible in light response curves is shown in Fig. 3 for a single species and plant age. Although this variation is large, Leverenz (1987) showed that the convexity of the curve was correlated with chlorophyll content (Fig. 4). It has also been established that the maximum rate of photosynthesis in saturating light conditions is a function of the Rubisco activity and end-product synthesis limitations (Stitt, 1991). This suggests that much of the variation in light response curves can be accounted for by these factors alone.

The relationship between photosynthesis and irradiance is not always as simple as that discussed above. Photoinhibition, i.e., a decrease in the optimum photon yield induced by light (Powles, 1984), reduces photosynthetic rates after exposure to light (Ögren and Evans, 1992). Photoinhibition is greater at higher light intensities, and is more severe when temperatures are extreme or water deficits are large, or both. Photoinhibition under mild conditions is reversible over a short period of time (minutes to hours), but reversibility of the greater photoinhibition that

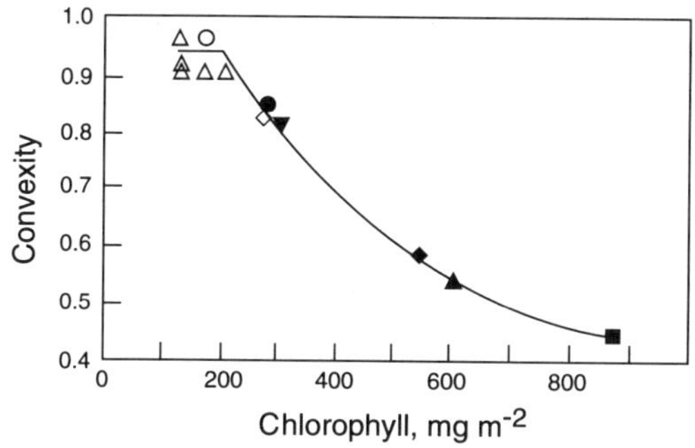

Figure 4 Convexity of the light response curve as a function of the chlorophyll content of foliage for five conifer species: (■) *Picea mariana*, (▲) *Pinus sylvestris*, (▼) *Picea glauca*, (♦) *Picea sitchensis*, and (●) *Picea abies*. Open symbols are measurements made in a light-integrating sphere of *Pinus sylvestris* (△) and *Picea abies* (○); closed symbols are measurements made using unilateral illumination (Data from Leverenz, 1987).

results from a more extreme environment takes longer (Ögren and Evans, 1992). It has been suggested that the rapidly reversible form of photoinhibition is a protective mechanism that dissipates excessive energy that could not be otherwise used and would result in tissue damage (Krause, 1988). Lowering of photosynthetic rates, with recovery times consistent with the rapidly and slowly reversible forms of photoinhibition, was observed by D. W. Sheriff (unpublished, 1994) in a laboratory study of photosynthesis in well-watered *P. radiata* at moderate and high temperatures, respectively. High temperatures and water deficits have induced symptoms consistent with photoinhibition in *P. radiata* in the field (D. W. Sheriff and E. K. S. Nambiar, unpublished, 1994).

The response of canopy photosynthesis to light is similar to that of individual leaves, except that because of mutual shading, the linear phase of the response of photosynthesis to light occurs over a wider range of irradiances for a canopy than for a leaf. Differences among species in leaf arrangement, as determined by the silhouette area ratio, alter light absorption and photosynthesis of the needles (Smolander *et al.*, 1987). The effect of leaf arrangement on light absorption affects the shade acclimation of individual needles in different orientations on the shoot (Leverenz and Jarvis, 1980). The effect of leaf arrangement on light absorption changes diurnally and seasonally, due to differences in sun altitude and azimuth (Smith *et al.*, 1991). With the exception of obligate shade species, and the possible exception of extreme environmental conditions, the photosynthetic productivity of a plant is positively associated with the total irradiance that impinges on the top of its canopy (e.g., Rook and Corson, 1978) (see Chapter 1, this volume).

Ultraviolet radiation can reduce photosynthesis by damage to enzymes and pathways (e.g., Tevini and Teramura, 1989). A negative effect of ultraviolet-B (UV-B) radiation on pine photosynthesis has been reported, but the effect was not consistent at all elevated UV-B levels (Sullivan and Teramura, 1989). Some species are able to produce UV-B-absorbing compounds that obviate negative effects of high UV levels (e.g., Sullivan and Teramura, 1989). The penetration of UV radiation into leaves differs among conifers due to optical and anatomical characteristics. The characteristics of the cuticle, epidermis, and hypodermis are particularly important (Bornman and Vogelmann, 1988).

B. Water

It is often difficult to separate direct and indirect effects of water deficits on photosynthesis, especially with diffusive conductance and photosynthesis approximately linearly related (e.g., Wong *et al.*, 1979; Matyssek, 1986). Among conifers the relationship between diffusive conductance and net photosynthesis has been reported for a number of species (Fig. 5). With few exceptions, the response is linear and consistent with respect to slope. The link between these variables may be hormonal (Cowan *et al.*, 1982), consistent with the theory of "general" control

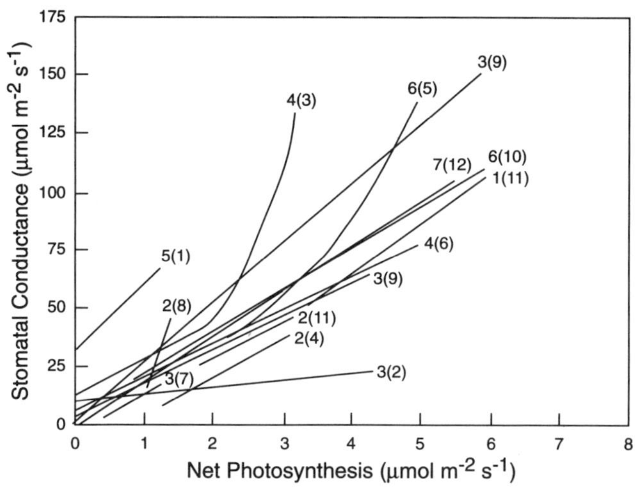

Figure 5 Relationship between stomatal conductance and net photosynthesis for conifer species; the first number next to each line refers to the species, the number in parentheses is the reference. Species: (1) *Larix decidua*; (2) *Picea abies*; (3) *Pinus radiata*; (4) *Pinus ponderosa*; (5) *Pinus sylvestris*; (6) *Pinus taeda*; (7) *Pinus contorta*. References: (1) Küppers and Schulze (1985); (2) Conroy *et al.* (1988); (3) DeLucia and Heckathorn (1989); (4) Wallin *et al.* (1990); (5) Day *et al.* (1991); (6) Monson and Grant (1989); (7) Squire *et al.* (1988); (8) Zimmermann *et al.* (1988); (9) Thompson and Wheeler (1993); (10) Teskey *et al.* (1986); (11) Matyssek (1986); (12) Benecke and Nordmeyer (1982).

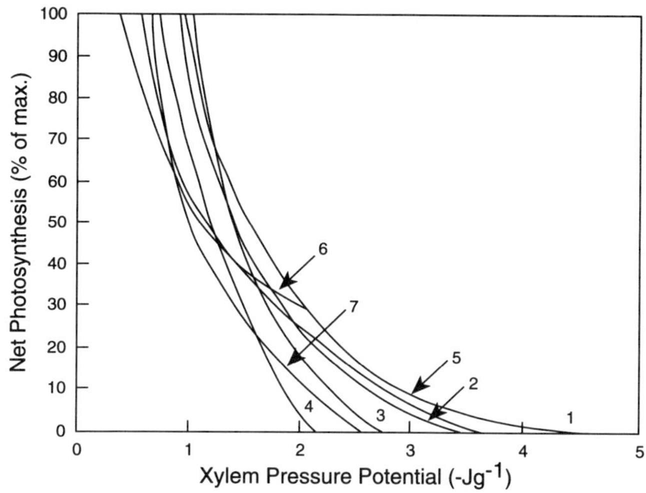

Figure 6 Influence of xylem pressure potential on the rate of net photosynthesis for selected conifers. (1) *Pseudotsuga menziesii*; (2) *Tsuga heterophylla*; (3) *Picea glauca*; (4) *Pinus contorta*; (5) *Picea rubens*; (6) *Pinus halepensis*; (7) *Pinus ponderosa*.

of physiological activity by plant hormones, some aspects of which are discussed below. The overall effect of water deficits on leaf photosynthesis will usually be qualitatively similar to that found by Sheriff and Whitehead (1984) for *P. radiata*. As foliar water potential (or content) declines, photosynthetic rate often remains unchanged initially, until a water potential is reached at which photosynthesis declines, at first steeply and then more gradually as water deficits increase, until there is a zero (or negative) carbon balance. For conifers, the foliar water status at which photosynthesis begins to decline and the rate of this decline can vary with species (Grieu *et al.*, 1988), plant age (Squire *et al.*, 1988), growth conditions (Lee *et al.*, 1990), and genotype (Monson and Grant, 1989), although the pattern among species is quite similar (Fig. 6). Direct effects of water deficits on photosynthesis operate mainly through effects on electron transport and phosphorylation (e.g., Kaiser, 1987). For example, Conroy *et al.* (1986) obtained results indicating that electron flow subsequent to photosystem II is adversely affected by water deficits in *P. radiata*. Bunce (1977) found an increase in mesophyll resistance with leaf water deficit in a range of species, including several conifers.

C. Temperature

Most conifers retain their leaves for 2 or more years. As a result the foliage is subjected to a wide range of temperatures as the seasons change. Rates of net photosynthesis vary with temperature such that

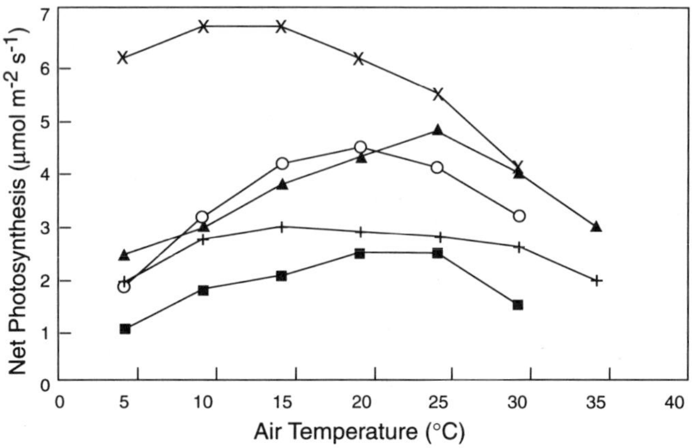

Figure 7 Response of net photosynthesis to air temperature for conifers adapted to a range of climates from cold to warm. Each line represents the highest seasonal rates when authors presented data from more than one time during the year. Species represented are (×) *Pinus ponderosa* (Data from Monson and Grant, 1989), (+) *Pinus sylvestris* (Data from Küppers and Schulze, 1985), (▲) *Pinus taeda* (Data from Strain *et al.*, 1976), (○) *Pinus strobus* (Data from Jurik *et al.*, 1988), and (■) *Picea engelmanii* (Data from Hadley and Smith, 1987).

they decline both above and below an optimum temperature (e.g., Wood and Brittain, 1973). Optimum temperatures for photosynthesis are generally between 15 and 30° C (Belous, 1986). The response of most conifers to temperature is relatively flat (Fig. 7) with photosynthesis decreasing gradually at higher or lower temperatures. At many times of the year the temperatures are far from optimum (Smith and Carter, 1988). For conifers in cold climates, this is especially true in winter, when frost and freezing temperatures greatly limit rates of photosynthesis (Troeng and Linder, 1982a). Positive net photosynthesis is possible at air temperatures below freezing, although the rates are very low (Guehl, 1985, Neilson *et al.*, 1972). Maintenance of photosynthetic capacity at low temperatures is a function of the light environment, because photoinhibition occurs in conifers in the winter when they are exposed to low temperatures and moderate to high light levels. Low, but above-freezing, temperatures that induce cold tolerance in plants also appear to cause photoinhibition (Öquist and Huner, 1991). Cold acclimation, however, provides no resistance to photoinhibition, as hardened and nonhardened *Pinus sylvestris* had the same degree of photoinhibition (Strand and Öquist, 1985; Öquist and Huner, 1991). This was also evident in *Pinus contorta* (Öquist and Malmberg, 1989). In winter, temperatures below freezing produce substantial changes in photosynthetic capacity by causing damage to the photosynthetic apparatus, including disruption of the electron transport chain, the chlorophyll–photosystem complex, and in particular photosystem II (Öquist *et al.*, 1980; Öquist, 1986).

After winter, damage to the chloroplasts is repaired. The rate of recovery is temperature dependent (Lundmark *et al.*, 1988a), with more rapid rates of recovery at higher temperatures. No difference in the rate of recovery was detected in a comparison of *P. contorta*, *P. sylvestris*, and *Picea abies*, and recovery took several months under field conditions (Lundmark *et al.*, 1988b). DeLucia and Smith (1987) reported that the photosynthetic capacity of *Picea engelmannii* foliage continued to increase until midsummer. In contrast, the recovery of *Pinus strobus* from depressed rates of photosynthesis in winter was very rapid, although it did not occur until April (Jurik *et al.*, 1988). Jurik *et al.* concluded that there was little benefit to carbon gain in the winter from evergreen foliage, but that carbon gain of this species would benefit from 3 to 6 weeks of favorable weather before deciduous foliage would emerge. A similar conclusion has been reached by others studying conifers in cold climates (Schulze *et al.*, 1977). In regions with mild winters substantial photosynthesis takes place in the late fall, winter, and early spring periods (Cropper and Gholz, 1992; Waring and Franklin, 1979), although photosynthetic capacity is still lower than in the summer (Reynolds *et al.*, 1980), as is the temperature optimum for net phtosynthesis (Strain *et al.*, 1976).

In summer, photosynthesis is still often limited by temperature. Significant reductions in the maximum rate of net photosynthesis are evident after nighttime frosts (Hällgren *et al.*, 1990). Temperatures below 0°C cause a reversible reduction in the mesophyll capacity for photosynthesis and a parallel decrease in stomatal conductance (Fig. 1) (Lundmark *et al.*, 1988a, DeLucia and Smith, 1987). At one time it was thought that stomatal limitations to diffusion of CO_2 were responsible for the reduction in photosynthesis, but more recent analyses of the relationship between net photosynthesis and intercellular CO_2 concentration and fluorescence techniques have confirmed that the primary effect is in the mesophyll (Strand and Lundmark, 1987; Hallgren *et al.*, 1990). Even above 0°C, air temperature is often limiting to photosynthesis, either by being below the photosynthetic optimum in cold climates (Smith and Carter, 1988), or above it in warm climates (Teskey *et al.*, 1987; Attiwill and Cromer, 1982). Summer air temperatures also reduce canopy carbon gain by substantially increasing nighttime respiration (Price and Black, 1990).

D. Carbon Dioxide Partial Pressure

At present atmospheric CO_2 levels, photosynthesis of all C_3 plants, including conifers, may be limited by the supply of CO_2. The reason for this limitation is that Rubisco, the carboxylating enzyme, also functions as an oxygenase and, therefore, there is competition between CO_2 and O_2 for active sites on the enzyme. Although enzymatic data on conifers are very limited, it appears that the kinetic characteristics of Rubisco, including the affinity for CO_2, are similar to those in spinach (Weidner and Kraus, 1987; Gezelius, 1975; Beadle *et al.*, 1983). Although the $K_m(CO_2)$, or concentration of CO_2 at which Rubisco is half saturated, is lower than the $K_m(O_2)$, the high ratio of O_2 partial pressure to CO_2 partial pressure in the atmosphere generates a substantial photorespiratory loss of carbon from the plant.

Atmospheric CO_2 reaches the chloroplast by diffusion and is thus described by Fick's law. This states that the rate of CO_2 diffusion is a function of the concentration gradient of CO_2 from outside the leaf to the active site of the carboxylase enzyme and the sum of the resistances to flow exhibited by the leaf along this path (Larcher, 1980). Resistances to CO_2 include the boundary layer of the needles, the stomatal pore, and the flow across the mesophyll of the leaf. In general, stomatal conductance to CO_2 in conifers is lower than that of most deciduous trees under similar environmental conditions. Very little is known about the mesophyll conductance of conifers. Mesophyll conductance to CO_2, however, is typically quite low in plants with low rates of photosynthesis (Loreto *et al.*, 1992). Additionally, low CO_2 diffusion rates from stomata to carboxylation sites have been demonstrated in evergreen species (Lloyd *et al.*, 1992).

At the leaf level, the measurement of stomatal conductance of CO_2 and CO_2 flux into the leaf can be used to evaluate the rate-limiting processes of photosynthesis, e.g., the rate of diffusion of CO_2 through the stomata versus biochemical limitations of photosynthesis. For example, Teskey *et al.* (1986) concluded that biochemical limitations, and not stomatal conductance, were primarily responsible for limiting photosynthesis under many environmental conditions, including changing light conditions and increasing water stress.

The partial pressure of atmospheric CO_2 has been steadily increasing since the middle of the nineteenth century. Since the late 1950s, when detailed measurements of CO_2 were begun in Mauna Loa, Hawaii, atmospheric levels of CO_2 have increased by about 12% (Keeling *et al.*, 1982). Several reviews have summarized the photosynthetic response of trees, including conifers, to increased atmospheric CO_2 (Kramer and Sionit, 1987; Eamus and Jarvis, 1989; Jarvis, 1989), and some generalities have emerged. For conifers grown under nonlimiting water and nutrient supply, photosynthesis is enhanced and stomatal conductance is reduced with long-term CO_2 enrichment (Eamus and Jarvis, 1989). However, higher foliage temperature and greater thermal stress in *Pinus ponderosa* may have resulted from CO_2-induced stomatal closure and consequent reductions in transpiration (Surano *et al.*, 1986). In addition, short-term increases in CO_2 often stimulate photosynthesis to a greater degree than long-term increases in CO_2, indicating the possibility of some acclimatory response (Fetcher *et al.*, 1988). When conifers are grown under conditions in which they naturally exist, including nutrient limitations or water stress, the photosynthetic response of CO_2 is less predictable (Tolley and Strain, 1984, 1985; Higginbotham *et al.*, 1985; Conroy *et al.*, 1988, 1990). For example, after two growing seasons loblolly pine maintained a higher photosynthetic nitrogen use efficiency when grown in elevated CO_2 than when grown in ambient CO_2 (R. B. Thomas, unpublished data, 1993). On the other hand, Conroy *et al.* (1988) found that a phosphate deficiency in *P. radiata* decreased the maximum photosynthetic rate in elevated CO_2 due to a reduction in the efficiency of energy trapping of photosystem II. Obviously, there is a critical need for more information on the physiological responses of conifers to CO_2 enrichment. Also, as Jarvis (1989) pointed out, there is a critical lack of information on the effects of elevated CO_2 at the stand and forest scale.

E. Atmospheric Pollution

Many of the effects of air pollution indirectly alter the rate of photosynthesis. For example, acid precipitation may affect nutrient availability (Havranek *et al.*, 1990), carbohydrate content (Barnes *et al.*, 1990; Mengel *et al.*, 1990; Paynter *et al.*, 1991), and growth (Thornton *et al.*, 1990). All of these factors can indirectly modify the rate of photosynthesis. Other gaseous pollutants, most notably, ozone, nitrogen dioxide, and sulfur dioxide, can pass through the stomata and damage mesophyll cells, which lowers the rate of photosynthesis on a whole-leaf basis (Fig. 1). Many examples are available for conifers (for example, Sasek and Richardson, 1989; Lorenc-Plucinska, 1988; Boyer *et al.*, 1986; Freer-Smith and Dobson, 1989; Katainen *et al.*, 1987; Reich *et al.*, 1987; Larsen *et al.*, 1990; Barnes *et al.*, 1990). See Matyssek *et al.* (1994) for additional information on the effect of pollution on conifer forests.

IV. Indirect Effects

The preceding discussion indicates the ways that the environment can directly affect photosynthesis of a plant. This effect is usually short term, and does not involve an active response. Here we will discuss responses to the environment that are active and that can occur over a wide range of time periods.

Controls that indirectly influence photosynthesis operate to maintain a degree of homeostasis within a plant. Their primary activity may not be the control of photosynthesis, although they can profoundly affect it. The controls operate in either of two ways, and although mechanisms and outcomes of these are often similar, their implications differ. The best known of these is feedback control, whereby a change in magnitude of an internal variable initiates a response. The response virtually always tends to maintain the system constant; that is, it is one of negative feedback. There is increasing evidence for the existence of feedforward control, originally discussed by Cowan (1977) and later by Farquhar (1978) in relation to stomatal responses to humidity. This form of response is also found in reactions to other external factors. A control loop of this form allows a plant to "anticipate" effects of the level of supply and demand of resources and of changes in these. The plant alters its activity accordingly so that impending deficits are minimized. Phenomena discussed in this section will include both feedback and feedforward effects.

If both feedback and feedforward responses operate when resource supply is limited, the feedforward response will be effective only for resources not actively accumulated by a plant, if it is initiated at a smaller resource deficit (higher "concentration") and if it is more sensitive to change than a feedback response.

A. Light

Light instantaneously affects the internal and external environments of a plant or plant organ. A plant cannot, therefore, often exhibit a feedforward response to a change in light. However, a response to an approaching change in irradiance by the photosystem in climates where irradiance varies with season is in effect a feedforward response. The mechanism for this may involve sensing a change in daylength, in absolute daylength, or in daily maximum or average irradiance. These and other responses to factors that affect the irradiance incident on a leaf or a plant may reduce damage by, for example, photoinhibition and increase resource use efficiency (see Chapter 5, this volume). The overall "aim" of these responses is probably to maximize the overall photosynthetic gain and the photosynthetic efficiency of a plant.

As a plant grows, previously well-exposed foliage often becomes increasingly shaded; sometimes a change in canopy structure results from growth of a faster or larger-growing neighbor, or from a neighbor's death. These result in a change in a leaf's light environment and often in the distribution of light within the plant's canopy (see Chapter 1, this volume). Shading will produce a lower rate of photosynthesis and at least initially a reduced efficiency of use of many resources by affected leaves (Chapter 5, this volume). Shade may cause changes in the distribution of resources in the leaf. An example of this is a change in partitioning of nitrogen to thylakoids versus ribulose bisphosphate carboxylase. Newly shaded leaves would increase their photosynthetic efficiency by investing a greater proportion of nitrogen in thylakoids, newly exposed leaves by investing a greater proportion in ribulose bisphosphate carboxylase (e.g., Evans, 1989). When a plant's light environment changes, or when previously sunlit leaves become shaded, the whole plant can also acclimate to optimize its photosynthetic capacity. It will do this by altering the distribution of resources within its canopy, for example, by redistributing nitrogen so that a greater proportion is in sunlit leaves (e.g., Field and Mooney, 1983). There may also be an altered allocation of newly partitioned photosynthate to foliage biomass so that areas of the canopy now sunlit or shaded will produce sun and shade leaves. These allocate biomass differently, to form leaves that are thicker or broader, respectively.

B. Water

1. Water in the Atmosphere A feedforward response to humidity causes leaf conductance to decline because the evaporative power of the air has increased (Cowan, 1977; Farquhar, 1978). This reduces the magnitude of water deficits that would develop within a plant as a result of increased transpiration into drier air. In some species, conductance is lowered sufficiently such that transpiration falls as the air becomes drier (e.g., Sheriff, 1977). This reduction of conductance limits diffusion of CO_2 into the leaf, and hence reduces photosynthesis. With well-watered cuttings of *P. radiata*, Bennett and Rook (1978) found that changes in stomatal conductance were the cause of the response of photosynthesis to ambient humidity. Stomatal responses to humidity vary between species (Sheriff, 1977; Sandford and Jarvis, 1986), and are under genetic control, as shown for *P. radiata* by Bennett and Rook (1978) (Fig. 8).

2. Water in the Soil Drying of soil around a plant's roots will cause a reduction in stomatal conductance independently of plant water status in some species (Blackman and Davies, 1985; Davies and Meizner, 1990; Gollan *et al.*, 1986; Zhang and Davies, 1990). The "message" that soil is drying is carried from roots to leaves by hormonal messenger(s) (e.g., Munns and King, 1988). The reduction in conductance will lower rates of transpiration, thereby conserving an increasingly scarce resource. It will also reduce photosynthesis (Fig. 1), and increase transpiration efficiency (Chapter 5). Because these hormones also inhibit growth (e.g., Passioura, 1988), water is clearly not conserved to enable turgor maintenance so that growth can continue. The consequence of this mechanism must be to provide a greater possibility of survival. Success of this tactic, however, depends on competitors not using the saved water. The

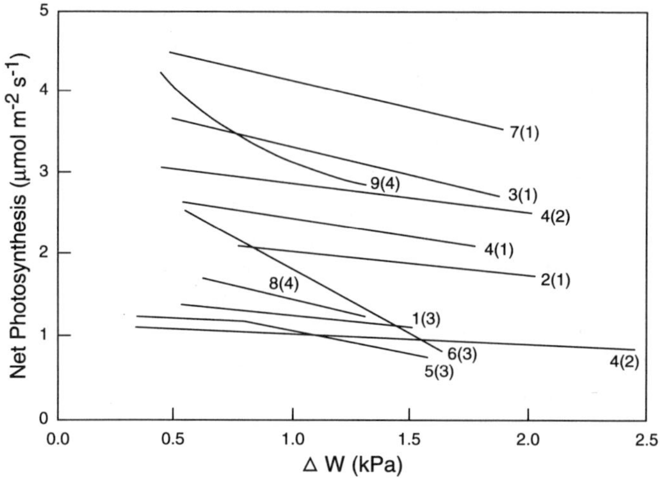

Figure 8 Effect of vapor pressure deficit on net photosynthesis of conifers. The first number represents the species, the second, in parentheses, is the reference. Species: (1) *Cedrus atlantica*; (2) *Larix eurolepis*; (3) *Picea sitchensis*; (4) *Pinus sylvestris*; (5) *Pseudotsuga macrocarpa*; (6) *Pseudotsuga menziesii*; (7) *Pinus contorta*; (8) *Pinus radiata* clone 456; (9) *Pinus radiata* clone 457. References: (1) Sandford and Jarvis (1986); (2) Küppers and Schulze (1985); (3) Grieu *et al.* (1988); (4) Bennett and Rook (1978).

evidence for a molecular signal of soil water deficits from root to shoot is mounting, but little work has been done on conifers. However, Blake and Ferrell (1977) demonstrated that the abscisic acid concentration in the xylem sap of *Pseudotsuga menziesii* increased in response to soil drying. This suggests that the mechanism is likely to exist in this species, and possibly in other conifers.

3. Water Loss Some species have the ability to limit the amount of water they use in a day. Hormones that inhibit physiological activity are secreted into roots, carried in the transpiration stream, and accumulate in leaves over the period of each day in proportion to the amount of water transpired. As foliar hormone concentrations increase, stomatal conductance is reduced and photosynthesis inhibited (e.g., Loveys, 1984a,b). By the following day these hormones have been metabolized or translocated to other parts of the plant. In some cases hormones that inhibit physiological activity originate in leaves, are translocated to roots, and later (the following day) transported to leaves in the transpiration stream (Loveys, 1984b). The result of this is that on days of high potential evaporation stomata are open only in the earlier part of the day, and that the period of opening is inversely related to the potential transpiration rate.

This phenomenon is a mechanism that integrates plant activity over time. A result is that some photosynthetic activity occurs on all days, but this is restricted when the weather is hot and dry and conditions do not favor high rates of photosynthesis. Thus more water remains stored in the soil until conditions are more favorable for photosynthesis.

C. Nutrients

A deficiency in supply of at least some nutrients to roots can produce a hormone-mediated response (e.g., Golbach *et al.*, 1975; Krauss, 1978) that reduces stomatal conductance. This response acts synergistically with the response to soil water content (e.g., Radin and Ackerson, 1981), although the interaction is difficult to investigate for plants with their roots in soil because availability of nutrients varies with soil moisture. As with the hormone-mediated response to soil moisture, changes in hormone balance that result from alteration of the root environment reduce shoot growth, as well as gas exchange (e.g., Krauss and Marschner, 1982). It is reasonable to conclude that plants adopt this tactic to conserve nutrients: it may be a way of limiting growth so that the requirement for nutrients does not surpass their supply.

Some nutrients are not directly involved in the photosynthetic pathway but are necessary for growth. If a shortage of one or more of these exists and does not cause a hormonal response from roots, its lack may nevertheless reduce rates of photosynthesis by source–sink feedback, whereby a lack of demand for photosynthate reduces photosynthesis (see Chapter 6, this volume).

A greater availability of nutrients may cause increased partitioning of carbon to foliage, in which case the effect of nutrition on photosynthesis will result from both a greater rate of photosynthesis per unit foliage and a larger amount of foliage. Over time these effects will accumulate by "compound interest" to produce a much larger biomass from plants with better nutrition, provided other potentially limiting factors are the same irrespective of nutrition. For example, Sheriff (1992) has demonstrated using a simple model and experimental data that an approximately 10% greater rate of photosynthesis will result in a 31% greater seedling biomass over a period of 80 days (taking into account greater respiration at higher nutrient concentration, but not any other loss of carbon). Over the same period a 3.3% per day greater partitioning of biomass to foliage at higher nutrient concentration caused a 22% difference in plant biomass.

D. Temperature

To some extent it is possible for a plant to alter both the temperature of its foliage and the response of photosynthesis to temperature. Conifers growing in cold climates with short growing seasons can enhance their rates of photosynthesis by producing shoots with high packing densities of needles. It is often the case in these areas that air temperatures are below the optimum for photosynthesis. When these needles are exposed to sunlight their temperature is raised above air temperature, and is closer to the optimum for photosynthesis (Smith and Carter, 1988). It has been suggested that the shape of krummholz vegetation not only aids in increasing the temperature of the foliage in summer, it also helps to protect the foliage from damage in winter by promoting the accumulation of a protective snow layer (Hadley and Smith, 1987).

In a way similar to that discussed above, plants usually exposed to high air temperatures and large insolation loads will have leaf temperatures not greatly above air temperature if their foliage has a low packing density and is composed of narrow leaves.

In many species leaf conductance increases with temperature (e.g., Hofstra and Hesketh, 1969; Mahan and Upchurch, 1988; Upchurch and Mahan, 1988), especially at temperatures that exceed those for normal enzyme function (Burke and Upchurch, 1989). This change in conductance enables increased transpirational cooling at higher air temperatures, which tends to keep leaf temperature constant. Conductance is also inversely related to humidity deficit, as discussed above, and there can be a nonlinear interaction between effects of temperature and humidity (e.g., Wilson, 1947). The amount of transpirational cooling will, therefore, not be a simple function of temperature or humidity deficit.

A change in the average ambient temperature will often cause acclimation of the photosynthetic system such that it operates more efficiently in, and has a temperature optimum nearer to that of, the new environment (Rook, 1969; Strain *et al.*, 1976). This adjustment indicates possibilities for at least limited acclimation to a changing climate.

E. Growth

Source–sink feedback can cause changes in photosynthetic activity, even in photosynthetic potential (Leverenz, 1981), that do not relate to a direct response of photosynthesis to either the external environment or to those parts of the internal environment that are directly altered by the external environment (see Chapter 6). For example, photosynthate demand by ectomycorrhizas has been linked to an increase in net photosynthetic rate in seedlings of *P. menziesii* (Dosskey *et al.*, 1990). Correlations between photosynthesis and growth can result from growth being stimulated by environmental triggers, and, in turn, the increase in growth then stimulates photosynthesis (Maier and Teskey, 1992). Alternatively, these correlations could result from a direct effect of the environment on photosynthesis, which makes available more carbohydrate and thereby enables greater growth. Which mechanism predominates will depend on a number of factors, and will involve the balance of source–sink relationships. In some cases, for example, when the soil environment indirectly controls activity, photosynthesis and growth may simultaneously respond to the same hormonal signals.

F. Simultaneous Control

Direct environmental effects on photosynthesis can often be detected over short time intervals, i.e., seconds to hours, and can dramatically alter the rates of photosynthesis. Changes in irradiance due to clouds or the normal diurnal cycle are excellent examples of a direct effect that can rapidly change the rate of photosynthesis. Indirect influences on photosynthesis tend to have effects over longer time intervals. Processes such as growth, or the synthesis and transport of hormones and carbohydrates, occur more slowly, yet have important consequences to photosynthesis. Indirect controls can be considered to be superimposed over the direct controls and operate simultaneously. For example, higher rates of photosynthesis may be possible during periods of growth, in comparison to when growth has stopped. In this case internal controlling factors, presumably sink demands for carbohydrate, define the limit within which environmental factors can influence photosynthesis. The effects of indirect controls are not always obvious yet they can substantially modify the response to direct regulating factors, with important consequences for growth and survival. An example of this is the control exerted by hormonal signals on stomatal apertures. In certain stress situations this control overrides the external environmental controls. Over even longer time intervals, growth, the retention period of foliage, and the development of a complex crown structure significantly modify the direct effects of environment on photosynthesis of individual leaves.

V. Consequences for Growth and Survival

Deficits of resources within a plant reduce its biological activity—e.g., photosynthesis, cell expansion, and cell division. Some species and ecotypes are better able to tolerate internal deficits than others, which must, in order to survive, avoid these deficits when the supply of one or more resources from the external environment declines beyond a particular level. If a plant is living in a resource-poor environment (less resources than are required for optimum activity at all stages of the life cycle), it can maintain its activity at the maximum level for that environment until either of the following events occurs:

1. Reduced availability of resource(s) in a plant's internal environment forces a reduction in activity. This may involve acclimation of the

component systems of a plant to increasing deficits with time. There are consequences to this: (a) The development of lethal or near-lethal deficits. These would cause a large loss of accumulated carbon through death of the plant or of some of its organs, and a subsequent loss of photosynthetic capacity. (b) A competitive advantage if the plant is better able to withstand (tolerate) these deficits than are its competitors. This will apply only if a plant is using the same resources as its competitors, and a similar or greater resource base: if cooccurring plants exploit a greater resource (e.g., are deeper rooted and therefore have greater access to water), this will not apply. (c) A demand on stored carbohydrate or on the photosynthetic system to provide energy, or carbohydrate, or both for acclimation. However, a plant does not know how large the deficits will become, or how long they will last, although an "average" over evolutionary time may be "known" to its genome.

2. It senses the onset of resource paucity and a resulting "signal" lowers activity. There are also consequences to this: (a) Prevention of the onset of lethal or near-lethal internal levels of the resource, e.g., by restricting growth while concentrations of nutrient or water near the roots are low. (b) Eking out of available resources, provided these are not consumed by a neighbor, to allow survival for long periods of deficiency, and more efficient use of resources. This could involve redistribution of resources within a plant, as well as more efficient use within an organelle or cell.

VI. Integration of Photosynthetic Rate over Time

A. Controlling Factors

When photosynthesis is limited simultaneously by more than one factor, the controlling mechanisms may interact to produce unexpected responses. For example, photosynthesis of plantation-grown *Pinus radiata* was negatively associated with foliar nitrogen concentration, and positively associated with foliar phosphorus concentration when temperatures were high and humidities low (Sheriff *et al.*, 1986). A positive association between photosynthesis and foliar nutrition at moderate temperatures, and a negative association at high temperatures, were found in a laboratory study of *P. radiata* seedlings (D. W. Sheriff, unpublished, 1994).

It is sometimes stated that plants integrate the effects of their environment. Individual plant processes, for example, rates of photosynthesis, respond simultaneously to, and may integrate the effects of, a range of variables in the internal and external environments of a plant. But, as we have seen for water, at least some plants have the ability to integrate effects of their environment over time. Although they are unable (we assume) to foresee the future, "information" in their genome determines tactics and strategies that would have been successful in responses to environmental change over evolutionary time. This is likely to provide a greater chance of survival during adverse periods, and of productivity summed over periods of months to years.

B. Time

Measurements of photosynthesis and respiration are made over very short time periods, with a resolution of seconds, minutes, or hours. This measurement interval is not very appropriate for examining growth or productivity because the time scales of these processes are of a higher order (Ford and Teskey, 1991). An approach for using physiological information to evaluate long-term trends (months, years) in carbon gain and allocation has been to use processed-based mathematical models that consider the factors limiting growth. Typically these models incorporate hourly or daily environmental data, and then estimate carbon gain and loss over the same time intervals. The net result is then summed over longer time intervals. The carbon gain of single trees and stands can be estimated by combining such models with the phenological and architectural characteristics of crowns and canopies (Wang *et al.*, 1992). Examples of "process models" used with conifers can be found in work by Dixon *et al.* (1990) and Irving (1991). For a comparison of two process models, BIOMASS and MAESTRO, which have been used to estimate carbon gain of a number of conifer species, the reader is referred to McMurtrie and Wang (1993).

Process modeling is a useful tool for integrating across time and space, because measurement techniques provide only extremely small subsets of the temporal and spatial variability in a tree or forest. The major advantage is that process modeling has the potential to provide an understanding of the mechanisms that control the growth of trees and the potential for prediction. We are far from reaching the full potential of these models for a number of reasons, including an incomplete understanding of the processes that are incorporated, for example, of the impact of environment, the interaction of stresses (e.g., water and nutrients), the control of carbon allocation, and the role of internal regulation (e.g., hormonal signals, source-sink regulation). A second factor limiting model utility is inadequate experimental sampling: spatial and temporal variability of photosynthesis within trees has often been ignored, and although there are exceptions (e.g., Fuchs *et al.*, 1977; Ginn *et al.*, 1991; Kull and Koppel, 1987; Rook and Corson, 1978; Troeng and Linder, 1982b), these data are scarce. An additional sampling problem is extrapolation across size and age of trees. It is well known that many characteristics of trees change with age. All aspects of growth and development, including phenology, anatomy, morphology, and physiology, change over time. Although there are few direct comparisons of plants at different stages of maturity (Cregg *et al.*, 1989), there are indications that gas exchange measurements made on young seedlings are quantitatively different from those on older plants (Cui and Smith, 1991; Squire *et al.*, 1988; Szaniawski and Wierzbicki, 1978). This is not surprising considering the differences between seedlings and older, larger trees with respect to carbohydrate availability, respirational demands of woody tissue, and proximity of sources and sinks. On the other hand, a very encouraging development has been the recent advances in eddy flux measurement techniques and instrumentation (see Section VII). Provided that models are constructed to give estimates of carbon flux in stands, carbon flux measurements have the potential to provide an independent estimate of model performance.

VII. Scaling Photosynthetic CO_2 Exchange from Foliage to Canopies and Regions

It is a paradox that although the greatest understanding of photosynthesis is at the leaf and biochemical level, some of the most pressing questions are at the canopy or landscape scale. These include the traditional concerns of forest growth and water use, and more recent issues of global ecology, such as the response of forests to a changing climate or their ability to sequester atmospheric carbon.

The previous discussion has shown the sensitivity of leaf photosynthesis to the environmental factors of light, humidity deficit, temperature, and carbon dioxide concentration. Successful scaling of leaf or shoot responses to the canopy or region depends on critically identifying how the factors change in a canopy and regionally. In essence, the problem is one of identifying how foliage affects its own environment at these different scales.

This was considered by Jarvis and McNaughton (1986) in the context of leaf transpiration. They developed the omega coefficient to show how transpiration becomes less sensitive to stomatal conductance as scale increases. An extension of their analysis up to the regional scale using control theory demonstrated how an increasing number of negative feedbacks reduce the sensitivity of transpiration to conductance (McNaughton and Jarvis, 1991).

Although the scaling of CO_2 exchange has not been considered in the same manner, the problem is similar although more simple. Within a canopy, the exchange of radiation and mass modifies the local environment of the foliage. In the daytime, canopies are sources of heat and water vapor and sinks for CO_2 and are thus warmer, moister, and depleted in CO_2 relative to the air above them. Because of the rough nature of coniferous forests, however, turbulent mixing in these canopies is large and variation between the canopy and the external environments is small. Denmead and Bradley (1985) found, for example, that at midday in a *P. ponderosa* forest, the maximum canopy air temperature was $<2°$C higher than that at a reference height above the forest, whereas water vapor density and CO_2 concentration differed by only about 0.4 g/kg (4%) and 1 ppm (0.3%), respectively. At the canopy level, the modification of the light environment by foliage and structural elements is of far greater importance than other environmental factors because leaf photosynthesis is a strong but nonlinear function of photosynthetically active radiation.

Extensive canopies also influence the temperature, water vapor content, and CO_2 concentration of the regional atmospheric boundary layer. Although generally minor, these influences feed back on the processes of leaf photosynthesis. For example, the CO_2 concentration above an extensive tropical forest was found to decrease by about 30 ppm (~8%) between 8 a.m. and noon (Wofsy *et al.*, 1988). With a CO_2–photosynthesis relationship typical of many conifers (Fig. 1), this would equate to a ~5% reduction in maximum photosynthetic uptake when photosynthesis is not otherwise severely limited.

Canopy photosynthesis can be inferred from micrometeorological measurements of net ecosystem CO_2 exchange, or can be estimated with models that couple a simulation of canopy environment to leaf gas exchange. To move from net CO_2 exchange above a forest (F_c) to canopy photosynthesis (P), the aboveground (R_a) and belowground (R_b) production of respiratory CO_2 must be accounted for. It is also necessary to allow for any changes in CO_2 storage within the air column below the flux measurement height ($F_{\Delta s}$) between measurement periods as

$$P = F_c + F_{\Delta s} - R_a - R_b. \quad (1)$$

This requirement for respiratory data complicates estimates of canopy photosynthesis and increases potential sources of error. However, for many questions (e.g., changes in stand net carbon balance), measurements of F_c may be all that is necessary.

A. Temporal Scaling of Coniferous Photosynthesis

The natural environment is dynamic. The irradiance incident on foliage may vary by a factor of 10 or even 100 over several seconds or minutes because of changing shading brought about by clouds or the movement of other foliage relative to the sun. Pearcy (1990) has reviewed the influence of sunflecks on canopy photosynthesis. Most of the information available relates to understory and crop species, although the general results probably apply to conifers as well. There are two main dynamic responses of photosynthesis to rapid changes in photosynthetically active photon flux density (PPFD). The first is that it takes some minutes for photosynthesis of foliage that has been in low light for an extended period to respond fully to an increase in light (e.g., Chazdon and Pearcy, 1986a). This limitation has both biochemical and stomatal components. A long period (~30 minutes) of high light levels or closely spaced shorter periods of high light levels are necessary to "induce" the photosynthetic system to function at peak efficiency (Pearcy, 1990). The second response is often observed in induced foliage following short-term (0.1 to 10 seconds) increases in PPFD and takes the form of enhanced CO_2 uptake that typically exceeds the mean of the steady-state rates at the same PPFDs (e.g., Laisk *et al.*, 1984). This is understood biochemically as a buildup of the ribulose-1,5-bisphosphate and triose phosphate pools during the sunfleck and their subsequent utilization in the postsunfleck period (Pearcy, 1990). The extra efficiency of utilization of such sunflecks may range from 30 to 80% (Chazdon and Pearcy, 1986b).

The ultimate efficiency of canopy photosynthesis in a varying light environment will depend on the biochemistry of the species, the acclimation of the foliage to sun or shade, and the temporal details of the light environment. This is undoubtedly a research area requiring additional attention.

B. Modeling Canopy Photosynthesis

Baldocchi (1993) has discussed canopy photosynthesis models and their use in scaling CO_2 exchange up to the canopy level. Conifers represent a class of vegetation wherein both turbulent transport and structural complexity are high. This means that simple submodels of transport are sufficient (e.g., constant atmospheric conditions), but that complex submodels of radiation penetration are required, such as those of Grace *et al.* (1987) or Wang and Jarvis (1990).

C. Measuring Canopy CO_2 Exchange

The eddy correlation and flux gradient techniques have both been used to measure canopy or regional scale CO_2 exchange directly. The scale "seen" by these techniques depends on the source "footprint" or flux region (Schmid and Oke, 1990), which depends on the height of instrumentation above the surface, and on the windspeed. As height and windspeed increase, so does the source region. Tower-based measurements may "see" 50% of the observed flux coming from an area of a few hectares (e.g., Hollinger *et al.*, 1994), whereas aircraft-based systems may integrate the exchange characteristics of dozens of square kilometers (Desjardins *et al.*, 1985).

Micrometeorological techniques measure the net movement of mass or energy across the plane of instrumentation above a canopy. The eddy correlation technique calculates the net flux of CO_2 (F_c) across this plane as the mean covariance between fluctuations in vertical wind velocity (w) and the density of CO_2 (c) (Baldocchi *et al.*, 1988):

$$F_c = \rho_a(\overline{w'c'}), \tag{2}$$

where ρ_a is the density of air, the primes denote deviations from the mean, and the overbar signifies a time average.

Flux gradient methods assume that turbulent transfer is analogous to molecular diffusion. The flux is proportional to the product of the mean vertical mixing ratio gradient and an eddy diffusivity (K). For CO_2, the flux is calculated as

$$F_c = K\,\partial c/\partial z, \tag{3}$$

where $\partial c/\partial z$ is the mean gradient in CO_2 density (Baldocchi *et al.*, 1988). The eddy diffusivity for CO_2 can be based on that for momentum (determined by aerodynamic methods) or energy (based on Bowen-ratio, energy-balance techniques). Because forest canopies are aerodynamically rough, the above-canopy gradients of CO_2 and other scalars are small (e.g., Jarvis *et al.*, 1976). The sensor resolution of gradient systems must be greater than that of eddy correlation systems to provide a comparable accuracy (Baldocchi *et al.*, 1988; Businger and Delany, 1990). The flux gradient method is not reliable within canopies because the source and sink distributions of other scalars are different from that of CO_2, and because the intermittent nature of turbulent transport means that fluxes can be transported counter to a mean gradient (Denmead and Bradley, 1985).

Several studies have investigated the net CO_2 exchange of coniferous forests (Table I). Net values varied by more than a factor of 5 between a low-productivity, juvenile *Pseudotsuga menziesii* stand and fast-growing *Pinus radiata* or *Picea sitchensis* stands. These midday values of net canopy CO_2 exchange are several times higher than comparable maximum leaf-level rates of photosynthesis (Figs. 1 and 7).

These studies did not deduct estimates of soil (R_b) and canopy (R_a) respiration from the measurements of net CO_2 exchange. Thus, rates of canopy photosynthesis will be considerably higher than the rates of net CO_2 exchange in Table I. Price and Black (1990) estimated that R_b in their Douglas fir stand was about 2 μmol/m^2/sec and found that F_c could drop to zero on warm days. This suggests that R_a at these times was in the range of ~5 μmol/m^2/sec. Jarvis (1992) found lower values of R_b in

Table I Measured Midday Net CO_2 Fluxes (F_c) above Various Coniferous Forest Canopies

Species	F_c (μmol/m^2 ground surface/sec)	Method	Reference
Pinus radiata	30–40	Gradient	Denmead (1969)
Pinus sylvestris	11–27	Gradient	Jarvis *et al.* (1976)
Pinus taeda	~25	Gradient	Lorenz and Murphy (1985)
Pinus strobus	9–21	Aircraft eddy correlation	Desjardins *et al.* (1985)
Pinus ponderosa	12	Eddy correlation	Denmead and Bradley (1985)
Picea sitchensis	20	Gradient	Jarvis *et al.* (1976)
Picea sitchensis	30	Gradient	Jarvis (1992)
Pseudotsuga menziesii	5–10	Gradient	Price and Black (1990)

Picea sitchensis (~0.8 μmol/m^2/sec at 10° C) and used a model to estimate R_a. In this fast-growing stand R_a was about 23 μmol/m^2/sec at 10° C and increased with a Q_{10} of 2. These studies both suggest that total respiratory losses are often similar in magnitude to net CO_2 uptake, meaning that canopy photosynthesis is of the order of twice the net rates observed in the studies tabulated in Table I. As pointed out by Jarvis (1992), careful measurements of woody tissue area and respiration rate are required if one is to estimate canopy photosynthesis from micrometeorological measurements.

D. Scaling with Present Models

In only a few instances have modeled results been compared with actual measurements. These comparisons, however, are of particular value because they demonstrate where present models accurately reproduce nature, and where they may be missing a process that only comes into existence, or changes in importance, at a larger scale.

Jarvis (1992) compared measurements of net CO_2 exchange of *Picea sitchensis* with estimates made by the canopy model MAESTRO (Wang and Jarvis, 1990). The model simulated the general pattern and magnitude of canopy CO_2 exchange. It also duplicated the observed decrease in net forest CO_2 uptake on warmer days, and suggested that more of this decrease was associated with temperature-enhanced respiration than with a vapor pressure deficit induced reduction in CO_2 uptake through stomatal closure.

Models can also reveal how the importance of environmental factors change with scale. For example, the directional characteristics of the

PPFD are not relevant at the scale of a needle. Zelawski *et al.* (1973) found, however, that shoot photosynthesis was enhanced by diffuse light compared to an equal flux of beam radiation. Models (e.g., Oker-Blom, 1985; Jarvis *et al.*, 1985) simulate this enhancement behavior for a whole canopy. An earlier version of MAESTRO, for example, suggested that canopy CO_2 uptake under diffuse PPFD should be about twice that attained with an equal direct beam flux (Jarvis *et al.*, 1985). In several studies (Price and Black, 1990; Hollinger *et al.*, 1994) net forest CO_2 uptake was found to be enhanced on overcast days.

References

Attiwill, P. M., and Cromer, R. N. (1982). Photosynthesis and transpiration of *Pinus radiata* D. Don under plantation conditions in southern Australia. I. Response to irrigation with waste water. *Aust. J. Plant Physiol.* 9:749–760.

Baldocchi, D. (1993). Scaling water vapor and carbon dioxide exchange from leaves to a canopy: Rules and tools. *In* "Scaling Physiological Processes: Leaf to Globe" (J. Ehleringer, ed.), pp. 77–114. Academic Press, New York.

Baldocchi, D. D., Hicks, B. B., and Meyers, T. P. (1988). Measuring biosphere–atmosphere exchanges of biologically related gases with micrometeorological methods. *Ecology* 69:1331–1340.

Barnes, J. D., Eamus, D., and Brown, K. A. (1990). The influence of ozone, acid mist and soil nutrient status on Norway spruce (*Picea abies* (L.) Karst.). *New Phytol.* 115:149–156.

Beadle, C. L., Hart, J. W., and Jarvis, P. G. (1983). Carboxylase activities in Sitka spruce and some other conifers. *Photosynthetica* 17:321–333.

Belous, V. K. (1986). Temperature dependence of photosynthesis in conifers. *Sov. Plant Physiol. (Engl. Transl.)* 33:367–372.

Benecke, U., and Nordmeyer, A. H. (1982). Carbon uptake and allocation by *Nothofagus solandri* var. cliffortiodes (Hook. f.) Poole and *Pinus contorta* Douglas ex Loudon spp. *contorta* at montane and subalpine altitudes. *In* "Carbon Uptake and Allocation in Subalpine Ecosystems as a Key to Management. Proceedings IUFRO Workshop," pp. 9–21. Oregon State Univ., Corvallis.

Bennett, K. J., and Rook, D. A. (1978). Stomatal and mesophyll resistances in two clones of *Pinus radiata* D. Don known to differ in transpiration and survival rate. *Aust. J. Plant Physiol.* 5:231–238.

Black, C. H. (1988). Interaction of phosphorus fertilizer form and soil medium on Douglas-fir seedling phosphorus content, growth and photosynthesis. *Plant Soil* 106: 191–199.

Blackman, P. G., and Davies, W. J. (1985). Root to shoot communication in maize plants of the effects of soil drying. *J. Exp. Bot.* 36:39–48.

Blake, J., and Ferrell, W. K. (1977). The association between soil and xylem water potential, leaf resistance and abscisic acid content in droughted seedlings of Douglas-fir (*Pseudotsuga menziesii*). *Physiol. Plant.* 39:106–109.

Bornman, J. F., and Vogelmann, T. C. (1988). Penetration of blue and UV radiation measured by fiber optics in spruce and fir needles. *Physiol. Plant.* 72:699–705.

Boyer, J. N., Houston, D. B. and Jensen, K. F. (1986). Impacts of chronic SO_2, O_3 and SO_2 + O_3 exposures on photosynthesis in *Pinus strobus* clones. *Eur. J. For. Pathol.* 16:293–299.

Brix, H. (1979). Effects of plant stress on photosynthesis and survival of four conifers. *Can. J. For. Res.* 9:160–165.

Brix, H. (1981). Effects of nitrogen fertilizer source and application rates on foliar nitrogen concentration, photosynthesis, and growth of Douglas-fir. *Can. J. For. Res.* 11: 775–780.

Bunce, J. A. (1977). Nonstomatal inhibition of photosynthesis at low water potentials in intact leaves of species from a variety of habitats. *Plant Physiol.* 59:348–350.

Burke, J. J., and Upchurch, D. R. (1989). Leaf temperature and transpirational control in cotton. *Environ. Exp. Bot.* 29:487–492.

Businger, J. A., and Delany, A. C. (1990). Chemical sensor resolution required for measuring surface fluxes by three common micrometeorological techniques. *J. Atmos. Chem.* 10:399–410.

Ceulemans, R. J., and Saugier, B. (1991). Photosynthesis. *In* "Physiology of Trees" (A. S. Raghavendra, ed.), pp. 21–50. Wiley, New York.

Chazdon, R. L., and Pearcy, R. W. (1986a). Photosynthetic responses to light variation in rainforest species. I. Induction under constant and fluctuating conditions. *Oecologia* 69:517–523.

Chazdon, R. L., and Pearcy, R. W. (1986b). Photosynthetic responses to light variation in rainforest species. II. Carbon gain and photosynthetic efficiency during lightflecks. *Oecologia* 69:524–531.

Conroy, J. P., Smillie, R. M., Küppers, M., Bevege, D. I., and Barlow, E. W. (1986). Chlorophyll *a* fluorescence and photosynthetic and growth responses of *Pinus radiata* to phosphorus deficiency, drought stress, and high CO_2. *Plant Physiol.* 81:423–429.

Conroy, J. P., Küppers, M., Küppers, B., Virgona, J., and Barlow, E. W. R. (1988). The influence of CO_2 enrichment, phosphorus deficiency and water use of *Pinus radiata* D. Don. *Plant, Cell Environ.* 11:91–98.

Conroy, J. P., Milham, P. J., Reed, M. L., and Barlow, E. W. (1990). Increases in phosphorus requirements for CO_2-enriched pine species. *Plant Physiol.* 92:977–982.

Cowan, I. R. (1977). Stomatal behavior and environment. *Adv. Bot. Res.* 4:117–228.

Cowan, I. R., Raven, J. R., Hartung, W., and Farquhar, G. D. (1982). A possible role for abscisic acid in coupling stomatal conductance and photosynthetic carbon metabolism in leaves. *Aust. J. Plant Physiol.* 9:489–498.

Cregg, B. M., Halpin, J. E., Dougherty, P. M., and Teskey, R. O. (1989). Comparative physiology and morphology of seedlings and mature forest trees. *In* "Air Pollution Effects on Vegetation" (R. D. Noble, J. L. Martin, and K. F. Jensen, eds.), pp. 111–118. USDA Forest Service, Broomall, Pennsylvania.

Cropper, W. P., Jr., and Gholz, H. L. (1993). Simulation of the carbon dynamics of a Florida slash pine plantation. *Ecol. Modell.* 66:231–249.

Cui, M., and Smith, W. K. (1991). Photosynthesis, water relations and mortality in *Abies lasiocarpa* seedlings during natural establishment. *Tree Physiol.* 8:37–46.

Davies, W. J., and Meizner, F. C. (1990). Stomatal responses of plants in drying soil. *Biochem. Physiol. Pflanz.* 186:357–366.

Day, T. A., Heckathorn, S. A., and DeLucia, E. H. (1991). Limitations of photosynthesis in *Pinus taeda* L. (loblolly pine) at low soil temperatures. *Plant Physiol.* 96:1246–1254.

DeLucia, E. H., and Heckathorn, S. A. (1989). The effect of soil drought on water-use efficiency in a contrasting Great Basin desert and Sierran montane species. *Plant, Cell Environ.* 12:935–940.

DeLucia, E. H., and Smith, W. K. (1987). Air and soil temperature limitations on photosynthesis in Engelmann spruce during summer. *Can. J. For. Res.* 17:527–533.

Denmead, O. T. (1969). Comparative micrometeorology of a wheat field and a forest of *Pinus radiata*. *Agric. For. Meteorol.* 6:357–371.

Denmead, O. T., and Bradley, E. F. (1985). Flux-gradient relationships in a forest canopy. *In* "The Forest–Atmosphere Interaction" (B. A. Hutchinson and B. B. Hicks, eds.), pp. 421–442. Reichel, Dordrecht, The Netherlands.

Desjardins, R. L., MacPherson, J. L., Alvo, P., and Schuepp, P. H. (1985). Measurements of turbulent heat and CO_2 exchange over forests from aircraft. *In* "The Forest–Atmosphere Interaction" (B. A. Hutchinson and B. B. Hicks, eds.), pp. 645–658. Reichel, Dordrecht, The Netherlands.

Dixon, R. K., Meldahl, R. S., Ruark, G. A., and Warren, W. G. (1990). "Process Modelling of Forest Growth Responses to Environmental Stress." Timber Press, Portland, Oregon.

Dosskey, M. G., Linderman, R. G., and Boersma, L. (1990). Carbon-sink stimulation of photosynthesis in Douglas-fir seedlings by some ectomycorrhizas. *New Phytol.* 115: 269–274.

Eamus, D., and Fowler, D. (1990). Photosynthetic and stomatal conductance responses to acid mist of red spruce seedlings. *Plant, Cell Environ.* 13:349–357.

Eamus, D., and Jarvis, P. G. (1989). The direct effects of increase in the global atmospheric CO_2 concentration on natural and commercial temperate trees and forests. *Adv. Ecol. Res.* 19:1–55.

Evans, J. R. (1989). Photosynthesis and nitrogen relationships in leaves of C_3 plants. *Oecologia* 78:9–19.

Farquhar, G. D. (1978). Feedforward responses of stomata to humidity. *Aust. J. Plant Physiol.* 5:787–800.

Farquhar, G. D. (1989). Models of integrated photosynthesis of cells and leaves. *Philos. Trans. R. Soc. London B* 323:357–367.

Fetcher, N., Jaeger, C. H., Strain, B. R., and Sionit, N. (1988). Long-term elevation of atmospheric CO_2 concentration and the carbon exchange rates of saplings of *Pinus taeda* L. and *Liquidambar styraciflua* L. *Tree Physiol.* 4:255–262.

Field, C., and Mooney, H. A. (1983). Leaf age and seasonal effects on light, water, nitrogen use efficiency in a California shrub. *Oecologia* 56:348–355.

Ford, E. D., and Teskey, R. O. (1991). The concept of closure in calculating carbon balance of forests: Accounting for differences in spatial and temporal scales of component processes. *Tree Physiol.* 9:307–324.

Freer-Smith, P. H., and Dobson, M. C. (1989). Ozone flux to *Picea sitchensis* (Bong.) Carr and *Picea abies* (L.) Karst during short episodes and the effects of these on transpiration and photosynthesis. *Environ. Pollut.* 59:161–176.

Fuchs, M., Schulze, E.-D., and Fuchs, M. I. (1977). Spatial distribution of photosynthetic capacity and performance in a mountain spruce forest of northern Germany. *Oecologia* 29:329–340.

Gerwick, C. (1982). Response of terrestrial plants to mineral nutrients. *In* "CRC Handbook of Biosolar Resources" (O. R. Zaborsky, ed.), pp. 213–222. The Chemical Rubber Company, Cleveland, Ohio.

Gezelius, K. (1975). Extraction and some characteristics of ribulose-1,5-diphosphate carboxylase from *Pinus sylvestris*. *Photosynthetica* 9:192–200.

Ginn, S. E., Seiler, J. R., Cazell, B. H., and Kreh, R. E. (1991). Physiological and growth responses of eight-year-old loblolly pine stands to thinning. *For. Sci.* 37:1030–1040.

Golbach, E., Golbach, H., Wagner, H., and Michael, G. (1975). Influence of N-deficiency on the abscisic acid content of sunflower plants. *Physiol. Plant.* 34:138–140.

Gollan, T., Passioura, J. B., and Munns, R. (1986). Soil water status affects the stomatal conductance of fully turgid wheat and sunflower leaves. *Aust. J. Plant Physiol.* 13: 459–464.

Grace, J. C., Jarvis, P. G., and Norman, J. M. (1987). Modelling the interception of solar radiant energy in intensively managed stands. *N.Z. J. For. Sci.* 17:193–209.

Grieu, P., Guehl, J. M., and Aussenac, G. (1988). The effects of soil and atmospheric drought on photosynthesis and stomatal control of gas exchange in three coniferous species. *Physiol. Plant.* 73:97–104.

Guehl, J. M. (1985). Comparative study of winter carbon dioxide assimilation potential of three temperate zone conifers (*Pseudotsuga menziesii*, *Abies alba*, and *Picea excelsa*). *Ann. Sci. For.* 42:23–38.

Hadley, J. L., and Smith, W. K. (1987). Influence of krummholz mat microclimate on needle physiology and survival. *Oecologia* 73:82–90.

Hällgren, J.-E., Lundmark, T., and Strand, M. (1990). Photosynthesis of Scots pine in the field after night frosts during summer. *Plant Physiol. Biochem.* 28:437–445.

Havranek, Von W. M., Pfeifhofer, H., and Grill, D. (1990). Pigmentgehalte und Gaswechsel von Tief- und Hochlagenfichten nach chronischer Ozonbelastung. *Forstwiss. Centralbl.* 109:200–209.

Higginbotham, K. O., Mayo, J. M., L'Hirondelle, S., and Drystofiak, D. K. (1985). Physiological ecology of lodgepole pine (*Pinus contorta*) in an enriched CO_2 environment. *Can. J. For. Res.* 15:417–421.

Hofstra, G., and Hesketh, J. D. (1969). The effect of temperature on stomatal aperture in different species. *Can. J. Bot.* 47:1307–1310.

Hollinger, D. Y., Kelliher, F. M., Byers, J. N., Hunt, J. E., McSeveny, T. M., and Weir, P. W. (1994). Carbon dioxide exchange between an undisturbed old-growth temperate forest and the atmosphere. *Ecology* 75:134–150.

Irving, P. M. (1991). "Acidic Deposition: State of Science and Technology. Volume III. Terrestrial, Materials, Health and Visibility Effects. U.S. National Acid Precipitation Assessment Program." Government Printing Office, Washington, D.C.

Jarvis, P. G. (1989). Atmospheric carbon dioxide and forests. *Philos. Trans. R. Soc. London B* **324**:369–392.

Jarvis, P. G. (1992). Resource capture by coniferous forest—A case study. *In* "Resource Capture by Crops" (J. L. Monteith, R. K. Scoti, and M. H. Unsworth, eds.), pp. 167–186. University of Notingham, Notingham, U.K.

Jarvis, P. G., and McNaughton, K. G. (1986). Stomatal control of transpiration: Scaling up from leaf to region. *Adv. Ecol. Res.* 15:1–49.

Jarvis, P. G., James, G. B., and Landsberg, J. J. (1976). Coniferous forest. *In* "Vegetation and the Atmosphere, Volume 2. Case Studies" (J. L. Monteith, ed.), pp. 171–240. Academic Press, London.

Jarvis, P. G., Miranda, H. S., and Muetzelfeldt, R. I. (1985). Modelling canopy exchanges of water vapor and carbon dioxide in coniferous forest plantations. *In* "The Forest–Atmosphere Interaction" (B. A. Hutchinson and B. B. Hicks, eds.), pp. 521–543. Reichel, Dordrecht, The Netherlands.

Jordan, D. N., and Smith, W. K. (1993). Simulated influence of leaf geometry on sunlight interception and photosynthesis in conifer needles. *Tree Physiol.* 13:29–39.

Jurik, T. W., Briggs, G. B., and Gates, D. M. (1988). Springtime recovery of photosynthetic activity of white pine in Michigan. *Can. J. Bot.* 66:138–141.

Kaiser, W. M. (1987). Effects of water deficit on photosynthetic capacity. *Physiol. Plant.* 77:142–149.

Katainen, H.-S., Mäkinen, E., Jokinen, J., Karjalainen, R., and Kellomäki, S. (1987). Effects of SO_2 on photosynthetic and respiration rates in Scots pine seedlings. *Environ. Pollut.* 46:241–251.

Keeling, C. D., Bacastow, R. B., and Whorf, T. P. (1982). Measurements of the concentration of carbon dioxide at Mauna Loa Observatory, Hawaii. *In* "Carbon Dioxide Review 1982" (W. C. Clark, ed.), pp. 377–383. Oxford University Press, New York.

Kramer, P. J., and Sionit, N. (1987). Effects of increasing carbon dioxide concentration on the physiology and growth of forest trees. *In* "The Greenhouse Effect, Climate Change, and U.S. Forests" (W. E. Shands and J. S. Hoffman, eds.), pp. 219–246. The Conservation Foundation, Washington, D.C.

Krause, G. H. (1988). Photoinhibition of photosynthesis. An evaluation of damaging and protective mechanisms. *Physiol. Plant.* 74:566–574.

Krauss, A. (1978). Tuberization and abscisic acid content in *Solanum tuberosum* as affected by nitrogen nutrition. *Potato Res.* 25:183–192.

Krauss, A., and Marschner, H. (1982). Influence of nitrogen nutrition, daylength and temperature on contents of gibberellic and abscisic acid on tuberization in potato plants. *Potato Res.* 25:13–21.

Kull, O., and Koppel, A. (1987). Net photosynthetic response to light intensity of shoots from different crown positions and age in *Picea abies* (L.) Karst. *Scand. J. For. Res.* 2:157–166.

Küppers, M., and Schulze, E.-D. (1985). An empirical model of net photosynthesis and leaf conductance for the simulation of diurnal courses of CO_2 and H_2O exchange. *Aust. J. Plant Physiol.* 12:513–526.

Laisk A., Kiirats, O., and Oja, V. (1984). Assimilatory power (postillumination CO_2 uptake in leaves). Measurement, environmental dependences and kinetic properties. *Plant Physiol.* 76:723–729.

Lange, O. L., Zellner, H., Gebel, J., Schramel, P., Kostner, B., and Czygan, F. C. (1987). Photosynthetic capacity, chloroplast pigments, and mineral content of the previous year's spruce needles with and without the new flush: Analysis of the forest-decline phenomenon of needle bleaching. *Oecologia* 73:351–357.

Larcher, W. (1980). "Physiological Plant Ecology." Springer-Verlag, New York.

Larsen, J. B., Yang, W., and Tiedemann, A. V. (1990). Effects of ozone on gas exchange, frost resistance, flushing and growth of different provenances of European silver fir (*Abies alba* Mill.). *Eur. J. For. Pathol.* 20:211–218.

Lee, W. S., Chevone, B. I., and Seiler, J. R. (1990). Growth and gas exchange of loblolly pine seedlings as influenced by drought and air pollutants. *Water, Air Soil Pollut.* 51: 105–116.

Leverenz, J. W. (1981). Photosynthesis and transpiration in large forest grown Douglas-fir: Interactions with apical control. *Can. J. Bot.* 59:2568–2576.

Leverenz, J. W. (1987). Chlorophyll content and the light response curve of shade-adapted conifer needles. *Physiol. Plant.* 71:20–29.

Leverenz, J. W. (1988). The effects of illumination sequence, CO_2 concentration, temperature and acclimation on the convexity of the photosynthetic light response curve. *Physiol. Plant.* 74:332–341.

Leverenz, J. W., and Jarvis, P. G. (1980). Photosynthesis in Sitka spruce (*Picea sitchensis* (Bong.) (Carr.) IX. The relative contribution made by needles in various positions on the shoot. *J. Appl. Ecol.* 17:59–68.

Linder, S., and Rook, D. A. (1984). Effects of mineral nutrition on carbon dioxide exchange and partitioning of carbon in trees. *In* "Nutrition of Plantation Forests" (G. D. Bowen, and E. K. S. Nambiar, eds.), pp. 211–236. Academic Press, London.

Lloyd, J., Syvertsen, J. P., Kriedemann, P. E., and Farquhar, G. D. (1992). Low conductances for CO_2 difussion from stomata to the sites of carboxylation in leaves of woody species. *Plant, Cell and Environ.* 15:873–899.

Long, S. P., and Hällgren, J.-E. (1985). Measurement of CO_2 assimilation by plants in the field and the laboratory. *In* "Techniques in Productivity and Biosynthesis" (J. Coombs, D. O. Hall, S. P. Long, and J. M. O. Scurlock, eds.), 2nd Ed., pp. 62–94. Pergamon, Oxford.

Lorenc-Plucinska, G. (1988). Effect of nitrogen dioxide on CO_2 exchange in Scots pine seedlings. *Photosynthetica* 22:108–111.

Lorenz, R., and Murphy, C. E., Jr. (1985). Sulfur dioxide, carbon dioxide, and water vapor flux measurements utilizing a microprocessor-controlled data acquisition system in a pine plantation. *In* "The Forest–Atmosphere Interaction" (B. A. Hutchinson and B. B. Hicks, eds.), pp. 133–147. Reichel, Doredrecht, The Netherlands.

Loreto, F., Harley, P. C., Marco, G. D., and Sharkey, T. D. (1992). Estimation of mesophyll conductance to CO_2 flux by three different methods. *Plant Physiol.* 98: 1437–1443.

Loveys, B. R. (1984a). Diurnal changes in water relations and abscisic acid in field-grown *Vitis vinifera* cultivars III. The influence of xylem-derived abscisic acid on leaf gas exchange. *New Phytol.* 98:563–573.

Loveys, B. R. (1984b). Abscisic acid transport and metabolism in grapevine (*Vitis vinifera* L.). *New Phytol.* 98:575–582.

Lundmark, T., Hallgren, J.-E., and Degermark, C. (1988a). Effects of summer frosts on the gas exchange of field-grown *Pinus sylvestris* L. seedlings. *Scand. J. For. Res.* 3: 441–448.

Lundmark, T., Hallgren, J.-E., and Heden, J. (1988b). Recovery from winter depression of photosynthesis in pine and spruce. *Trees* 2: 110–114.

McLaughlin, S. B., Andersen, C. P., Edwards, N. T., Roy, W. K. and Layton, P. A. (1990). Seasonal patterns of photosynthesis and respiration of red spruce saplings from two elevations in declining southern Appalachian stands. *Can. J. For. Res.* 20:485–495.

McMurtrie, R. E., and Wang, Y.-P. (1993). Mathematical models of the photosynthetic response of tree stands to rising CO_2 concentrations and temperatures. *Plant, Cell Environ.* 16: 1–13.

McNaughton, K. G., and Jarvis, P. G. (1991). Effects of spatial scale on stomatal control of transpiration. *Agric. For. Meteorol.* 54:279–301.

Mahan, J. R., and Upchurch, D. R. (1988). Maintenance of constant leaf temperature by plants—I. Hypothesis—Limited homeothermy. *Environ. Exp. Bot.* 28:351–357.

Maier, C. A., and Teskey, R. O. (1992). Internal and external control of net photosynthesis and stomatal conductance of mature eastern white pine (*Pinus strobus*). *Can. J. For. Res.* 22: 1387–1394

Matyssek, R. (1986). Carbon, water and nitrogen relations in evergreen and deciduous conifers. *Tree Physiol.* 2: 177–187.

Matyssek, R., Reich, P., Oren, R., and Winner, W. (1994). Response mechanisms of conifers to air pollutants. *In* "Ecophysiology of Coniferous Forests" (W. K. Smith and T. M. Hinckley, eds.), pp. 255–308. Academic Press, San Diego.

Melzack, R. N., Bravdo, B., and Riov, J. (1985). The effect of water stress on photosynthesis and related parameters in *Pinus halepensis*. *Physiol. Plant.* 64:295–300.

Mengel, K., Breininger, M. Th., and Lutz, H. J. (1990). Effect of simulated acidic fog on carbohydrate leaching, CO_2 assimilation and development of damage symptoms in young spruce trees (*Picea abies* L. Karst) *Environ. Exp. Bot.* 30: 165–173.

Monson, R. K., and Grant, M. C. (1989). Experimental studies on Ponderosa pine. III. Differences in photosynthesis, stomatal conductance and water-use efficiency between two genetic lines. *Am. J. Bot.* 76: 1041–1047.

Munns, R., and King, R. W. (1988). Abscisic acid is not the only stomatal inhibitor in the transpiration stream of wheat plants. *Plant Physiol.* 88:703–708.

Neilson, R. E., Ludlow, M. M., and Jarvis, P. G. (1972). Photosynthesis in Sitka spruce (*Picea sitchensis* (Bong.) Carr.). II. Response to air temperature. *J. Appl. Ecol.* 9:721–745.

Ögren, E., and Evans, J. R. (1992). Photoinhibition of photosynthesis *in situ* in six species of *Eucalpytus*. *Aust. J. Plant Physiol.* 19:223–232.

Oker-Blom, P. (1985). Photosynthesis of a scots pine shoot: Simulation of the irradiance distribution and photosynthesis of a shoot in different radiation fields. *Agric. For. Meteorol.* 34:31–40.

Öquist, G. (1986). Effects of winter stress on chlorophyll organization and function in Scots pine. *J. Plant Physiol.* 122:169–179.

Öquist, G., and Huner, N. P. A. (1991). Effects of cold acclimation on the susceptibility of photosynthesis to photoinhibition in Scots pine and in winter and spring cereals: A fluorescence analysis. *Funct. Ecol.* 5:91–100.

Öquist, G., and Malmberg, G. (1989). Light and temperature dependent inhibition of photosynthesis in frost-hardened and un-hardened seedlings of pine. *Photosynthe. Res.* 20:261–277.

Öquist, G., Brunes, L., Hallgren, J.-E., Gezelius, K., Hallen, M., and Malmberg, G. (1980). Effects of artifical frost hardening and winter stress on net photosynthesis, photosynthetic electron transport and RuBP carboxylase activity in seedlings of *Pinus sylvestris*. *Physiol. Plant.* 48:526–531.

Oren, R., Schulze, E.-D., Matyssek, R., and Zimmermann, R. (1986). Estimating photosynthetic rate and annual carbon gain in conifers from specific leaf weight and leaf biomass. *Oecologia* 70:187–193.

Passioura, J. B. (1988). Root signals control leaf expansion in wheat seedlings in drying soil. *Aust. J. Plant Physiol.* 15:687–693.

Paynter, V. A., Reardon, J. C., and Shelburne, V. B. (1991). Carbohydrate changes in shortleaf pine (*Pinus echinata*) needles exposed to acid rain and ozone. *Can. J. For. Res.* 21:666–671.

Pearcy, R. W. (1990). Sunflecks and photosynthesis in plant canopies. *Annu. Rev. Plant Physiol. Plant Mol. Biol.* 41:421–453.

Powles, S. B. (1984) Photoinhibition of photosynthesis induced by visible light. *Annu. Rev. Plant Physiol.* 35:15–44.

Price, D. T. , and Black, T.A. (1990). Effects of short-term variation in weather on diurnal canopy CO_2 flux and evapotranspiration of a juvenile Douglas-fir stand. *Agric. For. Meteorol.* 50:139–158.

Radin, J. W., and Ackerson, R. C. (1981). Water relations of cotton plants under nitrogen deficiency. III. Stomatal conductance, photosynthesis, and abscisic acid accumulation during drought. *Plant Physiol.* 67:115–119.

Reich, P. B., and Schoettle, A. W. (1988). Role of phosphorus and nitrogen in photosynthetic and whole plant carbon gain and nutrient use efficiency in eastern white pine. *Oecologia* 77:25–33.

Reich, P. B., Schoettle, A. W., Stroo, H. F., Troiano, J., and Amundson, R. G. (1987). Effects of ozone and acid rain on white pine (*Pinus strobus*) seedlings grown in five soils. I. Net photosynthesis and growth. *Can. J. Bot.* 65:977–987.

Reynolds, J. F., Strain, B. R., Cunningham, G. L., and Knoerr, K. R. (1980). Predicting primary productivity for forest and desert ecosystem models. *In* "Predicting Photosynthesis for Ecosystem Models. Volume II" (J. D. Hesketh and J. W. Jones, eds.), pp. 169–207. CRC Press, Boca Raton, Florida.

Rook, D. A. (1969). The influence of growing temperature on photosynthesis and respiration of *Pinus radiata* seedlings. *N. Z. J. Bot.* 7:43–55.

Rook, D. A., and Corson, M. J. (1978). Temperature and irradiance and the total daily photosynthetic production of the crown of a *Pinus radiata* tree. *Oecologia* 36:371–382.

Rook, D. A., Grace, J. C., Beets, P. N., Whitehead, D., Santantonio, D., and Madgwick, H. A. I. (1985). *In* "Attributes of Trees as Crop Plants" (M. G. R. Cannell and J. E. Jackson, eds.), pp. 507–524. Institute of Terrestrial Ecology, Huntingdon, England.

Rousseau, J. V. D., and Reid, C. P. P. (1990). Effects of phosphorus and ectomycorrhizas on the carbon balance of loblolly pine seedlings. *For. Sci.* 36:101–112.

Sandford, A. P., and Jarvis, P. G. (1986). Stomatal responses to humidity in selected conifers. *Tree Physiol.* 2:89–103.

Sasek, T. W., and Richardson, C. J. (1989). Effects of chronic doses of ozone on loblolly pine: Photosynthetic characteristics in the third growing season. *For. Sci.* 35:745–755.

Schmid, H. P., and Oke, T. R. (1990). A model to estimate the source area contributing to turbulent exchange in the surface layer over patchy terrain. *Q. J. R. Meteorol. Soc.* 116:965–988.

Schulze, E.-D., Fuchs, M., and Fuchs, M. I. (1977). Spatial distribution of photosynthetic capacity and performance in a mountain spruce forest of northern Germany. *Oecologia* 30:239–248.

Seiler, J. R., and Cazell, B. H. (1990). Influence of water stress on the physiology and growth of red spruce seedlings. *Tree Physiol.* 6:69–77.

Sheriff, D. W. (1977). The effect of humidity on water uptake by, and viscous flow resistance of, excised leaves of a number of species: Physiological and anatomical observations. *J. Exp. Bot.* 28:1399–1407.

Sheriff, D. W. (1992). Roles of carbon gain and allocation on growth at different N nutrition in *Eucalpytus camaldulensis* and *Eucalyptus globulus* seedlings. *Aust. J. Plant Physiol.* 19:637–652.

Sheriff, D. W., and Whitehead, D. (1984). Photosynthesis and wood structure in *Pinus radiata* D. Don during dehydration and immediately after rewatering. *Plant, Cell Environ.* 7:53–62.

Sheriff, D. W., Nambiar, E. K. S., and Fife, D. N. (1986). Relationships between nutrient status, carbon assimilation and water use efficiency in *Pinus radiata* (D. Don) needles. *Tree Physiol.* 2:73–88.

Smith, W. K., and Carter, G. A. (1988). Shoot structural effects on needle temperatures and photosynthesis in conifers. *Am. J. Bot.* 75:496–500.

Smith, W. K., Schoettle, A. W., and Cui, M. (1991). Importance of the method of leaf area measurement to the interpretation of gas exchange of complex shoots. *Tree Physiol.* 8:121–127.

Smolander, H., and Oker-Blom, P. (1989). The effect of nitrogen content on the photosynthesis of Scots pine needles and shoots. *Ann. Sci. For.* 46 (Suppl.): 473s–475s.

Smolander, H., Oker-Blom, P., Ross, J., Kellomaki, S., and Lahti, T. (1987). Photosynthesis of a Scots pine shoot: Test of a shoot photosynthesis model in direct radiation field. *Agric. For. Meteorol.* 39:67–80.

Squire, R. O., Neales, T. F., Loveys, B. R., and Attiwill, P. M. (1988). The influence of water deficits on needle conductance, assimilation rate and abscisic acid concentrations of seedlings of *Pinus radiata* D. Don. *Plant, Cell Environ.* 11:13–19.

Stitt, M. (1991). Rising CO_2 levels and their potential significance for carbon flow in photosynthetic cells. *Plant, Cell Environ.* 14:741–762.

Strain, B. R., Higginbotham, K. O., and Mulroy, J. C. (1976). Temperature preconditioning and photosynthetic capacity of *Pinus taeda* L. *Photosynthetica* 10:47–53.

Strand, M., and Lundmark, T. (1987). Effects of low night temperature and light on chlorophyll fluorescence of field-grown seedlings of Scots pine (*Pinus sylvestris* L.) *Tree Physiol.* 3:211–224.

Strand, M., and Öquist, G. (1985). Inhibition of photosynthesis by freezing temperatures and high light levels in cold-acclimated seedlings of Scots pine (*Pinus sylvestris*). I. Effects on the light-limited and light-saturated rates of CO_2 assimilation. *Physiol. Plant.* 64: 425–430.

Sullivan, J. H., and Teramura, A. H. (1989). The effects of ultraviolet-B radiation on loblolly pine. I. Growth, photosynthesis and pigment production in greenhouse-grown seedlings. *Physiol. Plant.* 77:202–207.

Surano, K. A., Daley, P. F., Houpis, J. L. J., Shinn, J. H., Helms, J. A., Palassou, R. J., and Costella, M. P. (1986). Growth and physiological responses of *Pinus ponderosa* Dougl. ex P. Laws to long-term elevated CO_2 concentrations. *Tree Physiol.* 2:243–259.

Szaniawski, R. K., and Wierzbicki, B. (1978). Net photosynthetic rate of some coniferous species at diffuse high irradiance. *Photosynthetica* 12:412–417.

Teskey, R. O., Fites, J. A., Samuelson, L. J., and Bongarten, B. C. (1986). Stomatal and non-stomatal limitations to net photosynthesis in *Pinus taeda* L. under different environmental conditions. *Tree Physiol.* 2:131–142.

Teskey, R. O., Bongarten, B. C., Cregg, B. M., Dougherty P. M., and Hennessey, T. C. (1987). Physiology and genetics of tree growth response to moisture and temperature stress: An examination of the characteristics of loblolly pine (*Pinus taeda* L.) *Tree Physiol.* 3:41–62.

Tevini, M., and Teramura, A. H. (1989). UV-B effects on terrestrial plants. *Photochem. Photobiol.* 50:479–487.

Thompson, W. A., and Wheeler, A. M. (1992). Photosynthesis by mature needles of field-grown *Pinus radiata*. *For. Ecol. Manage.* 52:225–242.

Thornton, F. C., Pier, P. A., and McDuffie, C., Jr. (1990). Response of growth, photosynthesis, and mineral nutrition of red spruce seedlings to ozone and acidic cloud deposition. *Environ. Exp. Bot.* 30:313–323.

Tolley, L. C. and Strain, B. R. (1984). Effects of CO_2 enrichment and water stress on growth of *Liquidambar styraciflua* and *Pinus taeda* seedlings under different irradiance levels. *Can. J. Bot.* 6:2135–2139.

Tolley, L. C., and Strain, B. R. (1985). Effects of CO_2 enrichment and water stress on gas exchange of *Liquidambar styraciflua* and *Pinus taeda* seedlings grown under different irradiance levels. *Oecologia* 65:166–172.

Troeng, E., and Linder, S. (1982a). Gas exchange in a 20-year-old stand of Scots pine. I. Net photosynthesis of current and one-year-old shoots within and between seasons. *Physiol. Plant.* 54:7–14.

Troeng, E., and Linder, S. (1982b). Gas exchange in a 20-year-old stand of Scots pine. II. Variation in net photosynthesis within and between seasons. *Physiol. Plant.* 54: 15–23.

Upchurch, D. R., and Mahan, J. R. (1988). Maintenance of constant leaf temperature by plants. II. Experimental observations in cotton. *Environ. Exp. Bot.* 28:359–366.

van den Driessche, R. (1987). Importance of current photosynthate to new root growth in planted conifer seedlings. *Can. J. For. Res.* 17:776–782.

Wallin, G., Skarby, L., and Sellden, G. (1990). Long-term exposure of Norway spruce, *Picea abies* (L.) Karst. to ozone in open-top chambers. I. Effects on the capacity of net photosynthesis, dark respiration and leaf conductance of shoots of different ages. *New Phytol* 115:335–344.

Wang, Y. P., and Jarvis, P. G. (1990). Description and validation of an array model—MAESTRO. *Agric. For. Meteorol.* 51:257–280.

Wang, Y.-P., McMurtrie, R. E., and Landsberg, J. J. (1992). Modelling canopy photosynthetic productivity. *In* "Crop Photosynthesis, Spatial and Temporial Determinants, Topics in Photosynthesis Volume 12" (N. R. Baker and H. Thomas, eds.), pp. 43–67. Elsevier, Amsterdam.

Waring, R. H., and Franklin, J. F. (1979). Evergreen forests of the Pacific Northwest. *Science* 204:1380–1386.

Weidner, M., and Kraus, M. (1987). Ribulose-1,5-bisphosphate carboxylase activity and the influence of air pollution in spruce. *Physiol. Plant.* 70:644–672.

Wilson, C. C. (1947). The effect of some environmental factors on the movements of guard cells. *Plant Physiol.* 23:5–37.

Wofsy, S. C., Harris, R. C., and Kaplan, W. A. (1988). Carbon dioxide in the atmosphere over the Amazon basin. *J. Geophys. Res.* 93(D2):1377–1387.

Wong, S. C., Cowan, I. R., and Farquhar, G. D. (1979). Stomatal conductance correlates with photosynthetic capacity. *Nature (London)* 282:424–426.

Wood, G. B., and Brittain, E. G. (1973). Photosynthesis, respiration and transpiration of radiata pine. *N. Z. J. For. Sci* 3:181–190.

Yushkov, V. I., and Zav'yalova, N. S. (1989). Structure and function of the assimilation apparatus of young siberian stone pine plants under different light regimes. *Sov. J. Ecol. (Engl. Transl.)* 19:194–200.

Zelawski, W., Szaniawski, R., Dybczynski, W., and Piechurowski, A. (1973). Photosynthetic capacity of conifers in diffuse light of high illuminance. *Photosynthetica* 7:351–357.

Zhang, J., and Davies, W. J. (1990). Changes in the concentration of ABA in xylem sap as a function of changing soil water status can account for changes in leaf conductance and growth. *Plant, Cell Environ.* 13:277–285.

Zimmermann, R., Oren, R., Schulze, E.-D., and Werk, K. S. (1988). Performance of two *Picea abies* (L.) Karst. stands at different stages of decline. *Oecologia* 76:513–518.

II

Resource Allocation and Utilization

5

Resource Use Efficiency

D. W. Sheriff, H. A. Margolis, M. R. Kaufmann, and P. B. Reich

I. Introduction

In this chapter we concentrate on the efficiency of use in fixing carbon and accumulating biomass of those resources most often of interest to ecophysiologists: light, nutrients, water, and carbon. A plant uses these in a variety of ways to enable it to survive, grow, and reproduce. Resources are used to aid in producing the necessary physical structure and in maintaining biochemical and photochemical machinery, to gather the energy that allows local reversal of entropy, and to accumulate carbon. Resources also have a role in the gathering of other resources.

In most natural environments, one or more resources will be in limited supply, at least for a portion of each year or for a part of the plant's life cycle, or both. This shortage will limit physiological activity and growth. Knowledge of how plants use resources is basic to an understanding of the biology of ecosystems and of competition between species. Efficiency of use of resources has traditionally been determined by calculating ratios of productivity per unit of resource [i.e., resource use efficiencies (RUEs); carbon use is often expressed as relative growth rate (RGR)]. Although RUEs and comparisons of values from different species can often be useful, they are frequently used incautiously. Comparisons have been made at a range of physical and temporal scales, often using the same terminology at each level of scale, frequently incorporating different meanings of productivity or different measures of resource use, sometimes with both different. This raises several questions: (1) What relationship could be expected between resource use efficiency and productivity? (2) Should RUEs measured at different scales use the same terminology? (3) How much similarity can be expected between RUEs determined for the same resource at different scales? (4) Is (carbon gain)/(resource), or (biomass increase)/(resource) a biologically appropriate measure of the efficiency of use of a resource?

When we obtain values for a RUE of a species, an ecotype, etc. we are often interested in the importance of differences or of the magnitude of that RUE in relation to growth and survival. We may compare RUEs between species within an ecosystem, or for different ecotypes of a species, or examine RUEs in relation to the environment in which the species has evolved or is now growing. Species will have evolved a higher RUE in response to a shortage of the resource in their environment. A resource shortage can result from an inherently poor supply of the resource in the environment and from competition for the resource by other plants. Mechanisms that allow plants to grow in an environment that is resource limited can be physiological, anatomical, morphological, ontogenetic, or phenological. RUEs of different resources will evolve separately, presumably in relation to the limitations of the resources in that environment and to other selective pressures. It is, perhaps, easiest to see this in plants native to extreme environments. For example, perennials from hot deserts usually use water, a very limited resource, very efficiently, but use the overabundant light less efficiently than do plants from mesic environments.

RUEs that relate to productivity are potentially important for many agronomic and tree crops, because production of plant biomass is a major aim. Here comparisons are usually between cultivars or species of similar morphology, or between simply contrasting treatments. The comparisons are, therefore, more straightforward than many analyses of natural ecosystems, and hence will probably more often show an effect of RUE on productivity. In any analysis, account must be taken of the fact that RUEs are ratios, and hence do not necessarily relate to absolute levels of either productivity or resource use: they show the relationship between these. It must also be remembered that a greater efficiency of use of a resource has potential to increase productivity only if supply or use of that resource is limiting in that environment, and that a limited supply of one resource can sometimes be overcome by use of another. An example of the latter is the compensation by long photoperiods for low light or low nitrogen nutrition (Küppers *et al.*, 1988). When considering natural ecosystems, account must be taken of the overriding importance of the ability of a species to survive and reproduce successfully. To do this it does not necessarily need to produce large amounts of biomass, either as individuals or as members of a community: factors other than a RUE or a combination of RUEs may be much more important. Thus, in relation to survival and reproduction, RUEs, as we usually define them, may have little importance in determining either productivity or the relative frequencies of occurrence of different species. Depending on the resource and how its RUE is defined, a RUE may be positively or negatively associated with productivity, or have little relationship with it. For example, efficiencies of use of transpired water and of foliar nitrogen (N) in assimilating carbon have opposing constraints, such that if they vary with an experimental treatment or between species they will often be negatively correlated (e.g., Farquhar and Kirschbaum, 1985). This has been found by Reich *et al.* (1989) in trees grown under a range of combinations of high and low water and nitrogen availability. The relationship was also observed by Field *et al.* (1983) over a range of species, but with a positive correlation between these for data from one species, which indicates complex interrelationships in resource use. Similarly, allocation of N to different components of the assimilatory system depends on the leaf's light environment (see later). Adjustment of this allocation for different light environments can increase the efficiency of use of both N and light. Data of Field *et al.* (1983) indicate little association between efficiencies of use of either water or nitrogen and assimilatory capacity. An example of little relationship between instantaneous RUEs and productivity is the greater competitive success of plants that are in positive carbon balance for longer portions of a year (Küppers, 1984).

Use of the same terminology at different scales can accentuate problems with scaling. For example, it may reduce the possibility of seeing relationships between RUEs and productivity because the terminology can constrain thought patterns and reduce communication, for example, between researchers working at different scales. Problems of scale are several, including the following considerations:

1. greater biomass infrastructure, the production and maintenance of which needs to be included at greater levels of physical scale; at the leaf level instantaneous rates of carbon assimilation are determined taking no account of its use, whereas at the whole-plant level assimilated carbon also has to produce and maintain the plant and its activities, and different availability of a resource may often change the use of carbon by altering its allocation and biomass partitioning.

2. Additional processes, for example, the absorption of resources, at greater levels of physical scale.

3. The different temporal scales for residence of different resources in a community, a plant, or a plant organ.

4. Viewing RUEs over different periods of time at the same physical scale, which raises questions of, for example, resource residence time.

Resource use efficiencies have most often been determined as carbon, or biomass gained per unit of resource in the biomass. This is more appropriate at levels of scale below that of a plant. At the plant level and above, the ability of plants to acquire resources (e.g., see Chapter 2, this volume), and the cost of that acquisition, are also very important. Here, the definition of RUE provides only part of the information about resource use and resource use efficiency. When working at the ecosystem level, for example, we might ask questions about the use and use efficiency of resources potentially available within the ecosystem.

In this chapter we attempt to address questions of the meaning of different RUEs, of RUE terminology, and of RUEs determined at different levels of scale, spending most time on those which seem important to us for improving our understanding of the RUE concept. To do this we broadly discuss current definitions of RUE, and do not attempt a detailed analysis of the importance of RUEs to community, ecosystem, or evolutionary questions.

II. Issues of Scaling

A. Hierarchy

A hierarchical approach places scales of space and time into perspective. Although the hierarchical approach has been developed to deal primarily with spatial patterns in landscapes (Urban *et al.*, 1987), it is useful for rationalizing how processes (including RUEs) relevant to structure and composition at one spatial or temporal scale relate to processes occurring at other scales. Space can be divided into discreet units such as an organ, a plant, a plant community, an ecosystem, and a landscape. Similarly, time can be divided into units such as a second, minute, day, year, or life cycle. In general, the relevant spatial and temporal scales are similar: processes occurring at small spatial scales normally have short time intervals, and processes occurring at large spatial scales require longer times.

A critical aspect of scale is that processes important at one level of scale are often different from those at other levels of scale. For example, carbon acquisition by a leaf includes photochemical and biochemical processes involved in carbon fixation in specialized tissues and organelles, whereas carbon fixation by a plant is the sum of fixation by all leaves and losses through respiration and loss of dry matter. Thus, differences in dry matter accumulation of species or treatments could result from either of two effects, or a combination of these. Responses to resource availability may be dominated by leaf area development, and this may be the main effect on primary productivity. For example, Linder and Rook (1984) found that fertilization of conifers affected leaf biomass more than rates of carbon assimilation. If differences in leaf RUEs are dominated by differences in carbon assimilation, differences in biomass accumulation might be expected, but if differences in RUEs are dominated by differences in resource availability, biomass accumulation is less likely to be different. Carbon acquisition by an ecosystem includes not only carbon fixation, but its redistribution within plants, and losses through respiration and decomposition of detritus and of carbon lost from roots. Scaling up involves not only more tissues and organs, more individual plants, and more time, it also involves more and different processes, and these cannot be represented by individual organs or plants. Thus, the behavior of systems at larger scales cannot be obtained by multiplying or summing only some processes from smaller scales (e.g, those used to determine RUEs at smaller scales). The behavior of each spatial level usually needs to be determined either by inclusion of *all* relevant and associated processes determined at smaller scales, or by studies of the relevant processes at the larger scale. We can see something of this from a more detailed examination of scaling from the leaf to the whole-plant level. The more of a resource a plant can gather and the less it uses for each net gain of carbon, the more carbon it will potentially be able to fix. This will increase supply of carbon-based compounds for structural material and metabolic activity, and will contribute to the plant's success. Thus, at the whole-plant level, carbon gain by all leaves (including those not operating under optimum conditions), resource use in gaining carbon, resource use in converting acquired carbon into biomass, and absorption of resources per carbon spent in getting them will all be important. For a particular efficiency of carbon gain, the less this carbon needs to be spent on activities other than growth, the greater will be the efficiency of its use for growth. Allocation of more carbon to biomass, and partitioning of more of this to leaves, will increase the efficiency of use of carbon in biomass production. Effects of greater leaf biomass are particularly important because of the "compound interest" effect of this in allowing a plant to have greater assimilatory capacity (see Chapter 4). Other constraints on productivity and survival have not been taken into account in this discussion.

The advantage of a hierarchical approach is that it is often not necessary to understand in detail all the smaller scale processes to determine behavior at the larger scales. Usually, processes relevant at one scale have significant influence at one or two scale levels up or down, but their influence declines beyond that.

An aspect of scaling particularly relevant to the study of RUEs is the feedback of processes at larger scales on processes at smaller scales. For example, stands may alter or create their own environment and affect processes occurring within individual trees. This is particularly important in relation to efficiencies of water use, because transpiration from an extensive stand causes the humidity of the air to be higher and the temperature to be lower than would otherwise be expected, thus potentially lowering transpiration. These effects can be large at the regional level, where a threefold change in canopy conductance may result in only a ~20% change in transpiration (Jarvis and McNaughton, 1986). Feedback from larger to smaller scales limits the use of measurements of smaller scale processes for estimating larger scale behavior.

Some other factors affect scaling: (1) the influence of stand structure on the environment of tree organs; (2) the consequences of responses of individual organs or plants are dependent on responses of all individuals; and (3) stand properties are not a simple sum of properties of its components.

These scaling issues are especially important with RUEs because at each level of scaling we are determining a ratio of two variables that each often involve different processes and time frames. In considering how different resources may contribute to development and growth of organs, individual trees, forest communities, and ecosystems, we need definitions of RUEs that have the same basis for all resources at each level. At the small spatial scale of a leaf, instantaneous RUEs are calculated from gas exchange data collected for individual leaves. These data are often expressed as rates with a timestep of 1 second, collected during measurements lasting no more than a few minutes. At this level, light and water have mean residence half-times of picoseconds and minutes. RUEs are defined as the flux of carbon (or CO_2), or gain in biomass per unit flux of the resource, and do not include a time component in their dimensions. For nutrients, where mean residence half-times are months to years for a particular leaf, the instantaneous RUE often has been defined as the flux of carbon (or CO_2) per unit of the resource resident in the organ at the time of measurement. This produces a RUE that is a rate.

In an ecosystem, we most often want information on biomass accumulation either per plant or per unit ground area over periods of months to years or decades, or longer. Mean residence half-times are nanoseconds for light, days for water in a tree, months to years for water in an ecosystem, years to decades for nutrients in a tree, and often hundreds of years for nutrients in an undisturbed ecosystem. This means that sometimes the time bases for resources needs to be adjusted so their magnitudes are similar for all resources.

B. Individual RUEs

Issues related to each RUE are mentioned briefly in this section and are discussed in more detail later.

1. Light Use Efficiency Assimilatory light use efficiency (ALUE) of a leaf, tree, canopy, or ecosystem can be similarly defined: the efficiency of use of light impinging on, or absorbed by, a leaf or canopy in assimilating carbon or accumulating biomass. Relationships between rates of carbon assimilation and ALUE, and questions of scaling to the canopy level, have been discussed in some detail by Sinclair and Horie (1989). ALUE, or biomass accumulation per unit light (LUE), at levels of greater scale can also be defined for light input per unit ground area. Although this has limited physiological meaning, or importance, unless the canopy is closed and intercepting all incident light, it includes an "efficiency of capture" term. This can be important to an understanding of changes with time in how ecosystems operate, and of differences between ecosystems. Particularly at the ecosystem level, values of LUE indicate only a part of the effect of light, which can also influence growth via hormonal effects, for example, photoperiodicity.

2. Nutrient Use Efficiencies Instantaneous ANUEs of foliage (flux of CO_2/foliar nutrient) (e.g., Sheriff, 1992) are useful for comparing different species or treatments from short-term experiments. ANUE of individual leaves, without reference to the whole plant, are of limited use to an ecophysiologically based understanding, because in this definition there is no indication of how nutrients contribute to carbon assimilation or growth over the period that they are in a plant or a plant organ. If nutrient turnover cannot be established, the ANUE can be defined as (gain in carbon, or biomass)/(nutrient concentration), where nutrient concentration is the mean concentration during a substantial period. At the whole-plant level and above, NUE has often been determined [for example, by Schlesinger *et al.* (1989) and by Sheriff and Nambiar (1991)] as (increase in biomass)/(nutrient content of plant biomass). At the ecosystem level, use of site total, or available nutrient rather than tissue nutrient, would indicate the capacity of plants to utilize site resources (Son and Gower, 1991).

3. Water Use Efficiency At smaller levels of scale, instantaneous water use efficiency (WUE) is better called transpiration efficiency [TE = (carbon assimilation)/(transpiration)], as is increasingly found in published papers (e.g., Virgona *et al.*, 1990). For consistency of nomenclature, and to avoid confusion with transpiration efficiencies determined for biomass production per unit of water transpired by whole plants (e.g., Herwitz and Gutterman, 1990), we use the term assimilatory transpiration efficiency (ATE) for instantaneous values. These are physiologically useful, but instantaneous values are usually determined only when conditions are suitable for carbon assimilation. They ignore water loss at other times, for example, at night. This definition also neglects other important functions of water in a plant, including effects of hydration or of turgor on growth. At the tree, canopy, or ecosystem levels, carbon (or biomass) gain is compared to total water use (Squire *et al.*, 1987). Depending on how this is assessed, water "use" may include transpiration, evaporation (including interception losses), runoff, and deep drainage, but usually it is limited to total evapotranspiration.

III. Resource Use Efficiencies at Various Spatial Scales

A. Definitions of Scales

For spatial scales ranging from individual leaves to ecosystems, RUEs can be determined at temporal scales ranging from instantaneous to long term. For a single leaf, losses of assimilated carbon, other than simultaneous respiratory losses within the leaf, are neglected for calculations of instantaneous RUEs, whereas for longer term RUEs these losses are included. At larger scales, losses include respiration of the living biomass (autotrophic respiration), production of detritus through shedding of plant parts, mortality, and consumption by animals and disease. For an ecosystem, total respiration is the sum of autotrophic respiration and respiration by heterotrophic organisms involved in decomposition of detritus and carbon lost from roots. The carbon balance of an ecosystem includes gross photosynthesis, autotrophic respiration, growth, production and heterotrophic decomposition of detritus, and growth of heterotrophs (Waring and Schlesinger, 1985). Typically, heterotroph production is a small component of the carbon balance, but heterotrophic activity eventually consumes most of the primary production.

Instantaneous rates of net primary productivity (NPP) may be determined for any spatial scale as the net carbon assimilation for all foliage being considered, whether this be a single leaf, branch, tree crown, or overstory and understory canopy. In practice, instantaneous values have been much easier to determine for single leaves and branches. Recent advances in eddy correlation techniques for CO_2 flux are making it easier to determine NPP for entire plant communities, but observations on single trees remain a problem because of difficulties in measuring or estimating belowground processes such as root respiration and fine root turnover. Estimates of gross primary productivity (GPP) are much more difficult, and are usually obtained by measuring NPP and adding an estimate of respiration (Waring and Schlesinger, 1985).

In the following sections, RUEs are discussed for spatial scales of leaves to ecosystems. At the smaller spatial scales, most discussion is on instantaneous efficiencies of net primary productivity or of efficiencies integrated for up to a day, whereas at the forest community or ecosystem level, discussions center primarily on annual timesteps.

B. Leaf Level

1. Light Use Efficiency Light use efficiency of a leaf, ALUE, is related to the rate of carbon assimilation (A) and the photosynthetic photon flux density (PPFD) as ALUE = A/PPFD; this is sometimes called quantum yield and can be determined either on the basis of light incident on a leaf (AL_iUE) or on the basis of the amount of light absorbed by a leaf (AL_aUE). The latter is physiologically more meaningful because it indicates the efficiency with which energy from absorbed light is utilized. However, this is a more difficult quantity to scale up to larger physical scales because it is difficult to measure light absorption by only the foliar canopy of a stand of vegetation, and to separate component species. Most frequently, leaf area index [LAI = (leaf area)/(ground area)] or a similar measure is determined. In this case light use efficiency can usually be related only to light incident on the foliage. Changes in factors that alter light absorption (e.g., foliar [N]) will most often modify AL_iUE, but will often not affect AL_aUE at PPFDs below that for light saturation.

Over short time periods, ALUE will remain constant in an unchanging environment unless environmental extremes cause damage to the assimilatory system (e.g., photoinhibition). Changes in light over short time periods will cause ALUE to change little at low light levels, where carbon assimilation increases linearly with PPFD. As carbon assimilation approaches and attains light saturation, ALUE will be negatively associated with PPFD. Over longer time periods, previously exposed leaves that have become shaded may take on the characteristics of shade leaves. This will alter their responses to PPFD such that ALUE will often then be lower at high PPFD and higher at low PPFD than it was previously.

Factors in the internal and external environments of leaves affect their ALUE. Effects of changes in these environments on carbon assimilation and light use are summarized in Table I.

2. Nutrient Use Efficiency Carbon assimilation is often measured in an environment expected to maximize carbon assimilation at current leaf nutrient concentrations. These data give the potential (maximum) rate

Table I Effects of Changes in the Numerator or the Denominator of Components of Instantaneous RUEs of Leaves[a]

RUE	Means of increasing RUE	Effect on NPP of leaf	Effect on quantitative use of resource
ALUE	Greater efficiency of net carbon assimilation	Increase	Nil
	Greater absorption of intercepted light[b]	Increase	Increase
	Greater light interception	Increase	Increase
ANUE	Greater efficiency of carbon assimilation	Increase	Nil
	Lower foliar nutrient concentration[c]	Nil	Reduce
ATE	Greater efficiency of carbon assimilation	Increase	Nil
	Lower transpiration by reduced stomatal conductance	Reduce	Reduce

[a] Net carbon assimilation can be affected by a range of factors, as discussed in Chapter 4 of this volume. The magnitude of these effects can vary with species, time of year, ontogeny, and phenology. Factors that affect the supply or use of the resource are given in Table II.

[b] Usually only below the PPFD at which carbon assimilation is light saturated.

[c] Usually only below the foliar nutrient concentration at which carbon assimilation is saturated for that nutrient.

of carbon assimilation, and the potential assimilatory nutrient use efficiency (PANUE). This may rarely be achieved in a natural environment because many combinations of nonnutrient variables (e.g., light, humidity, temperature, water status) can produce conditions not optimal for carbon assimilation.

Below, we will discuss nutrient use efficiencies in general terms, because we are interested here in the general concept, and because values are usually calculated in the same way for each nutrient. However, there is no generalized ANUE. It needs to be calculated separately for each nutrient, because constraints on carbon assimilation by one nutrient will often affect calculated efficiencies of use of other nutrients. ANUE has most frequently been derived for nitrogen because this is the nutrient required in largest quantity for carbon assimilation (Gerwick, 1982). Therefore, some examples given below will be based on assimilatory nitrogen use efficiency (AN_nUE) calculated as $A/[N]$. These examples are used to illustrate principles important for interpreting NUE data, not details of effects universal to all nutrients.

At the leaf level, ANUE can be determined either for total leaf content of a nutrient or for that portion of the nutrient involved in carbon assimilation. In the latter case efficiency of nutrient use depends principally on two factors: (1) whether there are limitations to carbon assimilation that are not associated with the nutrient in question, and (2) ways the nutrient is partitioned to different parts of the assimilatory pathway, if used in more than one of these. If ANUEs are calculated using total foliar nutrient concentration, it is also necessary to conceptually take into account partitioning to assimilatory activity versus other components: nonassimilatory biochemical processes, structure, and storage.

Values of ANUE may be different when expressed over the life of a leaf than when expressed on an instantaneous basis (Rundel, 1982). The importance of this can be seen from comparison of two leaves. One is long-lived, retains its nutrients for a long time, and has a small instantaneous ANUE; the other has a large ANUE and is short-lived. The former, and the nutrients it contains, may contribute more to carbon gain during its life than the latter. Leaf longevity is important to definitions of NUE, because it seems likely that slower leaf turnover may be a response, possibly the most important response in evolutionary time, to low nutrient availability (Reich *et al.*, 1992; van der Werf, 1993). Leaf longevity allows longer retention of the carbon and nitrogen that are hard-won under nutrient-limiting conditions. To take account of this, a measure of nutrient turnover can be included in calculation of ANUEs at all spatial levels, as discussed, for example, by Rundel (1982) and by Berendse and Aerts (1987). The definition of ANUE is then similar to that for light and water, i.e., (net flux of CO_2 or carbon)/(flux of resource), and allows determination of the contribution the nutrient makes to carbon gain during the time it is in a plant or a plant organ. It also makes more simple integration of ANUE over different levels of organization. To use this definition at levels of scale greater than that of a leaf, it is necessary to take account of internal retranslocation of nutrients from senescent and green nonsenescent older leaves to new foliage (e.g., Nambiar and Fife, 1991). New, more exposed, foliage needs a good supply of nutrients to enable rapid carbon assimilation. Older foliage becomes more shaded and is unable to utilize large nutrient supply, so carbon gain by a plant is optimized by appropriate arrangement of nutrient concentrations in differently exposed leaves (DeJong and Doyle, 1985; Field, 1983). This is particularly relevant to conifers, most of which have long-lived foliage, and to hypotheses about greater nutrient use efficiency. Of importance to this is the carbon gain per unit of foliar nutrient during the time this is in foliage. Carbon gain assessed in this way is often greater for species with long-lived foliage (e.g., Monk, 1966; Small, 1972). Although conifers as a group tend to have foliage that is longer-lived than foliage of the temperate woody angiosperms with which they are most frequently compared, coniferous species vary widely in needle longevity (see Reich *et al.*, 1994). This raises questions of how the NUE of conifers compares with that of angiosperms, and how NUE differs among conifers. Instantaneous values of potential assimilatory nitrogen use efficiency (PAN_nUE) are greater in species with

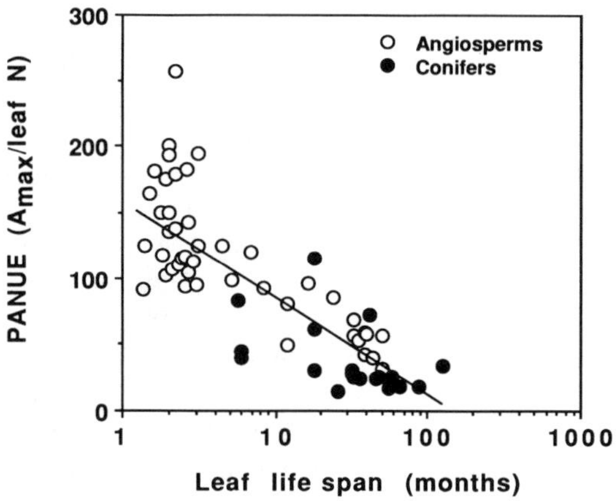

Figure 1 Variation in potential assimilatory nitrogen use efficiency with life span of leaves of conifer and angiosperm arboreal species.

shorter leaf life spans and higher A_{max} per leaf mass (Fig. 1). They are, therefore, on average greater in deciduous hardwood trees than in evergreen conifers. Data for coniferous species fall roughly along the same line as data for angiosperm herbs and trees. This leads us to two conclusions: (1) there is nothing intrinsically different about the AN_nUEs of conifers and angiosperms—no differences are noted when species with similar ecological traits are considered, and (2) conifers vary widely in their PAN_nUEs in a way consistent with general variation among species of differing leaf life span (Reich *et al.*, 1992). Differences between conifers, and between conifers and angiosperms, result from species-specific differences in leaf life span, A_{max}, and foliar [N], which are unique to every species.

Internal recycling of nutrients and longevity of foliage were not significantly associated with each other or with leaf [N] in tree species investigated by Reich *et al.* (1992). Comparison of conifer and nonconifer groups (Fig. 2) shows no difference in the proportion of N retranslocated. However, *Larix* species appear to retranslocate a greater proportion of foliar N compared to evergreen conifers (Gower and Richards, 1990). These data are consistent with previous reports of no general differences in the proportion of foliar N retranslocated by temperate evergreen conifers and deciduous hardwood trees (Small, 1972; Chapin and Kedrowski, 1983). Because most conifers have long-lived foliage with low [N], and the proportion of N retranslocated does not vary consistently among species, it follows that litterfall [N] should decrease as leaf life span increases or as [N] of green foliage declines. A litterfall-based NUE (Vitousek, 1982) can be calculated as the inverse of litterfall

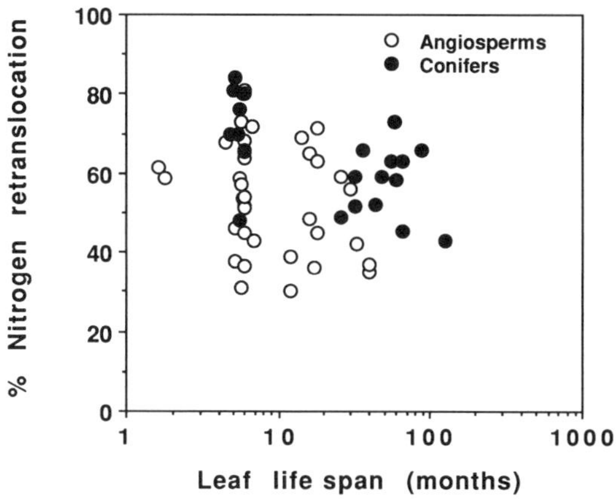

Figure 2 Variation in the proportion of foliar nitrogen that is retranslocated from leaves of arboreal conifers and angiosperms with the life span of leaves.

[N]. Among species of all plant life forms this NUE was weakly positively related to leaf life span if data points for two species with the shortest leaf life spans were included in the data set. No relationship was found if these two species were omitted (Reich *et al.*, 1992). Overall, conifers did not differ from angiosperm trees.

Trends for litterfall-nitrogen use efficiency (LN_nUE) vis-à-vis leaf life span differ from those for PAN_nUE. To analyze this question we must view the system more holistically. In some niches, carbon consumed (i.e., lost to the plant) by frequent leaf shedding, or the time taken for nutrients to be recycled through decomposition, will potentially limit plant activity. A crude relative index of maximum time-integrated potential carbon gain per unit leaf N, based on PAN_nUE, leaf lifetime, and percent N retranslocation, increased with increasing leaf life span ($p < 0.001$, $r^2 = 0.65$) (Reich *et al.*, 1992), with a similar relationship for conifers and broadleaf tree species. Definitive analyses of integrated long-term NUE for conifers and other species, based on physiological measurements, remain to be made.

a. Amount of Nutrient in Assimilatory Pathway Compared with Amount in Other Compounds Maximum (potential) rates of carbon assimilation are positively associated with leaf nitrogen concentration because most nitrogen in a leaf is incorporated into the photosynthetic pathway (e.g., Evans, 1989). For a leaf, the relationship between A_{max} and [N] is linear, with zero A_{max} at a positive value of [N], at all but high nitrogen concentrations, where carbon assimilation saturates for [N]. As a consequence the relationship between foliar nitrogen concentration and PAN_nUE has four phases (e.g., Field and Mooney, 1986; Sheriff, 1992): (1) a positive association between foliar nitrogen and PAN_nUE at lower concentrations, (2) a curvilinear relationship over a range of [N] beyond which PAN_nUE saturates, (3) saturation of PAN_nUE, and (4) a negative relationship between foliar nitrogen and PAN_nUE at higher concentrations. This occurs because with increasing foliar [N] a larger proportion of nitrogen is incorporated into compounds in the assimilatory pathway. Beyond a limit, more N cannot be readily incorporated into these compounds and is stored. Factors not related to incorporation of N into the assimilatory pathway may limit carbon assimilation at higher nitrogen concentrations. An example of this is limitations to diffusion of CO_2 from ambient air to sites of carbon assimilation. Relationships between foliar [N] and carbon assimilation hold in a general sense for all leaves, but vary quantitatively among species and for leaves of the same species produced under different conditions. Leaf specific weight (LSW = weight/area) is often negatively related to PAN_nUE, probably because leaves of high LSW have more nitrogen in structural compounds and in substances involved with metabolism not directly linked to carbon assimilation (Field and Mooney, 1986). The relationship between foliar nutrient concentration, carbon assimilation, and PANUE may be different for other nutrients.

b. Partitioning of Nutrient within Assimilatory Pathway Within the photosynthetic pathway, nitrogen is principally in Calvin cycle proteins, mainly ribulose bisphosphate carboxylase, and in thylakoid membranes of chloroplasts. The proportion of foliar nitrogen in each of these components varies both with species and with a leaf's microenvironment. A major factor of the microenvironment that influences this is the light regime under which a leaf develops (Evans, 1989). Usually, there is a negative relationship between the "average" PPFD incident on a leaf and the proportion of nitrogen partitioned to the thylakoids in its chloroplasts; at lower light more thylakoids are needed for light capture. There is a corresponding positive relationship between PPFD and the proportion of nitrogen partitioned to ribulose bisphosphate carboxylase: at high light levels more is needed to utilize the captured energy. When nitrogen is partitioned to maximize carbon assimilation in a leaf's usual light environment, leaf AN_nUE will be improved compared to that resulting from a fixed partitioning of nitrogen. Thus, PAN_nUEs of leaves from different light environments need to be measured under light regimes similar to those they normally experience. In low light environments foliar nitrogen concentrations do not usually affect carbon assimilation, so under these conditions values of PAN_nUE or of AN_nUE have little meaning (e.g., Gulmon and Chu, 1981). Changes in assimilatory capacity not caused by nutrition, and changes in foliar nutrient concentration, will usually both affect a leaf's ANUE. The general effects of these changes are shown in Table I.

Plants that live in an environment normally deficient in a nutrient may optimize use of the nutrient partly by maximizing its ANUE. However, this can have a deleterious effect on the efficiency of water use (see below).

c. Limitation of Carbon Assimilation by Factors Other Than Nutrients Instantaneous nutrient use efficiencies are reduced by nonnutrient factors that limit assimilatory capacity, for example, supply of light or CO_2. When carbon assimilation is affected by a number of environmental factors not related to nutrition, their effects on NUE are usually, at least qualitatively, independent of foliar nutrient concentrations. For example, very low photosynthetic photon flux density, sub- or supraoptimal temperatures, and desiccation will reduce the supply of energy to the assimilatory system, affect the relationship between rates of gross carbon assimilation and respiration, or reduce stomatal conductance. A reduction in stomatal conductance, also often caused by an increase in vapor deficit of the air, will limit the supply of CO_2. These effects will lower assimilatory activity compared to that occurring under optimal conditions. At least in the shorter term these limitations will not affect either foliar nutrient concentrations or the apportioning of nutrients to different compounds, and will reduce ANUE in proportion to the effect they have on carbon assimilation.

3. Water Use Efficiency Carbon assimilation depends on the assimilatory capacity of the mesophyll and on gas- and liquid-phase resistances to CO_2 diffusion from the bulk atmosphere to assimilation sites in the mesophyll. Transpiration depends on the difference in partial molar concentration of water vapor between the bulk air and sites of evaporation from a leaf, and on gas-phase resistances of that pathway. Transpiration efficiency will vary between zero and infinity for periods of net CO_2 uptake to when carbon assimilation is zero, and will be negative for periods of net CO_2 efflux. ATE will be altered by physiological responses that cause changes in assimilatory capacity or in stomatal conductance. Effects of these changes on carbon assimilation and on water use are summarized in Table I. ATE is also inversely proportional to evaporative demand if the stomata do not respond to this.

Effects on ATE of a change in carbon assimilation or in transpiration are clear. However, both of these are controlled by other variables, some of which affect both carbon assimilation and transpiration. The magnitude of effects on ATE of changes in these other variables is not always obvious, but can be easily evaluated by simple algebraic expressions. The rate of transpiration (E) can be defined in terms of the concentration of water vapor in ambient air (w_a) and at evaporation sites in a leaf (w_i), as well as diffusion resistances of the leaf boundary layer (r_b) and stomata (r_s):

$$E = \frac{w_i - w_a}{r_b + r_s}. \tag{1}$$

This does not take into account cuticular transpiration, which is usually insignificant when stomata are open and which is generally low in conifers.

Similarly, A can be defined in terms of the CO_2 concentrations of ambient air (c_a) and of intercellular air spaces in the leaf (c_i), as well as the above gas-phase resistances:

$$A = \frac{c_a - c_i}{1.35r_b + 1.6r_s}. \tag{2}$$

The values 1.35 and 1.6 allow for CO_2 having a smaller gas-phase diffusion coefficient compared to water vapor.

Thus,

$$\text{ATE} = \frac{(c_a - c_i)(r_b + r_s)}{(1.35r_b + 1.6r_s)(w_i - w_a)}. \tag{3}$$

It is clear that an increase in $w_i - w_a$ will lower ATE by 1.35 to 1.6 times more than the same change in $c_a - c_i$. Usually, $w_i - w_a$ is much greater than $c_a - c_i$ and has a greater possible range of values (0 to 0.01227 at 10°C and 0 to 0.04243 at 30°C for H_2O versus 0 to 0.00015 for CO_2). This means that atmospheric factors affecting humidity will often alter ATE more than those affecting CO_2.

Between sites where CO_2 dissolves in water and sites of carbon assimilation there is an appreciable resistance (r_m) to diffusion in the liquid phase, so

$$\text{ATE} = \frac{(c_a - c_m)(r_b + r_s)}{(1.35r_b + 1.6r_s + r_m)(w_i - w_a)}, \tag{4}$$

where c_m is the concentration of CO_2 at assimilatory sites. Because of this additional resistance, a change in gas-phase resistances will have a smaller effect on carbon assimilation than on transpiration. Thus, a change in r_b or in r_s (if not caused through a feedback loop by a change in carbon assimilation) will change ATE in the same direction as the change in resistance, but by less than the change in transpiration.

The technique of $\delta^{13}C$ analysis (Farquhar *et al.*, 1982) has potential for providing a reliable, simply obtained, integrated measurement of ATE for scales of leaf to whole plant. The stable isotope content of carbon in plant tissue is a useful indicator of ATE in certain agricultural crops (Farquhar and Richards, 1984; Hall *et al.*, 1990) and halophytes (Guy and Reid, 1986; Guy *et al.*, 1989), although the technique is unreliable for some species. Because of faster diffusion, and preferential uptake by ribulose bisphosphate carboxylase (Rubisco) of $^{12}CO_2$, $^{13}CO_2$ is used for carbon assimilation in smaller proportion than its concentration in the atmosphere, but the proportion increases as c_i declines. Thus, there is a positive correlation between the ratio of ^{13}C to ^{12}C in plant tissues and the diurnal ATE integrated over time. The technique is potentially most useful when aerodynamic conductances are large compared with those of the stomata, and when rates of carbon assimilation are high. However, the method does not take account of nocturnal transpiration, or properly account for nonstomatal factors that affect carbon assimilation. It therefore requires separate calibration for each species, and possibly for the same species in widely different environments.

Many plants grow in environments that, for at least a portion of a year, experience a shortage of water. However, they need to assimilate carbon to survive, grow, and reproduce. One way plants can enhance carbon assimilation under these conditions is to maximize carbon gained per unit water lost. Thus, it has been suggested that leaves operate to (tend to) maximize ATE, on both an instantaneous basis and a daily basis (e.g., Cowan, 1977). This has been examined further by Cowan and Farquhar (1977) and Farquhar and Sharkey (1982). It is apparent from field data that some species of conifers, for example, *Pinus sylvestris* (Berninger and Hari, 1993), optimize in this way, whereas others, for example, *Pinus taeda* (Fites and Teskey, 1988), do not.

4. Interactions between Resource Use Efficiencies When the environment is changing, ALUE and ANUE of a leaf will usually change in the same direction, except at low PPFDs, where carbon assimilation may be little affected by foliar nitrogen (see above). However, leaves of different species, those with different foliar [N], or those produced in different environments, especially in different light regimes, may have different relationships between PALUE and PANUE. They will respond differently to changes in PPFD because of differences in partitioning of nitrogen both within a leaf and within assimilatory pathways. For example, over the range of PPFD that produces a linear relationship between *A* and PPFD, ALUE determined on the basis of absorbed light does not vary greatly with foliar [N], and AN_nUE is lower at high foliar [N] (Fig. 3). However, foliage with greater [N] becomes light saturated, with a greater rate of carbon assimilation, at a higher PPFD. Thus, as PPFD increases, AN_nUE continues to increase and ALUE remains unchanged for longer in leaves with high [N]. The detail of the relationship between the ALUEs and the AN_nUEs of leaves with low and high [N]

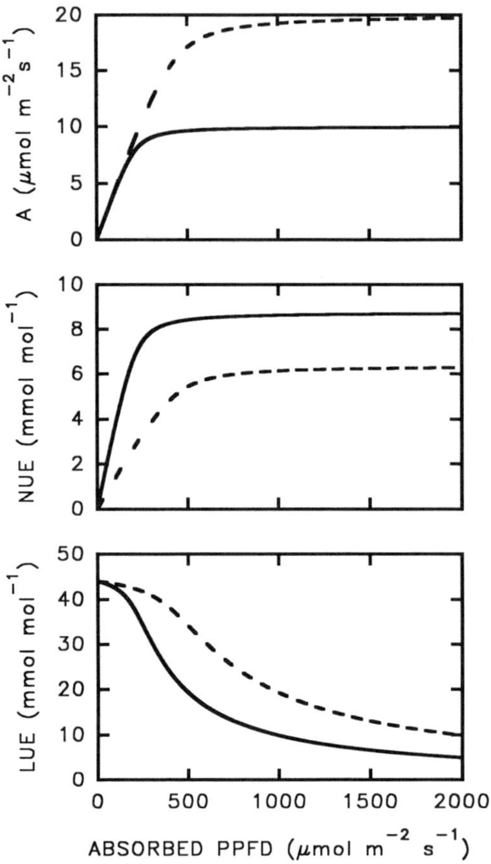

Figure 3 Simulated comparison of carbon assimilation (A), instantaneous nitrogen use efficiency (NUE), and instantaneous light use efficiency (LUE) of (———) a leaf with low [N] and (- - - -) a leaf with high [N]. The relationship between PPFD (I) and A was simulated using the nonrectangular hyperbola (Thornley and Johnson 1990): $A = \{\alpha I + A_{max} - [(\alpha I + A_{max})^2 - 4\theta\alpha I A_{max}]^{0.5}\}/(2\theta)$, with a quantum efficiency (α) of 0.044 mol (CO_2) mol^{-1} (PPFD), and θ of 0.95.

depends on the difference in [N] and on the response of A to [N], but a typical, simulated, scenario is shown in Fig. 3.

With a constant external environment, assimilatory capacity is determined primarily by foliar [N], whereas assimilation is affected by the supply of CO_2, and therefore altered by any effect of r_s on c_i (see Chapter 4, this volume). At constant foliar [N] an increase in r_s will increase ATE, but cause ANUE to decline.

If a leaf adjusts r_s to maximize ANUE, c_i will be low and near constant, the reciprocal of r_s will be proportional to leaf assimilatory capacity (related to nutrient concentration), and ATE will be smaller at greater [N]. A negative correlation of this kind has been observed between PATE and PAN_nUE for four of five species (Field *et al.*, 1983). On the other hand, if r_s was constant irrespective of foliar [N], c_i would be lower at greater [N]. Then, PATE and PAN_nUE would be positively associated over the range of N concentrations for which there was a negative relationship between [N] and c_i. Most leaves probably operate between these two extremes.

In some species stomatal conductance exhibits a strong negative response to a change in leaf-air vapor pressure difference (VPD), and this reduces the impact on transpiration of a change in VPD. An increase in VPD then increases r_s, so tending to maintain transpiration constant. The change in r_s reduces c_i or carbon assimilation, or both, and tends to reduce ANUE and maintain ATE.

A high RUE can "maximize" carbon assimilation in an environment wherein that resource is consistently limiting, although when more than one resource or other factors are limiting, there may be conflict between requirements for optimization. However, this is not necessarily a general rule. For example, a plant that has a large ATE as a result of a high rate of carbon assimilation is not necessarily a good water saver in a dry environment: the high ATE may result from a high assimilatory potential rather than from a low rate of transpiration. Also, as found with some annual and ephemeral species, high spending of water when it is available will enhance productivity and reproductive success. In a mixed community there is likely to be little advantage to a plant that is frugal in use of a resource if this makes the resource more available to neighbors that are able to use the resource.

C. Branch and Plant Levels

Instantaneous RUEs are not often determined at these levels, but can be obtained in a way similar to those at the leaf level. These RUEs will often relate more realistically to physiological activity under natural conditions than do values obtained for individual leaves, because the canopy environment is often more natural than is used for measurements on individual leaves.

Carbon assimilation, transpiration, and light interception are affected by changes in environmental variables over short periods of time and over short distances. To relate experimentally determined instantaneous RUEs of a branch or plant to those of leaves and to the environment in which measurements are made, effective values for variables in the environment of the branch or plant are needed. Because new levels of heterogeneity are added, this requires a different approach for a branch or tree than for a leaf. If there are linear relationships between compo-

Table II Factors That Affect the Availability or Demand for a Resource

RUE	Resource	Influencing factors
ALUE	Quantity of light intercepted	Leaf angle, shading, time of day, time of year
	Quantity of light absorbed	Quantity of chlorophyll per leaf area, quantity of light intercepted, reflectance of waxes, hairs, etc. on leaf surface
ANUE	Foliar nutrient concentration	Ability of plant to absorb nutrients, ability of site to supply nutrients, conflicting demands from other nutrient sinks within the plant
	Nutrient use in carbon assimilation	Partitioning of nutrient within the leaf
ATE	Transpiration	Amount of energy available to evaporate water, diffusive conductance of boundary layer, diffusive conductance of stomatal antechambers, diffusive conductance of stomata, permeability of leaf cuticle and waxes, vapor pressure difference between leaf and ambient air

nents of a RUE (e.g., carbon assimilation, transpiration, nutrient concentration, and incident light) and values of an environmental variable, effective values of components of the RUE are equivalent to those occurring in the mean branch environment. If this relationship is nonlinear, allowance must be made for the nonlinearity.

The environment of a branch or plant is different from that of a leaf, especially when the latter is artificially maintained in an environment that will maximize carbon assimilation. Differences will particularly affect the carbon assimilation, diffusive conductance, and light environment of foliage. Some generalizations can be made about these effects (see Table II), but physiological responses to the microenvironment will vary with species. Stomatal conductance may be affected, for example, by the influence of a branch versus a leaf's boundary layer on humidity, whereas the average light environment of foliage will depend on the amount of mutual shading on leaf angle, and on the transmittance and reflectance of light by individual leaves. The arrangement of foliage on a branch may affect leaf temperature, and this will affect physiological processes and potential rates of transpiration. In particular, closer packing of foliage will raise leaf temperature on illumination, and can shift this temperature toward the assimilatory optimum (see Chapter 4). Looser packing of foliage can similarly lower leaf temperature without expenditure of water in hot climates. Effects of scale discussed above apply to all RUEs, and are additional to those specific to any one RUE.

1. Light Use Efficiency At the branch and plant levels instantaneous ALUEs are obtained for foliage with very heterogeneous light environments. A measure of AL_aUE is difficult to interpret in terms of measurements on individual leaves, especially for conifer foliage, because of variation in light levels. Values of AL_iUE will be similar to those of the tree crown, taking into account the position of the branch within the crown (i.e., whether it has sun or shade foliage), the average leaf angle, and the angle of the sun.

2. Nutrient Use Efficiency Values of ANUE for a branch or plant will differ from those determined at the leaf level for a number of reasons. On a branch some leaves are shaded by others, so the average light intensity will be lower. This will often cause measured rates of carbon assimilation to be lower than those expected from values of incident light and measures of incident light. Derived values of ANUE at PPFDs that would saturate A of a leaf will thus be lower than PANUE of a single leaf. When the relationship between nutrient concentration and carbon assimilation is not linear, a measured mean foliar nutrient concentration may not indicate assimilatory capacity if nutrient concentrations of leaves vary appreciably. If total nutrient content of a branch or plant is used to derive ANUE, because average values of nutrient concentration are usually lower than the average of the foliage, ANUE will be larger than if calculated on the basis of foliar nutrient concentration. This will often be the case despite respiration of nonassimilatory tissue of a branch or plant reducing the effective value of net carbon assimilation. At levels of scale greater than that of a leaf, carbon assimilation by leaves has to provide the structure and maintenance of nonassimilating tissues. This will lower RUEs at these larger scales, and will often make more similar RUEs measured under different conditions (e.g., Sheriff and Nambiar, 1991).

3. Water Use Efficiency The WUE of a branch or plant can be determined as an ATE, similar to that discussed earlier for a leaf. Alternatively, it can be calculated as a WUE that includes evaporation of water intercepted by and condensed on leaves. Depending on the energy balance, which will usually differ in a gas-exchange chamber from that under ambient conditions, a greater boundary layer resistance of a foliage mass than of a leaf will tend to produce a greater ATE for a branch or plant than for an isolated leaf under otherwise similar conditions: the effect on the numerator of Eq. (4) will be greater than the effect on its denominator. Self-shading within a branch or plant will tend to reduce WUE or ATE compared with isolated leaves at the same incident PPFD.

4. Interactions between Resource Use Efficiencies Interactions at the branch and plant levels are qualitatively similar to those found for individual leaves, taking into account effects of variation in leaf microenvironment and, when integrating from the leaf level, overall effects of nonlinearities. An important factor, especially in natural environments (as opposed to gas-exchange cuvettes, where air is moved very rapidly over the foliage), is the higher resistance of the boundary layer of branches compared to that of leaves. This will lower rates of gas exchange and will lessen effects of a change in stomatal resistance, with considerably more effect on transpiration than on carbon assimilation because of the constraint on the latter of the series resistance r_m [see Eq. (4)].

5. Whole-Plant Resource Use Efficiency by Growth Analysis Growth analysis can be readily used to determine RUE at the whole-plant level, where resource use efficiency for plant growth is a function of a variety of factors interacting over significant periods of time. Growth analysis can provide integrated measures of plant performance using measurements of changes in biomass, plant morphology, and plant nutrient content.

Absolute growth indicates little about the vigor of plants, because this is greatly influenced by plant size (Brand *et al.*, 1987). A large plant may have greater growth than a smaller plant, even though the smaller plant is more vigorous and uses available resources more efficiently. To allow for effects of plant size, measures of relative growth (growth rate per unit of biomass) have been developed. The most common of these is the relative growth rate (RGR), generally expressed in grams/gram/time or as percent/time.

RGR can be divided into a series of terms that show different aspects of plant acclimation to changes in environmental conditions. The simplest case, that of classical growth analysis, divides RGR into a term indicating the efficiency of foliage for producing biomass under a given set of environmental conditions, unit leaf assimilation rate (ULAR); a term indicating biomass partitioning between photosynthetic and nonphotosynthetic tissues, the leaf weight ratio [LWR = (leaf biomass)/(plant biomass)]; and a term indicating foliar morphology, specific leaf area (SLA; m^2/gram):

$$\mathrm{RGR} = (1/W)(dW/dt) \tag{5a}$$
$$= \mathrm{ULAR} \times \mathrm{LWR} \times \mathrm{SLA} \tag{5b}$$
$$= (1/\mathrm{LA})(dW/dt)(\mathrm{LW/W})(\mathrm{LA/LW}), \tag{5c}$$

where W is plant biomass (dry weight), dW/dt is the instantaneous rate of change of plant biomass, LA is plant leaf area (fresh), and LW is leaf biomass (dry weight).

Alternatively, RGR can be divided into terms indicating resource use efficiency, resource availability, and biomass partitioning as described in Hunt *et al.* (1984). Such analyses can be used for determining ecophysiological responses of plants to changing environmental conditions. In the case of light, we have:

$$\begin{aligned} \mathrm{RGR} &= \mathrm{LUE(PPFD/LA)(SLA)} && \text{(6a)} \\ &= (\mathrm{LA}/W)(\mathrm{PPFD/LA})(1/\mathrm{PPFD}\ dW/dt), && \text{(6b)} \end{aligned}$$

where PPFD is in millimoles/m^2/time, LUE is in grams of growth/millimole (PPFD), and PPFD/LA = millimoles of light availability to a plant/m^2 of leaf area.

A similar analysis can be conducted for mineral nutrients:

$$\begin{aligned} \mathrm{RGR} &= \mathrm{NUE(Nu/RW)RWR} && \text{(7a)} \\ &= (1/\mathrm{Nu}\ dW/dt)(\mathrm{Nu/RW})(\mathrm{RW/W}), && \text{(7b)} \end{aligned}$$

where NUE is in grams of growth/gram of foliar nutrient/time, Nu/RW is an index of nutrient availability (grams of nutrient in foliage/gram of root), and RWR is root weight ratio (grams of root/gram of plant). Details on calculation of these indices are in the appendix of Margolis and Brand (1990).

Similar analysis for TE often requires long-term, laborious monitoring of plant water balance as well as growth. However, ATE can often be determined by the $\delta^{13}C$ technique, as discussed earlier. Alternatively, plant water balance is relatively easy to determine for potted plants, where losses in weight not due to drainage or mineral nutrient accumulation are assumed to show transpiration. When plants are not in pots, plant water balance may be calculated by modeling transpiration using the Penman–Monteith equation, although responses of stomatal conductance to many environmental variables must be well-known if accurate simulations are to be performed. This approach can give a value for transpiration or evapotranspiration (Monteith, 1965, 1973), and thus allows calculation of TE or WUE. The latter term includes water transpired by plants and that evaporated from surfaces of leaves, litter, soil, etc. wetted by precipitation.

A growth analysis equation can be constructed as follows:

$$\begin{aligned} \mathrm{RGR} &= \mathrm{TE(H_2O/RW)RWR} && \text{(8a)} \\ &= (1/\mathrm{H_2O}\ dW/dt)(\mathrm{H_2O/RW})(\mathrm{RW}/W), && \text{(8b)} \end{aligned}$$

where TE is grams of growth/gram of water transpired and H_2O/RW is grams of water transpired/gram of root.

Although Eqs. (5)–(8) are not identical, their examination quickly reveals strong similarities in the importance to plant growth of the resources carbon, light, nutrients, and water.

a. Resource Use Efficiency of Conifer Seedlings in Response to Silvicultural Practices Resource use efficiencies of conifer plants are of considerable economic importance to improving silvicultural treatments designed to improve growth and survival. Conifer seedlings are the main stock for commercial plantations in many climates. Large-scale reforestation programs have also been seriously considered as a means of reducing the increase in atmospheric CO_2 (Price, 1990; Moulton and Richards, 1990).

We here examine indices of resource use efficiency for an experimental white pine (*Pinus strobus* L.) plantation in eastern Canada (Margolis and Brand, 1990) by growth analysis of whole-plant performance. Treatments designed to manipulate availability of environmental resources to seedlings were applied in a replicated factorial design. Three levels of soil surface modification were undertaken to increase soil temperature. Two levels are reported here: T_0 = no site preparation; T_1 = scarification of organic horizons and installation of black fiberglass sheets around the seedlings to further increase soil temperature. There were two levels of fertility: F_0 = no fertilization and F_1 = fertilized with slow-release fertilizer at 30 g/seedling; there were also two levels of brush control: B_0 = no brush control and B_1 = all competing vegetation removed from around the seedlings. Further details are in Brand and Janas (1988), Brand (1990), Margolis and Brand (1990), and Brand (1991).

The greatest response of RGR was to soil surface modification, an increase of 81% (Table III). Fertilization and brush control resulted in only small increases in RGR. Lower soil temperature was associated with a lower LUE, even though light availability (PPFD/LA) did not change. Lower LUE was accompanied by a greater SLA. Negative effects of low soil temperature on RGR and LUE were buffered by adjustments in plant morphology. Low soil temperatures have been shown to affect carbon assimilation adversely (Jurik *et al.*, 1988; Day *et al.*, 1989), and tend to result in symptoms associated with carbon limitation, e.g., lower starch levels, higher shoot-to-root ratios, and thinner leaves.

The primary effect of fertilizer was to increase LUE, likely caused by the positive effect of increased foliar nitrogen on photosynthesis (Table III). Brush control resulted in a 20% increase in light availability and an increase in water and nutrient availability (Brand and Janas, 1988), but had no effect on LUE or on the amount of carbon allocated to foliage (SLA) (Table III). The growth response to brush control can be attributed solely to an increase in the availability of light, and not to an increase in its efficiency of use.

Table III Growth Analysis of Light Use Efficiency and Associated Indices for White Pine[a]

Treatment[b]	Parameter[c] RGR (%/yr)	LUE (g/mol/yr)	PAR/LA (mol/m²)	SLA (mm²/g)
T_0	0.42 ± 0.07	1.62 ± 0.26	23.01 ± 1.44	12,300 ± 210
T_1	0.76 ± 0.06	3.26 ± 0.29	22.54 ± 0.89	10,900 ± 170
F_0	0.56 ± 0.05	2.25 ± 0.21	23.39 ± 1.06	11,600 ± 180
F_1	0.64 ± 0.07	2.78 ± 0.30	21.86 ± 0.73	11,400 ± 200
B_0	0.56 ± 0.07	2.53 ± 0.30	2058 ± 81	11,700 ± 220
B_1	0.64 ± 0.05	2.50 ± 0.31	2468 ± 83	11,300 ± 150

[a] Data are for seedlings one growing season after planting following manipulation of environmental conditions. Adapted from Margolis and Brand (1990), with permission.

[b] T_0, No site preparation; T_1, scarification of organic horizons and installation of black fiberglass sheets around the seedlings to further increase soil temperature; F_0, no fertilization; F_1, fertilized with 30 g/seedling of slow-release fertilizer; B_0, no brush control; B_1, all competing vegetation removed from around the seedlings. Means ± one standard error; $n = 48$ for T_0 and T_1, $n = 72$ for F_0, F_1, B_0, and B_1.

[c] RGR, Relative growth rate; LUE, light use efficiency; PAR/LA, light availability (PAR radiation per unit leaf area); SLA, specific leaf area.

The nitrogen response of these white pine seedlings is shown in Table IV. Seedlings growing in the unmodified soil surface had lower NUE and less relative allocation to root growth (RWR) than those in the warmer soil environment. However, this decrease in allocation to root growth was buffered by an increase in their relative ability per unit of root to supply N to foliage (N/RW). Although there were positive effects

Table IV Growth Analysis of Nitrogen Use Efficiency and Associated Indices for White Pine[a]

Treatment[b]	Parameter[c] RGR (%/yr)	NUE (g/g/yr)	N/RW (mg/g)	RWR (g/g)
T_0	0.42 ± 0.07	53 ± 0.8	36 ± 1	0.25 ± 0.001
T_1	0.76 ± 0.06	102 ± 0.9	31 ± 1	0.30 ± 0.001
F_0	0.56 ± 0.05	77 ± 0.7	32 ± 1	0.28 ± 0.006
F_1	0.64 ± 0.07	80 ± 0.8	34 ± 1	0.28 ± 0.007
B_0	0.56 ± 0.07	80 ± 0.9	31 ± 1	0.28 ± 0.007
B_1	0.64 ± 0.05	78 ± 0.5	34 ± 1	0.27 ± 0.005

[a] Data are for seedlings one growing season after planting, following manipulation of environmental conditions. Adapted from Margolis and Brand (1990), with permission.

[b] As in Table III.

[c] RGR, relative growth rate; NUE, nitrogen use efficiency; N/RW, index of nitrogen availability (foliar nitrogen per unit root weight); RWR, root weight ratio.

of scarification on N availability in the first year, after 4 years there was a decrease in nutrient availability (Munson *et al.*, 1993). The fertilization and brush control treatments did not affect NUE, although a small increase in the ability of unit root to supply nitrogen to foliage was apparent.

Thus, effects of environmental conditions on resource use efficiency at the whole-plant level cannot be understood without considering the effects of environmental changes on resource availability and morphological acclimation. A decrease or increase in the availability of a resource may or may not result in changes in the use efficiency of that resource, because other resources may have an overriding effect. A decrease in resource availability that affects resource use efficiency may be buffered by changes such as morphological acclimations of the plant, e.g., thinner leaves, relatively less foliage biomass, relatively more root biomass. Different species or genotypes can exhibit greatly different capacities to buffer changes in the availability of environmental resources through morphological and physiological acclimation (Brand, 1990). These differences are the most important factors determining the relative ability of different species or genotypes to resist increasing environmental stress, as well as their ability to respond to silvicultural manipulations designed to reduce such stress levels.

D. Forest Community/Ecosystem Levels

1. Ecosystem Efficiency Although the emphasis of this chapter is on RUEs at levels of individual trees and smaller, several aspects of RUEs at forest community and ecosystem levels are important. Prominent among these is the increase in spatial and temporal scales of processes and driving variables that parallels any increase in these scales of the system being investigated. RUEs of complex systems include overstory and understory canopy layers and all stages of tree development. They involve not only the utilization of resources for growth of individual trees and understory plants, but also processes such as mortality of plants and plant organs, and decomposition of detritus. Thus, RUEs must be examined in the broader context of integration, with the behavior of the system as a whole reflecting all of the organisms and processes that make up the system. However, it is not always necessary to understand the detail of all these processes.

A measure of ecosystem efficiency is the ratio of net ecosystem productivity (NEP) to GPP (Waring and Schlesinger, 1985). NEP is determined by subtracting heterotrophic respiration from NPP. The condition of a forest ecosystem is reflected in this ratio. In rapidly growing forests the ratio of NEP/GPP is high, whereas in degrading forests the ratio is lower, as discussed below. Nutrient use efficiencies may be expressed in several ways. When reported as NPP per unit of nutrient lost from the vegetation (which is replaced by nutrient uptake), conifer ecosystems tend to have high efficiencies compared with deciduous ecosystems (Waring and Schlesinger, 1985). This difference is attributed at least in part to conifers having a much lower foliage turnover rate and reduced nutrient cycling.

2. Exchanging Abundant Light for Limiting Water: Physiological Adjustments in Resource Use Efficiency in Response to Thinning One of the most abrupt changes in environmental conditions in managed conifer forests are those resulting from thinning. In such cases, levels of certain environmental resources, such as light, increase dramatically, while other resources, such as the supply of water to foliage of each tree to meet an increase in transpirational demand, may become more limiting. Bloom *et al.* (1985) suggest that plant response to resource limitation follows an economic analogy whereby plants have a tendency to exchange more abundant environmental resources in order to obtain a greater supply of more limiting environmental resources.

Thinned and unthinned balsam fir [*Abies balsamea* (L.) Mill.] saplings showed a negative exponential relationship between light availability (PPFD) and LUE (Fig. 4) (Pothier and Margolis, 1991). At light levels less than 200 MJ/m^2/sec, LUE increased dramatically with PPFD. On the other hand, thinned trees exposed to light levels greater than 400 MJ/m^2/sec had low LUE. These increased radiation loads and the consequent changes in vapor pressure deficits increased transpirational demand on foliage (Whitehead and Jarvis, 1981; Pothier and Margolis, 1990). Thus, for thinned trees, water becomes a more limiting environ-

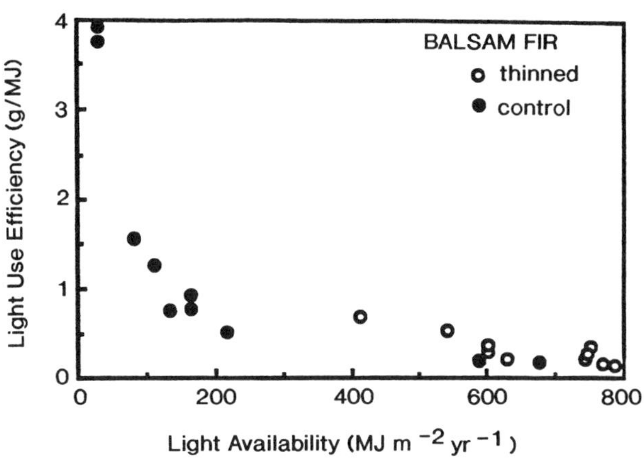

Figure 4 Relation between light use efficiency and light availability of thinned and control balsam fir trees. Reprinted from Pothier and Margolis (1991), with permission.

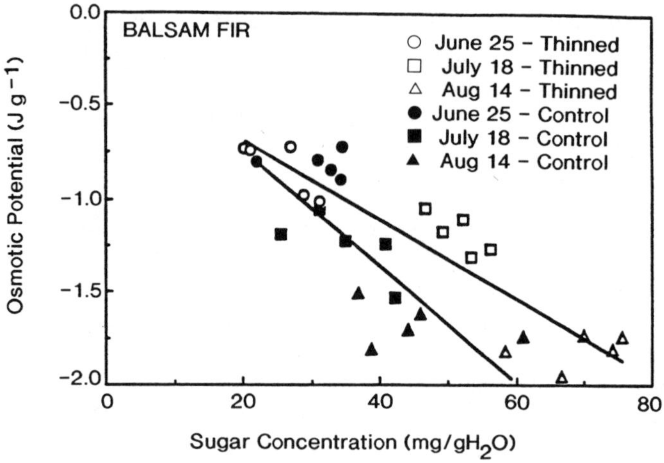

Figure 5 Relation between osmotic potential of needles at full saturation ($y_{p,sat}$) and sugar concentration (per gram of H_20 in needles) in needles for thinned and control balsam fir trees for three sampling dates during 1989. Lines correspond to simple linear regressions between $y_{p,sat}$ and sugar concentration for thinned and control trees ($R^2 = 0.85$ and 0.48, respectively). Reprinted from Pothier and Margolis (1990), with permission.

mental resource than light. One mechanism by which these trees can barter abundantly available light to obtain a greater supply of more limiting water is to use the increased light levels to increase carbon assimilation, and hence leaf sugar concentration. Greater sugar concentrations result in lower leaf osmotic potentials, and thus increase the ability of leaves to extract water from the soil (Fig. 5). The different slopes for the relationship between osmotic potential and sugar concentration of thinned and unthinned trees suggest that the contribution of sugars to the osmotic potential was larger in the thinned than in the unthinned trees at osmotic potentials below −1.0 J/g. Given that 40 mmol sugar/g of H_2O is equivalent to 0.1 J/g (Munns and Weir, 1981), sugars accounted for 53% of the osmotic potential in thinned trees, but for only 38% in unthinned trees in August. Thus, by increasing sugar concentrations to lower leaf osmotic potentials, thinned trees seem capable of exchanging abundantly available light for a greater supply of more limiting water.

3. Growth Efficiency The ratio of (biomass of wood produced)/(foliage area) has been widely used as an index of "growth efficiency." This is a holistic index of the efficiency of resource use by foliage to produce stem biomass. It will vary geographically as availability of resources changes, and with age both of an individual and of a single-aged stand as competition between individuals changes. Especially in the single-aged stands usually found in plantations, growth efficiency declines with time (e.g., Waring and Schlesinger, 1985). Kaufmann and Ryan (1986) examined growth in relation to potential absorbed radiation (irradiance × projected leaf area) for Engelmann spruce, subalpine fir, and lodgepole pine. They found that growth efficiency declined from the youngest to oldest ages in all species. Subsequent work by Ryan and Waring (1992) on lodgepole pine indicated that only a portion of the decline could be attributed to increased maintenance respiration in larger trees, and recent work by B. Yoder *et al.* (personal communication) shows that maximum net carbon assimilation is lower in old ponderosa and lodgepole pine trees than in younger trees, largely as a result of lower leaf conductance.

Reanalysis of data of Beets and Pollock (1987) for a *Pinus radiata* plantation in the north island of New Zealand shows a steady initial decline in growth efficiency that is reversed suddenly by thinning (Fig. 6). These effects parallel the greater limitations in availability of some resources that would accompany an increase in stand biomass of a vigorous plantation, and the great increase in availability per tree of some resources that would follow thinning. It is probable that shortage of one or more resources will decrease some RUEs. For example, the average PPFD in-

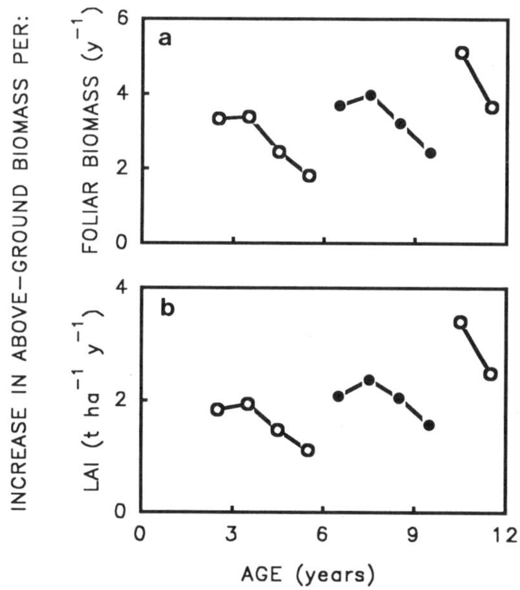

Figure 6 Increase in aboveground biomass per unit of foliage (a) mass (b) area (growth efficiency) of a *Pinus radiata* stand. The change from open to filled and from filled to open symbols indicates times of thinning. Calculated from data of Beets and Pollock (1987).

cident on unit area of foliage declines as a canopy closes, and increases dramatically on thinning.

A more specific example in forests of limited supply of some resources affecting the efficiency of use of others is the limitation of aboveground LUE of *P. radiata* by restricted water and nutrient supplies (Raison and Myers, 1992).

IV. Discussion

Values for RUEs have undoubted use in increasing our understanding of ecosystems and of how members of a species respond to biotic and to abiotic factors in their environment. However, their incautious use in attempting an understanding of parameters such as productivity, and in scaling in space and time, can be misleading. For example, increasing one or more RUEs in the way we currently define them may not be a very important factor for survival or reproduction of a plant. For example, a greater instantaneous ANUE at the leaf level may result in a higher ANUE for the plant, unless foliage area is much reduced. However, effects at the plant level will be smaller than for a leaf because of "dilution" by other factors (e.g., Sheriff and Nambiar, 1991). Maximizing ANUE may not be an overriding consideration for either productivity or competitive success. At the canopy level it may be advantageous to have low foliar nutrient concentrations, and thus to sacrifice leaf ANUE, in order to increase foliage area. An advantage of this would be greater light interception, which would produce greater shading of competitors. This could increase a plant's ability to compete for light, or make it better able to live in a low-light environment, although often at the cost of incorporating more carbon in leaf structure. This would also often increase both carbon assimilation by a plant and its ALUE if these are calculated on the basis of ground area.

Use of the same terminology for RUEs at different scales of space or time, or both, can be misleading and may result in poor communication. Further, at different scales of space and time the quantities being measured differ, so that using the same terminology at different scales can result in poor definition.

Scaling of values of RUEs over several levels of scale of space or time can result in misleading conclusions unless great care is taken to vary appropriately the number of variables included in an analysis. Many more factors have to be included to determine values successfully at several levels greater of scale or time.

Although current definitions of RUEs are often useful, many of them do not clearly address the response by a plant to availability of a resource, even though, as pointed out earlier, availability may alter RUEs over evolutionary time. In the shorter term, low availability of, for example, nitrogen in the rooting medium can, by changing physiological activity, reduce gas exchange (e.g., Radin and Ackerson, 1981; Krauss, 1978) and shoot growth (e.g., Krauss and Marschner, 1982) independently of any effects of foliar nitrogen concentration. Similarly, gas exchange and plant growth can be reduced by dry soil, independently of plant water content or water potential (e.g., Munns, 1992; Passioura, 1988; Turner, 1986). In the longer term, shortage of one or more resources in the environment can change allocation and partitioning of carbon, and probably of other resources, and this will make the plant's ability to assimilate carbon different, usually smaller, from that of a plant with adequate resources. For example, shortage of water or nutrients in the soil will often increase the partitioning of biomass to root growth (e.g., Sands and Mulligan, 1990). Alterations to RUEs and to carbon assimilation can also be expected from different stages of ontogeny and phenology.

Notation of Symbols

A	Rate of carbon assimilation per leaf area or per leaf mass
A_{max}	Rate of carbon assimilation per leaf area or per leaf mass when carbon assimilation is not limited
ALUE	Assimilatory light use efficiency (rate of carbon assimilation per unit incident or absorbed light)
AL_aUE	Assimilatory use efficiency of absorbed light (rate of carbon assimilation per unit absorbed light)
AL_iUE	Assimilatory use efficiency of incident light (rate of carbon assimilation per unit incident light)
ANUE	Assimilatory nutrient use efficiency (rate of carbon assimilation per unit nutrient content of foliage)
AN_nUE	Assimilatory nitrogen use efficiency (rate of carbon assimilation per unit nitrogen of foliage)
ATE	Assimilatory transpiration efficiency (rate of carbon assimilation per unit water transpired)
c_a	Concentration of carbon dioxide in ambient air
c_m	Concentration of carbon dioxide at assimilatory sites in the leaf mesophyll
c_i	Concentration of carbon dioxide in leaf air spaces
E	Transpiration rate
GPP	Gross primary productivity (carbon assimilation)
LA	Leaf area
LN_nUE	Litterfall-nitrogen use efficiency (inverse of litterfall [N])
LSW	Leaf specific weight (leaf weight per unit leaf area)
LUE	Light use efficiency (rate of biomass increase per unit incident or absorbed light)
LW	Leaf weight

LWR	Leaf weight ratio (leaf biomass per plant biomass)
NEP	Net ecosystem productivity (primary productivity of an ecosystem − heterotrophic respiration)
NPP	Net primary productivity (carbon assimilation − respiration)
NUE	Nutrient use efficiency (rate of biomass increase per unit nutrient content of plant or foliage)
Nu	Nutrient
Nu/R	Index of nutrient availability (nutrient in foliage per root biomass)
PANUE	Potential assimilatory nutrient use efficiency; rate of carbon assimilation when variables, other than possibly nutrient concentrations, are not limiting
PAN_nUE	Potential assimilatory nitrogen use efficiency; rate of carbon assimilation when variables, other than possibly nitrogen concentrations, are not limiting
PPFD	Photosynthetic photon flux density
r_b	Boundary layer diffusion resistance
r_m	Mesophyll resistance to CO_2
r_s	Stomatal diffusion resistance
RGR	Relative growth rate (increase in mass per unit mass per time)
RUE	Resource use efficiency (rate of carbon assimilation or biomass increase per unit resource "used")
RW	Root biomass
RWR	Root weight ratio (root biomass per plant biomass)
SLA	Specific leaf area (leaf area per unit leaf weight)
TE	Transpiration efficiency (rate of biomass increase per unit water transpired)
ULAR	Unit leaf assimilation rate
W	Plant biomass
w_a	Concentration of water vapor in ambient air
w_i	Concentration of water vapor at evaporation sites in a leaf
WUE	Water use efficiency (rate of biomass increase per unit water "used")

Acknowledgments

We thank Dr. Paul Kriedemann, Division of Forestry, CSIRO, Canberra, ACT, Australia, and Dr. Chris Beadle, Division of Forestry, CSIRO, Hobart, Tasmania, Australia, for their helpful comments on the manuscript. Particular thanks to Dr. Bob Teskey, Visiting Fellow, Plantation Forest Research Centre, Division of Forestry, CSIRO, Mount Gambier, South Australia, for helpful comments on the manuscript and discussions with D.W.S., assisting in efficient use of intellectual resources.

References

Beets, P. N., and Pollock, D. S. (1987). Accumulation and partitioning of dry matter in *Pinus radiata* as related to stand age and thinning. *N. Z. J. For. Sci.* 17:246–271.

Berendse, F., and Aerts, R. (1987). Nitrogen-use-efficiency: A biologically meaningful definition? *Funct. Ecol.* 1:293–296.

Berninger, F., and Hari, P. (1993). Optimal regulation of gas exchange: Evidence from field data. *Ann. Bot.* 71:135–140.

Bloom, A. J., Chapin III, F. S., and Mooney, H. A. (1985). Resource limitations in plants—An economic analogy. *Annu. Rev. Plant Physiol.* 11:141–146.

Brand, D. G. (1990). Growth analysis of responses by planted white pine and white spruce to changes in soil temperature, fertility, and brush competition. *For. Ecol. Manage.* 30: 125–138.

Brand, D. G. (1991). The establishment of boreal and sub-boreal conifer plantations: An integrated analysis of environmental conditions and seedling growth. *For. Sci.* 37: 68–100.

Brand, D. G., and Janas, P. S. (1988). Growth and acclimation of planted white pine and white spruce to changes in response to environmental conditions. *Can. J. For. Res.* 18: 320–329.

Brand, D. G., Weetman, G. F., and Rehsler, P. R. (1987). Growth analysis of perennial plants: The relative production rate and its yield components. *Ann. Bot.* 59: 45–53.

Chapin III, F. S., and Kedrowski, R. A. (1983). Seasonal changes in nitrogen and phosphorus fractions and autumnal retranslocation in evergreen and deciduous taiga trees. *Ecology* 64: 376–391.

Cowan, I. R. (1977). Stomatal behaviour and environment. *Adv. Bot. Res.* 4: 117–228.

Cowan, I. R., and Farquhar, G. D. (1977). Stomatal function in relation to leaf metabolism and environment. *Soc. Exp. Bot. Symp.* 31:471–505.

Day, T. A., DeLucia, E. H., and Smith, W. K. (1989). Influence of cold soil and snowcover on photosynthesis and leaf conductance in two Rocky Mountain conifers. *Oecologia* 80: 546–552.

DeJong, T. M., and Doyle, J. F. (1985). Seasonal relationships between leaf nitrogen content (photosynthetic capacity) and leaf canopy light exposure in peach (*Prunus persica*). *Plant, Cell Environ.* 8: 701–706.

Evans, J. R. (1989). Photosynthesis and nitrogen relationships in leaves of C_3 plants. *Oecologia* 78: 9–19.

Farquhar, G. D., and Kirschbaum, M. U. F. (1985). Environmental constraints on carbon assimilation. *In* "Regulation of Sources and Sinks in Crop Plants" (B. Jeffcoat, A. F. Hawkins, and A. D. Stead, eds.), pp. 87–97. British Plant Growth Regulator Group Monograph No. 12, Long Ashton, Bristol.

Farquhar, G. D., and Richards, R. A. (1984). Isotopic composition of plant carbon correlates with water use efficiency of wheat genotypes. *Aust. J. Plant Physiol.* 11: 539–552.

Farquhar, G. D., and Sharkey, T. D. (1982). Stomatal conductance and photosynthesis. *Annu. Rev. Plant Physiol.* 33: 317–345.

Farquhar, G. D., O'Leary, M. H., and Berry, J. A. (1982). On the relationship between carbon isotope discrimination and the intercellular carbon dioxide concentration in leaves. *Aust. J. Plant Physiol.* 9: 121–137.

Field, C. (1983). Allocating leaf nitrogen for the maximisation of carbon gain: Leaf age as a control on the allocation program. *Oecologia* 56: 341–347.

Field, C., and Mooney, H. A. (1986). The photosynthesis nitrogen relationship in wild plants. *In* "On the Economy of Plant Form and Function" (T. J. Givnish, ed.), pp. 25–55. Proceedings of the Sixth Maria Moors Cabot Symposium, Cambridge Univ. Press, Cambridge.

Field, C., Merino, J., and Mooney H. A. (1983). Compromises between water-use efficiency in five species of California evergreens. *Oecologia* 60: 384–389.

Fites, J. A., and Teskey, R. O. (1988). CO_2 and water vapour exchange of *Pinus taeda* in relation to stomatal behaviour: Test of an optimization hypothesis. *Can. J. For. Res.* 18: 150–157.

Gerwick, B. C. (1982). Response of terrestrial plants to mineral nutrients. *In* "C. R. C. Handbook of Biosolar Resources" (O. R. Zaborsky, ed.), pp. 213–222. The Chemical Rubber Company, Cleveland, Ohio.

Gower, S. T., and Richards, J. H. (1990). Larches: Deciduous conifers in an evergreen world. *BioScience* 40: 818–826.

Gulmon, S. L., and Chu, C. C. (1981). The effects of light and nitrogen on photosynthesis, leaf characteristics, and dry matter allocation in the chaparral shrub, *Diplacus aurantiacus*. *Oecologia* 49:207–212.

Guy, R. D., and Reid, D. M. (1986). Photosynthesis and the influence of CO_2-enrichment on $\delta^{13}C$ values in a C_3 halophyte. *Plant, Cell Environ.* 9:65–72.

Guy, R. D., Warne, P. G., and Reid, D. M. (1989). Stable carbon isotope ratio as an index of water-use efficiency in C_3 halophytes—Possible relationship to strategies for osmotic adjustment. *In* "Stable Isotopes in Ecological Research: Ecological Studies 68. Proceedings of an International Workshop, UCLA, April 1986" (P. W. Rundel, J. R. Ehleringer, and K. A. Nagy, eds.), pp. 55–75. Springer-Verlag, Berlin.

Hall, A. E., Mutters, R. G., Hubick, K. T., and Farquhar, G. D. (1990). Genotypic differences in carbon isotope discrimination by cowpea under wet and dry field conditions. *Crop Sci.* 30:300–305.

Herwitz, S. R., and Gutterman, Y. (1990). Biomass production and transpiration efficiencies of eucalypts in the Negev Desert. *For. Ecol. Manage.* 31:81–90.

Hunt, R., Warren-Wilson, J., Hand, D. W., and Sweeney, D. G. (1984). Integrated analysis of growth and light interception in winter lettuce. I. Analytical methods and environmental indices. *Ann. Bot.* 54:743–757.

Jarvis, P. G., and McNaughton, K. G. (1986). Stomatal control of transpiration: Scaling up from leaf to region. *Adv. Ecol. Res.* 15:1–49.

Jurik, T. W., Briggs, G. M., and Gates, D. M. (1988). Springtime recovery of photosynthetic activity in white pine in Michigan. *Can. J. Bot.* 66:138–141.

Kaufmann, M. R. and Ryan, M. G. (1986). Physiographic, stand, and environmental effects on individual tree growth and growth efficiency in subalpine forest. *Tree Physiol.* 2:47–59.

Krauss, A. (1978). Tuberization and abscisic acid content in *Solanum tuberosum* as affected by nitrogen nutrition. *Potato Res.* 21:183–192.

Krauss, A., and Marschner, H. (1982). Influence of nitrogen nutrition, daylength and temperature on contents of gibberellic and abscisic acid on tuberization in potato plants. *Potato Res.* 25:13–21.

Küppers, M. (1984). Carbon relations and competition between woody species in a Central European hedgerow. II. Stomatal responses, water use, and hydraulic conductivity in the root/leaf pathway. *Oecologia* 64:344–354.

Küppers, M., Koch, G., and Mooney, H. A. (1988). Compensating effects to growth of changes in dry matter allocation in response to variation in photosynthetic characteristics induced by photoperiod, light and nitrogen. *Aust. J. Plant Physiol.* 15:287–297.

Linder, S., and Rook, D. A. (1984). Effects of mineral nutrition on carbon dioxide exchange and partitioning in trees. *In* "Nutrition of Plantation Forests" (G. D. Bowen and E. K. S. Nambiar, eds.), pp. 211–236. Academic Press, London.

Margolis, H. A., and Brand, D. G. (1990). An ecophysiological basis for understanding plantation establishment. *Can. J. For. Res.* 20:375–390.

Monk, C. D. (1966). An ecological significance of evergreeness. *Ecology* 47:504–505.

Monteith, J. L. (1965). Evaporation and environment. *Symp. Soc. Exp. Biol.* 19:205–236.

Monteith, J. L. (1973). "Principles of Environmental Physics." Arnold, London.

Moulton, R. J., and Richards, K. R. (1990). Costs of sequestering carbon through tree planting and forest management in the United States. USDA For. Service Gen. Tech. Rep. WO-58.

Munns, R. (1992). A leaf elongation assay detects an unknown growth inhibitor in xylem sap from wheat and barley. *Aust. J. Plant Physiol.* 19:127–135.

Munns, R., and Weir, R. (1981). Contribution of sugars to osmotic adjustment in elongating and expanding zones of wheat leaves during moderate water deficits at two light levels. *Aust. J. Plant Physiol.* 8:93–105.

Munson, A. D., Margolis, H. A., and Brand, D. G. (1993). Intensive silvicultural treatment: Impacts on soil fertility and planted conifer response. *Soil Sci. Soc. Am. J.* 57:246–255.

Nambiar, E. K. S., and Fife, D. N. (1991). Nutrient retranslocation in temperate conifers. *Tree Physiol.* 9:185–207.

Passioura, J. B. (1988). Root signals control leaf expansion in wheat seedlings growing in drying soil. *Aust. J. Plant Physiol.* 15:687–693.

Pothier, D., and Margolis, H. A. (1990). Changes in the water relations of balsam fir and white birch saplings after thinning. *Tree Physiol.* 6:371–380.

Pothier, D., and Margolis, H. A. (1991). Analysis of growth and light interception of balsam fir and white birch saplings following precommercial thinning. *Ann. Sci. For.* 48:123–132.

Price, C. (1990). The allowable burn effect: Does carbon-fixing offer a new escape from the bogey of compound interest. *For. Chron.* 66:572–578.

Radin, J. W., and Ackerson, R. C. (1981). Water relations of cotton plants under nitrogen deficiency. III. Stomatal conductance. *Plant Physiol.* 67:115–119.

Raison, R. J., and Myers, B. J. (1992). The Biology of Forest Growth experiment: Linking water and nitrogen availability to the growth of *Pinus radiata*. *For. Ecol. Manage.* 52: 279–308.

Reich, P. B., Walters, M. B., and Tabone, T. J. (1989). Response of *Ulmus americana* seedlings to varying nitrogen and water status. II. Water and nitrogen-use efficiency in photosynthesis. *Tree Physiol.* 5:173–184.

Reich, P. B., Walters, M. B., and Ellsworth, D. S. (1992). Leaf life-span in relation to leaf, plant and stand characteristics among diverse ecosystems. *Ecol. Monogr.* 62:365–392.

Reich, P. B., Koike, T., Gower, S. T., and Schoettle, A. W. (1994). Causes and Consequences of Variation in Conifer Leaf Life-Span. *In* "Ecophysiology of Coniferous Forests" (W. K. Smith and T. M. Hinckley, eds.). Academic Press, San Diego. In press.

Rundel, P. W. (1982). Nitrogen utilization efficiencies in Mediterranean-climate shrubs of California and Chile. *Oecologia* 55:409–413.

Ryan, M. G., and Waring, R. H. (1992). Maintenance respiration and stand development in a subalpine lodgepole pine forest. *Ecology* 73:2100–2108.

Sands, R., and Mulligan, D. R. (1990). Water and nutrient dynamics and tree growth. *For. Ecol. Manage.* 30:91–111.

Schlesinger, W. H., DeLucia, E. H., and Billings, W. D. (1989). Nutrient-use efficiency of woody plants on contrasting soils in the Western Great Basin, Nevada. *Ecology* 70: 105–113.

Sheriff, D. W. (1992). Nitrogen nutrition, growth and gas exchange of *Eucalyptus camaldulensis* and *Eucalyptus globulus* seedlings. *Aust. J. Plant Physiol.* 19:637–652.

Sheriff, D. W., and Nambiar, E. K. S. (1991). Nitrogen nutrition, growth and gas exchange in *Eucalyptus globulus* Labill. seedlings. *Aust. J. Plant Physiol.* 18:37–52.

Sinclair, R., and Horie, T. (1989). Leaf nitrogen, photosynthesis, and crop radiation use efficiency: A review. *Crop Sci.* 29:90–98.

Small, E. (1972). Photosynthetic rates in relation to nitrogen recycling as an adaptation to nutrient deficiency in peat bog plants. *Can. J. Bot.* 50:2227–2233.

Son, Y., and Gower, S. T. (1991). Aboveground nitrogen and phosphorus use by five plantation-grown trees with different leaf longevities. *Biogeochemistry* 14:167–191.

Squire, R. O., Attiwill, P. M., and Neales, T. F. (1987). Effects of changes of available water and nutrients on growth, root development and water use of *Pinus radiata* seedlings. *Aust. For. Res.* 17:99–111.

Thornley, J. H., and Johnson, I. R. (1990). "Plant and Crop Modelling. A Mathematical Approach to Plant and Crop Physiology." Clarendon Press, Oxford.

Turner, N. C. (1986). Adaptation to water deficits: A changing perspective. *Aust. J. Plant Physiol.* 13:175–190.

Urban, D. L., O'Neil, R. V., and Shugart, H. H., Jr. (1987). Landscape ecology—A hierarchical perspective can help scientists understand spatial patterns. *BioScience* 37: 119–127.

Van der Werf, A. (1993). "Fast- and slow-growing monocots. The effect of N-supply on respiration, photosynthesis and optimization of carbon and nitrogen allocation." Ph. D. Thesis, University of Utrecht, The Netherlands.

Virgona, J. M., Hubrick, K. T., Rawson, H. M., Farquhar, G. D., and Downes, R. W. (1990). Genotype variation in transpiration efficiency, carbon-isotope discrimination and carbon allocation during early growth in sunflower. *Aust. J. Plant Physiol.* 17:207–214.

Vitousek, P. M. (1982). Nutrient cycling and nutrient use efficiency. *Am. Nat.* 119: 553–572.

Waring, R. H., and Schlesinger, W. H. (1985). The carbon balance of trees. *In* "Forest Ecosystems Concepts and Management" (R. H. Waring and W. H. Schlessinger, eds.), pp. 7–37. Academic Press, Orlando, Florida.

Whitehead, D., and Jarvis, P. G. (1981). Coniferous forests and plantations. *In* "Water Deficits and Plant Growth Volume 6" (T. T. Kozlowski, ed.), pp. 49–152. Academic Press, New York.

6

Source–Sink–Storage Relationships of Conifers

R. J. Luxmoore, R. Oren, D. W. Sheriff, and R. B. Thomas

I. Introduction

Irradiance, air temperature, saturation vapor pressure deficit, and soil temperature vary in association with Earth's daily rotation, inducing significant hourly changes in the rates of plant physiological processes. These processes include carbon fixation in photosynthesis, sucrose translocation, and carbon utilization in growth, storage, and respiration. The sensitivity of these physiological processes to environmental factors such as temperature, soil water availability, and nutrient supply reveals differences that must be viewed as an interactive whole in order to comprehend whole-plant responses to the environment. Integrative frameworks for relationships between plant physiological processes are needed to provide syntheses of plant growth and development. Source–sink–storage relationships, addressed in this chapter, provide one framework for synthesis of whole-plant responses to external environmental variables.

To address this issue, some examples of carbon assimilation and utilization responses of five conifer species to environmental factors from a range of field environments are first summarized. Next, the interactions between sources, sinks, and storages of carbon are examined at the leaf and tree scales, and finally, the review evaluates the proposition that processes involved with carbon utilization (sink activity) are more sensitive to the supply of water and nutrients (particularly nitrogen) than are the processes of carbon gain (source activity) and carbon storage. The terms "sink" and "source" refer to carbon utilization and carbon gain, respectively. The physiological processes determining these are referred to as "sink activity" (carbon storage, tissue growth) and "source activity" (photosynthesis), as appropriate. The relative roles of stored carbon reserves and of current photosynthate in meeting sink demand are addressed. The activities and vigor of various sources and sinks of a species (including symbionts) depend on the microclimate of a plant or organ and on seasonal changes in climate (Cannell 1985, 1989). Discussions focus on source–sink–storage relationships within the diurnal, wetting–drying, and annual cycles of conifer growth and development, and some discussion of life cycle aspects is also presented.

II. Source–Sink Relationships for Five Conifer Species

Experimental data for conifer species from a latitudinal gradient are summarized to provide examples of source–sink interactions for a wide range of seasonality and growth conditions. The gradient ranges from a montane climate in Germany to a warm temperate environment on the coastal plain of Florida. The observations suggest that source–sink interactions tend to be more strongly expressed in more favorable environments, such as those with higher mean annual temperature. In areas with low mean annual temperature, rates of growth and photosynthesis seem to be independent, whereas, in warmer environments of lower latitudes, photosynthesis tends to be high in spring when growth demand is high.

A. Norway Spruce (*Picea abies*) and Related Species

Seasonal patterns of maximum photosynthesis and growth rates of an 89-year-old Norway spruce forest in a mountainous climate (mean annual temperature of 6.4°C) in the Solling, Germany (latitude 52°N) appear to follow somewhat unsynchronized patterns, suggesting that source and sink activities are independent. The daily sums of net photosynthesis of 1-year-old needles from a mature *Picea abies* tree in the stand followed nearly a bell-shaped annual trend, beginning in March, terminating in October, and peaking in July and August (Schulze *et al.*, 1977a). These patterns of photosynthesis did not match the growth pattern of *Picea abies* in the study areas. Twig and needle growth began in May. Needle production was nearly completed by the end of June, at which time stem growth increased and was completed at the end of September. Spring growth rate was high before photosynthesis reached its

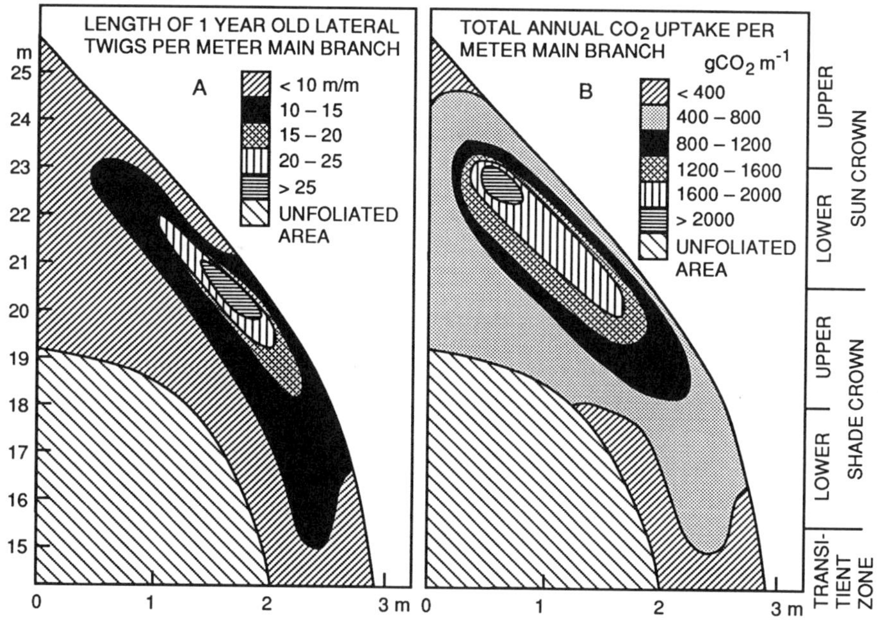

Figure 1 Distribution of regions of similar rates of (A) growth of 1-year-old twigs and (B) annual CO_2 gain per meter of main branch in a cross-section of a *Picea abies* crown; *x* axis, length of main branches; *y* axis, tree height. (From Schulze *et al.*, 1977b.)

highest rate in mid summer. Shoot extension growth and needle development of *Picea sitchensis* (Sitka spruce) also began in early May and lasted for about 9 weeks, although nutrient deficiency may delay bud break and shorten the development period by about 3 weeks (Chandler and Dale, 1990).

Spatial variation of photosynthesis and growth in different regions within a crown of an 89-year-old Norway spruce showed a weak association. The highest rate of annual CO_2 assimilation occurred in the central part of the upper sun crown where favorable microclimatic conditions coincided with a high needle biomass (Fig. 1); the highest rate of growth was found at the border between the sun and shade crown zones, toward the periphery of the crown, where growth was promoted by favorable temperature and light (Schulze *et al.*, 1977b). Similar results were also demonstrated in an earlier study with *P. sitchensis* by Norman and Jarvis (1974). Further, according to the analysis of Cannell and Morgan (1990), with a branching pattern favoring lateral twig production over branch elongation, *P. sitchensis* branches can continue to grow slowly, even at the base of the crown, while still exporting assimilates to support branches in zones of rapid growth.

Variations in concentrations of soluble carbohydrates in the stem wood of 80-year-old Norway spruce trees seemed to coincide with changes in sinks for, rather than the sources of, carbohydrates (Oren and Zimmermann, 1989). A decline in carbohydrate concentration, probably associated with an increase in the root sink in late April or early May, was followed by a quick recharge by the end of May, then by a decline during the period of canopy growth during June, and an increase following a cold period, which may have reduced growth more than photosynthesis in July (Fig. 2, top). After the cold period, carbohydrate concentration declined to its lowest level of about 50% of the maximum, where it remained until growth was completed at the end of September. After growth ceased some recharge of carbohydrate storage occurred during October. On a ground area basis, total nonstructural carbohydrates were a maximum in May (Fig. 2, bottom).

Ecotypic variation in measured rates of CO_2 uptake of *P. abies* progenies are often not correlated with growth. *Picea abies* seedlings of a

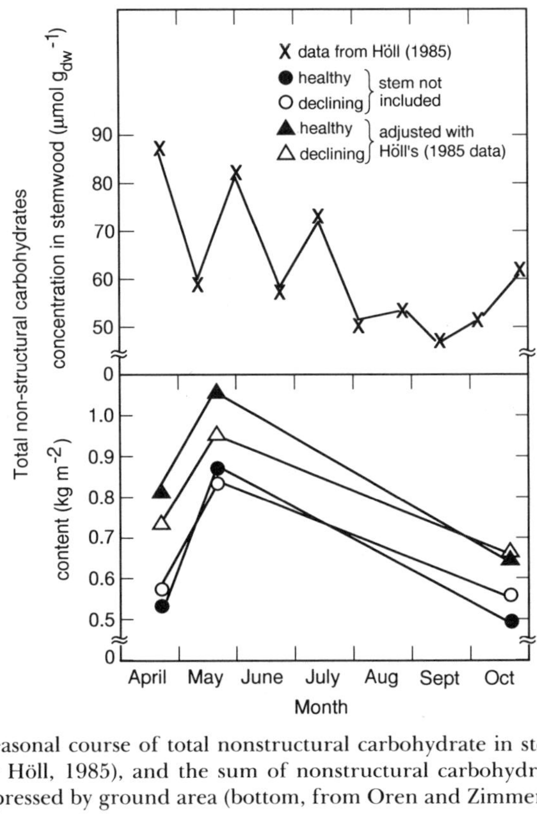

Figure 2 Seasonal course of total nonstructural carbohydrate in stem wood of *Picea abies* (top; after Höll, 1985), and the sum of nonstructural carbohydrate in all biomass components expressed by ground area (bottom, from Oren and Zimmermann, 1989).

southern provenance grew in height threefold faster than seedlings of northern populations (~90 mm vs. 30 mm) during the first year of a 2-year experiment (Pelkonen and Luukonen, 1974). At the end of the second growing season, a season in which height growth by the southern provenance was twice that of northern populations (~55 mm vs. 26 mm), net CO_2 uptake rate per seedling, measured after shoot growth was completed, was nearly fourfold higher in southern relative to northern provenances (~2.8 vs. 0.7 mg CO_2/hr/plant). Net photosynthetic rate on a needle basis was lower in the southern relative to northern progenies (~4.6 vs. 5.1 mg CO_2/g/hr), despite slightly lower photorespiration and dark respiration. A larger investment in foliage must have been made by the southern provenance on a per plant basis to account for greater growth and lower net photosynthesis than northern populations. Scotch pine (*Pinus sylvestris*) provenances also displayed an inverse relationship between CO_2 assimilation rate per unit leaf weight and whole-plant growth (Gordon and Gatherum, 1968) with southern provenances having lower rates of photosynthesis and greater leaf weight than northern populations.

B. Douglas Fir (*Pseudotsuga menziesii*)

Brix (1981a) reported growth and photosynthetic responses of Douglas fir to nitrogen (N) fertilization in 30-year-old codominant trees. The stand was on Vancouver Island, British Columbia (latitude 50° N), growing in a maritime climate with a mean annual temperature of about 9° C. The fertilizer was applied as ammonium nitrate on three occasions during a 4-year period with annual application rates ranging up to 896 kg N/ha. Foliar N concentration increased with increase in fertilizer application, reaching a maximum of 17.4 mg/g at about 448 kg N/ha treatment and declining at higher rates. The rate of net photosynthesis at the optimum foliar N concentration was 22% higher than the control treatment, which had a mean foliar N concentration of 9.5 mg/g. Stem diameter growth increased at all fertilizer application rates and was greater by an average of 76% over 4 years for the annual treatments of 448 kg N/ha relative to the control. Stem volume and dry matter production thus showed a much greater response to fertilization than did net photosynthesis per unit of leaf. In a related study (Brix, 1983), increased growth was shown to be due to a large increase in leaf area and, to a lesser extent, to an increase in net assimilation rate. Both studies show the greater relative response of sink activity compared to source activity to nitrogen supply, because leaf growth is initially a sink response to fertilization. An increase in leaf area resulted in increased interception of photosynthetically active radiation, which then led to further carbon gain.

Seedlings of *Pseudotsuga menziesii* transported less carbohydrates to roots in May during a period of high shoot and needle growth and more in November, after bud set, during the time of maximum root elongation (Hermann, 1977; Webb, 1977). Thus, phenological development and changing environmental conditions affect growth and result in changes in nonstructural carbohydrate content, particularly when photosynthesis rates are low. Dosskey *et al.* (1990) grew Douglas fir seedlings with three different ectomycorrhizal species and found one that induced higher photosynthesis compared to nonmycorrhizal controls; the other two mycorrhizal fungi had no effect on photosynthesis. There was no increase in nutrient concentration of leaves in seedlings that had higher photosynthesis, indicating a nonnutritional basis for mycorrhizal effects on photosynthesis. Dosskey *et al.* concluded that the sink strength generated by the fungal hyphae increased leaf photosynthesis.

There is a significant influence of apical control on the rate of photosynthesis shown in the measurements of Leverenz (1981) in a 37-year-old Douglas fir stand. Terminal shoots had higher stomatal and mesophyll conductances, transpiration rates, and photosynthesis rates than neighboring lateral shoots. Measurements of photosynthesis on terminal and lateral shoots as a function of irradiance show higher rates with terminal shoots (Fig. 3). Similar findings were obtained when measurements of photosynthesis were plotted against air temperature. There was no significant difference in the xylem water potential or microclimate of terminal and lateral shoots. Internal factors associated with apical dominance were controlling photosynthesis; however, these effects were not apparent at the fourth whorl from the terminal shoot, showing that apical control cannot be exerted over long distances from the terminal shoot. Terminal shoots seem to show a greater sink strength for photosynthate. Ross (1972) determined that a smaller fraction of photoassimilated ^{14}C was exported from terminal shoots than from current foliage on adjacent lateral shoots. The photosynthesis measurements of Leverenz (1981) also showed large variability that could not be attributed to irradiance (Fig. 3) or temperature, indicating the possibility of an additional role for internal regulation of photosynthesis to account for the variability.

C. Radiata Pine (*Pinus radiata*)

There are a few studies with *Pinus radiata* in which experimental manipulation of source–sink relationships has been conducted, whereas other studies provide inferential information. Many measurements indicate the possibility of source–sink interactions without a clear demonstration that they occur. For example, field data from near Mt. Gambier in southeastern South Australia (latitude 38° S) show a positive correlation

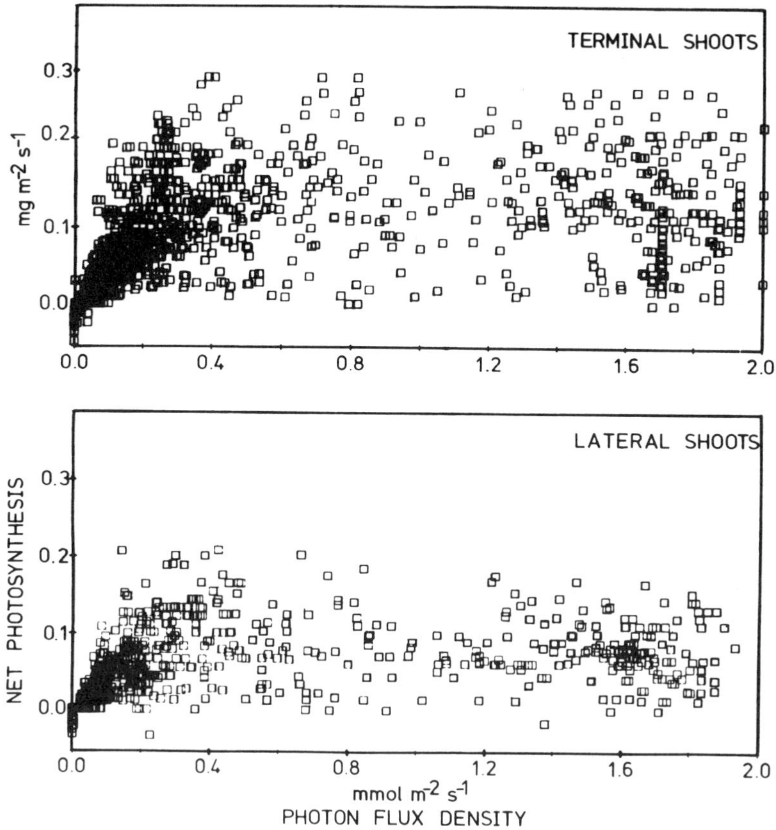

Figure 3 Relationships between field measurements of net photosynthesis and photon flux density for current-year foliage of terminal and lateral shoots from a 37-year-old Douglas fir stand. (From Leverenz, 1981.)

between carbon assimilation and growth, with stems growing mainly in spring and early summer, and carbon assimilation having a distinct maximum in spring (D. W. Sheriff and E. K. S. Nambiar, unpublished, 1992). This pattern is probably a response, at least in part, to the Mediterranean climate, where dry, warm to hot summers, sometimes dry autumns, and cool moist winters limit physiological activity. The mean annual temperature of the study area is 13.2°C. To understand the interrelationships between carbon sources and sinks it is necessary to appreciate how demands of different sinks interact, as well as how changes in sink activity affect sources and vice versa.

The interacting effects of vegetative growth at the apex of a shoot (sink) and the amount of foliage (source) on the shoot of radiata pine seedlings were examined by Sweet and Wareing (1966) and by Wood

(1968). Sink activity was reduced by removal of actively growing buds: debudding reduced carbon assimilation of foliage on the shoot. Removal of a part of the foliage caused the rate of carbon assimilation of the remaining foliage to increase. This can explain the findings that removal of small amounts of foliage has little effect on growth, whereas removal of large portions can reduce growth (Rook and Whyte, 1976). When a small amount of foliage is removed, carbon assimilation of remaining foliage increases in an apparent compensation for the loss of assimilatory surface area. Beyond some level of foliage removal, assimilation cannot compensate because leaves have a finite maximum assimilation rate.

Day length has also been shown to be associated with change in plant growth patterns. Jenkins *et al.* (1977) found that stem and branch elongation were greatest during periods with increasing day length (spring), with lesser elongation rates during sustained long days (summer). Stem diameter growth showed smaller photoperiod effects although there was a tendency for greater diameter increment with decreasing day length (autumn). Diameter growth generally has a lower priority for carbon allocation than does elongation growth of shoots.

Light quality and quantity directly influence carbon gain of radiata pine. Carbon assimilation and dry matter accumulation were both greater under conditions of simulated shade than at the same light intensity but with light quality similar to that of daylight (Warrington *et al.*, 1988). This effect of light quality on carbon assimilation was greater with higher photosynthetic photon flux density (PPFD). Light quality thus affected the assimilatory source and, directly or indirectly, altered at least one sink. A greater PPFD, with the same light quality, caused greater carbon assimilation (Warrington *et al.*, 1988). This indicates that growth may have been source limited at low light, a hypothesis that would explain the positive associations between photosynthesis and PPFD and between growth and PPFD, and the lower partitioning of assimilate to roots (further removed than other sinks) at low than at high PPFD (Sheriff and Rook, 1990).

D. Loblolly Pine (*Pinus taeda*)

Loblolly pine covers a wide geographic range of the southeastern portion of the United States, and the species experiences a wide variety of environmental conditions (Boltz *et al.*, 1986; Teskey *et al.*, 1987). Ecotypic differentiation is an important adaptive component in growth and carbon allocation in the species.

Carbon assimilation, allocation, and growth of loblolly pine vary significantly between seasons (Chung and Barnes, 1980a,b; Adams *et al.*,

1990; Kuhns and Gjerstad, 1991) due to changes in environmental conditions, such as temperature (Strain *et al.*, 1976; Drew and Ledig, 1981). Loblolly pine acclimates to seasonal changes in temperature, and has the potential to grow considerably during the winter season (Strain *et al.*, 1976). Birk and Matson (1986) found that nonstructural carbohydrate accumulated in the needles of loblolly during the winter months prior to bud break and dissipated quickly during the spring as shoots began to elongate. Phototsynthetic activity during periodic growth flushes may differ from that observed between flushes. For example, net photosynthesis of the previous year's needles often showed a late spring maximum during the first growth flush whereas maximum photosynthesis of the current year's needles occurred later in autumn (Adams *et al.*, 1990). Respiratory losses are often highest in the spring and decline sharply during the summer when photosynthate allocation to new leaves increases. Carbon allocation shifted in autumn to fine roots at the expense of the leaves (Drew and Ledig, 1981; Adams *et al.*, 1990).

Environmental stresses can have a large impact on source–sink activity of loblolly pine. For example, water deficit commonly limits the number of shoot flushes and bud development (Zahner, 1962). Seedlings stop growing at mild levels of water stress and are slow to begin growth again following rewatering (Kaufmann, 1968; Brix, 1962; Stransky and Wilson, 1964). Diurnal changes in water potential probably account for the fact that most radial and extension growth occurs at night, as noted in the review by Zahner (1968). Mild water stress increased starch production in source needles and strong water stress decreased starch production (Kuhns and Gjerstad, 1988). Bongarten and Teskey (1987) found that dry matter allocation to roots increased when loblolly seedlings were exposed repeatedly to drought cycles.

Nitrogen availability often regulates growth of loblolly through leaf area production, but can also affect photosynthetic rates (Penning de Vries *et al.*, 1975). Carbohydrates accumulate in needles in response to nutrient deficiency. This is associated with a decrease in the demand by carbohydrate sinks for growth and metabolism. Loblolly pine with high foliar N used starch reserves earlier in the growing season, and accumulated more starch after the growing season compared to pines with low foliar N (Birk and Matson, 1986; Adams *et al.*, 1986).

E. Slash Pine (*Pinus elliottii*)

Slash pine is an important commercial conifer that grows on the coastal plains of the southeastern United States (~ latitude 30° N) in soils that generally have shallow water tables. Water seems to be less of a limiting factor to stand growth than nutrient (nitrogen) supply (Gholz *et al.*,

1985). Stand fertilization with N, P, K, and ground limestone applied quarterly over 4 years caused an increase both in the leaf area index and stem growth of slash pine in a 21-year-old stand (Gholz *et al.*, 1991). The mean annual air temperature for the site was 21.7° C. The growth response of slash pine to fertilization contrasted with the photosynthetic responses measured on the two foliar age classes of the trees. No stimulation of photosynthesis was found with the fertilization treatment even though leaf nitrogen and phosphorus concentrations were significantly higher than for control trees (Teskey *et al.*, 1994). In addition, there was no significant relationship between net photosynthesis rate and air temperature, or with vapor pressure deficit. Net photosynthesis was related to photosynthetically active radiation (PAR) for the two age classes of foliage; however, large variability in the relationships was obtained. Current foliage had a range of net photosynthesis rates from about 1 to 3 μmol/m^2/sec at a PAR of 1000 μmol/m^2/sec (Fig. 4). The 1-year-old foliage had a somewhat smaller range of net photosynthesis at the same PAR (not shown). These wide ranges of values, obtained over a 2-year observation period, cannot be accounted for by the dominant external variables (light, temperature, vapor pressure deficit). These data provide strong evidence for internal regulation of net photosynthesis.

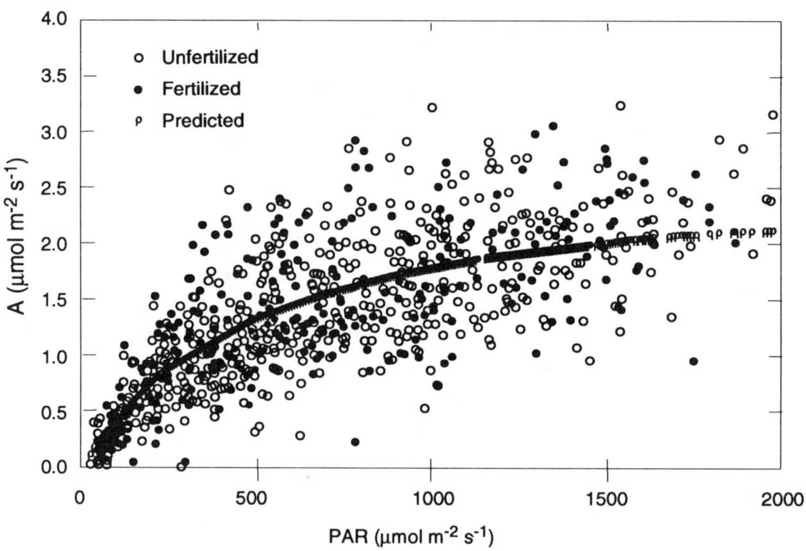

Figure 4 Relationship between net photosynthesis (A) and photosynthetically active radiation (PAR) measured on current foliage in fertilized and unfertilized stands of mature *Pinus elliottii* trees. A regression function fitted to the data is shown. (From Teskey *et al.* 1994.)

III. Source–Sink–Storage Interactions

The contrasting behavior in the relationship between growth activity and photosynthesis of Norway spruce and of slash pine suggests two possible extremes in source–sink relationships of conifers. The spruce, growing in a montane climate, had a low rate of photosynthesis even during midsummer and evidence of source–sink interactions was not apparent. There was evidence for internal regulation of photosynthesis in slash pine growing in warm temperate conditions. Other species, from temperate climates, have shown a correlation between high rates of growth and photosynthesis during spring. Maier and Teskey (1992) observed that springtime rates of photosynthesis of mature eastern white pine (*Pinus strobus*) were higher when shoot growth was high than during the nonshoot growth period (Fig. 5). Results from experiments with Douglas fir and radiata pine seedlings, described above, also provide evidence of interaction between sink activity and photosynthesis perhaps favored by close proximity of source and sink in small plants, as well as by low levels of storage reserves. An electrical analog is outlined next to summarize these general relationships.

A. Electrical Analog of Source–Sink–Storage Relationships

Consider the following relationship in which the rate of carbon assimilation (source), partitioning of this carbon in different plant organs

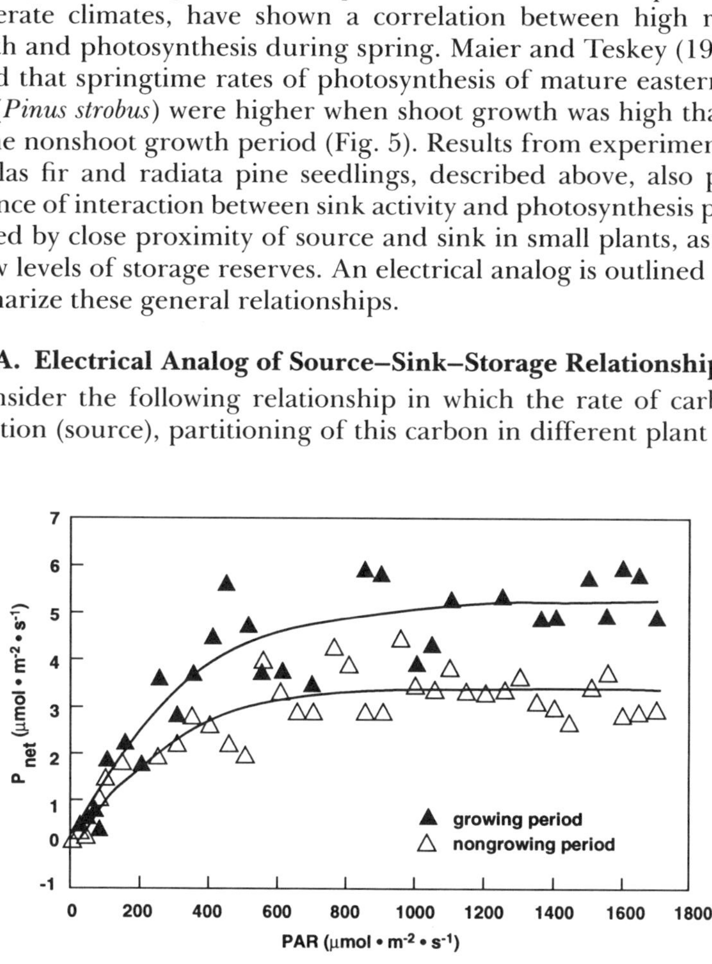

Figure 5 Measurements of net photosynthesis (P_{net}) in relation to photosynthetically active radiation (PAR) for mature *Pinus strobus* showing greater photosynthesis during the spring shoot-growth (21 April–20 June) period (▲) than during the following summer (20 June–15 October) period (△). (From Maier and Teskey, 1992.)

(translocation), input to and output from storage (e.g., starch reserves), and utilization by sink organs all vary in response to a variety of internal and external controls. When sink activity increases, source activity increases, and sink activity can be limited if insufficient carbon is assimilated. This leads to competition between sinks for the limited carbon supply. Surplus assimilate may be stored in plant organs and buffer the internal carbon dynamics by becoming a source or sink for assimilate, depending on the regulation of internal supply and demand. Holistically, we can view most of these relationships as resulting (1) from various sources and sinks with differing "strengths," (2) from the "buffering power" of carbon storage, and (3) from transport pathway "resistances" between sources, sinks, and storage.

An analog diagram for these relationships using electronic circuit symbols is given in Fig. 6 with the following components: (1) a generator

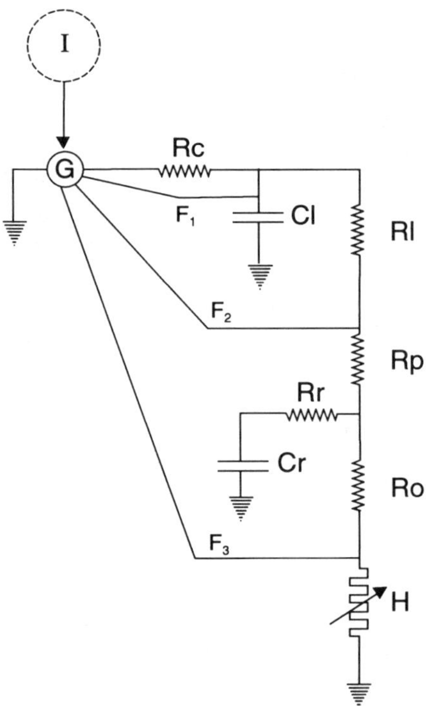

Figure 6 Analogs for source–sink–storage relationships using a generator to represent carbon assimilation, resistance (R) for the transport pathway, a capacitor (C) for assimilate storage, and a heating element (H) for sink activity. Solid lines define pathways of component linkage, and other lines represent negative feedback control. The three examples represent feedback (F) from leaf starch storage (F_1), from phloem loading (F_2), and from remote sink activity (F_3). I, Energy input (sun); G, generator (source); Rc, chloroplast; Rl, leaf; Rp, phloem; Rr, ray cell; Ro, organ.

for carbon fixation; (2) resistances for cell to cell transfer, transfer to storage, and movement though phloem; (3) a capacitor for assimilate storage; and (4) heating elements for utilization of assimilate in growth and respiration by plant sinks. In its simplest form, the analog includes one of items (1), (3), and (4), and two of item (2). More detailed analogs can be formulated to represent multiple sources and sinks. The influence of feedback regulation, whereby sink strength can modify source activity, can take several forms. Examples of feedback control from leaf carbohydrate storage, phloem loading, and remote sink activity are identified by labels F_1, F_2, and F_3, respectively in Fig. 6. Further discussion of these feedback controls is given in later sections.

At levels of organization smaller than a plant organ, the analog is unlikely to be useful. However, at the organ scale and larger the analog can qualitatively account for a variety of observations, such as competition between sinks and the large influence of sink changes on nearby sources (Watson and Casper, 1984). Partitioning of carbohydrates appears to be determined by the activity and storage of importing sinks (Daie, 1985; Wolsvinkel, 1985). The analog may not be useful for quantitatively describing source and sink relationships; however, it provides a unifying concept for evaluating the outcomes of source, sink, and storage interactions in situations wherein environmental factors have differential effects on source and sink activity.

The actual mechanism(s) of feedback regulation of photosynthesis are not completely understood for C_3 plants at the present time; however, there is extensive empirical evidence for the effects of feedback regulation of photosynthesis. There are several scales over which feedback effects of sink activity on photosynthesis can occur. Some effects occur within the leaf, and others operate through sink activity at distal growth centers.

B. Source–Sink Feedback within Leaves

Feedback regulation within a leaf has been variously attributed to depletion of the mineral phosphate supply, deactivation of ribulose 1,5-bisphosphate carboxylase/oxygenase (Rubisco), sucrose accumulation, and/or decline in phloem loading rates. At the cellular scale, Sharkey and Vanderveer (1989) identified the occurrence of low phosphate concentrations in the stroma of chloroplasts during feedback-limited photosynthesis. Inorganic phosphate in the chloroplast is required for ATP synthesis and it also regulates starch synthesis. In the cytosol, phosphate is required for sucrose synthesis. Phosphate incorporation into enzyme-mediated reactions for carbon fixation may lead to a local phosphorus deficiency, inducing feedback inhibition of photosynthesis during some parts of the diurnal cycle. Woodrow and Berry (1988) concluded that product feedback mainly influences metabolite partitioning (starch, sucrose) and may have little *direct* effect on CO_2 fixation, except under water stress or low temperature whereby CO_2 fixation may decline. There are a number of possible stages in feedback regulation of photosynthesis. Woodrow and Berry (1988) suggest that the activation state of Rubisco may impose the greatest control on photosynthesis. Stitt (1991) reviewed evidence for feedback regulation of photosynthesis in herbaceous plants. He attributed feedback inhibition of photosynthesis to low demand for carbohydrate in distal parts of the plant. Stitt further suggested that carbohydrate accumulation could *directly* inhibit photosynthesis by mechanical damage from large starch grains or from low levels of inorganic phosphate resulting from inhibition of sucrose synthesis. These effects were considered to be short term, whereas long-term regulation of photosynthesis was considered to be an *indirect* response involving change in the quantity and activation state of Rubisco.

Low phloem loading rates can lead to a buildup in leaf sucrose, favoring starch formation in chloroplasts. Evidence that starch accumulation in leaves is correlated with a decline in CO_2 fixation was reported by Guinn and Mauney (1980). Conversely, rapid depletion of leaf starch was accompanied by recovery in photosynthetic rate in the investigation of Sasek *et al.* (1985). Stitt (1991) showed reasonable evidence for starch synthesis being stimulated by feedback regulation of sucrose production. However, stimulation of starch synthesis can occur without photosynthesis being inhibited. Stitt (1991) further showed examples wherein sink inhibition of photosynthesis was not associated with low levels of internal phosphate concentration. Stitt concluded that a hierarchy of interacting regulatory mechanisms can generate a very complex and flexible response at the pathway level. His review provides significant evidence suggesting that this is the case for feedback regulation of photosynthesis. Trewavas (1986) has considered network responses and emphasizes the importance of *context*, whereby the effect of change in any individual constituent of a network is also determined by the state of the network. A critical property of a network is that external and internal effects on one or more constituents impact the whole network and not just those constituents. Chapter 4 of this volume provides further discussion of photosynthetic responses to external and internal factors.

C. Photosynthetic Response to Remote Sink Activity

There are many and varied examples in which an increase in sink activity is associated with an increase in photosynthesis. Reid *et al.* (1983) showed that needles of ectomycorrhizal loblolly pine had consistently higher photosynthesis rates than did plants without the symbiont. Similarly, Paul and Kucey (1981) demonstrated, with ^{14}C experiments, greater carbon fixation rates in plants with enhanced sink activity due to nitrogen fixation and mycorrhizal symbionts.

Photosynthetic rates have been altered experimentally by changing the ratio of sink to source. Partial defoliation of *Pinus contorta* was followed by a significant increase in photosynthesis in remaining needles, even at saturating light intensities (Sweet and Wareing, 1968; Wareing *et al.*, 1968). Remaining needles became greener and larger and their protein content increased (Wareing *et al.*, 1968). This was attributed to increased protein synthesis caused by an increased supply of cytokinins to the remaining leaves from roots. Increased photosynthesis after partial defoliation was prevented by removal of part of the roots (Wareing *et al.*, 1968). On the other hand, removal of buds, a large sink for carbon, reduced photosynthetic rates of *Pinus radiata* (Sweet and Wareing, 1966). Similar results were found with Scotch pine, whereby bud removal caused lower photosynthetic rates despite having greater Rubisco activity (Gezelius *et al.*, 1981). In addition, bud removal caused an increase in needle starch accumulation, presumably because carbohydrates could not be used in new shoot growth (Gezelius *et al.*, 1981). Troeng and Langström (1991) showed that pruning of current shoots from the upper crown of a young Scotch pine caused an immediate reduction of about 50% in the carbon assimilation and mesophyll conductance of remaining foliage, although both recovered within 10 days.

Reproductive growth can provide a strong sink for carbohydrates. First-year and second-year cones of red pine (*Pinus resinosa*) mobilized almost three times more carbohydrates from old needles than from shoots (Dickmann and Kozlowski, 1970a). Dick *et al.* (1991) have shown greater assimilation rates of *P. contorta* needles on branches bearing male cones than on vegetative branches or branches bearing female cones. This response was attributed to fewer current-year needles on branches bearing male cones and to development of the male cone buds.

Production of cones also has a potential to reduce the source strength of carbon assimilation, because these structures take space that would otherwise be occupied by foliage. The large sink strength of cones, and their displacement of foliage, might be expected to reduce plant growth, but there is often a positive association between tree growth and presence of female cones (e.g., Chalupka and Giertych, 1975; Simpson and Powell, 1981). Dick *et al.* (1990) found that in *P. contorta* effects of cones on carbon balance of the tree were small and the magnitude of these effects depended on environmental conditions. Respiratory loss by female cones was at most 3% of carbon assimilated by the tree. Although male cones reduced the number of needles by 33%, during sunny weather their assimilation increased, partially compensating for smaller needle numbers in a high-density stand and largely compensating in a low-density stand. In cloudy weather the carbon balance of trees with male cones was better than that of vegetative trees because of lower respiration rates of cones than of foliage. However, the observation that cone-bearing trees have less nonstructural carbohydrates than do trees without cones (Grochowska, 1973; Lenz and Küntzel, 1974) indicates that the carbohydrates necessary for cone production are not solely from current assimilate.

D. Sink Activity in Relation to Storage and Current Photosynthate

Kozlowski (1992) has provided a very comprehensive review on the topic of carbohydrate sources and sinks in woody plants. A large spatial and temporal variation in the use of reserve and currently produced carbohydrates was noted for cell metabolism and the growth of plant structures. The perennial woody plant was viewed as a highly integrated system of sinks that competed for the carbohydrate produced by foliage. Internal competition for carbohydrates was shown in ^{14}C tracer experiments by changes in the rate of carbohydrate transport as the relative sink strengths of various organs changed (Kozlowski, 1992). Cropper and Gholz (1990) showed that the labile carbon pool in slash pine, defined as the sum of soluble sugar and starch levels, was small relative to the magnitudes of gross primary production and total respiration estimated in simulation of whole-plant function. Simulated growth dynamics were very sensitive to the model parameters representing the labile carbon pool.

Many temperate conifer species exhibit determinate leaf, twig, and flower growth (preformed bud control) and the early spring growth flush is supported with stored reserves and current photosynthate from old needles. Stem thickening and root growth are usually indeterminate, but have periodic behavior due to reduction in root growth during periods of shoot growth. Deans and Ford (1986) determined the patterns of shoot and root growth for a 9-year-old Sitka spruce tree (Fig. 7) and noted the cooccurrence of shoot elongation and root growth in spring. This was followed by reduced root growth and onset of trunk thickening during summer, and by an increase in root growth in autumn. Measured changes in starch reserves indicated that root growth in close proximity to the stem was dependent on current photosynthate or on imported reserves. Distal roots were shown to deplete starch reserves in root bark to a greater extent than could account for distal root growth alone, indicating carbohydrate export from root bark (presumably to other sinks). Asynchrony between shoot and root growth allows the same storage reserves to serve different parts of a plant during a growing season (Chapin, 1980).

Tree species exhibiting indeterminate growth tend to proliferate leaf area readily on sites with favorable nutrition, whereas determinate species may take several months to a year to express a leaf area response to increased nutrient supply (Linder and Rook, 1984). Dickson (1989) re-

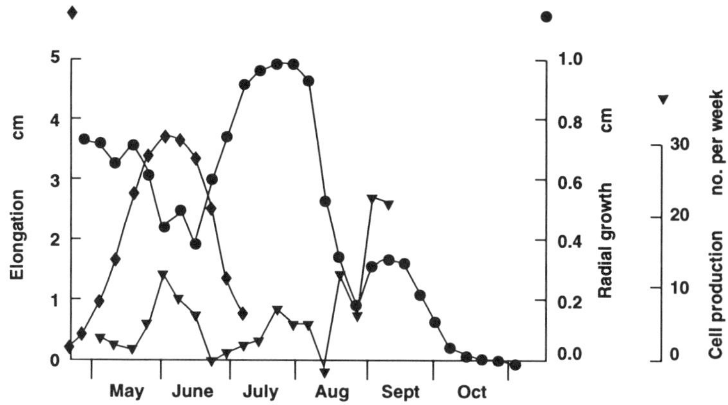

Figure 7 Weekly increments in shoot elongation (♦), stem radial growth (●), and root cell production (▼) determined at 25 cm from the stem in a 9-year-old *Picea sitchensis* stand in Scotland. (From Deans and Ford, 1986.)

viewed carbon and nitrogen allocation in trees and noted that during the spring growth flush, both stored and recently fixed carbohydrates contributed to new shoot growth in conifers. In the case of late-season diameter growth, only current photosynthate was used. Thus, current photosynthate is an important contributor to many plant growth activities, and any environmental stress that inhibits height or diameter growth could lead to feedback inhibition of photosynthesis. Allocation of carbohydrate to storage is a relatively low-priority function according to Dickson (1989). However, Cannell and Dewar (1994) suggest in a review of carbon allocation in trees that storage reserves may act as high-priority sinks.

Whereas on an annual basis photosynthesis may not limit growth in some environments, on a shorter temporal scale photosynthesis and growth do not maintain a constant ratio. Before bud break, growth rate is lower than the photosynthetic rate, resulting in an accumulation of nonstructural carbohydrates (Fig. 2). During intensive growth after bud break the rate of growth is greater than that of the photosynthetic rate. Stored carbohydrates are rapidly used until July and at a slower rate until October (Oren and Zimmermann, 1989). In *Picea abies* forests, cone and seed production of about 80 g/m^2/year may reduce the net primary production by all other biomass components by about 400 g/m^2/year (after Glazov, cited in Walter and Breckle, 1989). In Douglas fir, grand fir (*Abies grandis*), and western white pine (*Pinus monticola*) competition for photosynthate was indicated in years of prolific cone production by a decrease in annual stem wood increment (Eis *et al.*, 1965). Cone count records were compared with annual diameter

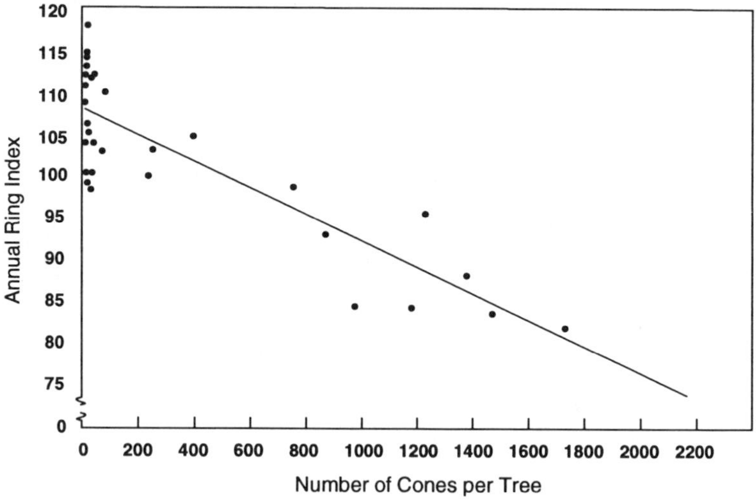

Figure 8 Relationship between annual diameter increment and the annual cone production per tree for *Pseudotsuga menziesii* determined over a 28-year observation period on 80 mature trees. (From Eis *et al.*, 1965.)

increment obtained from two increment cores taken at right angles from each of 80 Douglas fir, 14 grand fir, and 9 western white pine trees that were growing on a range of sites in British Columbia, Canada. The Douglas fir trees were from 68 to 104 years of age at the end of the 28-year observation period. A decline in annual stem increment occurred only in the year of abundant cone production (Fig. 8). Greater cone production was associated with a smaller annual increment. No relationship between cone production and ring increment was found for the year preceding, or the year following, a year of prolific flowering. This led Eis *et al.* (1965) to suggest that photosynthate produced in the current year, rather than stored reserves from previous years, was the dominant source of carbohydrate for prolific cone production.

Increasing atmospheric CO_2 to three times ambient concentration resulted in a greater and longer lasting growth response when applied during the more active growth stage of *Picea mariana* seedlings (Campagna and Margolis, 1989). The concentration of total nonstructural carbohydrate was unaffected by CO_2 enrichment, indicating that CO_2 uptake kept pace with the demand for growth. In *P. radiata* seedlings, elevated CO_2 concentration enhanced carbon assimilation and growth (Conroy *et al.*, 1990), including the thickness of tracheid walls, which increased wood density. Carbon allocation patterns of *P. radiata* families were modified by CO_2 enrichment; there was relatively less allocation to stems than to roots (Conroy *et al.*, 1990).

IV. Conifer Development and Source–Sink Relationships

A. Leaf

From germination to senescence, the source of fixed carbon shifts from cotyledons to primary needles and to secondary fascicled needles. Any plant tissue that cannot meet its carbon requirement autonomously is a potential carbon sink, and allocation of carbon to a particular sink is controlled by interrelated physiological processes. Actively expanding leaves are significant sinks for carbohydrate, and become carbohydrate sources as they grow. Carbon allocated to leaf biomass is partitioned to structure, physiologically active materials, defense, secretions, and storage. Leaf structure is affected by the environment in which it develops. For example, leaves produced under high light, low nutrition, and water deficit usually have high leaf specific weight (Linder and Rook, 1984). This results from increased leaf thickness, or elevated carbohydrate storage in the leaf, or both, and can allow for efficient use of available resources for carbon assimilation (Chapters 1 and 4). Photosynthate from leaves with a positive carbon balance is allocated to respiration, defense, repair, and maintenance, and is exported to other organs.

B. Shoot

Developing shoots are strong sinks for carbohydrates for a growth period that depends on species, genotype, and climate (Kozlowski, 1962; Kozlowski *et al.*, 1991). Shoot development following bud burst first requires the import of carbohydrates both from current photosynthesis (Dickmann and Kozlowski, 1968) and from storage (Krueger, 1967; Gordon and Larson, 1970; Schier, 1970), and much of these carbohydrates may be respired (Kozlowski and Gentile, 1958; Clark, 1961). This phase is followed by carbohydrate self-sufficiency, and then by a potential to store and export carbohydrates (Olofinboba and Kozlowski, 1973). To a much smaller degree than foliage, branches and stems are an assimilatory source of carbon. This latter source is often not sufficient to replace respiratory losses by these tissues, but can be useful in improving the carbon budget of plants and organs, especially of leafless branches and stems [e.g., *Larix decidua* (Keller, 1973)].

Microclimate effects on carbon allocation to stems are frequently observed as a result of silvicultural treatments. The most extreme effect may be seen after the removal of a part of the standing biomass. Thinning usually increases diameter growth at the base of stems more than near the top (Waring *et al.*, 1981). This happens at least in part because carbon assimilation is increased more near the base than near the top of the tree, and the proximity of this greater source strength to the stem base has the greater effect on growth there. Thus stem taper increases.

Pruning of lower branches has an opposite effect on taper (Larson, 1963a,b), at least in part, because removal of lower foliage causes the ratio of upper and lower assimilatory sources of the tree to increase.

C. Tree

Allocation to various structural components of trees is affected mainly by growth form and climate (Whittaker and Woodwell, 1968), thus allocation patterns are usually determined by climate and stand conditions (Benecke and Nordmeyer, 1982). For example, greater transpirational demand lowers the ratio leaf area to sapwood cross-sectional area (Waring and Schlesinger, 1985) by limiting foliage production, or by enhancing sapwood production and maintenance, or both. A high priority for carbohydrate allocation is the production of new "local" plant structure (Quinlan, 1965). Export of carbohydrate often has lower priority. For example, branch growth is directly proportional to carbon assimilation by its foliage, and little carbon is exported by shaded branches to other parts of a tree (Linder and Axelsson, 1982). This may also explain the bimodal peak sometimes observed in annual root production (Krueger and Trappe, 1967). During active shoot growth there is less assimilate available for root growth. This suggestion is supported by observations that indicate a dependence of fine root growth on current assimilation (Ritchie and Dunlap, 1980). Carbohydrate storage may have a low priority in carbon allocation (Dickson, 1989) and Cranswick (1979) found that the amount of carbohydrate stored was negatively associated with shading and defoliation.

Data indicate that plants may grow in relation to the quantity of supply of soil resources, and in inverse relation to constriction of roots by, for example, a layer of soil with high penetration resistance. These responses are at least partly mediated by hormonal root–shoot communication (Passioura, 1988; Pereira, 1990), and can affect carbon allocation and biomass partitioning. There is often an association between resource supply and partitioning aboveground, or to foliage, or both (Linder and Rook, 1984).

Deficiencies of nitrogen or phosphorus increase biomass partitioning to roots (Nambiar, 1980; Linder and Axelsson, 1982), and a deficit of nitrogen will increase extension growth of tap roots in *P. radiata* seedlings (Nambiar, 1980). The effect of nutrition on carbon allocation to roots is also found in the negative relationship between the biomass ratio of fine root to foliage and site quality, a measure of availability of soil resources (Santantonio, 1989). This is consistent with a positive association between plant nutrition and leaf size (Brix, 1981b; Will and Hodgkiss, 1977) and leaf number (Brix and Ebell, 1969). Some or all of these effects involve changes in hormone balance, but other mechanisms can be described more simply. For example, greater biomass partitioning to roots at low nutrition can occur because roots are nearest to the soil nutrient source. At nutritionally poor sites roots use much of the absorbed nutrients and leave little for translocation to aboveground tissues and shoot growth (Clarkson, 1985). This mechanism will also explain the proliferation of roots at sites of high resource availability in normally infertile soil. Allocation to biomass can also be influenced by stand characteristics. For example, a greater proportion of carbon is allocated to tree stems and used for maintenance respiration in an unthinned than in a thinned stand (Lavigne, 1991). Further aspects of carbon acquisition and allocation during stand development are given by Gower *et al.* in Chapter 7 of this volume.

D. Life-Cycle Stages

Carbon assimilation and allocation vary in response to stages of ontogeny and phenology. Structures and organs begin as carbohydrate sinks during growth and plant development. Some sinks, especially foliage, develop into and continue as sources until they die, whereas the role of others may often alternate between storage, sink, and source, depending on photosynthate supply and demand. The capacity for carbohydrate supply and use develops as a plant grows. Carbohydrate supply to an embryo during germination is from storage in the seed. Sasaki and Kozlowski (1968a,b) showed that initiation and expansion of cotyledons can rely entirely on the limited seed storage. Further growth depends on carbohydrates from photosynthesis (Sasaki and Kozlowski, 1970; Kozlowski and Borger, 1971), thus growth of young seedlings is limited by their capacity to assimilate carbon. At these early stages, there is little carbohydrate storage and most current assimilate will be used to produce new plant structures, particularly roots.

Plant development is accompanied by changing patterns of allocation. For example, carbon assimilation of forest stands increases with age, reaching a constant value, but the ratio of maintenance respiration to growth respiration increases with age (Cannell, 1989). Carbon turnover often increases in the roots as stands age. Vogt *et al.* (1982) found an increase, of about 1.5 times, in the proportion of carbon allocated to mycorrhizae and fine roots of mature, compared to young, trees. Therefore, the rate of biomass production declines with age in response to increased root sinks, root turnover, and respiratory load (Cannell, 1989).

Once plants reach sexual maturity, reproductive structures, usually produced on a yearly basis, become a new sink for carbohydrate. Growing cones contain chlorophyll and have potential to assimilate carbon, but usually at a rate lower than that of their respiration (Dickmann and Kozlowski, 1970b, 1973; Kozlowski *et al.*, 1991).

V. Environmental Influences on Sources and Sinks

A. Some Temperature Considerations

Photosynthesis and cell growth have differing responses to temperature. This can influence source–sink interactions. Low temperatures in the soil reduced root and shoot growth of conifers, especially in subalpine environments (Barney, 1951; Bowen, 1970; Lopushinsky and Kaufmann, 1984; Andersen *et al.*, 1986; Vapaavuori *et al.*, 1992). Low root temperature can decrease stomatal conductance and transpiration (Teskey *et al.*, 1984; Day *et al.*, 1989). In addition, DeLucia (1986) has shown that reduced photosynthesis of Engelmann spruce (*Picea engelmannii*), due to low root temperatures, is related to increased biochemical limitations of photosynthesis. However, although root chilling (0.7° C) for 7 days progressively reduced the rate of photosynthesis by about 50%, soluble carbohydrate concentrations increased slightly, indicating that growth was reduced by a similar or somewhat greater proportion. This suggests that the supply of current assimilates may not limit growth under low-temperature conditions. This is an important consideration in evaluation of source–sink relationships of conifers along latitudinal and altitudinal gradients.

Warm daytime temperatures favor shoot growth, and cool night temperatures may restrict growth. Shoot growth rates have been observed to lag 2 to 5 hours behind air temperature changes in the diurnal cycle and, in the studies of Kanninen (1985), a 2-hour phase shift between air temperature and elongation rate of *Pinus sylvestris* shoots was shown. There is also a lag of several hours in the diurnal maximum soil temperature. This may favor root growth during the late afternoon–evening period. Plant water potential and temperature changes during the diurnal cycle can have large effects on cell growth and thus influence the timing of sink demand in relation to source activity.

B. Water Supply

Wetting–drying cycles vary greatly in duration (week to months) with climate and can have a marked effect on source–sink relationships. As soil drying occurs, root growth is impeded and a chemical signal (abscisic acid) is released from roots and transported to foliage, where it induces stomatal closure (Gollan *et al.*, 1989). This mechanism provides feedback from the root sink that reduces the rates of photosynthesis and transpiration by impeding gas exchange.

Sink activity is determined by the irreversible increase in cell size, as well as by carbon storage. Gholz *et al.* (1990) noted that restricted leaf growth is a dominant response of forests, including conifers, to water stress. Grier and Running (1977) compared the leaf area index values (LAI) of coniferous forests in relation to a range of local water budgets in western Oregon and showed lower LAI on sites with lower plant available water. Generally, reduction of water potential (or water content) of tissues has a much larger effect on cell expansion and growth (sink activity) than on photosynthesis (source activity) as shown by Rook *et al.* (1977) in *P. radiata*. In a review of water effects on plant processes, Hsiao (1973) identified cell growth, wall synthesis, and protein synthesis as very sensitive to stress, whereas stomatal opening, CO_2 assimilation, and respiration were less sensitive. Boyer (1970) showed that leaf expansion of three herbaceous species was reduced more than 80% by a decrease in leaf xylem water pressure to -0.5 MPa, whereas the accompanying net photosynthesis was largely unaffected. Teskey *et al.* (Chapter 4, Fig. 6) show similar responses of net photosynthesis to xylem pressure for seven conifer species. Some herbaceous plants show a daytime buildup of solutes in cells that results in increased cell turgor, allowing cell expansion even during conditions of high evaporative demand (Acevedo *et al.*, 1979). Osmotic adjustment seems to be a major regulatory mechanism for leaf expansion (Dale, 1988). Roots also have a significant ability for osmotic adjustment that assists root sinks in overcoming soil mechanical resistance during growth (Russell, 1977).

The influence of environmental conditions on shoot growth is well demonstrated in the shoot elongation measurements reported by Cremer (1976) for *P. radiata*. The daily patterns of shoot elongation were determined during a range of weather conditions, and Cremer showed that a considerable part of shoot elongation can occur at night following hot sunny conditions (Fig. 9). The results also showed that, under rainy conditions or moderate sunshine, elongation may occur during the daytime. Under the latter conditions, shoot sink activity is high at the time when photosynthesis is proceeding, but under conditions of high evaporative demand (even with favorable soil water supply) aboveground sink activity may be low during the photoperiod. Shoot elongation growth was greatest on a moderately sunny day, with the majority of growth occurring in the late afternoon and evening (Fig. 9). Cremer (1976) showed a slight increase in growth on a hot sunny day, which was half of that measured on a rainy day. Elongation occurred throughout the day

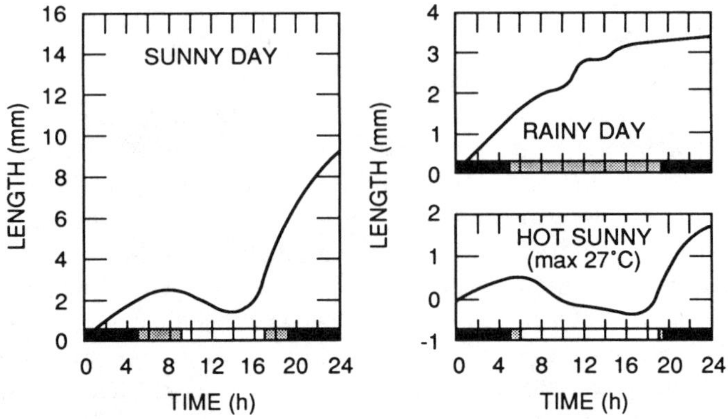

Figure 9 Daily patterns of shoot elongation in *Pinus radiata* for several differing environmental conditions (from Cremer, 1976). The dark, grey, and white sections of the time axes represent night, daytime cloudy conditions, and sunny periods, respectively.

under the cloudy conditions and only at night for the hot and sunny conditions. Water deficits increase carbon partitioning to branches rather than to stems (Snowdon, 1990; Snowdon and Benson, 1992).

There are two main feedback effects on photosynthesis induced by water stress, one associated with sucrose accumulation resulting from reduced sink activity, and the other due to stomatal closure. These effects operate in diurnal and wetting–drying cycles. Diurnal variation in leaf water pressures (potentials) can often range between −0.2 and −2.0 MPa, resulting in daytime conditions favoring photosynthesis but not necessarily favoring cell growth (Rook *et al.*, 1977). Conversely, conditions that favor growth during the day may result in enhanced photosynthesis due to reduced feedback inhibition. Restricted foliar sink activity during the day favors photosynthate accumulation in leaf storage, as well as sucrose translocation to roots. Translocation is generally insensitive to water stress (Dale and Sutcliffe, 1986), and can proceed under conditions that inhibit leaf cell expansion. Internal resistances to water flow cause a water potential gradient through a plant when water is flowing. Formation of this gradient is buffered by the hydraulic capacitance of tissues (change in water content per unit change in water potential, Hinckley *et al.*, 1978). Greater carbon allocation to roots may be favored in plants with large internal water flow resistances if shoot sink activity is decreased to a relatively greater extent than root sink activity by a water potential lower in shoots than in roots. Several review articles report the occurrence of greater tissue growth during periods when photosynthesis rates were small or nil; however, exceptions to this tendency have also been reported in the same review articles (Hsiao, 1973; Hinckley *et al.*, 1981).

C. Nutrition

Leaf area adjustment is usually the major plant response to nutrient supply under field conditions (Terry *et al.*, 1983; Linder and Rook, 1984). As mentioned, early leaf growth and expansion result from sink activity. Linder (1985) showed that the increase in aboveground production of *P. radiata* irrigated with nutrient-rich water could be accounted for by increased foliar interception of solar radiation. This indicates a much greater adjustment of source size than of activity per unit of source to favorable nutrition and water supply. In a modeling analysis of results from another field experiment, McMurtrie *et al.* (1990) showed that the majority (67%) of the increase in *P. radiata* growth resulting from irrigation and fertilization could be accounted for by a lengthening of the active growth period. Greater source size and enhanced activity per unit of source accounted for 23 and 10% of the production gain, respectively. Thus, an increase in photosynthetic activity per unit leaf area was a minor response to increased water and nutrient supply. Greater availability of nutrients often changes the relative strength of sinks, for example, increasing the allocation of aboveground biomass to branches and foliage with less going to stems in *P. radiata* (Mead *et al.*, 1984; Snowdon and Benson, 1992). In one case nutrition had no effect on allocation of carbon to aboveground plant parts (Cromer *et al.*, 1985).

An increase in light-saturated photosynthetic rate per unit leaf area can result from higher leaf nitrogen concentration (Evans, 1989). This allows plants, particularly determinate species, to increase source activity with nitrogen supply. Canopy photosynthesis may be less sensitive to nitrogen because only a part of the canopy is at light saturation. Brix (1981a) found an increase in photosynthesis of 30-year-old Douglas fir trees with nitrogen fertilization; however, this response was small relative to the increase in diameter growth with the fertilization treatment.

There is considerable evidence of plants increasing starch storage when there is little growth sink activity. During a winter period (February), Birk and Matson (1986) found in a 30-year-old loblolly pine stand greater starch storage in needles with higher nitrogen concentrations. This can be attributed to enhanced photosynthetic rates in needles with high N and the lack of a growth sink. During the growing season (June), the needles with high N were low in starch, whereas trees with low N status accumulated starch. This would result from the greater growth sink for photosynthate in trees with higher N status, illustrating the greater sensitivity of growth than of photosynthesis to N stress.

Nitrogen nutrition of conifers involves uptake and remobilization from stored nitrogen reserves, and the level of stored N reserves can influence the source–sink relationships for carbon. The principal soluble forms of nitrogen storage in woody plants are arginine, which has a high N:C ratio, and asparagine (Titus and Kang, 1982), with glutamine being significant in some conifers. Proteins are probably the most significant insoluble reserves in trees (van den Driessche, 1984). Titus and Kang (1982) cite several examples (for apple trees) wherein high levels of stored N are highly correlated with new shoot growth during the following spring. Miller (1984) noted that the relative demands made by forests for soil minerals decrease markedly after canopy closure as trees increasingly depend on internal storage and retranslocation to meet nutrient requirements for annual growth. Meier *et al.* (1985) also estimated an increased contribution of retranslocation in meeting the nutritional requirements of aging *Abies amabilis* (silver fir) stands. Following canopy closure additional phytomass accumulates largely as low-nutrient wood. Nambiar and Fife (1991) have clearly demonstrated the role of current needles of radiata pine as sources of nitrogen that can be readily exported to new growth sinks during growth flushes.

Removal of sinks for carbon and nitrogen may lead to a more favorable nutrient status of remaining foliage. For example, removal of buds and prevention of new twig and foliage production in a nutritionally balanced *Picea abies* tree caused a significant increase in maximum photosynthesis rate, determined at 2.5 ml/liter of CO_2 (A_{2500}), in May, even though less assimilate was necessary to support growth (Weikert *et al.*, 1989). However, a month later maximum photosynthetic rate declined to a rate similar to that of unaltered shoots, indicating the possibility of feedback regulation. Except for 1-year-old needles, all investigated needle age classes showed a midseason depression in A_{2500}. This midseason depression was temporally associated with the period in which the concentrations of mobile elements and chlorophyll pigments were lowest due to nutrient retranslocation to support new twig and needle growth (Oren and Schulze, 1989; Oren *et al.*, 1993). The net photosynthetic rate of 1-year-old foliage of a nutritionally-limited *P. abies* tree measured at 340 μl/liter CO_2 (A_{340}) was similar in May for branches whereon new growth was controlled by bud removal in comparison to others whereon growth proceeded normally. However, in October, at the end of the growing season, midday A_{340} of foliage on twigs in which growth was not modified was lower relative to A_{340} values of twigs from which buds were removed (~1.5 vs. 2.5 μmol/m^2/sec). This resulted in an estimated increase in daily and annual CO_2 uptake of 25% for foliage of the bud-removal treatment. Thus, branches whose growth was reduced by the treatment had significantly higher maximum and actual photosynthetic rates than branches whose growth was not curtailed. Similar results were obtained with *P. sylvestris* (Gezelius *et al.*, 1981). These studies demonstrate that under conditions in which nutrient supply to foliage is limited, improved foliar nutrition due to bud removal may enhance photosynthesis.

Leaf and stem growth have been shown to be more sensitive to nitrogen stress than is photosynthesis. Any plants that have the opposite response, so that photosynthesis is the more sensitive, would quickly become both N and carbon limited. Such plants are unlikely to be a significant component of any present-day flora because the double deficiency of N and photosynthate supply would be a major disadvantage in competition with neighbors. On the other hand, when growth is more sensitive than photosynthesis to nutrient deficit there is potential for carbohydrate storage, which can be used in growth if nutrition improves.

VI. General Discussion

A. Synthesis Framework

A number of photosynthesis models represent the process as a function of external environmental variables, such as light, temperature, vapor pressure deficit, soil water status, and nitrogen supply (e.g., Tenhunen *et al.*, 1980), without consideration of the internal environment of whole-plant function. This can be a serious limitation, particularly when simulating tree growth for temperate or warmer climates. Confounding of internal feedback regulation with responses to external variables has occurred in the estimation of model variables when no consideration has been given to feedback effects on photosynthesis. This has also resulted from limitations in data sets that do not include measures of feedback parameters.

Incorporation of a suitable feedback framework in plant growth models is a challenge because the mechanism(s) has not been determined. A similar situation exists for simulation of stomatal behavior wherein empirical relationships are used in simulate responses. Luxmoore (1991) offered a framework for the effects of water and nutrition (nitrogen) on source–sink relationships with the following attributes:

1. Growth is more sensitive than photosynthesis to water stress and nutrient deficiency.
2. Restricted growth leads to reduced sucrose translocation from foliage.
3. Buildup of stored carbohydrates in foliage induces (or correlates with) feedback inhibition of photosynthesis.

4. The tissue receiving the most limiting resource (water, nutrient) from the environment is favored to have "first use," leading to preferential growth relative to the other parts of the plant (Brouwer, 1966).

There is extensive evidence for the first attribute, and several examples were reviewed above; however, it is noted that in cold climates source–sink relationships may not be apparent, perhaps due to relatively low rates of biological activity. The second attribute follows as a consequence of a reduced sucrose gradient induced by attribute 1; however, the effect can be buffered by buildup of stored reserves throughout the plant. Evidence for attribute 3 was given in the section on source–sink feedback within leaves. The fourth attribute is consistent with empirical plant growth responses to limited environmental resources. For example, shaded plants show preferential shoot growth. Preferential root growth occurs with soil water and nutrient stresses. It is appropriate to note that nutrient deficiency does not always result in a relative increase in root mass. Ericsson *et al.* (1992) showed a relative decline in root weight of birch seedlings with reduced Mg, Mn, or K supply, whereas for other elements (N, P, S, Fe) the proportion of root mass increased with decrease in supply. Chapin (1991) has proposed a centralized system of physiological responses to environmental stress operating through declines in growth rate and in rates of resource acquisition. A hormonally mediated response system was suggested to provide an integrated change in nutrient, water, and carbon balances within the plant. Source–sink interactions were suggested to control photosynthesis most strongly under nutrient-limited conditions.

B. Summary, Conclusions, and Developments

Went (1974) concluded from his professional experience in plant physiological research that "growth limits photosynthesis rather than photosynthesis limiting growth." It is evident that growth and photosynthesis are interdependent and their interactions can be buffered by storage. The physiology of growth sinks for carbon is much more sensitive to environmental factors such as water stress and nitrogen deficiency than is photosynthesis, favoring feedback regulation of photosynthesis under such conditions. Source–sink relationships have been poorly understood or ignored in many photosynthesis and plant growth simulators. Accounting for these relationships in computer models will lead to advances in understanding of whole-plant physiological responses to environmental conditions. The examples of source–sink relationships from a latitudinal gradient suggest that source–sink interactions may be less significant in cooler climates. The possibility of global warming may increase the significance of source–sink interactions in conifers of boreal regions.

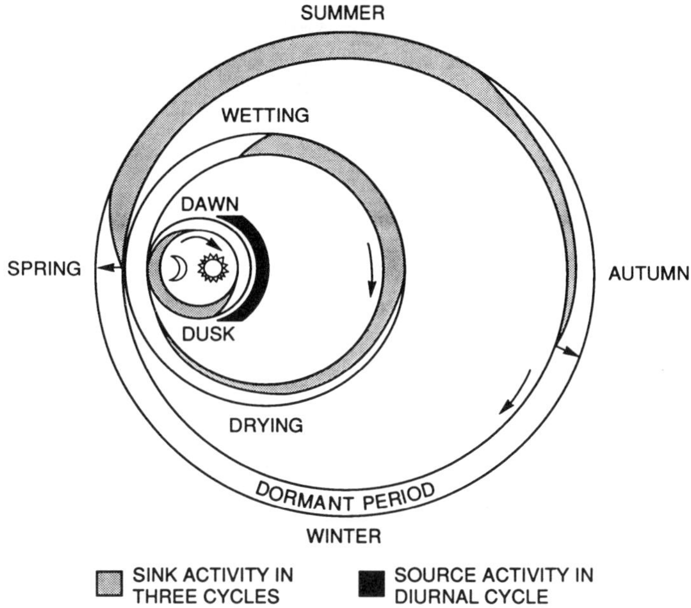

Figure 10 Schematic representation of the diurnal period of photosynthesis with nested cycles of sink activity for diurnal, wetting–drying, and annual cycles (from Luxmoore, 1991). Each of the outer cycles is composed of a variable number of the lower scale cycles, providing an essentially infinite set of possible water, carbon, and nutrient interactions in vegetation. These interactions are constrained by each upper scale with annual cycles being constrained by the plant life cycle.

The oftentimes subtle interactions of plant responses to water, carbon, nutrients, and weather during the diurnal cycle lead to a daily increment in carbon gain and growth that contributes to plant responses in subsequent diurnal cycles. The annual cycle of plant growth is a consequence of at least 365 diurnal cycles and a variable number of wetting–drying cycles. Viewing annual growth as a nested set of cycles (Fig. 10) with varying interactions of carbon source physiology and sink physiology with water, nutrients, and weather leads to the realization that an almost infinite number of outcomes is possible. Nevertheless, plants respond with patterns that may be generalized. There are certain periods within each cycle in which sink activity is favored, and these cyclic patterns have some similarity across temporal scales.

In the annual cycle, source–sink relationships are constrained by the phenological controls on plant processes (Cannell, 1990), and the aging of plant tissues (Sestak, 1985). Bud break is often determined by daylength and thermal period in the spring, whereas senescence and frost hardening processes are accompanied by a decline in photosynthesis. Three time scales of interactions between a plant and its environment (hours, weeks/months, seasons) are constrained by the length of the growing season.

Source–sink relationships can be used for developing hypotheses about whole-plant function. For example, greater root resistance to water flow may favor carbon allocation to roots due to greater water stress in shoots than in roots, which preferentially restricts shoot growth; growth may limit photosynthesis under some conditions and vice versa under others; and nutrient-deficient plants may show a growth response to atmospheric CO_2 enrichment by stimulation of belowground processes (Luxmoore, 1981). Methods of increasing sink activity (e.g., genetic selection) may be a much more effective means of increasing conifer timber production than by directly seeking increased rates of photosynthesis.

Continuing research on the leaf scale and whole-plant mechanism(s) of feedback regulation of photosynthesis will aid in developing appropriate algorithms to represent the phenomenon in photosynthesis and plant growth models. Feedback control may involve biochemical processes such as protein synthesis, as shown in the study of Van Oosten and Besford (1994). They measured the down-regulation of small subunit transcripts of Rubisco in tomato plants with elevated atmospheric CO_2 or with feeding of sucrose or glucose, suggesting that photosynthetic end products may contribute to feedback regulation through synthesis of Rubisco. Many aspects of needle physiology, phloem loading and unloading, translocation, and carbon utilization and storage in conifers need to be quantified in a whole-plant context to provide a quantitative understanding of source–sink feedback regulation of carbon dynamics.

Acknowledgments

This research was sponsored in part by the Carbon Dioxide Research Program, Environmental Sciences Division, Office of Health and Environmental Research, U.S. Department of Energy under Contract Number DE-AC05–84OR21400 with Martin Marietta Energy Systems, Inc., and by contract with Duke University. Additional support was provided by the Division of Forestry, CSIRO, Australia, and the School of the Environment, Duke University, Durham, North Carolina.

References

Acevedo, E., Hsiao, T. C., and Henderson, D. W. (1979). Diurnal growth trends, water potential and osmotic adjustment of maize and sorghum leaves in the field. *Plant Physiol.* 64:476–480.

Adams, M. B., Allen, H. L., and Davey, C. B. (1986). Accumulation of starch in roots and foliage of loblolly pine (*Pinus taeda* L.): Effect of season, site and fertilization. *Tree Physiol.* 2:35–46.

Adams, M. B., Edwards, N. T., Taylor, G. E. Jr., and Skaggs, B. L. (1990). Whole plant ^{14}C-photosynthate allocation in *Pinus taeda*: Seasonal patterns at ambient and elevated ozone levels. *Can. J. For. Res.* 20:152–158.

Andersen, C. P., Sucoff, E. I., and Dixon, R. I. (1986). Effects of root zone temperature on root initiation and elongation in red pine seedlings. *Can. J. For. Res.* 16:696–700.

Barney, C. W. (1951). Effect of soil temperature and light intensity on root growth of loblolly pine. *Plant Physiol.* 26:146–163.

Benecke, U., and Nordmeyer, A. H. (1982). Carbon uptake and allocation by *Nothofagus solandri* var. *cliffortioidies* (Hook. f.) Poole and *Pinus contorta* Dougl. ex. Loundon ssp. *contorta* at montane and subalpine altitudes. *In* "Carbon Uptake and Allocation in Subalpine Ecosystems as a Key to Management" (R. H. Waring, ed.), pp. 9–21. IUFRO Workshop, Forest Research Laboratory, Oregon State Univ., Corvallis.

Birk, E. M., and Matson, P. A. (1986). Site fertility affects seasonal carbon reserves in loblolly pine. *Tree Physiol.* 2:17–27.

Boltz, B. A., Bongarten, B. C., and Teskey, R. O. (1986). Seasonal patterns of net photosynthesis of loblolly pine from diverse origins. *Can. J. For. Res.* 16:1063–1068.

Bongarten, B. C., and Teskey, R. O. (1987). Dry weight partitioning and its relationship to productivity in loblolly pine seedlings from seven sources. *For. Sci.* 33:255–267.

Bowen, G. D. (1970). Effects of soil temperature on root growth and on phosphate uptake along *Pinus radiata* roots. *Aust. J. Soil Res.* 8:31–42.

Boyer, J. S. (1970). Leaf enlargement and metabolic rates in corn, soybean, and sunflower at various water potentials. *Plant Physiol.* 46:233–235.

Brix, H. (1962). The effects of water stress on the role of photosynthesis and respiration in tomato plants and loblolly pine seedlings. *Physiol. Plant.* 15:10–20.

Brix, H. (1981a). Effects of nitrogen fertilizer source and application rates on foliar nitrogen concentration, photosynthesis, and growth of Douglas-fir. *Can. J. For. Res.* 11: 775–780.

Brix, H. (1981b). Effects of thinning and nitrogen fertilization on growth, leaf area, and photosynthesis rate in Douglas-fir. *For. Sci.* 15:189–196.

Brix, H. (1983). Effects of thinning and nitrogen fertilization on growth of Douglas-fir: Relative contribution of foliage quantity and efficiency. *Can. J. For. Res.* 13:167–175.

Brix, H., and Ebell, L. F. (1969). Effects of nitrogen fertilization on growth, leaf area, and photosynthesis rate in Douglas Fir. *For. Sci.* 15:189–196.

Brouwer, R. (1966). Root growth of grasses and cereals. *In* "The Growth of Cereals" (F. L. Milthorpe and D. J. Ivins, eds.), pp. 153–166. Butterworth, London.

Campagna, M. A., and Margolis, H. A. (1989). Influence of short-term atmospheric CO_2 enrichment on growth, allocation patterns, and biochemistry of black spruce seedlings at different stages of development. *Can. J. For. Res.* 19:773–782.

Cannell, M. G. R. (1985). Dry matter partitioning in tree crops. *In* "Attributes of Trees as Crop Plants" (M. G. R. Cannell and J. E. Jackson, eds.), pp. 160–193. Titus Wilson & Sons, Kendal, Cumbria, U. K.

Cannell, M. G. R. (1989). Physiological basis of wood production: A review. *Scand. J. For. Res.* 4:459–490.

Cannell, M. G. R. (1990). Modelling the phenology of trees. *Silva Carelica* 15:11–27.

Cannell, M. G. R., and Dewar, R. C. (1994). Carbon allocation in trees: A review of concepts for modelling. *Adv. Ecol. Res.* 25:59–104.

Cannell, M. G. R., and Morgan, J. (1990). Theoretical study of variables affecting the export of assimilates from branches of Picea. *Tree Physiol.* 6:257–266.

Chalupka, W., and Giertych, M. (1975). The effect of growth on cone crops in Norway spruce (*Picea abies* L. Karst.). *Arbor. Kornickie* 20:193–200.

Chandler, J. W., and Dale, J. E. (1990). Needle growth in Sitka spruce (*Picea sitchensis*):

Effects of nutrient deficiency and needle position within shoots. *Tree Physiol.* 6:41–56.
Chapin, F. S. (1980). The mineral nutrition of wild plants. *Annu. Rev. Ecol. Syst.* 11: 233–260.
Chapin, F. S. (1991). Integrated responses of plants to stress. *BioScience* 41:29–36.
Chung, H. H., and Barnes, R. L. (1980a). Photosynthate allocation in *Pinus taeda.* II. Seasonal aspects of photosynthate allocation to different biochemical fractions in shoots. *Can. J. For. Res.* 10:338–347.
Chung, H. H., and Barnes, R. L. (1980b). Photosynthate allocation in *Pinus taeda.* III. Photosynthate economy: its production, consumption, and balance in shoots during the growing season. *Can. J. For. Res.* 10:348–356.
Clark, J. (1961). Photosynthesis and respiration in white spruce and balsam fir. Univ. Tech. Bull. 85. College of Forestry, State Univ. New York, Syracuse.
Clarkson, D. T. (1985). Factors affecting mineral nutrient acquisition by plants. *Annu. Rev. Plant Physiol.* 36:77–115.
Conroy, J. P., Milham, P. J., Mazur, M., and Barlow, E. W. R. (1990). Growth, dry weight partitioning and wood properties of *Pinus radiata* D. Don after 2 years of CO_2 enrichment. *Plant, Cell Environ.* 13:329–337.
Cranswick, A. M. (1979). Food reserves of Douglas fir trees defoliated by Lepidopterous larvae in Kaingaroa SF., F.R.I. Intern. Rep. No. 13. Prod. For. Div., Forest Research Institute, Rotorua, New Zealand.
Cremer, K. W. (1976). Daily patterns of shoot elongation in *Pinus radiata* and *Eucalyptus regnans. New Phytol.* 76:459–468.
Cromer, R. N., Barr, N. J., Williams, E. R., and McNaught, A. M. (1985). Response to fertilizer in a *Pinus radiata* plantation. 1. Above-ground biomass and wood density. *N. Z. J. For. Sci.* 15:59–70.
Cropper, W. P., and Gholz, H. L. (1990). Modelling the labile carbon dynamics of a Florida slash pine plantation. *Silva Carelica* 15:121–130.
Daie, J. (1985). Carbohydrate partitioning and metabolism. *Hortic. Rev.* 7:69–108.
Dale, J. E. (1988). The control of leaf expansion. *Annu. Rev. Plant Physiol. Plant Mol. Biol.* 39:267–295.
Dale, J. E., and Sutcliffe, J. F. (1986). Phloem transport. *In* "Plant Physiology a Treatise Volume IX: Water and Solutes in Plants" (F. C. Steward, ed.), pp. 455–549. Academic Press, New York.
Day, T. A., DeLucia, E. H., and Smith, W. K. (1989). Influence of cold soil and snowcover on photosynthesis and leaf conductance in two Rocky mountain conifers. *Oecologia* 80:546–552.
Deans, J. D., and Ford, E. D. (1986). Seasonal patterns of radial root growth and starch dynamics in plantation-grown Sitka spruce trees of differing ages. *Tree Physiol.* 1: 241–251.
DeLucia, E. H. (1986). Effect of low root temperature on net photosynthesis, stomatal conductance, and carbohydrate concentration in Engelmann spruce (*Picea engelmannii* Parry ex Engelm.). *Tree Physiol.* 2:143–154.
Dick, J. M., Jarvis, P. G., and Barton, C. V. M. (1990). Influence of male and female cones on assimilate production of *Pinus contorta* trees within a forest stand. *Tree Physiol.* 7:49–63.
Dick, J. M., Jarvis, P. G., and Leakey, R. R. B. (1991). Influence of male and female cones on needle CO_2 exchange rates of field-grown *Pinus contorta* Doug. trees. *Funct. Ecol.* 5:422–432.
Dickmann, D. I., and Kozlowski, T. T. (1968). Mobilization by *Pinus resinosa* cones and shoots of ^{14}C-photosynthate from needles of different ages. *Am. J. Bot.* 55:900–906.
Dickmann, D. I., and Kozlowski, T. T. (1970a). Mobilization and incorporation of photo-assimilated ^{14}C by growing vegetative and reproductive tissues of adult *Pinus resinosa* Ait. trees. *Plant Physiol.* 45:284–288.

Dickmann, D. I., and Kozlowski, T. T. (1970b). Photosynthesis by rapidly expanding green strobili of *Pinus resinosa*. *Life Sci.* 9 (Part 2, No. 10):549–552.

Dickmann, D. I., and Kozlowski, T. T. (1973). Water, nutrient and carbohydrate relations in growth of *Pinus resinosa* ovulate strobili. *In* "Proceedings of the First All Union Symposium on Sexual Reproduction in Conifers," pp. 195–209. Novosibirsk, U.S.S.R.

Dickson, R. E. (1989). Carbon and nitrogen allocation in trees. *Ann. Sci. For.* 46 (Suppl.):631s–647s.

Dosskey, M. G., Linderman, R. G., and Boersma, L. (1990). Carbon-sink stimulation of photosynthesis in Douglas-fir seedling by some ectomycorrhizas. *New Phytol.* 115: 269–274.

Drew, A. P., and Ledig, F. T. (1981). Seasonal patterns of CO_2 exchange in the shoot and root of loblolly pine seedlings. *Bot. Gaz.* 142:200–205.

Eis, S., Garman, E. H., and Ebell, L. F. (1965). Relation between cone production and diameter increment of douglas fir (*Pseudotsuga menziesii* (Mirb.) Franco), grand fir (*Abies grandis* (Dougl.) Lindl.), and western white pine (*Pinus monticola* Dougl.). *Can. J. Bot.* 43:1553–1559.

Ericsson, T., Rytter, L., and Linder, S. (1992). Nutritional dynamics and requirements of short rotation forests. *In* "Ecophysiology of Short Rotation Forest Crops" (C. P. Mitchell, J. B. Ford-Robertson, T. Hinckley, and L. Sennerby-Forsse, eds.), pp. 35–65. Elsevier, London.

Evans, J. R. (1989). Photosynthesis and nitrogen relationships in leaves of C_3 plants. *Oecologia* 78:9–19.

Gezelius, K., Ericsson, A., Hällgren, J.-E., and Brunes, L. (1981). Effects of bud removal in Scots pine (*Pinus sylvestris*) seedlings. *Physiol. Plant.* 51:181–188.

Gholz, H. L., Fisher, R. F., and Pritchett, W. L. (1985). Nutrient dynamics in slash pine plantation ecosystems. *Ecology* 66:647–659.

Gholz, H. L., Ewel, K. C., and Teskey, R. O. (1990). Water and forest productivity. *For. Ecol. Manage.* 30:1–18.

Gholz, H. L., Vogel, S. A., Cropper, W. P., McKelvey, K., Ewel, K. C., Teskey, R. O., and Curran, P. J. (1991). Dynamics of canopy structure and light interception in *Pinus elliottii* stands, north Florida. *Ecol. Monogr.* 61:33–51.

Gollan, T., Davies W. J., Schurr, U., and Zhang, J. (1989). Control of gas exchange: Evidence for root-shoot communication on drying soil. *Ann. Sci. For.* 46 (Suppl.): 393s–400s.

Gordon, J. C., and Gatherum, G. E. (1968). Photosynthesis and growth of selected Scotch pine populations. *Silva Fenn.* 2:183–194.

Gordon, J. C., and Larson, P. R. (1970). Redistribution of ^{14}C-labeled reserve food in young red pines during shoot elongation. *For. Sci.* 16:14–20.

Grier, C. C., and Running, S. W. (1977). Leaf area of mature northwestern coniferous forest: Relation to site water balance. *Ecology* 58:893–899.

Grochowska, M. J. (1973). Comparative studies on physiological and morphological features of bearing and nonbearing spurs of the apple tree. I. Changes in starch content during growth. *J. Hortic. Sci.* 48:347–356.

Guinn, G., and Mauney, J. R. (1980). Analysis of CO_2 exchange assumptions: Feedback control. *In* "Predicting Photosynthesis for Ecosystem Models" (J. D. Hesketh and J. W. Jones, eds.), Vol. 2, pp. 1–16. CRC Press, Boca Raton, Florida.

Hermann, R. K. (1977). Growth and production of tree roots: A review. *In* "The Belowground Ecosystem: A Synthesis of Plant-Associated Processes" (J. K. Marshall, ed.), pp. 7–28. Range Sci. Dept. Sci. Ser. No. 26, Colorado State Univ., Fort Collins, Colorado.

Hinckley, T. M., Lassoie, J. P., and Running, S. W. (1978). Temporal and spatial variations in the water status of forest trees. *For. Sci. Monogr.* 20:1–72.

Hinckley, T. M., Teskey, R. O., Duhme, F., and Richter, H. (1981). Temperate hardwood forests. *In* "Water Deficits and Plant Growth Volume VI Woody Plant Communities" (T. T. Kozlowski, ed.), pp. 153–208. Academic Press, New York.

Höll, W. (1985). Seasonal fluctuation of reserve materials in the trunkwood of spruce [*Picea abies* (L.) Karst.]. *J. Plant Physiol.* 117:355–362.

Hsiao, T. C. (1973). Plant response to water stress. *Annu. Rev. Plant Physiol.* 25:519–570.

Jenkins, P. A., Hellmers, H., Edge, E. A., Rook, D. A., and Burdon, R. D. (1977). Influence of photoperiod on growth and wood formation of *Pinus radiata*. *N. Z. J. For. Sci.* 7:172–191.

Kanninen, M. (1985). Shoot elongation in Scots pine: Diurnal variations and response to temperature. *J. Exp. Bot.* 36:1760–1770.

Kaufmann, M. R. (1968). Water relations of pine seedlings in relation to root and shoot growth. *Plant Physiol.* 43:281–288.

Keller, T. (1973). CO_2 exchange of bark of deciduous species in winter. *Photosynthetica* 7:320–324.

Kozlowski, T. T. (1962). Photosynthesis, climate, and growth of trees. *In* "Tree Growth" (T. T. Kozlowski, ed.), pp. 149–164. Ronald, New York.

Kozlowski, T. T. (1992). Carbohydrate sources and sinks in woody plants. *Bot. Rev.* 58:107–222.

Kozlowski, T. T., and Borger, G. A. (1971). Effect of temperature and light intensity early in ontogeny on growth of *Pinus resinosa* seedlings. *Can. J. For. Res.* 1:57–65.

Kozlowski, T. T., and Gentile, A. C. (1958). Respiration of white pine buds in relation to oxygen availability and moisture content. *For. Sci.* 4:147–152.

Kozlowski, T. T., Kramer, P. J., and Pallardy, S. G. (1991). "The Physiological Ecology of Woody Plants." Academic Press, San Diego.

Krueger, K. W. (1967). Nitrogen, phosphorus, and carbohydrates in expanding and year-old douglas-fir shoots. *For. Sci.* 13:352–356.

Krueger, K. W., and Trappe, J. M. (1967). Food reserves and seasonal growth of Douglas-fir seedlings. *For. Sci.* 13:352–356.

Kuhns, M. R., and Gjerstad, D. H. (1988). Photosynthate allocation in loblolly pine (*Pinus taeda*) seedlings as affected by moisture stress. *Can. J. For. Res.* 18:285–291.

Kuhns, M. R., and Gjerstad, D. H. (1991). Distribution of ^{14}C labelled photosynthate in loblolly pine (*Pinus taeda*) seedlings as affected by season and time of exposure. *Tree Physiol.* 8:259–271.

Larson, P. R. (1963a). Stem form and silviculture. *Proc. Soc. Am. Foresters*, 103–107.

Larson, P. R. (1963b). Stem form and development of forest trees. *For. Sci. Monogr.* No. 5.

Lavigne, M. B. (1991). Effects of thinning on the allocation of growth and respiration in young stands of balsam fir. *Can. J. For. Res.* 21:186–192.

Lenz, F., and Küntzel, U. (1974). Carbohydrate content of citrus as affected by fruit load. *Gartenbauwissenschaft* 39:99–101.

Leverenz, J. W. (1981). Photosynthesis and transpiration in large forest-grown Douglas-fir: Interactions with apical control. *Can. J. Bot.* 59:2568–2576.

Linder, S. (1985). Potential and actual production in Australian forest stands. *In* "Research for Forest Management" (J. J. Landsberg and W. Parsons, eds.), pp. 11–35. CSIRO, Melbourne, Australia.

Linder, S., and Axelsson, B. (1982). Changes in carbon uptake and allocation patterns as a result of irrigation and fertilization in a young *Pinus sylvestris* stand. *In* "Carbon Uptake and Allocation: Key to Management of Subalpine Forest Ecosystems" (R. H. Waring, ed.), pp. 38–44. IUFRO Workshop, For. Res. Lab., Oregon State Univ., Corvallis.

Linder, S., and Rook, D. A. (1984). Effects of mineral nutrition on carbon dioxide exchange and partitioning of carbon in trees. *In* "Nutrition of Plantation Forests" (G. D. Bowen and E. K. S. Nambiar, eds.), pp. 211–236. Academic Press, New York.

Lopushinsky, W., and Kaufmann, M. R. (1984). Effects of cold soil on water relations and spring growth of Douglas-fir seedlings. *For. Sci.* 30:628–634.

Luxmoore, R. J. (1981). CO_2 and phytomass. *BioScience* 31:626.

Luxmoore, R. J. (1991). A source-sink framework for coupling water, carbon, and nutrient dynamics of vegetation. *Tree Physiol.* 9:267–280.

McMurtrie, R. E., Benson, M. L., Linder, S., Running, S. W., Talsma, T., Crane, W. J. B., and Myers, B. J. (1990). Water/nutrient interactions affecting the productivity of stands of *Pinus radiata*. *For. Ecol. Manage.* 30:415–423.

Maier, C. A., and Teskey, R. O. (1992). Internal and external control of net photosynthesis and stomatal conductance of mature eastern white pine (*Pinus strobus*). *Can. J. For. Res.* 22:1387–1394.

Mead, D. J., Draper, D., and Madgwick, H. A. I. (1984). Dry matter production of a young stand of *Pinus radiata*: Some effects of nitrogen fertilizer and thinning. *N. Z. J. For. Sci.* 14:97–108.

Meier, C. E., Grier, C. C., and Cole, D. W. (1985). Below and above ground N and P use by *Abies amabilis* stands. *Ecology* 66:1928–1942.

Miller, H. G. (1984). Dynamics of nutrient cycling in plantation ecosystems. *In* "Nutrition of Plantation Forests" (G. D. Bowen and E. K. S. Nambiar, eds.), pp. 53–78. Academic Press, New York.

Nambiar, E. K. S. (1980). Root configuration and root regeneration in *Pinus radiata* seedlings. *N. Z. J. For. Sci.* 10:249–263.

Nambiar, E. K. S., and Fife, D. N. (1991). Nutrient retranslocation in temperate conifers. *Tree Physiol.* 9:185–207.

Norman, J. M., and Jarvis, P. G. (1974). Photosynthesis in sitka spruce (*Picea sitchensis* (Bong.) Carr.). III. Measurements of canopy structure and interception of radiation. *J. Appl. Ecol.* 11:375–398.

Olofinboba, S., and Kozlowski, T. T. (1973). Accumulation and utilization of carbohydrate reserves in shoot growth of *Pinus resinosa*. *Can. J. For. Res.* 3:346–353.

Oren, R., and Schulze, E.-D. (1989). Nutritional disharmony and forest decline: A conceptual model. *In* "Forest Decline and Air Pollution: A Study on Spruce" (*Picea abies*) on acid soils. (E.-D.Schulze, O. L. Lange, and R. Oren, eds.), Ecol. Studies Ser. Vol. 77, pp. 425–443. Springer-Verlag, Berlin.

Oren, R., and Zimmermann, R. (1989). CO_2 assimilation and the carbon balance of healthy and declining Norway spruce stands. *In* "Forest Decline and Air Pollution: A Study on Spruce (*Picea abies*) on Acid Soils" (E.-D. Schulze, O. L. Lange, and R. Oren, eds.), Ecol. Studies. Ser. Vol. 77, pp. 352–369. Springer-Verlag, Berlin.

Oren, R., Werk, K. S., Buchmann, N., and Zimmermann, R. (1993). Chlorophyll-nutrient relationships identify nutritionally caused decline in *Picea abies* stands. *Can. J. For. Res.* 23:1187–1195.

Passioura, J. B. (1988). Root signals control leaf expansion in wheat seedlings growing in drying soil. *Aust. J. Plant Physiol.* 15:687–693.

Paul, E. A., and Kucey, R. M. N. (1981). Carbon flow in plant microbial associations. *Science* 213:473–474.

Pelkonen, P., and Luukonen, O. (1974). Gas exchange in three populations of Norway spruce. *Silvae Genet.* 23:160–164.

Penning de Vries, F. W. T., Murphy, C. E., Jr., Wells, C. G., and Jorgensen, J. R. (1975). Simulation of nitrogen distribution and its effect on productivity in even aged loblolly pine plantations. *In* "Mineral Cycling in Southeastern Ecosystems" (F. G. Howell, J. B. Gentry, and M. H. Smith, eds.), ERDA Symposium Ser. (Conference 740513), pp. 70–83. National Technical Information Center, Springfield, Virginia.

Pereira, J. S. (1990). Whole-plant regulation and productivity in forest trees. *In* "Importance of Root to Shoot Communication in the Responses to Environmental Stress" (W. J. Davies and B. Jeffcoat, eds.), Monograph 21, pp. 237–250. British Society for Plant Growth Regulation, Bristol.

Quinlan, J. D. (1965). The pattern of distribution of ^{14}C in a potted rootstock following assimilation of $^{14}CO_2$ by a single leaf. *Annu. Rep. East Malling Res. Stn. Kent,* 117–118.

Reid, C. P. P., Kidd, F. A., and Ekwebelam, S. A. (1983). Nitrogen nutrition, photosynthesis and carbon allocation in ectomycorrhizal pine. *Plant Soil* 71:415–432.

Ritchie, G. A., and Dunlap, J. R. (1980). Root growth potential: Its development and expression in forest tree seedlings. *N. Z. J. For. Sci.* 10:218–248.

Rook, D. A., and Whyte, A. G. D. (1976). Partial defoliation and growth of 5-year-old radiata pine. *N. Z. J. For. Sci.* 6:40–56.

Rook, D. A., Swanson, R. H., and Cranswick, A. M. (1977). Reaction of radiata pine to drought. *In* "Proceedings of the Soil and Plant Water Symposium," pp. 55–67. Massey Univ., Palmeston North, New Zealand.

Ross, S. D. (1972). The seasonal and diurnal source–sink relationships of photoassimilated ^{14}C in the Douglas-fir branch. Ph.D. Thesis, University of Washington, Seattle.

Russell, R. S. (1977). "Plant Root Systems: Their Function and Interaction with the Soil," pp. 169–192. McGraw-Hill, London.

Santantonio, D. (1989). Dry-matter partitioning and fine-root production in forests—New approaches to a difficult problem. *In* "Biomass Production by Fast-Growing Trees" (J. S. Pereira and J. J. Landsberg, eds.), pp. 57–72. Kluwer Academic Publ., Dordrecht, The Netherlands.

Sasaki, S., and Kozlowski, T. T. (1968a). The role of cotyledons in early development of pine seedlings. *Can. J. Bot.* 46:135–140.

Sasaki, S., and Kozlowski, T. T. (1968b). Effects of herbicides on seed germination and early seedling development of *Pinus resinosa. Bot. Gaz.* 129:238–246.

Sasaki, S., and Kozlowski, T. T. (1970). Effects of cotyledon and hypocotyl photosynthesis on growth of young pine seedlings. *New Phytol.* 69:493–500.

Sasek, T. W., DeLucia, E. H., and Strain, B. R. (1985). Reversibility of photosynthetic inhibition in cotton after long-term exposure to elevated CO_2 concentrations. *Plant Physiol.* 78:619–622.

Schier, G. A. (1970). Seasonal pathways of ^{14}C-photosynthate in red pine labelled in May, July, and October. *For. Sci.* 16:2–13.

Schulze, E.-D., Fuchs, M. I., and Fuchs, M. (1977a). Spacial distribution of photosynthetic capacity and performance in a mountain spruce forest of northern Germany. I. Biomass distribution and daily CO_2 uptake in different crown layers. *Oecologia* 29:43–61.

Schulze, E.-D., Fuchs, M., and Fuchs, M. I. (1977b). Spacial distribution of photosynthetic capacity and performance in a mountain spruce forest of northern Germany. III. The significance of the evergreen habit. *Oecologia* 30:239–248.

Sestak, Z. (ed.) (1985). "Photosynthesis during Leaf Development." Dr. W. Junk Publ., Dordrecht, The Netherlands.

Sharkey, T. D., and Vanderveer, P. J. (1989). Stromal phosphate concentration is low during feedback limited photosynthesis. *Plant Physiol.* 91:679–684.

Sheriff, D. W., and Rook, D. A. (1990). Wood density and above-ground growth in high and low density clones of *Pinus radiata* D. Don. *Aust. J. Plant Physiol.* 17:615–628.

Simpson, J. D., and Powell, G. R. (1981). Some factors influencing cone production in young black spruce in New Brunswick. *For. Chron.* 57:267–269.

Snowdon, P. (1990). Effect of water on growth of *Pinus radiata* D. Don. *In* "Forest Soils and Modern Forest Management" (Bo-Qun Lin, ed.), pp. 3–8. Publishing House of Northeast Forestry Univ., Harbin, China.

Snowdon, P., and Benson, M. L. (1992). Effects of combinations of irrigation and fertilization on the growth and above ground biomass production of *Pinus radiata*. *For. Ecol. Manage.* 52:87–116.

Stitt, M. (1991). Rising CO_2 levels and their potential significance for carbon flow in photosynthetic cells. *Plant, Cell Environ.* 14:741–762.

Strain, B. R., Higginbotham, K. O., and Mulroy, J. C. (1976). Temperature preconditioning and photosynthetic capacity of *Pinus taeda* L. *Photosynthetica* 10:47–53.

Stransky, J. J., and Wilson, D. R. (1964). Terminal elongation of loblolly and shortleaf pine seedlings under soil moisture stress. *Soil Sci. Soc. Am. Proc.* 28:439–440.

Sweet, G. B., and Wareing, P. F. (1966). Role of plant growth in regulating photosynthesis. *Nature (London)* 210:77–79.

Sweet, G. B., and Wareing, P. F. (1968). A comparison of the rates of growth and photosynthesis in first-year seedlings of four provenances of *Pinus contorta* Dougl. *Ann. Bot.* 32:735–751.

Tenhunen, J. D., Hesketh, J. D., and Gates, D. M. (1980). Leaf photosynthesis models. *In* "Predicting Photosynthesis for Ecosystem Models" (J. D. Hesketh and J. W. Jones, eds.), Vol. 1, pp. 123–181. CRC Press, Boca Raton, Florida.

Teskey, R. O., Hinckley, T. M., and Grier, C. C. (1984). Temperature-induced change in the water relations of *Abies amabilis* (Dougl.) Forbes. *Plant Physiol.* 74:77–80.

Teskey, R. O., Bongarten, B. C., Cregg, B. M., Dougherty, P. M., and Hennessey, T. C. (1987). Physiology and genetics of tree growth response to moisture and temperature stress: An examination of the characteristics of loblolly pine (*Pinus taeda* L.). *Tree Physiol.* 3:41–61.

Teskey, R. O., Gholz, H. L., and Cropper, W. P., Jr. (1994). Influence of climate and nutrient availability on net photosynthesis of mature slash pine. *Tree Physiol.*, in press.

Terry, N., Waldron, L. J., and Taylor, S. E. (1983). Environmental influences on leaf expansion. *In* "The Growth and Functioning of Leaves" (J. E. Dale and F. L. Milthorpe, eds.), pp. 179–205. Cambridge Univ. Press, Cambridge.

Titus, J. S., and Kang, S. M. (1982). Nitrogen metabolism, translocation, and recycling in apple trees. *Hortic. Rev.* 4:204–246.

Trewavas, A. (1986). Understanding the control of plant development and the role of growth substances. *Aust. J. Plant Physiol.* 13:447–457.

Troeng, E., and Langström, B. (1991). Gas exchange in young Scots pine following pruning of current shoots. *Ann. Sci. For.* 48:359–366.

van den Driessche, R. (1984). Nutrient storage, retranslocation, and relationship of stress to nutrition. *In* "Nutrition of Plantation Forests" (G. D. Bowen and E. K. S. Nambiar, eds.), pp. 181–209. Academic Press, New York.

Van Oosten, J.-J., and Besford, R. T. (1994). Sugar feeding mimics effect of acclimation to high CO_2—Rapid down regulation of RuBisCO small subunit transcripts but not of the large subunit transcripts. *J. Plant Physiol.* 143:306–312.

Vapaavuori, E. M., Rikala, R., and Ryyppo, A. (1992). Effects of root temperature on growth and photosynthesis in conifer seedlings during shoot elongation. *Tree Physiol.* 10:217–230.

Vogt, K. A., Grier, C. C. Meier, C. E., and Edmonds, R. L. (1982). Mycorrhizal role in net primary production and nutrient cycling in *Abies amabilis* ecosystems in western Washington. *Ecology* 63:370–380.

Walter, H., and Breckle, S.-W. (1989). "Ecological Systems of the Geobiosphere. 3. Temperate and Polar Zonobiomes of Northern Eurasia." Springer-Verlag, Berlin.

Wareing, P. F., Khalifa, M. M., and Treharne, K. J. (1968). Rate limiting processes in photosynthesis at saturating light intensities. *Nature (London)* 220:453–457.

Waring, R. H., Newman, K., and Bell, J. (1981). Efficiency of tree crowns and stemwood production at different canopy leaf densities. *Forestry* 54:129–137.

Waring, R. H., and Schlesinger, W. H. (1985). The carbon balance of trees. *In* "Forest Ecosystems Concepts and Management" (R. H. Waring and W. H. Schlesinger, eds.), pp. 7–37. Academic Press, Orlando, Florida.

Warrington, I. J., Rook, D. A., Morgan, D. C., and Turnbull, H. L. (1988). The influence of simulated shadelight and daylight on growth, development and photosynthesis of *Pinus radiata, Agathis australis* and *Dacrydium cupressium. Plant, Cell Environ.* 12:343–356.

Watson, M. A., and Casper, B. B. (1984). Morphogenetic constraints on patterns of carbon distribution in plants. *Annu. Rev. Ecol. Syst.* 15:233–258.

Webb, W. L. (1977). Rate of current photosynthate accumulation in roots of Douglas-fir seedlings: Seasonal variation. *In* "The Belowground Ecosystem: A Synthesis of Plant-Associated Processes" (J. K. Marshall, ed.), Range Sci. Dept. Sci. Ser. No. 26., pp. 149–152. Colorado State Univ., Fort Collins, Colorado.

Weikert, R. M., Welder, M., Lippert, M., Schramel, P., and Lange, O. L. (1989). Photosynthetic performance, chloroplast pigments, and mineral content of various needle age classes of spruce (*Picea abies*) with and without the new flush: An experimental approach for analyzing forest decline phenomena. *Trees* 3:161–172.

Went, F. W. (1974). Reflections and speculations. *Annu. Rev. Plant Physiol.* 25:1–26.

Whittaker, R. H., and Woodwell, G. M. (1968). Dimension and production relations of trees and shrubs in the Brookhaven forest, New York. *J. Ecol.* 56:1–25.

Will, G. M., and Hodgkiss, P. D. (1977). Influence of nitrogen and phosphorus stresses on the growth and form of radiata pine. *N. Z. J. For. Sci.* 7:307–320.

Wolsvinkel, P. (1985). Phloem unloading and turgor sensitive transport: Factors involved in sink control of assimilate partitioning. *Physiol. Plant.* 65:331–339.

Wood, G. B. (1968). Photosynthesis and growth in *Pinus radiata* D. Don as affected by environmental factors and inherent qualities. Ph. D. Thesis, Australian National University, Canberra.

Woodrow, I. E., and Berry, J. A. (1988). Enzymatic regulation of photosynthetic CO_2 fixation in C_3 plants. *Annu. Rev. Plant Physiol. Plant Mol. Biol.* 39:533–594.

Zahner, R. (1962). Terminal growth and wood formation by juvenile loblolly pine under two soil moisture regimes. *For. Sci.* 8:345–352.

Zahner, R. (1968). Water deficits and growth of trees. *In* "Water Deficits and Plant Growth" (T. T. Kozlowski, ed.), pp. 99–110. Academic Press, New York.

7

Carbon Allocation and Accumulation in Conifers

Stith T. Gower, J. G. Isebrands, and David W. Sheriff

I. Introduction

Forests cover approximately 33% of the land surface of the earth, yet they are responsible for 65% of the annual carbon (C) accumulated by all terrestrial biomes (Schlesinger, 1991). In general, total C content and net primary production rates are greater for forests than for other biomes, but C budgets differ greatly among forests. Despite several decades of research on forest C budgets, there is still an incomplete understanding of the factors controlling C allocation. Yet, if we are to understand how changing global events such as land use, climate change, atmospheric N deposition, ozone, and elevated atmospheric CO_2 affect the global C budget, a mechanistic understanding of C assimilation, partitioning, and allocation is necessary. Numerous abiotic and biotic factors influence C allocation patterns, which in turn affect the capacity of plants to obtain resources from the atmosphere and soil.

Although reviews on various components of conifer forest C budgets, such as photosynthesis (see Chapter 4, this volume) and detritus production (Vogt *et al.*, 1986), are available, a synthesis of the influence of abiotic and biotic factors on leaf, canopy, and stand-level C budgets of conifer forests is lacking. The objective of this chapter is to review the major factors that influence C allocation and accumulation in conifer trees and forests. In keeping with the theme of this book, we will focus primarily on evergreen conifers. However, even among evergreen conifers, leaf, canopy, and stand-level C and nutrient allocation patterns differ, often as a function of leaf development and longevity (Gower and Richards, 1990; Gower *et al.*, 1993a; Reich *et al.*, 1994).

The terminology related to C allocation literature is often inconsistent, confusing and inadequate for understanding and integrating past and current research. For example, terms often used synonymously to describe C flow or movement include translocation, transport, distribution, allocation, partitioning, apportionment, and biomass allocation. A common terminology is needed because different terms have different meanings to readers. In this paper we use C allocation, partitioning, and accumulation according to the definitions of Dickson and Isebrands (1993). Partitioning is the process of C flow into and among different chemical, storage, and transport pools. Allocation is the distribution of C to different plant parts within the plant (i.e., source to sink). Accumulation is the end product of the process of C allocation.

II. Distribution of Conifer Forests

Evergreen conifers occur from tropical to boreal forests and from temperate rainforests to dry woodlands (Walter, 1979; Kikuzawa, 1991). Conifers commonly dominate in temperate regions only after disturbance or where soil infertility is low. The dominance of evergreen conifers in the Pacific Northwest of the United States can be explained partly by the unique environmental conditions of dry, warm days and cool nights in the summer and mild, moist conditions in the winter (Waring and Franklin, 1979). Although conifers occur on many of the major soil orders, conifers, especially pines, are more abundant on nutrient-poor soils (Miller *et al.*, 1979). Nitrogen availability commonly limits net primary production (NPP) in temperate and boreal conifer forests whereas phosphorus availability limits NPP in subtropical and tropical environments (Ballard, 1984). In summary, temperature, moisture, and nutrient conditions differ greatly among evergreen conifer forests.

To help understand assimilation and allocation patterns of conifer forests, we briefly contrast the major differences in C and nutrient cycles of evergreen conifer and deciduous forests. Perhaps the most striking difference between evergreen conifers and broadleaf deciduous species is leaf longevity, which is correlated to a number of leaf, canopy, and ecosystem structural and functional characteristics (Gower *et al.*, 1993a; Reich *et al.*, 1994). Even among conifer species leaf longevity ranges from less than a year for *Larix* spp. and *Taxodium* spp. to greater than 40 years for *Pinus longaeva*. Leaf litterfall nitrogen content is often less for evergreen conifers than for broadleaf deciduous forests (Vogt *et al.*,

1986). The litter of evergreen conifers decomposes slower than that from deciduous trees due to its lower litter quality (i.e., higher lignin and lower nitrogen concentrations), resulting in greater forest floor mass beneath conifer forests compared to broadleaf deciduous forests. Soil nitrogen availability is commonly lower in evergreen conifer forests than in deciduous forests due to the positive feedback of leaf litterfall quality on soil nitrogen dynamics (Gower and Son, 1992). In addition, evergreen conifer forests have lower annual nitrogen uptake rates compared to deciduous forests (Gosz, 1981; Nadelhoffer *et al.*, 1984; Son and Gower, 1991).

III. Controls on Carbon Assimilation

Net primary production of a seedling, tree, or forest is the balance between total canopy photosynthesis (gross primary production; GPP) and the amount of C lost via respiration (autotrophic respiration). There are numerous direct and indirect feedbacks on C assimilation and allocation in forests (Fig. 1). For example, net canopy photosynthesis is the product of net photosynthetic rate and photosynthetic surface area integrated over selected daily and seasonal time periods, and net primary production is the difference between net canopy photosynthesis and autotrophic respiration of woody tissues. Clearly, net primary production cannot be estimated without a complete understanding of the influence of biophysical controls on physiological processes such as photosynthesis and respiration (Chapter 4) and on the allocation of C to components such as leaf area and live woody biomass

Under optimal environmental conditions, NPP is linearly related to intercepted photosynthetically active radiation (IPAR) for crop plants and tree seedlings (Monteith, 1977) and forests (Linder, 1985; Landsberg, 1986). Factors that influence the relationship include (1) canopy reflectance, (2) canopy architecture and persistance of the canopy during the year, and (3) light use efficiency (Cannell, 1989). The first factor that affects the relationship between NPP and IPAR is canopy albedo, which averages 15–20% for deciduous forests versus 10–15% for conifer forests (Rosenberg *et al.*, 1983). Canopy characteristics related to architecture include rate of leaf area development, maximum leaf area, and leaf area duration. The rate of leaf area development in the spring, although important for deciduous species, is gradually less important for evergreen conifers that retain their needles for a greater number of years because the new foliage comprises an increasingly smaller fraction of the total leaf area (Gower *et al.*, 1993a). Leaf area duration positively influences annual IPAR. For example, Cannell *et al.* (1987) reported a

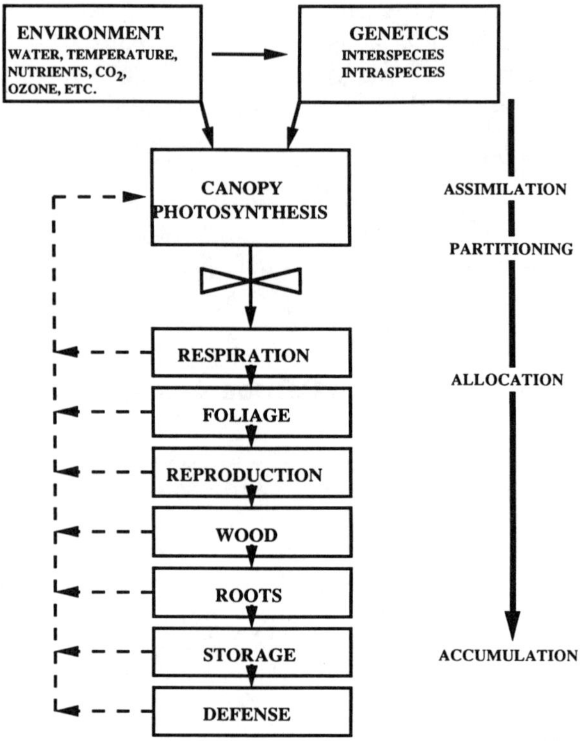

Figure 1 Conceptual model illustrating the major components of a forest carbon budget as they relate to carbon assimilation, partitioning, allocation, and accumulation.

linear decrease in IPAR with later dates of canopy development for a *Salix viminalis* stand. Another important factor influencing IPAR is the leaf area index (LAI) of the tree or forest. IPAR is curvilinearly related to LAI, with 95% of the incoming radiation being absorbed in a forest with a LAI of 6 (Fig. 2), assuming radiation attenuation approximates the Beer–Lambert Law. Major factors governing LAI include environmental constraints (see Section V,B) and biotic factors such as leaf longevity (Gower *et al.*, 1993a; Reich *et al.*, 1994).

Crown architecture also influences radiation attenuation, which affects IPAR. Light extinction coefficients (k) for conifers range from 0.4 to 0.6 (Jarvis and Leverenz, 1983) and for a constant LAI, IPAR generally increases with k, especially for forests with LAIs from 1 to 6 (Fig. 2). Factors that influence k include shoot angle, clumping of foliage on the shoot, and spectral reflectance. As conifer shoots are held more erect (i.e., perpendicular to the soil surface), k decreases (Jordan and Smith, 1993). Clumped foliage such as needles on conifer shoots allows more

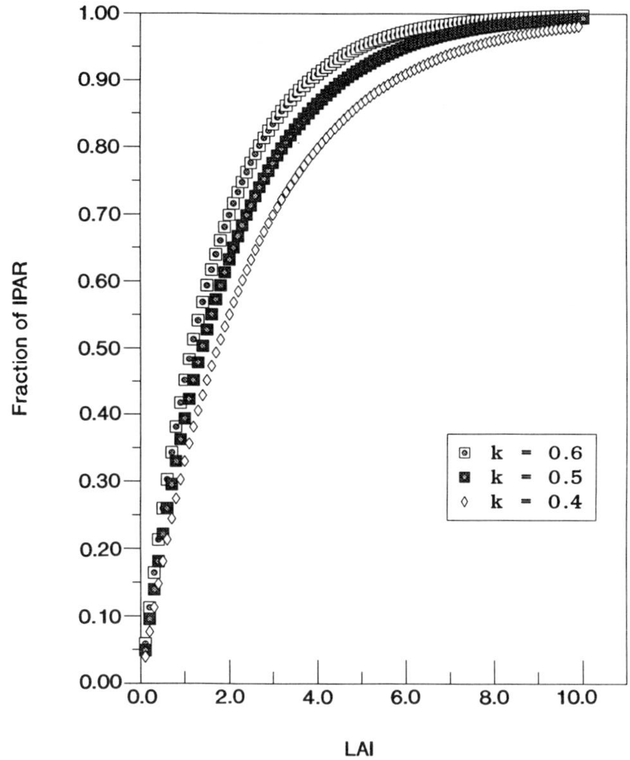

Figure 2 Relationship between intercepted photosynthetic active radiation (IPAR) and leaf area index (LAI) for three light extinction coefficients (0.4, 0.5, 0.6) common for conifers.

radiation to penetrate deeper in the canopy. In fact, Leverenz (1980) speculated that the shoot morphology of conifers maximizes radiation use efficiency within the canopy and later demonstrated that the ratio of shoot silhouette area: total projected needle area of shade shoots was positively correlated to aboveground net primary production (ANPP) of evergreen conifer forests (Leverenz and Hinckley, 1990; Leverenz, 1992).

The third major factor influencing the relationship between NPP and IPAR is light use efficiency, or ε, defined as dry biomass accumulated per unit of intercepted radiation (Monteith, 1977). Values of ε from field studies range from 0.5 to 1.4 (Cannell, 1989), falling well short of the theoretical maximum value of 6.0 calculated by Jarvis and Leverenz (1983). Most estimates of ε do not include belowground NPP, but including belowground dry matter production would increase ε by 20–50% for conifer forests (see Section V,C). Multiple constraints on photosyn-

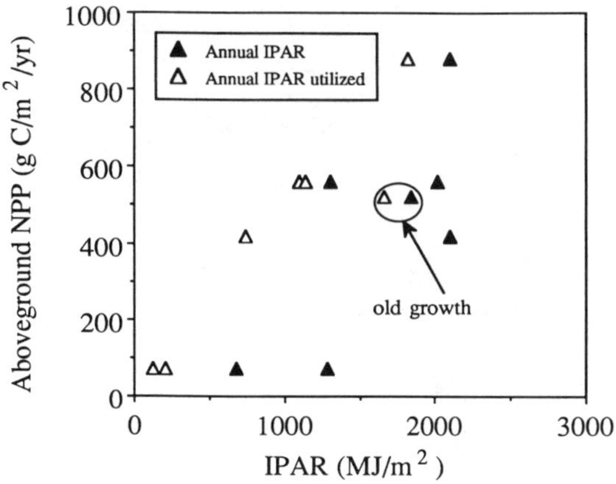

Figure 3 Relationship between aboveground net primary production (NPP) and intercepted photosynthetic active radiation (IPAR) for seven conifer forests along a precipitation gradient in Oregon. Closed symbols are for total IPAR and open symbols are for utilized IPAR after subtracting the IPAR that was not utilized by the trees due to environmental constraints (cold temperature, vapor pressure deficit, and water stress) on photosynthesis. (Adapted from Runyon *et al.*, 1994.)

thesis imposed by environmental stresses may in part explain the inefficient use of radiation (Teskey and Whitehead 1994). For example, Runyon *et al.* (1994) reported a strong positive linear relationship between ANPP and IPAR for seven conifer forests located along a large environmental gradient in western Oregon only if they excluded the proportion of IPAR that was not utilized because of environmental constraints on net canopy photosynthesis (Runyon *et al.*, 1994) (Fig. 3). Efforts to model environmental constraints cannot simply use leaf responses because of differences in physical structure and nonlinear relationships between light and photosynthesis (Norman, 1993). For example, changes in photosynthetic photon flux density (PPFD) have similar effects on C assimilation of a canopy and of its constituent needles; however, the canopy is light saturated at a higher PPFD than are needles at the top of the canopy because C assimilation of needles lower in the canopy increases with PPFD (Chapter 4).

IV. Carbon Allocation Patterns in Conifers

Carbon allocation patterns within conifers are highly integrated and are determined at early stages of development. These patterns are regulated by complex interactions between competing sources and sinks within the plant (Dickson, 1989), where a source is a net exporter of photosynthate and a sink is a net importer of photosynthate (Dickson, 1991). Moreover, an intimate interdependency between the shoot and root exists throughout the life of a plant (Kozlowski, 1971a,b). Carbon allocation patterns in all woody plants are closely related to the source–sink interactions determined by shoot growth (Fig. 4). Sink strength varies markedly by species, genotype (clone), shoot type, age of plant, location within the plant, season, and environmental conditions (Kozlowski, 1992).

A. Plant Genetic Factors

There are three basic phenological patterns of C allocation to shoots in woody plants (Dickson, 1991; Kozlowski, 1992). The first pattern is typical of many conifers and is associated with determinate (or fixed) shoot growth that is characterized by a single, short burst of shoot growth in the late spring and early summer followed by a long lag period of budset. Distribution of assimilate is dependent on the flush cycle, with most of the assimilate (i.e., >90%) allocated to shoots during the flushing episode, but is largely directed (i.e., >95%) to the lower stem and roots during the lag stage (Fig. 4). Conifers with this pattern include many cold-temperate and boreal genera (or species) such as *Abies*, *Picea*, *Pseudotsuga*, *Tsuga*, *Pinus ponderosa*, *Pinus resinosa*, *Pinus strobus*, and *Pinus sylvestris*. The second pattern is associated with indeterminate (or free) shoot growth, characterized by continuous shoot growth over most of the growing season. Distribution of assimilate in these plants varies with the stage of leaf development. Developing needles are net importers of assimilate until they become fully expanded, at which time they export both acropetally to developing shoots and basipetally to the stem and roots. Mature leaves export assimilate almost exclusively to the lower stem and roots. Conifers exhibiting this pattern include *Ginkgo*, *Larix*, and some tropical pines. The third pattern is associated with semideterminate (or recurrent flushing) shoot growth, characterized by periodic flushes of shoot growth with intermediate lag stages. Assimilate is transported to developing shoots during a flush and downward to the stem and roots during the lag stage; the cycle is repeated during each consecutive flushing episode. Conifer species with this pattern include warm-temperate pines such as *Pinus echinata*, *Pinus elliottii*, *Pinus palustris*, and *Pinus taeda*, and subtropical pines such as *Pinus caribaea* and *Pinus radiata*.

B. Mechanisms of Carbon Allocation

To understand fully C partitioning and allocation in conifers, it is important to understand their patterns and the underlying mechanisms. There have been numerous reviews of the mechanisms of C partitioning

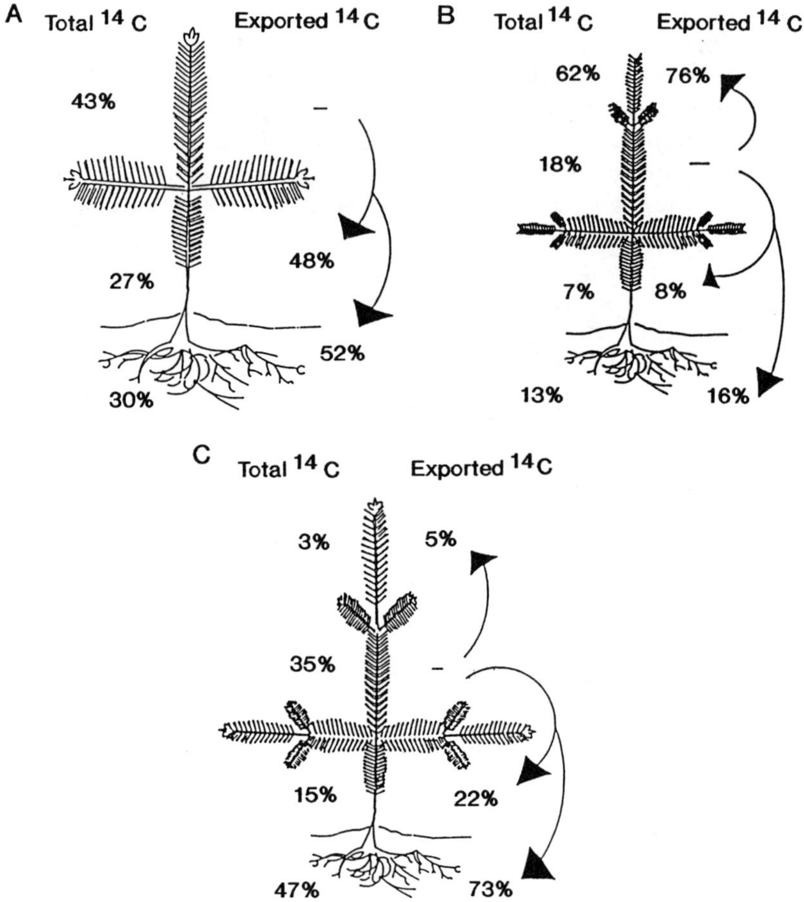

Figure 4 Allocation patterns of ^{14}C-labeled photosynthate from the second flush of mature needles of *Pinus resinosa*. Total ^{14}C (%) recovered is on the left and exported ^{14}C (%) is on the right. (A) A second flush seedling after budset; (B) during active growth of the third flush; and (C) at end of third flush after budset. (Adapted from Gordon and Larson, 1968, 1970.)

and allocation in herbaceous plants (Gifford and Evans, 1981; Geiger, 1986; Wyse, 1986; Ho, 1988; Stitt and Quick, 1989) and entire books have been written on the subject (Cronshaw *et al.*, 1986). Less is known about woody plants, especially conifers, but there is little reason to believe that the controls for C allocation will differ between herbaceous or crop plants and woody plants if similar plant ontogenetic stages and temporal factors are considered (Dickson, 1991) (see Table I). Unfortunately, the controlling mechanisms of C allocation still remain largely unknown, although some of the genetic, biochemical, and physiological

Table I List of Carbon Allocation and Partitioning Studies in Conifers[a]

Genus and species	Subject	Reference
Abies balsamea	Partitioning, seasonal	Little (1970)
Abies spp.	Seasonal patterns	Loach and Little (1973)
Juniperus virginiana	Juvenile–mature patterns	Miller *et al.* (1990)
Larix spp.	4- to 5-year old trees	Schneider and Schmitz (1989)
Picea rubens	Branches, elevation	Anderson *et al.* (1991)
Pinus banksiana	Allocation, seedlings	Balantinecz *et al.* (1966)
Pinus banksiana	Cuttings–seedlings	Haissig (1984)
Pinus elliottii	Reserves, saplings	Allen (1964)
Pinus ponderosa	Field trees	Smith and Paul (1988)
Pinus resinosa	Reserves	Kozlowski and Winget (1964)
Pinus resinosa	Allocation, seedlings	Larson (1964)
Pinus resinosa	Reproduction	Dickmann and Kozlowski (1968)
Pinus resinosa	Ontogeny	Gordon and Larson (1968)
Pinus resinosa	Partitioning	Rangnekar and Forward (1969)
Pinus resinosa	Branches	Rangnekar *et al.* (1969)
Pinus resinosa	Reproduction	Dickmann and Kozlowski (1970)
Pinus resinosa	Ontogeny	Gordon and Larson (1970)
Pinus resinosa	Seasonal patterns	Schier (1970)
Pinus resinosa	Partitioning	Rangnekar and Forward (1972, 1973)
Pinus strobus	Allocation, seedlings	Shroya *et al.* (1966)
Pinus strobus	Seasonal patterns	Ursino and Paul (1973)
Pinus sylvestris	Photosynthesis–growth	Rutter (1957)
Pinus sylvestris	20-year-old tree	Ericsson (1978)
Pinus sylvestris	Ontogeny	Makela *et al.* (1980)
Pinus sylvestris	Partitioning, seedlings	Ericsson (1984)
Pinus taeda	Partitioning	Chung and Barnes (1977)
Pinus taeda	Partitioning	Chung and Barnes (1980)
Pinus taeda	Partitioning–fertilization	Adams *et al.* (1986)
Pinus taeda	Allocation	Kuhns and Gjerstad (1991)
Pseudotsuga menziesii	Ontogeny	Webb (1977)
Pseudotsuga menziesii	Ectomycorrhizae	Dosskey *et al.* (1990)
Pseudotsuga menziesii	Dark respiration, 3-year-old tree	Webb (1991)

[a] List includes representative references in each genus, published in English only.

aspects of the regulation of transport from sources and sinks are becoming more clear (Cronshaw *et al.*, 1986; see also Chapter 6, this volume).

Among the various conclusions that apply to C allocation in conifer trees and forests, there is increasing evidence that the structural and functional diversity among categories of plants is greater than once thought. Therefore, the mechanisms controlling C allocation and partitioning may not be universal and may be specific to species, genotypes, tissue type, or ontogentic stage (Cronshaw *et al.*, 1986). For example, there are differences in C partitioning and allocation patterns in seedlings and in trees, making extrapolation from seedlings to trees problematic (Cregg *et al.*, 1989). Genetic control is more important in C assimilation and allocation processes than once thought (Haissig *et al.*, 1992; Gower *et al.*, 1994). Also, the root and shoot interdependently produce the essential growth substances needed to maintain a functional balance in plants (Kozlowski, 1971a). The main control of C allocation between roots and shoots is related to sink activity, including sucrose synthesis (Stitt and Quick, 1989; Dickson, 1991), respiration (Gifford and Evans, 1984), phloem loading and unloading (Ho, 1988), hormone synthesis (Gifford and Evans, 1981), and cell turgor and genetic control (Wyse, 1986). Another important factor is the proximity of the source to the sink and their vascular connections (Gifford *et al.*, 1984; Dickson, 1989). Thus, the key to understanding the mechanisms of C allocation in plants seems to be with the properties of sinks, including the initial establishment of the sink at an early developmental stage, duration of sink growth, as well as sink effects on photosynthesis (Gifford and Evans, 1981).

V. Controls on Carbon Allocation

Cost–benefit theory suggests that plants have evolved to maximize resource use efficiency (Bloom *et al.*, 1985); therefore, C allocation patterns should theoretically vary depending on the availability of resource(s) that vary at different temporal and spatial scales. With few exceptions, environmental conditions differ enough during the year such that different resources control C assimilation and allocation during different times of the year. Futhermore, C assimilation and allocation patterns vary within and among conifers due to changes in the magnitude of resource limitations. Below, we briefly examine controls on C allocation.

A. Respiration

Autotrophic respiration is a poorly understood component of forest C budgets. Respiration is commonly separated into construction and maintenance components. Ion uptake respiration is also recognized (Amthor, 1986; Ryan, 1991), but so few estimates exist for plants that our focus is on the former two components. Together, construction and maintenance respiration are a significant C cost. For example, Ryan and Waring (1992) estimated that maintenance respiration comprised about 18% of GPP for a subalpine *Pinus contorta* forest; Linder and Axelsson (1982) estimated that autotrophic respiration comprised 34% of annual photosynthesis for a *P. sylvestris* tree; Kinerson *et al.* (1977) estimated that autotrophic respiration comprised 34% of GPP for a young *P. taeda* forest; and Kira (1975) reported that autotrophic respiration could approach 65% of GPP in a tropical forest.

The quantity of C allocated to respiration in conifer forests is strongly influenced by temperature. Based on C budgets for four conifer forests in contrasting climates, the percentage of net canopy photosynthesis allocated to autotrophic respiration is positively correlated to mean annual temperature (Ryan *et al.*, in press). The amount of accumulated living biomass, which changes with stand age and environmental conditions (see Section VI), is also an important factor influencing respiration. Runyon *et al.* (1994) speculated that an old-growth conifer forest allocated a disproportionately greater fraction of GPP to maintenance respiration compared to younger stands (Fig. 3); however, Ryan and Waring (1992) found only a slight increase in sapwood maintenance respiration in comparing a 40- and a 245-year-old *P. contorta* forest. One of the most critical questions related to forest C budgets pertains to how anticipated climate change will affect net canopy photosynthesis (GPP − autotrophic respiration) versus heterotrophic respiration, because the difference reflects the capacity of the forest to store C.

B. Foliage

Carbon allocated to foliage is partitioned to chemical fractions for structure, physiologically active materials, defense, and storage. Leaf structure or specific leaf area (SLA) is affected by numerous environmental factors. The maximum net photosynthetic rate under optimal conditions is positively correlated to SLA; however, low nutrient availability, increased needle longevity, and greater water vapor deficit reduce SLA (Linder and Rook, 1984; Reich *et al.*, 1991, 1992; Gower *et al.*, 1993a). Therefore, the actual SLA is a compromise between maximizing instantaneous C gain and long-term C, water, and nutrient use efficiency.

Water availability strongly influences the allometry between sapwood cross-sectional area and leaf area and the allocation of C to new foliage production. Leaf area (LA) is positively correlated to sapwood cross-sectional area (SA) for conifers (Grier and Waring, 1974), with the ratio being greater for conifers in more mesic environments (Fig. 5). Water availability also exerts a strong control on C allocation to new foliage production. Gower *et al.* (1992) demonstrated that for Douglas fir (*Pseudotsuga menziesii*), irrigated trees allocated more C to new foliage production than did similar-diameter control trees. Severe drought adversely affects the xylem water potential in the meristem at the base of needles (Zahner, 1968), which reduces needle length (Linder *et al.*, 1987).

Because nitrogen concentration is greater for foliage than for many

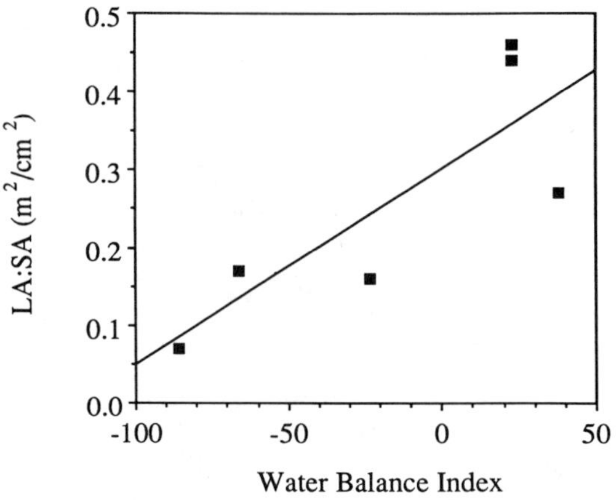

Figure 5 Leaf area:cross-sectional sapwood area ratios (LA:SA) for evergreen conifers along a water balance index gradient in Oregon. Water balance index (WB) is defined WB = P − E + SWC, where P and E are precipitation and evaporation for the study period, respectively, and SWC is soil water storage capacity. (Adapted from Waring *et al.*, 1982; Gholz, 1982.)

other tree tissues, and conifers commonly grow on nutrient limited soil, nitrogen fertilization increases the annual allocation of C to foliage (Brix and Mitchell, 1983; Grier *et al.*, 1984; Gholz *et al.*, 1991; Gower *et al.*, 1992, 1993b). The greater allocation of C to new foliage production results in greater (1) needle length, (2) production of needles per shoot, and (3) production of new shoots by fertilized than by control trees (Brix and Ebell, 1969; Brix, 1981; Gower *et al.*, 1992, 1993b). Greater nutrient availability often increases the new:total foliage mass ratio (Fig. 6). Because net photosynthesis is intrinsically greater for current than for older needles (Chabot and Hicks, 1982; Teskey *et al.*, 1984; Sheriff *et al.*, 1986), reduced needle longevity increases net canopy photosynthesis and production efficiency. Nutrient availability can also influence the allometry between total foliage mass or leaf area and stem diameter (Brix and Mitchell, 1983; Grier *et al.*, 1984; Gower *et al.*, 1993b); however, increased nutrient availability may not increase the total foliage mass of a tree if water or light availability is more limiting (Vose and Allen, 1988; Gower *et al.*, 1993b).

Many of the influences of water and nutrient availability on C allocation at the tree level also occur at the stand level. Water availability also governs new foliage production at the stand level (Gower *et al.*, 1992; Raison *et al.*, 1992a,b; Snowdon and Benson, 1992). Grier and Running (1977) and Gholz (1982) noted that total foliage mass or area was positively correlated to water availability for different conifer species. Irrigation studies have also confirmed a similar relationship (Gower *et al.*,

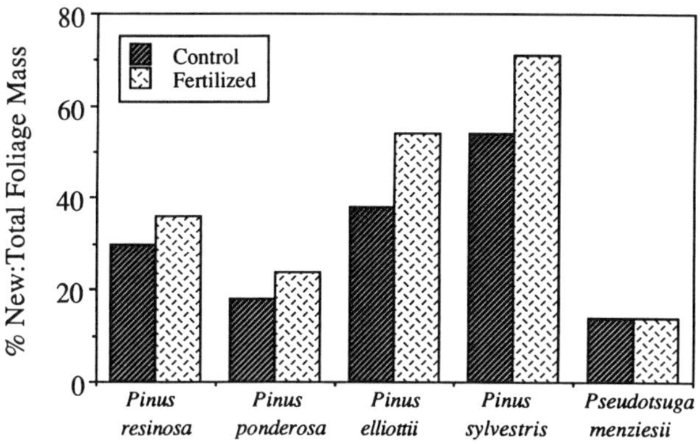

Figure 6 New: total foliage mass ratio for control and fertilized conifer forests. (Data are from Gholz *et al.*, 1991; Gower *et al.*, 1992; Linder and Axelsson, 1982; S. T. Gower unpublished data.)

1992; Raison *et al.*, 1992a,b; Snowdon and Benson, 1992), although the absolute increase is often smaller within than among conifer species. Extreme cold temperatures may reduce the amount of C accumulated in foliage below the physiological limit defined by water availability (Gholz, 1982). Fertilization often increases C accumulation in foliage as a result of increased new foliage production (Linder and Axelsson, 1982; Vose and Allen, 1988; Gholz *et al.*, 1991; Gower *et al.*, 1992; Raison *et al.*, 1992a,b; Snowdon and Benson, 1992; Fassnacht *et al.*, 1994).

C. Fine Roots

The ability of plant roots to acquire nutrients and water affects C assimilation and allocation (Fig. 1). Plant growth is positively correlated to the availability of soil resources, therefore physical barriers that constrict root growth adversely affect plant growth (Thomas and Strain, 1991). These responses are at least partially mediated by hormonal root–shoot communication (Davies *et al.*, 1986; Schulze, 1986; Turner, 1986; Passioura, 1988; Pereira, 1990) and can affect C allocation and accumulation. The above-mentioned effects of nutrition on C allocation involve changes in hormonal balance, but other mechanisms may be described more simply. For example, greater allocation of C to roots in nutrient-deficient soils can occur because roots are nearest to the nutrients. In nutrient-poor soils, roots use most of the limited absorbed nutrients, leaving little for translocation to aboveground tissues, thereby reducing their capacity to grow (Clarkson, 1985). This also explains the proliferation of roots at microsites of high resource availability in nutrient-poor soils (St. John, 1983; St. John *et al.*, 1983).

A second poorly understood component of forest C budgets is the

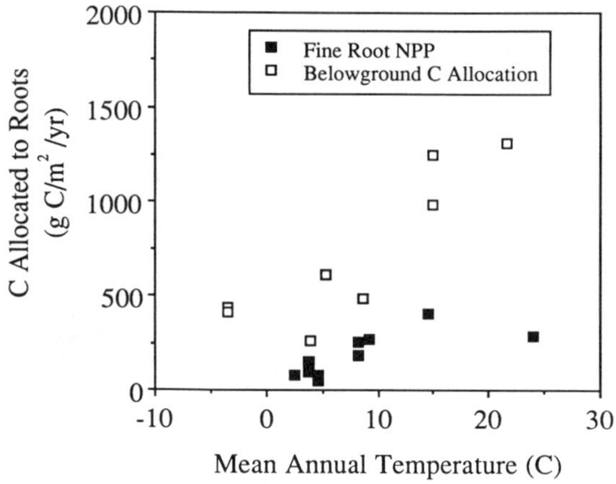

Figure 7 Fine root net primary production and total allocation of carbon belowground (see Raich and Nadelhoffer, 1989) for conifer forests in relation to mean annual temperature.

allocation of C to fine roots and mycorrhizae. A significant amount of C is allocated to fine roots annually, even though fine roots comprise a small fraction of the C content of forest trees. Based on a review of the world literature, annual allocation of C to fine root production for conifer forests ranges from 30 g C/m^2/yr for a *Picea rubens* forest in Ontario, Canada, to 710 g C/m^2/yr for a subalpine *Abies amabilis* forest in Washington state (Vogt *et al.*, 1986) (Table II). On a global scale, fine root NPP and total allocation of C to roots (i.e., root production + root respiration) are positively correlated to mean annual temperature (Fig. 7). Gower *et al.* (1994) also reported a strong positive correlation ($r^2 = 0.75$) between fine root NPP and mean annual temperature for pine forests. The difference between the two regression lines in Fig. 7 should approximate the quantity of C allocated to coarse root production and root respiration.

Water availability also influences belowground C allocation patterns. Annual C allocation to fine roots is greater in xeric than in mesic conifer forests (Santantonio and Hermann, 1985; Comeau and Kimmins, 1989). Gower *et al.* (1992) also reported that fine root production (absolute basis) was lower in irrigated and irrigated + wood chip forests (high water availability but low nitrogen availability) than in control Douglas fir forests, suggesting that water availability directly controls C allocation to fine roots and mycorrhizae.

Another important factor influencing fine root NPP is soil nutrient availability; however, there is uncertainty surrounding both the magnitude and direction of the effect. Both comparative and experimental studies have reported greater (absolute basis) fine root NPP for infertile than fertile *Pseudotsuga menziesii* forests in the western United States (Keyes and Grier, 1981; Kurz, 1989; Vogt *et al.*, 1990; Gower *et al.*, 1992), whereas other scientists have reported only a relative decrease in the amount of C allocated to fine roots as soil nitrogen availability increases (Linder and Axelsson, 1982; Comeau and Kimmins, 1989). Fertilization also decreases C allocation to mycorrhizae (Menge *et al.*, 1977; Menge and Grand, 1978). However, Aber *et al.* (1985) and Nadelhoffer *et al.* (1985) reported that fine root NPP is positively correlated to annual N mineralization for different forest types (including conifers) in Wisconsin. The disparity in the results may be related to the different methods used to estimate fine root NPP or may reflect differences in the relative importance of nutrition in controlling C assimilation and allocation in different climates.

Few studies have attempted to quantify how much C is allocated to mycorrhizae or to determine if the benefits provided by mycorrhizae exceed the C costs to the plant. Vogt *et al.* (1982) reported that 150 g C/m^2/yr, or 15% of total net primary production, was allocated to mycorrhizae net production in an *Abies amabilis* stand. Bevege *et al.* (1975) found that 15 times more ^{14}C-labeled assimilate was allocated to mycorrhizal-infected versus nonmycorrhizal roots. Respiration costs of mycorrhizae are largely unknown. Despite the large amount of C allocated to mycorrhizae, they are critical to the growth and survival of conifers. Growth rates are often 5- to 10-fold greater for innoculated versus nonmycorrhizal conifer seedlings (Lamb and Richards, 1971).

D. Woody Tissue

At the tree level, stem wood, stem bark, branch, and coarse root biomass are strongly correlated to stem diameter, although these relationships differ both within and among species. Differences in allometric relationships for stem wood and stem diameter within a species may reflect influences of environmental factors such as water and nutrient availability on tree form, whereas differences in allometric relationships between stem bark and stem diameter among species maybe an adaptation to fire (Gower *et al.*, 1987). Of all the woody components, the allometry between coarse root mass and stem diameter differs the least within and among conifer species (see Santantonio *et al.*, 1977) and may be explained by physical principles (Coutts, 1983).

Wood NPP rates for conifer forests range from 84 to 2105 g C/m^2/yr and commonly increase from cold to warm climates (Table II). When under stress, conifers allocate less C to woody tissue such as stem, branch, and coarse roots than to foliage and fine roots (Waring and Pitman, 1985; Gower *et al.*, 1992). Relative allocation of C to woody tissue

Table II Net Primary Production by Component for Evergreen Conifer Forests

Species	Age (yr)	Location	Biome[a]	Net primary production (g C/m²/yr) Woody[b]	Foliage	Fine roots	Source
Abies amabilis	23	Washington, U.S.A.	CT	270	57	311	Grier *et al.* (1981)
Abies amabilis	180	Washington, U.S.A.	CT	144	46	585	Grier *et al.* (1981)
Picea abies		Wisconsin, U.S.A.	CT	240	121	80	Nadelhoffer *et al.* (1985)
Pinus contorta[c]	70	British Columbia, Canada	CT	130	54	245	Comeau and Kimmins (1989)
Pinus contorta[d]	74	British Columbia, Canada	CT	275	108	210	Comeau and Kimmins (1989)
Pinus elliottii	27	Florida, U.S.A.	ST/WT	520	161	175	Gholz *et al.* (1986)
Pinus radiata	12	Puruki, New Zealand	ST/WT	1205	216	100	Beets and Pollock (1987)
Pinus radiata	12	Puruki, New Zealand	ST/WT	2105	288	100	Beets and Pollock (1987)
Pinus resinosa	>100	Wisconsin, U.S.A.	CT	201	113	99	Nadelhoffer *et al.* (1985)
Pinus strobus	>100	Wisconsin, U.S.A.	CT	275	129	129	Nadelhoffer *et al.* (1985)
Pinus sylvestris	28	Tammela, Finland	B	84	40	41[i]	Mälkönen (1974)
Pinus sylvestris	45	Tammela, Finland	B	175	76	70[i]	Mälkönen (1974)
Pinus sylvestris	47	Tammela, Finland	B	143	63	74[i]	Mälkönen (1974)
Pinus sylvestris	>100	Vilppula, Finland	B	161	54	74[i]	Paavilainen (1980)
Pinus sylvestris	20[e]	Jädraås, Sweden	B	157	73	200–236	Linder and Axelsson (1982)
Pinus sylvestris	20[f]	Jädraås, Sweden	B	497	220	106–124	Linder and Axelsson (1982)
Pinus taeda	16	North Carolina, U.S.A.	ST/WT	711	265	466	Kinerson *et al.* (1977)

Pseudotsuga menziesii	40[c]	Washington, U.S.A.	CT	301	90	329	Keyes and Grier (1981)
Pseudotsuga menziesii	40[d]	Washington, U.S.A.	CT	568	144	118	Keyes and Grier (1981)
Pseudotsuga menziesii	40[e]	Washington, U.S.A.	CT	356	168	237	Vogt *et al.* (1990)
Pseudotsuga menziesii	40[g]	Washington, U.S.A.	CT	542	162	143	Vogt *et al.* (1990)
Pseudotsuga menziesii	50[e]	New Mexico, U.S.A.	CT	650	87	210	Gower *et al.* (1992)
Pseudotsuga menziesii	50[g]	New Mexico, U.S.A.	CT	828	108	108	Gower *et al.* (1992)
Pseudotsuga menziesii	50[h]	New Mexico, U.S.A.	CT	922	92	132	Gower *et al.* (1992)
Biome averages[j]							
Subtropical/warm temperate (ST/WT)				1135 (72)	232 (15)	210 (13)	
Cold temperate (CT)				310 (47)	102 (16)	244 (37)	
Boreal (B)				144 (48)	61 (20)	95 (32)	

[a]Climate: ST/WT, subtropical/warm temperate; CT, cold temperate; B, boreal.
[b]Woody is equal to stem plus branch plus coarse root.
[c]Low fertility.
[d]High fertility.
[e]Control.
[f]Irrigated plus fertilized.
[g]Fertilized.
[h]Irrigated.
[i]Fine root NPP estimated using fine : total NPP ratio from Linder and Axelsson (1981).
[j]Values in parentheses are percent of total net primary production.

is greater in subtropical/warm-temperate than in cold-temperate and boreal conifer forests. Increased nutrient availability increases the amount of C allocated to woody tissue (Keyes and Grier, 1981; Gholz *et al.*, 1991; Gower *et al.*, 1992; Snowdon and Benson, 1992) and can be explained by the shift in C allocation from fine roots to stem production and perhaps greater leaf area.

E. Storage

Stored C is used to construct new tissue and repair damaged tissues. The vigor of the seedling or tree is related to the amount of stored carbohydrates, suggesting that trees with depleted carbohydrate reserves are disposed to die (Sprugel, 1976; Puttonen, 1986). In contrast, vigorous trees store sufficient carbohydrates such that they can survive environmental stresses such as insect attack, pathogen infestations, or drought (Waring, 1991). Starch is the main storage form for carbohydrates in conifers (Adams *et al.*, 1986; Ford and Deans, 1977). Other important storage forms include quinic and shikimic acids (Chung and Barnes, 1977; Cranswick *et al.*, 1987) and hemicellulose (Kimura, 1969); of lesser importance are sucrose, cyclitols, and monosaccharides (Kozlowski, 1992). The importance of the storage forms varies with both tissue type and species. Interested readers are referred to Kozlowski (1992) for more detailed information on other storage forms of carbohydrates.

Carbohydrate storage patterns in conifer and deciduous tree species differ in two ways: (1) conifers store larger amounts of carbohydates in foliage and (2) conifers tend to store less carbohydrates and accumulate them later in the season (Kozlowski, 1992). Few studies have examined seasonal patterns of starch storage for all major tissues, making the construction of stand-level storage budgets difficult; however, several patterns are consistent among conifers. The amount of carbohydrates in conifer needles varies with seasonal duration of shoot growth; maximum starch concentrations occur in pines with the shortest growing season (Kozlowski, 1992). A substantial amount of carbohydrates is stored in fine roots. Ericsson and Persson (1980) reported maximum starch concentrations of 30% by weight for fine roots of *P. sylvestris*. Although the concentration of starch in fine roots may be fixed for conifer seedlings (Marshall and Waring, 1985), data for mature *P. elliottii* suggest that starch is transported into fine roots after they are formed (Gholz and Cropper, 1991). Seasonal patterns of starch storage appear to be similar for all tissues, with the maximum occurring shortly before new growth begins in the spring (Ericsson, 1979; Ericsson and Persson, 1980; Deans and Ford, 1986; Cranswick *et al.*, 1987; Oren *et al.*, 1988; Gholz and Cropper, 1991). The relative importance of the various tissues in terms of whole-tree C storage varies with age because biomass components change with age. Needles and fine roots are the primary storage organs for conifer seedlings (Glerum, 1980), but coarse roots and stems are the primary carbohydrate storage sites for mature trees (Cranswick *et al.*, 1987; Gholz and Cropper, 1991).

F. Secondary Defense Compounds

Herbivores generally consume from 5 to 15% of the foliage in temperate forests during nonoutbreak years, but can consume all of the foliage in outbreak years (Schowalter *et al.*, 1986). Although small levels of herbivory do not decrease tree growth and may increase growth (Mattson and Addy, 1975), moderate to high levels of herbivory reduce photosynthetic surface area, deplete carbohydrate storage reserves, and increase the susceptibility of trees to other environmental stresses. As a result, trees allocate a portion of their annual C gain to construct and maintain defense compounds and physical barriers to deter herbivory. Clancy *et al.* (1994) provide a thorough discussion on herbivory and secondary defense compounds, therefore our discussion is restricted to the quantity of C allocated to defense compounds and the indirect effects of defense compounds on C assimilation and allocation.

Most evergreen conifers rely on C-based defense compounds (Swain, 1977). Little information is available on the amount of C generally allocated to defense compounds in woody plants. In general, the concentration of defense compounds appears to be inversely related to the leaf life span (Coley, 1988). Also, mild water stress increases the partitioning of C to defense compounds because cell division is adversely affected by water stress before photosynthesis. For example, Horner (1987) reported greater lignin concentrations in control than in irrigated Douglas fir during a drought year but little difference in lignin concentration between the two treatments during a wetter year. Futhermore, increased nitrogen availability decreased the amount of C partitioned to lignin whereas decreased nitrogen availability tended to increase lignin concentration in new foliage (Fig. 8).

Partitioning of C to volatile trace gases (e.g., terpenes, isoprenes) may comprise a significant component of the leaf carbon budget; however, we have a poor understanding of the environmental and metabolic controls over their production, despite the important influence they have on the chemistry of the atmosphere (Brasseur and Chatfield, 1991). Moderate water stress and elevated temperatures increase the production of monoterpenes (Hodges and Lorio, 1975; Tingey *et al.*, 1980). Efforts to better quantify the partitioning of C to trace gases will require an understanding of how environmental factors and genetic variability affect whole-plant C allocation patterns and what factors control the partitioning of C to the various species of trace gases (Lerdau, 1991).

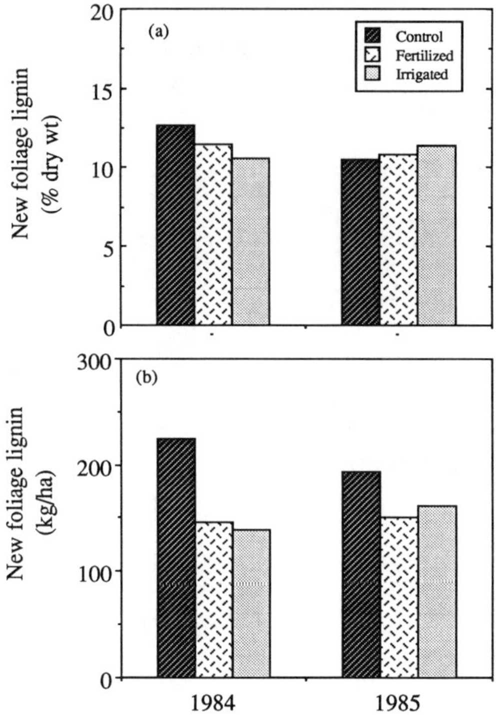

Figure 8 Needle lignin concentration and content for control, fertilized, and irrigated *Pseudotsuga menziesii* trees in New Mexico. (Adapted from Horner, 1987; Gower *et al.*, 1992.)

Defense compounds are also important because they indirectly affect C assimilation and allocation patterns. First, greater partitioning of C to defense compounds reduces stored C. Second, constitutive defenses such as lignin reduce the net photosynthetic rate, resulting in lower net C canopy gain. Third, defense compounds such as polyphenolics retard decomposition (Handley, 1961; Benoit and Starkey, 1964a,b; Davies *et al.*, 1964; Horner *et al.*, 1988). Foliage lignin : N ratios exert a strong control on N mineralization rates in forests (Melillo *et al.*, 1982; Gower and Son, 1992), which in turn influence net primary primary production and C allocation patterns.

G. Reproduction

Reproductive structures such as flowers, fruits, strobili, and seeds are strong C sinks and can consume 6–10% of the annual net canopy photosynthesis. Many trees produce more flowers than they can sustain to maturity, consequently when adverse environmental conditions restrict

C assimilation, flowers are aborted (Kozlowski, 1992). The time span during which strobili are strong carbohydrate sinks for conifers ranges from one season (i.e., *Abies*, *Larix*, *Picea*, and *Pseudotsuga*) to over two growing seasons and one winter for *Pinus* (Kozlowski, 1992). The C required to construct reproductive structures is commonly derived from stored carbohydrates during the early spring when net canopy photosynthesis is very low; however, later in the spring most of the carbohydrates are obtained from current photosynthate produced largely by 1-year-old needles (Kozlowski, 1992). Few data are available on the proportion of C allocated to reproduction for conifer forests. Linder and Troeng (1981) estimated that a 14-year-old *P. sylvestris* tree allocated 6% of its canopy photosynthate to cones and Fielding (1960) estimated that the amount of C that *P. radiata* allocated to cones was about 16% of stem production. Cone production has been thought to decrease stem production (Fielding, 1960; Linder and Rook, 1984); however, more recent analyses for *P. contorta* suggest that cone production does not affect stem production (Dick *et al.*, 1991).

H. Leaf Leachates and Root Exudates

Leaching of C from foliage is strongly influenced by wetting properties. Leaves with thick waxy cuticles, such as those on evergreen conifers, are not readily leached. However, damage to the epicuticular wax covering needles increases needle permeability and leaching loss of carbohydrates (Turunen and Huttunen, 1990). For example, increased needle permeability and leaching of carbohydrates have been reported for foliage subjected to artificial acidic misting (Scherbatskoy and Klein, 1983; Mengel *et al.*, 1988). In general, C leaching losses do not appear to be a major C flux.

Root exudation sites are predominantly restricted to regions of cell elongation and the zone of meristematic activity directly behind the root tip (Rovira, 1969; Rovira and Davey, 1974; Hale *et al.*, 1978). Although the type and amount of root exudate are influenced by species, atmospheric CO_2 concentration, and other environmental conditions (Bowen, 1969; Rovira, 1969; Norby *et al.*, 1987), the loss of C in exudates is a small flux relative to other fluxes discussed in this chapter. Rovira (1969) estimated that C contained in root exudates rarely exceeds 0.4% of the amount of C photosynthesized and Norby *et al.* (1987) estimated that loss of ^{14}C by exudation from *P. echinata* roots was about 0.5% of the total carbon allocated to roots. Despite the small amount of C lost in root exudates, root exudates play an important role in nutrient cycling because they influence nutrient availability and uptake indirectly through their effects on microbial activity in the soil (Uren and Reisenauer, 1988).

VI. Changes in Carbon Dynamics during Stand Development

Carbon allocation and accumulation patterns change as trees age. Most notably, the relative allocation of C to foliage reaches a maximum early in stand development and then remains stable or declines in older stands (Gower *et al.*, 1994). The total amount of C allocated to foliage reaches an asymptote earlier in more favorable climates, as previously noted for pine forests; however, the ratio of foliage:wood decreases with stand age regardless of species or climate (Fig. 9). Aboveground NPP commonly reaches steady state or declines in older pine forests

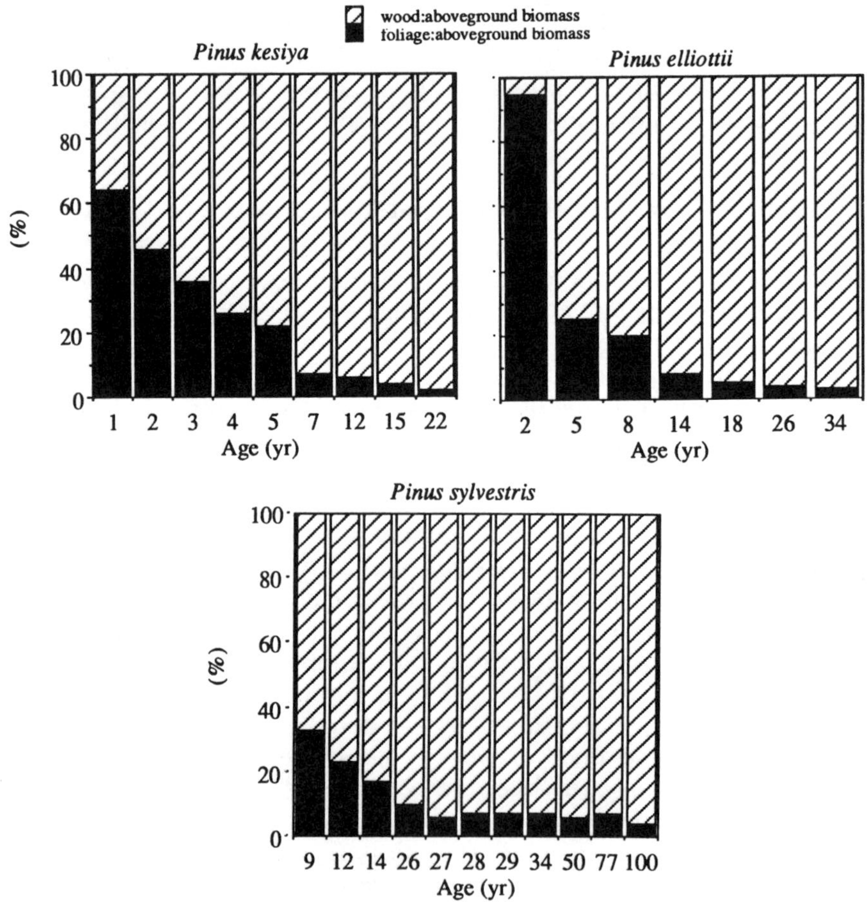

Figure 9 Foliage and wood aboveground C accumulation ratios for tropical (*Pinus kesiya*), warm-temperate (*Pinus elliottii*), and boreal (*Pinus sylvestris*) pines in contrasting climates. (Adapted from Gower *et al.*, 1994.)

(Gower *et al.*, 1994). The imbalance in C accumulation of foliage versus respiring tissue is the basis for the often-cited (but not rigorously tested) hypothesis that NPP decreases with stand age due to an imbalance between the accumulation of photosynthetic and respiring tissue (Kira and Shedei, 1967). However, this explanation for decreased NPP has been questioned (Sprugel, 1984; Ryan and Waring, 1992). Other explanations for decreased aboveground NPP in older conifer forests include decreased nutrient availability in older stands (Van Cleve *et al.*, 1981; Gholz *et al.*, 1985) and reduced stomatal conductance in mature trees (Ryan and Waring, 1992).

Few studies have compared belowground carbon accumulation and production patterns across a conifer forest age chronosequence. Vogt *et al.* (1987) found that maximum fine root NPP occurs earlier in high versus low site quality coastal *Pseudotsuga menziesii* forests. However, irrespective of site quality, fine root production and growth efficiency often peak simultaneously. This pattern suggests that maximum fine root and mycorrhizal NPP may occur during stand development when the ratio of the mass of photosynthesizing to respiring tissue is greatest, although additional data are needed to test this hypothesis rigorously.

VII. Multiple Resource Limitations

Thus far, we have largely focused on the effect of a single resource limitation on C assimilation and allocation; however, conifer forests routinely experience more than one environmental constraint concurrently. Unfortunately, little is known about the effects of concurrent multiple biotic and abiotic stresses on carbon allocation in plants (Dickson and Isebrands, 1991). Below, we provide several examples illustrating the nonlinear response of physiological processes controlling C budgets due to multiple resource limitations.

Fertilization commonly increases new foliage production by conifers (Fig. 10a), but the vertical distribution of increased foliage production varies depending on the light environment of the canopy (Fig. 10b). However, the increased new foliage production resulting from fertilization may not increase total foliage mass (Fig. 10c) if water is the primary resource limiting the amount of foliage or leaf area a site can support. Therefore, it is not surprising that the greatest increase in foliage mass (+40%) occurs for the nutrient-poor *P. resinosa* forests in a cool, moist climate, whereas little or no increase in total foliage mass occurs for fertilized *P. ponderosa* forests in an extremely dry climate. Other studies have also noted that the combined effect of fertilization and irrigation on foliage production and accumulation exceeded the additive effect of

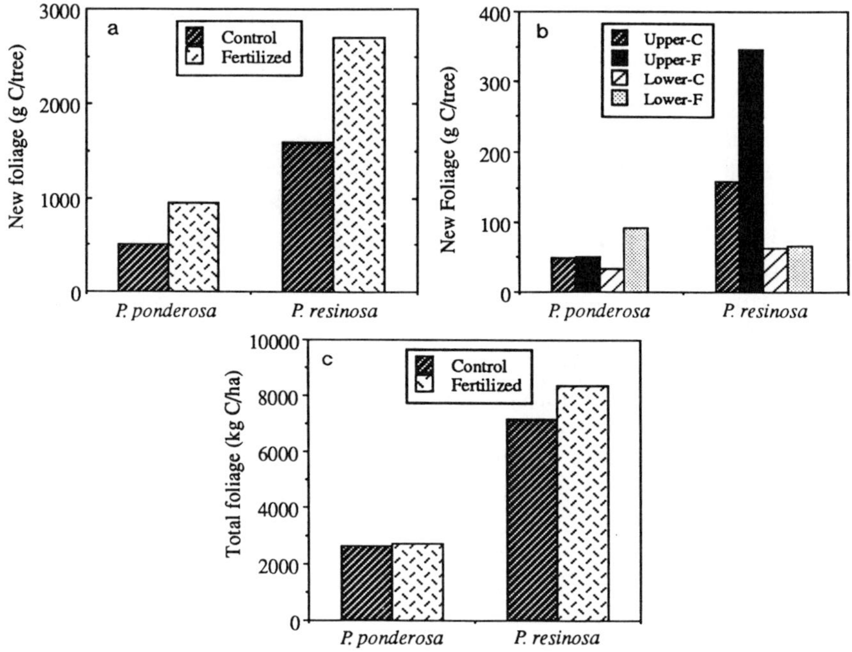

Figure 10 Influence of fertilization on (a) relationship between new foliage mass (g/tree) and stem diameter; (b) relationship between new foliage mass (g/tree) in the upper and lower third of the canopy and stem diameter; and (c) total foliage mass (kg/ha). (Data from Gower *et al.* 1993a; S. T. Gower, unpublished data.)

each resource when applied by itself (Linder, 1987; Raison *et al.*, 1992; Snowdon and Benson, 1992).

Air pollutants also influence C assimilation, but much less information is available on the effects on C partitioning and allocation (Kozlowski and Constantinidou, 1986; Pye, 1988). There have been numerous studies of air pollutants on carbon assimilation in conifers (Kozlowski and Constantinidou, 1986). Potential effects of elevated atmospheric CO_2 concentrations on C assimilation, allocation, partitioning, and accumulation are reviewed by Eamus and Jarvis (1989), Bazzaz (1990), Fararr and Williams (1991), and Woodward *et al.* (1991). Most of what is known is from studies on seedlings with a single pollutant such as sulfur dioxide, ozone, acid precipitation, and CO_2 (Table III). Air pollutants, however, are more likely to act concurrently (i.e., ozone and CO_2) or with other environmental stresses, to influence C budgets of conifers (Peterson *et al.*, 1989; McLaughlin and Kohut, 1992). For example, the magnitude of seedling or tree growth in response to elevated CO_2 varies depending on successional status, light environment, and soil nutrient availabil-

Table III List of Studies of the Effects of Air Pollutants on Carbon Allocation and Partitioning in Conifers[a]

Genus and species	Pollutant	Reference
Abies fraseri	Ozone	Tseng *et al.* (1988)
Picea abies	Ozone–acid precipitation	Barnes *et al.* (1990)
Picea rubens	Ozone	Alscher *et al.* (1989)
Picea rubens	Ozone	Amundson *et al.* (1991)
Picea rubens	Acid precipitation	McLaughlin *et al.* (1993)
Picea rubens	Ozone–acid precipitation	Taylor *et al.* (1986)
Pinus banksiana	SO_2	Malhotra and Sarkar (1979)
Pinus contorta	SO_2	Kahn and Malhotra (1977)
Pinus ponderosa	Ozone	Miller *et al.* (1969)
Pinus ponderosa	Ozone	Tingey *et al.* (1976)
Pinus taeda	Ozone	Adams *et al.* (1990)
Pinus taeda	Ozone–acid precipitation	Edwards *et al.* (1992)
Pinus taeda	Ozone–acid precipitation	Friend *et al.* (1992)
Pinus taeda	Ozone	Meier *et al.* (1990)
Pinus taeda	Ozone	Sasek *et al.* (1991)
Pinus taeda	Ozone	Tjoelker and Luxmoore (1991)
Pinus spp.	Ozone	Barnes (1972a,b)
Pinus sylvestris	Ozone	Skarby *et al.* (1987)
Pinus strobus	Ozone	Mann *et al.* (1980)
Pinus strobus	Ozone	McLaughlin *et al.* (1982)
Pinus strobus	Ozone	Wilkinson and Barnes (1973)

[a] List includes representative references in each genus, published in English only.

ity (Eamus and Jarvis, 1989; Bazzaz *et al.*, 1990). Ecosystem responses to elevated CO_2 are also likely to vary depending on the availability of other resources. Using a forest ecosystem process model FOREST-BGC (Running and Coughlan, 1988; Running and Gower, 1991), Running and Nemani (1991) simulated gross net primary production, net primary production, and respiration for two conifer forests in contrasting climates (short growing season/dry summer and long growing season/moist summer). Increased temperature (+4°C) and precipitation (+10%) decreased net canopy photosynthesis at both sites, although the decrease was substantially greater for the conifer forest in the subtropical environment due to excessive respiration costs throughout the year (Fig. 11). Assuming elevated CO_2 will increase water use efficiency (see Woodward *et al.*, 1991), NPP may increase for conifer forests that experience water stress whereas increased water use efficiency may have little or no effect on NPP for subtropical forests due to the greater respiration costs. Moreover, the effects of pollutants on C processes can vary with respect to genetics (Sasek *et al.*, 1991). These examples illustrate the non-

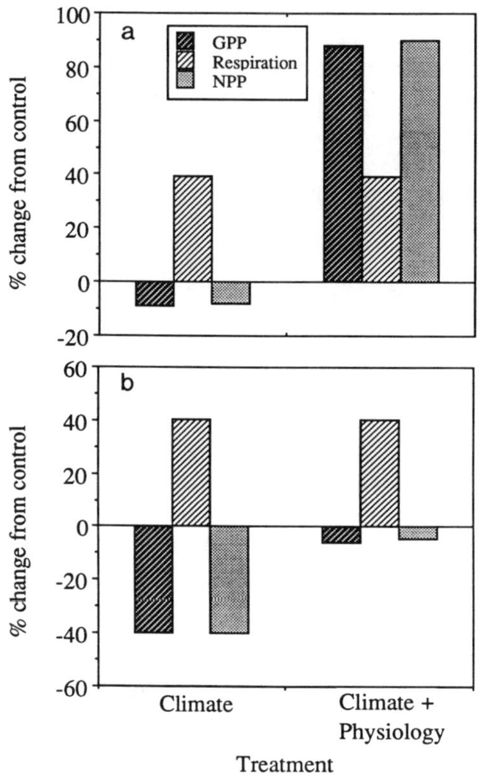

Figure 11 Simulated influence of climate change (+4° C and +10% precipitation) and climate change plus water use efficiency adjustment (+30%) due to elevated CO_2 on annual net canopy photosynthesis and net primary production for generic conifer forests in (a) short growing season, water stressed and (b) long growing season, nonwater stressed climates. (From Running and Nemani, 1991.)

linearity of C dynamics for forests that experience multiple resource limitations and some of the difficulties of extrapolating results from a single study to larger scales.

VIII. Future Research

Ecophysiological research during the last decade has contributed to a better understanding of C allocation and forest productivity. Yet, to answer scientific questions (Will fertilization increase forest net primary production? Are northern latitude forests a C sink?) requires a unified theory for understanding and predicting the effect of multiple environmental stresses on C assimilation and allocation. Knowledge of C assimilation, partitioning, and allocation patterns has greatly increased for a few well-studied conifers (*P. elliottii*, *Pseudotsuga menziesii*, *P. radiata*, and *P. sylvestris*), but still too often the "first principles" derived from these species do not completely apply to other conifers. This problem may be due to differences in methodology or an incomplete understanding of multiple stress interactions and their interactions on C allocation.

What approaches may be useful in developing a mechanistic understanding of C partitioning and allocation for conifers? Controlled-environment chamber experiments using seedlings will help elucidate basic controls on C partitioning and allocation and can be used to manipulate the availability of a number of resources. However, the results may have limited application to forests because seedlings have inherently different C partitioning and allocation patterns compared to mature trees (Cregg *et al.*, 1989; Woodward *et al.*, 1991). Stand-level, multiple resource manipulation experiments are expensive and difficult to design, especially for forests. One useful approach to increasing our understanding is through the use of physiological process models in conjunction with experimentation (Isebrands and Burk, 1992). Simple, process-based models can be used to determine if field measurements of C fluxes are "realistic" and to design the necessary measurements to test hypotheses related to controls on C allocation. Coupling physiologically based models of different temporal and spatial scale will likely play an important role in the future, although implementation of such models is in its infancy (Ehleringer and Field, 1993). Studies involving modelers, physiological ecologists, and micrometeorologists will provide the best opportunity to develop and test unifying theories on C partitioning and allocation. Eddy correlation, a technique traditionally used by micrometeorlogists to measure net CO_2, water, and energy flux from an ecosystem, should help resolve uncertainties related to scaling instantaneous CO_2 fluxes from chamber measurements to annual stand-level CO_2 fluxes and allow scientists to partition net CO_2 flux into C sinks and sources (Schimel, 1993). This decomposition is important because soil respiration and photosynthesis are controlled by different processes and differ in sensitivity to environmental stresses. Furthermore, such experiments provide physiological ecologists and modelers an opportunity to determine if the scaling approaches used to develop stand-level C budgets are valid, and if not, why.

Acknowledgments

The senior author acknowledges support from NSF Grants BSR-8918022 and BSR-9107419 during the preparation of this manuscript. The authors acknowledge the thoughtful reviews by Ian Woodward, John Raison, and Brent Haynes.

References

Aber, J. D., Melillo, J. M., Nadelhoffer, K. J., McClaugherty, C. A., and Pastor, J. (1985). Fine root turnover in forest ecosystems in relation to quantity and form of nitrogen availability: A comparison of two methods. *Oecologia* 66:317–321.

Adams, M. B., Allen, H. L., and Davey, C. B. (1986). Accumulation of starch in roots and foliage of loblolly pine (*Pinus taeda* L.): Effects of season, site and fertilization. *Tree Physiol.* 2:35–46.

Adams, M. B., Edwards, N. T., Taylor, G. E., Jr., and Skaggs, B. L. (1990). Whole-plant ^{14}C-photosynthate allocation in *Pinus taeda*: Seasonal patterns at ambient and elevated ozone levels. *Can. J. For. Res.* 20:152–158.

Allen, R. M. (1964). Contributions of roots, stems, and leaves to height growth of longleaf pine. *For. Sci.* 10:14–16.

Alscher, R. G., Cumming, J. R., Fellows, S., Fincher, J., Rubin, G., van Leuken, P., and Weinstein, L. H. (1989). Seasonal changes in the pigments, carbohydrates, and growth of red spruce as affected by ozone. *New Phytol.* 113:211–233.

Amthor, J. S. (1986). Evolution and applicability of a whole plant respiration model. *J. Theor. Biol.* 122:473–490.

Amundson, R. G., Alscher, R. G., Fellows, S., Rubin, G., Fincher, J., van Leuken, P., and Weinstein, L. H. (1991). Seasonal changes in the pigments, carbohydrates, and growth of red spruce as affected by exposure to ozone for two growing seasons. *New Phytol.* 118:127–138.

Andersen, C. P., McLaughlin, S. B., and Roy, W. K. (1991). A comparison of seasonal patterns of photosynthate production and use in branches of red spruce saplings at two elevations. *Can. J. For. Res.* 21:455–461.

Balantinecz, J. J., Forward, D. F., and Bidwell, R. G. S. (1966). Distribution of photoassimilated $^{14}CO_2$ in young jack pine seedlings. *Can. J. Bot.* 44:362–364.

Ballard, R. (1984). Fertilization of plantations. *In* "Nutrition of Plantation Forests" (G. D. Bowen and E. K. S. Nambiar, eds.), pp. 327–360. Academic Press, Orlando, Florida.

Barnes, R. L. (1972a). Effects of chronic exposure to ozone on photosynthesis and respiration of pines. *Environ. Pollut.* 3:133–138.

Barnes, R. L. (1972b). Effects of chronic exposure to ozone on soluble sugar and ascorbic acid contents of pine seedlings. *Can. J. Bot.* 50:215–219.

Barnes, J. D., Eamus, D., and Brown, K. A. (1990). The influence of ozone, acid mist and soil nutrient status on Norway spruce (*Picea abies* (L.) Karst.). II. Photosynthesis, dark respiration and soluble carbohydrates of trees during late autumn. *New Phytol.* 115: 149–156.

Bazzaz, F. A. (1990). The response of natural ecosystems to the rising global CO_2 levels. *Annu. Rev. Ecol. Syst.* 21:167–196.

Bazzaz, F. A., Coleman, J. S., and Morse, S. R. (1990). Growth response of seven major co-occurring tree species of the northeastern United States to elevated CO_2. *Can. J. For. Res.* 20:1479–1484.

Beets, P. N., and Pollock, D. S. (1987). Accumulation and partitioning of dry matter in *Pinus radiata* as related to stand-age and thinning. *N. Z. J. For. Sci.* 17:246–271.

Benoit, R. E., and Starkey, R. L. (1968a). Enzyme inactivation as a factor in the inhibition of decomposition of organic matter by tannins. *Soil Sci.* 105:203–208.

Benoit, R. E., and Starkey, R. L. (1968b). Inhibition of decomposition of cellulose and some other carbohydrates by tannin. *Soil Sci.* 105:291–296.

Bevege, D. I., Bowen, G. D., and Skinner, M. F. (1975). Comparative carbohydrate physiology of ecto- and endomycorrhizas. *In* "Endomycorrhizas" (F. E. Sanders, B. Mosse, and P. B. Tinker, eds.), pp. 149–174. Academic Press, New York.

Bloom, A. J., Chapin III, F. S., and Mooney, H. A. (1985). Resource limitations to plants—an economic analogy. *Annu. Rev. Ecol. Syst.* 16:363–392.

Bowen, G. D. (1969). Nutrient status effects on loss of amides and amino acids from pine roots. *Plant Soil* 30:139–142.

Brasseur, G. P., and Chatfield, R. B. (1991). The fate of biogenic trace gases in the atmosphere. *In* "Trace Gas Emissions by Plants" (T. D. Sharkey, E. A. Holland, and H. A. Mooney, eds.), pp. 1–27. Academic Press, San Diego.

Brix, H. (1981). Effects of nitrogen fertilization source and application rates on foliar nitrogen concentration, photosynthesis and growth of Douglas-fir. *Can. J. For. Res.* 11:775–780.

Brix, H., and Ebell, L. F. (1969). Effects of nitrogen fertilization on growth, leaf area and photosynthesis rate in Douglas-fir. *For. Sci.* 15:189–196.

Brix, H., and Mitchell, A. (1983). Thinning and fertilization effects on sapwood development and relationships of foliage quantity to sapwood area and basal area in Douglas-fir. *Can. J. For. Res.* 13:384–389.

Cannell, M. G. R. (1989). Physiological basis of wood production: A review. *Scand. J. For. Res.* 4:459–490.

Cannell, M. G. R., Milne, R., Sheppard, L. J., and Unsworth, M. H. (1987). Radiation interception and productivity of willows. *J. Appl. Ecol.* 24:261–278.

Chabot, B. F., and Hicks, D. J. (1982) The ecology of leaf life spans. *Annu. Rev. Ecol. Syst.* 13:229–259.

Chung, H. H., and Barnes, R. L. (1977). Photosynthate allocation in *Pinus taeda*. I. Substrate requirement for synthesis of shoot biomass. *Can. J. For. Res.* 7:106–111.

Chung, H. H., and Barnes, R. L. (1980). Photosynthate allocation in *Pinus taeda*. II. Seasonal aspects of photosynthate allocation to different biochemical fractions in shoots. *Can. J. For. Res.* 10:338–347.

Clancy, K. M., Wagner, M. R., and Reich, P. B. (1994). Ecophysiology and insect herbivory. *In* "Ecophysiology of Coniferous Forests" (W. K. Smith and T. M. Hinckley, eds.), pp. 125–180. Academic Press, San Diego.

Clarkson, D. T. (1985). Factors affecting mineral nutrient acquisition by plants. *Annu. Rev. Plant Physiol.* 36:77–115.

Coley, P. D. (1988). Effects of plant growth rate and leaf lifetime on the amount and type of anti-herbivore defense. *Oecologia* 74:531–536.

Comeau, P. G., and Kimmins, J. P. (1989). Above- and below-ground biomass and production of lodgepole pine on sites with differing soil moisture. *Can. J. For. Res.* 19: 447–454.

Coutts, M. P. (1983). Root architecture and tree stability. *Plant Soil* 71:171–188.

Cranswick, A. M., Rook, D. A., and Zabkiewicz, J. A. (1987). Seasonal changes in carbohydrate concentration and composition of different tissue types of *Pinus radiata* trees. *N. Z. J. For. Sci.* 17:229–245.

Cregg, B. M., Halpin, J. E., Dougherty, P. M., and Teskey, R. O. (1989). Comparative physiology and morphology of seedling and mature forest trees. *In* "Air Pollution Effects on Vegetation Including Forest Ecosystems" (R. D. Noble, J. L. Martin, and K. F. Jensen, eds.), pp. 111–118. *Proc. of the 2nd U.S.-U.S.S.R. Symposium.* Northeast Forest Exp. Station. Broomall, PA.

Cronshaw, J., Lucas, W. J., and Giaquinta, R. T. (eds.) (1986). "Phloem Transport." Alan R. Liss, New York.

Davies, R. I., Coulson, C. B., and Lewis, D. A. (1964). Polyphenols in plant, humus, and soil. III. Stabilization of gelatin by polyphenol tanning. *J. Soil Sci.* 15:299–309.

Davies, W. J., Metcalf, J., Lodge, T. A., and DaCosta, A. R. (1986). Plant growth substance and the regulation of growth under drought. *Aust. J. Plant Physiol.* 13:105–125.

Dick, J. M., Jarvis, P. G., and Leakey, R. R. B. (1991). Influence of male and female cones on needle CO_2 exchange rates of field-grown *Pinus contorta* Doug. trees. *Funct. Ecol.* 5:422–432.

Dickmann, D. I., and Kozlowski, T. T. (1968). Mobilization by *Pinus resinosa* cones and shoots of C14-photosynthate from needles of different ages. *Am. J. Bot.* 55:900–906.

Dickmann, D. I., and Kozlowski, T. T. (1970). Mobilization and incorporation of photoassimilated ^{14}C by growing vegetative and reproductive tissues of adult *Pinus resinosa* Ait. trees. *Plant Physiol.* 45:284–288.

Dickson, R. E. (1989). Carbon and nitrogen allocation in trees. *Ann. Sci. For.* 46 (Suppl.): 631–647.

Dickson, R. E. (1991). Assimilate distribution and storage. *In* "Physiology of Trees" (A. S. Raghavendra, ed.), pp. 51–85. Wiley, New York.

Dickson, R. E., and Isebrands, J. G. (1991). Leaves as regulators of stress response. *In* "Response of Plants to Multiple Stresses" (H. A. Mooney, W. E. Winner, E. J. Pell, and E. Chu, eds.), pp. 3–34. Academic Press, San Diego.

Dickson, R. E., and Isebrands, J. G. (1993). Carbon allocation terminology: Should it be more rational? *Bull. Ecol. Soc. Am.* 74:175–177.

Deans, J. D., and Ford, E. D. (1986). Seasonal patterns of radial root growth and starch dynamics in plantation-grown Sitka spruce trees of different ages. *Tree Physiol.* 1:241–251.

Dosskey, M. G., Linderman, R. G., and Boersma, L. (1990). Carbon-sink stimulation of photosynthesis in Douglas fir seedlings by some ectomycorrhizas. *New Phytol.* 115: 269–274.

Eamus, D., and Jarvis, P. G. (1989). The direct effects of increase in the global atmospheric CO_2 concentration on natural and commercial temperate trees and forests. *Adv. Ecol. Res.* 19:1–55.

Edwards, G. S., Friend, A. L., O'Neill, E. G., and Tomlinson, P. T. (1992). Seasonal patterns of biomass accumulation and carbon allocation in *Pinus taeda* seedlings exposed to ozone, acidic precipitation, and reduced soil Mg. *Can. J. For. Res.* 22:640–646.

Ehleringer, J. R., and Field, C. B. (1993). "Scaling Physiological Processes: Leaf to Globe." Academic Press, San Diego.

Ericsson, A. (1978). Seasonal changes in translocation of ^{14}C from different age-classes of needles on 20-year-old Scots pine trees (*Pinus sylvestris*). *Physiol. Plant.* 43:351–358.

Ericsson, A. (1979). Effects of irrigation and fertilization on the seasonal changes of carbohydrate reserves in different age-class needles on 20-year-old Scots pine trees (*Pinus sylvestris*). *Physiol. Plant.* 45:270–280.

Ericsson, A. (1984). Effects of low temperature and light treatment, following winter cold storage, on starch accumulation in Scots pine seedlings. *Can. J. For. Res.* 14:114–118.

Ericsson, A., and Persson, H. (1980). Seasonal changes in starch reserves and growth of fine roots of 20-year-old Scots pine. *Ecol. Bull.* 32:239–250.

Farrar, J. F., and Williams, M. L. (1991). The effects of increased atmospheric carbon dioxide and temperature on carbon partitioning, source-sink relations and respiration. *Plant, Cell Environ.* 14:819–830.

Fassnacht, K. S., Gower, S. T., Norman, J. M., and McMurtrie, R. E. (1994). Comparison of direct and three rapid optical methods of estimating surface area index in forests. *Agric. For. Meteorol.* accepted.

Fielding, J. M. (1960). "Branching and Flowering Characterisitcs of Monterey Pine," Bull. 37. Forestry and Timber Bureau, Canberra, Australia.

Ford, E. D., and Deans, J. D. (1977). Growth of Sitka spruce plantation: Spatial distribution and seasonal flucuations of lengths, weights, and carbohydrate concentrations of fine roots. *Plant Soil* 47:463–485.

Friend, A. L., Tomlinson, P. T., Dickson, R. E., O'Neill, E. G., Edwards, N. T., and Taylor, G. E., Jr. (1992). Biochemical composition of loblolly pine reflects pollutant exposure. *Tree Physiol.* 11:35–47.

Geiger, D. R. (1986). Processes affecting carbon allocation and partitioning among sinks. *In* "Phloem Transport" (J. Cronshaw, W. J. Lucas, and R. T. Giaquinta, eds.), pp. 375–388. Alan R. Liss, New York.

Gholz, H. L. (1982). Environmental limits on above-ground net primary production, leaf area, and biomass in vegetation zones of the Pacific Northwest. *Ecology* 63:469–481.

Gholz, H. L., and Cropper, W. P., Jr. (1991). Carbohydrate dynamics in mature *Pinus elliottii* var. *elliottii* trees. *Can. J. For. Res.* 21:1742–1747.

Gholz, H. L., Fisher, R. F., and Pritchett, W. L. (1985). Nutrient dynamics in slash pine plantation ecosystems. *Ecology* 66:647–659.

Gholz, H. L., Hendry, L. C., and Cropper, W. P., Jr. (1986). Organic matter dynamics of fine roots in plantations of slash pine (*Pinus elliottii*) in north Florida. *Can. J. For. Res.* 16:529–538.

Gholz, H. L., Vogel, S. A., Cropper, W. L., Jr., McKelvey, K., Ewel, K. C., Teskey, R. O., and Curran, P. J. (1991). Dynamics of canopy structure and light interception in *Pinus elliottii* stands of north Florida. *Ecol. Monogr.* 61:33–51.

Gifford, R. M., and Evans, L. T. (1981). Photosynthesis, carbon partitioning and yield. *Annu. Rev. Plant Physiol.* 32:485–509.

Gifford, R. M., Thorne, J. H., Hitz, W. D., and Giaquinta, R. T.. (1984). Crop productivity and photoassimilate partitioning. *Science* 225:801–808.

Glerum, C. (1980). Food sinks and food reserves of trees in temperate climates. *N. Z. J. For. Sci.* 10:176–185.

Gordon, J. C., and Larson, P. R. (1968). Seasonal course of photosynthesis, respiration, and distribution of ^{14}C in young *Pinus resinosa* trees as related to wood formation. *Plant Physiol.* 43:1617–1624.

Gordon, J. C., and Larson, P. R. (1970). Redistribution of ^{14}C-labeled reserve food in young red pines during shoot elongation. *For. Sci.* 16:14–20.

Gosz, J. R. (1981). Nitrogen cycling in coniferous ecosystems. *Ecol. Bull.* 33:405–426.

Gower, S. T., and Richards, J. H. (1990). Larches: Deciduous conifers in an evergreen world. BioScience 40:818–826.

Gower, S. T., and Son, Y. (1992). Differences in soil and leaf litterfall nitrogen dynamics for five forest plantations. *Soil Sci. Soc. Am. J.* 56:1959–1966.

Gower, S. T., Grier, C. C., Vogt, D. J., and Vogt, K. A. (1987). Allometric relations of deciduous (*Larix occidentalis*) and evergreen (*Pinus contorta* and *Pseudotsuga menziesii*) of the Cascade Mountains in central Washington. *Can. J. For. Res.* 17:630–634.

Gower, S. T., Vogt, K. A., and Grier, C. C. (1992). Carbon dynamics of Rocky Mountain Douglas-fir: Influence of water and nutrient availability. *Ecol. Monogr.* 62:43–65.

Gower, S. T., Reich, P. B., and Son, Y. (1993a). Canopy dynamics and aboveground production for five tree species with different leaf longevities. *Tree Physiol.* 12:327–345.

Gower, S. T., Gholz, H. L., Nakane, K., and Baldwin, V. C. (1994). Production and carbon allocation patterns of pine forests. *Ecol. Bull. (Stockholm)* 43. In press.

Gower, S. T., Haynes, B. E., Fassnacht, K. S., Running, S. W., and Hunt, E. R., Jr. (1993b). Influence of fertilization on the allometric relations for two pines in contrasting environments. *Can. J. For. Res.* 23:1704–1711.

Grier, C. C., and Waring, R. H. (1974). Conifer foliage mass related to sapwood area. *For. Sci.* 20:205–206.

Grier, C. C., and Running, S. W. (1977). Leaf area of mature northwestern coniferous forests: relation to site water balance. *Ecology* 58:893–899.

Grier, C. C., Vogt, K. A., Keyes, M. R., and Edmonds, R. L. (1981). Biomass distribution and above- and belowground production in young and mature *Abies amabilis* zone ecosystems on the Washington Cascades. *Can. J. For. Res.* 11:155–167.

Grier, C. C., Lee, K. M., and Archibald, R. M. (1984). Effect of urea fertilization on allometric relations in young Douglas-fir trees. *Can. J. For. Sci.* 15:900–904.

Hale, M. G., Moore, L. D., and Griffin, G. J. (1978). Root exudates and exudation. *In*

"Interactions between Nonpathogenic Soil Microorganisms and Plants" (Y. R. Dommergues and S. V. Krupa, eds.), pp. 163–203. Elsevier, New York.

Haissig, B. E. (1984). Carbohydrate accumulation and partitioning in *Pinus banksiana* seedlings and seedling cuttings. *Physiol. Plant.* 61:13–19.

Haissig, B. E., Davis, T. D., and Riemenschneider, D. E. (1992). Researching the controls of adventitious rooting. *Physiol. Plant.* 84:310–317.

Handley, W. R. C. (1961). Further evidence for the importance of residual leaf protein complexes in litter decomposition and the supply of nitrogen for plant growth. *Plant Soil* 15:37–73.

Ho, L. C. (1988). Metabolism and compartmentation of imported sugars in sink organs in relation to sink strength. *Annu. Rev. Plant Physiol. Mol. Biol.* 39:355–378.

Hodges, J., and Lorio, P. (1975). Moisture stress and xylem oleoresin in loblolly pine. *For. Sci.* 21:283–290.

Horner, J. D. (1987). The effects of manipulation of nitrogen and water availability on the polyphenol content of Douglas-fir foliage: Implications for ecosystem theory. Ph.D. Dissertation. University of New Mexico, Albuequerque.

Horner, J. D., Gosz, J. R., and Cates, R. G. (1988). The role of carbon-based secondary metabolites in decomposition in terrestrial ecosystems. *Am. Nat.* 132:869–883.

Isebrands, J. G., and Burk, T. E. (1992). Ecophysiological growth process models of short rotation forest crops. *In* "Ecophysiology of Short Rotation Forest Crops" (C. P. Mitchel, J. B. Ford-Robertson, T. M. Hinckley, and L. Sennerby-Forsse, eds.), pp. 231–266. Elsevier, New York.

Jarvis, P. G., and Leverenz, J. W. (1983). Productivity of deciduous and evergreen forests. *In* "Encyclopedia of Plant Physiology" (O. L. Lange, P. S. Nobel, C. B. Osmond, and H. Ziegler, eds.), New series Vol. 12D, IV, pp. 233–280. Springer-Verlag, Berlin.

Jordan, D. N., and Smith, W. K. (1993). Simulated influence of leaf geometry on sunlight interception and photosynthesis in conifer needles. *Tree Physiol.* 13:29–39.

Kahn, A. A., and Malhotra, S. S. (1977). Effects of aqueous sulphur dioxide on pine needle glycolipids. *Phytochemistry* 16:539–543.

Keyes, M. R., and Grier, C. C. (1981). Above- and below-ground net production in 40-year-old Douglas-fir stands on high and low productivity sites. *Can. J. For. Res.* 11: 599–605.

Kikuzawa, K. (1991). A cost-benefit analysis of leaf habit and leaf longevity of trees and their geographic pattern. *Am. Nat.* 138:1250–1263.

Kimura, M. (1969). Ecological and physiological studies on the vegetation of Mt. Shimagare. VII. Analysis of production processes of young *Abies* stand based on carbohydrate economy. *Bot. Mag.* 82:6–19.

Kinerson, R. S., Ralston, C., and Wells, C. (1977). Carbon cycling in a loblolly pine plantation. *Oecologia* 29:1–10.

Kira, T. (1975). Primary production of forests. *In* "Photosynthesis and Productivity in Different Environments" (J. P. Cooper, ed.), pp. 5–40. Cambridge Univ. Press, Cambridge.

Kira, T., and Shedei, T. (1967). Primary production and turnover of organic matter in different forest ecosystems of the western Pacific. *Jpn. J. Ecol. (Nippon Seitai Gakkaishi)* 17:70–87.

Kozlowski, T. T. (1971a). "Growth and Development of Trees. Volume I. Seed Germination, Ontogeny and Shoot Growth." Academic Press, New York.

Kozlowski, T. T. (1971b). "Growth and Development of Trees. Volume II. Cambial Growth, Root Growth, and Reproductive Growth." Academic Press, New York.

Kozlowski, T. T. (1992). Carbohydrate sources and sinks in woody plants. *Bot. Rev.* 58: 107–223.

Kozlowski, T. T., and Constantinidou, H. A. (1986). Responses of woody plants to environmental pollution. I. Sources and types of pollutants and plant responses. *For. Abstr.* 47:5–51.

Kozlowski, T. T., and Winget, C. H. (1964). The role of reserves in leaves, branches, stems, and roots on shoot growth of red pine. *Am. J. Bot.* 51:522–529.

Kuhns, M. R., and Gjerstad, D. H. (1991). Distribution of ^{14}C-labeled photosynthate in loblolly pine (*Pinus taeda*) seedlings as affected by season and time after exposure. *Tree Physiol.* 8:259–271.

Kurz, W. A. (1989). Net primary production, production allocation, and foliage efficiency in second growth Douglas-fir stands with differing site quality. Ph.D. Dissertation. University of British Columbia, British Columbia, Canada.

Lamb, R. J., and Richards, B. N. (1971). Effect of mycorrhizal fungi on the growth and nutrient status of slash and radiata pine seedlings. *Aust. For.* 35:1–7.

Landsberg, J. J. (1986). "Physiological Ecology of Forest Production." Academic Press, New York.

Larson, P. R. (1964). Contribution of different aged needles to growth and wood formation in young red pines. *For. Sci.* 10:224–238.

Lerdau, M. T. (1991). Plant function and biogenic terpene emission. *In* "Trace Gas Emissions by Plants" (T. D. Sharkey, E. A. Holland, and H. A. Mooney, eds.), pp. 121–134. Academic Press, San Diego.

Leverenz, J. W. (1980). Shoot structure and productivity in conifers. *In* "Understanding and Predicting Tree Growth" (S. Linder, ed.), pp. 135–137. Swedish Coniferous Forest Project Tech. Rpt. 25. Swedish University of Agricultural Sciences, Uppsala.

Leverenz, J. W. (1992). Shade shoot structure and productivity of evergreen conifer stands. *Scand. J. For. Res.* 7:345–353.

Leverenz, J. W., and Hinckley, T. M. (1990). Shoot structure, leaf area and productivity of evergreen conifer stands. *Tree Physiol.* 6:135–149.

Linder, S. (1985). Potential and actual production in Australian forest stands. *In* "Research for Forest Management" (J. J. Landsberg and W. Parsons, eds.), pp. 11–35. CSIRO, Melbourne, Australia.

Linder, S. (1987). Response to water and nutrients in coniferous ecosystems. *In* "Potentials and Limitations of Ecosystem Analysis" (E.-D. Schulze and H. Zwolfer, eds.), *Springer Verlag Ecological Studies* 61:180–202.

Linder, S., and Axelsson, B. (1982). Changes in carbon uptake and allocation patterns as a result of irrigation and fertilization in a young *Pinus sylvestris* stand. *In* "Carbon Uptake and Allocation: Key to Management of Subalpine Forest Ecosystems" (R. H. Waring, ed.), pp. 38–44. International Union Forest Research Organization (IUFRO) Workshop, Forest Research Laboratory, Oregon State Univ., Corvallis, Oregon.

Linder, S., and Rook, D. A. (1984). Effects of mineral nutrition on carbon dioxide exchange and partitioning of carbon in trees. *In* "Nutrition of Plantation Forests" (G. D. Bowen and E. K. S. Nambiar, eds.), pp. 211–236. Academic Press, London.

Linder, S., and Troeng, E. (1981). The seasonal variation in stem and coarse root respiration of a 20-year-old Scots pine (*Pinus sylvestris* L.). *Mitt. Forstl. Bundesversuchsanst. Wien* 142:125–139.

Linder, S. L., Benson, M. L., Myers, B. J., and Raison, R. J. (1987). Canopy dynamics and growth of *Pinus radiata*. I. Effects of irrigation and fertilization during drought. *Can. J. For. Res.* 17:1157–1165.

Little, C. H. A. (1970). Seasonal changes in carbohydrate and moisture content in needles of balsam fir (*Abies balsamea*). *Can. J. Bot.* 48:2021–2028.

Loach, K. and Little, C. H. A. (1973). Production, storage, and use of photosynthate during shoot elongation in balsam fir (*Abies balsamea*). *Can. J. Bot.* 51:1161–1168.

McLaughlin, S. B., and Kohut, R. J. (1992). The effects of atmospheric deposition and ozone on carbon allocation and associated physiological processes in red spruce. In

"Ecology and Decline of Red Spruce in the Eastern United States" (C. Eager and M. B. Adams, eds.), pp. 338–382. Springer-Verlag, New York.

McLaughlin, S. B., McConathy, R. K., Duvick, D., and Mann, L. K. (1982). Effects of chronic air pollution stress on photosynthesis, carbon allocation, and growth of white pine trees. *For. Sci.* 28:60–70.

McLaughlin, S. B., Tjoelker, M. J., and Roy, W. K. (1993). Acid deposition alters carbon allocation and growth of red spruce. Greenhouse studies reproduce field measurements. *Can. J. For. Res.* 23:380–386.

Makela, A., Kellomaki, S., and Hari, P. (1980). Ecophysiological studies of young Scots pine stands. III. Photosynthate allocation for needle growth and wood formation in current year shoots. *Silvae Fennica* 14:258–263.

Malhotra, S. S., and Sarkar, S. K. (1979). Effects of sulfur dioxide on sugar and free amino acid content of pine seedlings. *Physiol. Plant.* 47:223–228.

Malkonen, E. (1974). Annual primary production and nutrient cycle in some Scots pine stands. *Commun. Inst. For. Fenn. (Helsinki)*, No. 84.

Mann, L. K., McLaughlin, S. B., and Shriner, D. S. (1980). Seasonal physiological responses of white pine under chronic air pollution stress. *Environ. Exp. Bot.* 20:99–105.

Marshall, J., and Waring, R. H. (1985). Predicting fine root production and turnover by monitoring root starch and soil temperature. *Can. J. For. Res.* 15:791–800.

Mattson, W. J., and Addy, N. D. (1975). Phytophagous insects as regulators of forest primary production. *Science* 190:515–521.

Meier, S., Grand, L. F., Schoeneberger, M. M., Reinert, R. A., and Bruck, R. I. (1990). Growth, ectomycorrhizae and nonstructural carbohydrates of loblolly pine seedlings exposed to ozone and soil-water deficit. *Environ. Pollut.* 64:11–27.

Melillo, J. M., Aber, J. D., and Muratore, J. F. (1982). Nitrogen and lignin control of hardwood leaf litter decomposition dynamics. *Ecology* 63:621–626.

Menge, J. A., and Grand, L. F. (1978). Effect of fertilization on production of epigeous basidiocarps by mycorrhizal fungi in loblolly pine plantations. *Can. J. Bot.* 56: 2357–2362.

Menge, J. A., Grand, L. F., and Haines, L. W. (1977). The effect of fertilization on growth and mycorrhizae numbers in 11-year-old loblolly pine plantations. *For. Sci.* 23:37–44.

Mengel, K., Breininger, M. Th., and Lutz, H. J. (1988). Effect of acidic mist on nutrient leaching, carbohydrate status and damage symptoms of *Picea abies*. *In* "Air Pollution and Ecosystems" (P. Mathy, ed.), pp. 312–320. Reidel, Dordrecht, The Netherlands.

Miller, H. G., Cooper, J. M., Miller, J. D., and Pauline, O. J. L. (1979). Nutrient cycles in pine and their adaptation to poor soils. *Can J. For. Res.* 9:19–26.

Miller, P. M., Eddleman, L. E., and Kramer, S. (1990). Allocation patterns of carbon and minerals in juvenile and small-adult *Juniperus occidentalis*. *For. Sci.* 36:734–747.

Miller, P. R., Parmeter, J. R., Jr., Flick, B. H., and Martinez, C. W. (1969). Ozone dosage response of ponderosa pine seedlings. *J. Air Pollut. Control Assoc.* 19:435–538.

Monteith, J. L. (1977). Climate and efficiency of crop production in Britain. *Philos. Trans. R. Soc. London B* 281:277–294.

Nadelhoffer, K. J., Aber, J. D., and Melillo, J. M. (1984). Seasonal patterns of ammonium and nitrate uptake in nine temperate forest ecosystems. *Plant Soil* 80:321–335.

Nadelhoffer, K. J., Aber, J. D., and Melillo, J. M. (1985). Fine root production in relation to total net primary production along a nitrogen mineralization gradient in temperate forests: A new hypothesis. *Ecology* 66:1377–1390.

Norby, R. J., O'Neill, E. G., Hood, W. G., and Luxmoore, R. J. (1987). Carbon allocation, root exudation and mycorrhizal colonization of *Pinus echinata* seedlings grown under CO_2 enrichment. *Tree Physiol.* 3:203–210.

Norman, J. M. (1993). Scaling processes between leaf and canopy levels. *In* "Scaling Physiological Processes: Leaf to Globe" (J. R. Ehleringer and C. B. Field, eds.), pp. 41–76. Academic Press, New York.

Oren, R., Schulze, E.-D., Werk, K. S., Meyer, J. S., Schneider, B. U., and Heilmeier, H. (1988). Performance of two *Picea abies* (L.) Karst. stands at different stages of decline. I. Carbon relations and stand growth. *Oecologia* 75:25–37.

Paavilainen, E. (1980). Effect of fertilization on plant biomass and nutrient cycle on a drained dwarf shrub pine swamp. *Comm. Inst. For. Fenn. (Helsinki)*, No. 98.

Passioura, J. B. (1988). Root signals control leaf expansion in wheat seedlings growing in drying soil. *Aust. J. Plant Physiol.* 15:687–693.

Pereira, J. S. (1990). Whole-plant regulation and productivity in forest trees. *In* "Importance of Root to Shoot Communication in the Responses to Environmental Stress Monograph 21" (W. J. Davies and B. Jeffcoat, eds.), pp. 237–250. British Society for Plant Growth Regulation, Bristol.

Peterson, C. E., Mattson, K. G., and Mickler, R. A. (1989). Seedling response to sulfur, nitrogen and associated pollutants. U.S. Environ. Protection Agency Report 600-3-89-081. Corvallis, Oregon.

Puttonen, P. (1986). Carbohydrate reserves in *Pinus sylvestris* seedling needles as an attribute of seedling vigor. *Scand. J. For. Res.* 1:181–193.

Pye, J. M. (1988). Impact of ozone on the growth and yield of trees: A review. *J. Environ. Qual.* 17:347–360.

Raich, J. W., and Nadelhoffer, K. J. (1989). Belowground carbon allocation in forest ecosystems: Global trends. *Ecology* 70:1346–1354.

Raison, R. J., Myers, B. J., and Benson, M. L. (1992a). Dynamics of *Pinus radiata* foliage in relation to water and nitrogen stress: I. Needle production and properties. *For. Ecol. Manag.* 52:43–64.

Raison, R. J., Myers, B. J., and Benson, M. L. (1992b). Dynamics of *Pinus radiata* foliage in relation to water and nitrogen stress: I. Needle loss and temporal changes in total foliage mass. *For. Ecol. Manage.* 52:159–178.

Rangnekar, P. V., and Forward, D. F. (1969). Foliar nutrition and growth in red pine: the fate of photoassimilated carbon in a seedling tree. *Can. J. Bot.* 47:897–906.

Rangnekar, P. V., and Forward, D. F. (1972). Foliar nutrition and growth in red pine: distribution of photoassimilated carbon in seedlings during bud expansion. *Can. J. Bot.* 50:2053–2061.

Rangnekar, P. V., and Forward, D. F. (1973). Foliar nutrition and wood growth in red pine: Effects of darkening and defoliation on the distribution of ^{14}C-photosynthate in young trees. *Can. J. Bot.* 51:103–108.

Rangnekar, P. V., Forward, D. F., and Nolan, N. J. (1969). Foliar nutrition and wood growth in red pine: Distribution of radiocarbon photoassimilated by individual branches of young trees. *Can. J. Bot.* 47:1701–1711.

Reich, P. B., Uhl, C., Walters, M. B., and Ellsworth, D. S. (1991). Leaf life-span as a determinant of leaf structure and function among 23 species in Amazonian forest communities. *Oecologia* 86:16–24.

Reich, P. B., Walters, M. B., and Ellsworth, D. S. (1992). Leaf life-span in relation to leaf, plant and stand characteristics among diverse ecosystems. *Ecol. Monogr.* 62:365–392.

Reich, P. B., Koike, T., Gower, S. T., and Schoettle, A. W. (1994). Causes and consequences of variation in conifer leaf life span. *In* "Ecophysiology of Coniferous Forests" (W. K. Smith and T. M. Hinckley, eds.), pp. 225–254. Academic Press, San Diego.

Rosenberg, N. L., Blad, B. L., and Verma, S. B. (1983). "Microclimate: The Biological Environment," 2nd Ed. Wiley, New York.

Rovira, A. D. (1969). Plant root exudates. *Bot. Rev.* 35:35–57.

Rovira, A. D., and Davey, C. B. (1974). Biology of the rhizosphere. *In* "The Plant Root and Its Environment" (F. W. Carson, ed.), pp. 154–208. Univ. Press of Virginia, Charlottesville.

Running, S. W., and Coughlan, J. C. (1988). A general model of forest ecosystem processes for regional applications. I. Hydrologic balance, canopy gas exchange and primary production processes. *Ecol. Modell.* 42:125–154.

Running, S. W., and Gower, S. T. (1991). FOREST-BGC, a general model of forest ecosystem processes for regional applications. II. Dynamic carbon allocation and nitrogen budgets. *Tree Physiol.* 9:147–160.

Running, S. W., and Nemani, R. R. (1991). Regional hydrologic and carbon balance responses of forests resulting from potential climate change. *Clim. Change* 19:349–368.

Runyon, J., Waring, R. H., Goward, S. N., and Welles, J. M. (1994). Environmental limits on above-ground production: Observations from the Oregon transect. *Ecol. Appl.* 4:226–237.

Rutter, A. J. (1957). Studies in the growth of young plants of *Pinus sylvestris* L. I. The annual cycle of assimilation and growth. *Ann. Bot.* 21:299–425.

Ryan, M. G. (1991). Effects of climate change on plant respiration. *Ecol. Appl.* 1:157–167.

Ryan, M. G., and Waring, R. H. (1992). Maintenance respiration and stand development in a subalpine lodgepole pine forest. *Ecology* 73:2100–2108.

Ryan, M. G., Gower, S. T., Hubbard, R. M., Waring, R. H., Gholz, H. L., Cropper, W. L., Jr., and Running, S. W. (1994). Stem maintenance respiration of conifer stands in contrasting climates. *Oecologia.* In press.

Santantonio, D., and Hermann, R. K. (1985). Standing crop, production and turnover of fine roots on dry, moderate and wet sites of mature Douglas-fir in western Oregon. *Ann. Sci.* 42:113–142.

Santantonio, D., Hermann, R. K., and Overton, W. S. (1977). Root biomass studies in forest ecosystems. *Pedobiologia* 17:1–31.

Sasek, T. W., Richardson, C. J., Fendick, E. A., Bevington, S. R., and Kress, L. W. (1991). Carryover effects of acid rain and ozone on the physiology of multiple flushes of loblolly pine seedlings. *For. Sci.* 37:1078–1098.

Scherbatskoy, T., and Klein, R. M. (1983). Response of spruce and birch foliage to leaching by acidic mists. *J. Environ. Qual.* 12:189–195.

Schier, G. A. (1970). Seasonal pathways of ^{14}C-photosynthate in red pine labeled in May, July, and October. *For. Sci.* 16:2–13.

Schimel, D. S. (1993). New technologies for physiological ecology. *In* " Scaling Physiological Processes: Leaf to Globe" (J. R. Ehleringer and C. B. Field, eds.), Academic Press, San Diego.

Schlesinger, W. H. (1991). "Biogeochemistry: An Analysis of Global Change." Academic Press, New York.

Schneider, A., and Schmitz, K. (1989). Seasonal course of translocation and distribution of ^{14}C-labelled photoassimilate in young trees of *Larix decidua*. *Trees* 4:185–191.

Schowalter, T. D., Hargrove, W. W., and Crossley, D. A., Jr. (1986). Herbivory in forested ecosystems. *Annu. Rev. Entomol.* 31:177–196.

Schulze, E.-D. (1986). Whole-plant responses to drought. *Aust. J. Plant Physiol.* 13: 127–141.

Sheriff, D. W., Nambiar, E. K. S., and Fife, D. N. (1986). Relationships between nutrient status, carbon assimilation and water use in *Pinus radiata* (D. Don) needles. *Tree Physiol.* 2:73–88.

Shroya, T., Lister, G. R., Slankis, V., Krotkov, G., and Nelson, C. D. (1966). Seasonal change in respiration, photosynthesis, and translocation of the ^{14}C labelled products of photosynthesis in young *Pinus strobus* L. plants. *Ann. Bot.* 30:81–91.

Skarby, L., Troeng, E., and Bostrom, C. A. (1987). Ozone uptake and effects on transpiration, net photosynthesis, and dark respiration in Scots pine. *For. Sci.* 33:801–808.

Smith, J. L., and Paul, E. A. (1988). Use of an *in situ* labeling technique for the determination of seasonal ^{14}C distribution in ponderosa pine. *Plant Soil* 106:221–229.

Snowdon, P., and Benson, M. L. (1992). Effects of combinations of irrigation and fertilization on the growth and above-ground biomass production of *Pinus radiata*. *For. Ecol. Manage.* 52:87–116

Son, Y., and Gower, S. T. (1991). Aboveground nitrogen and phosphorus use by five plantation-grown tree species with different leaf longevities. *Biogeochemistry* 14:167–191.

Sprugel, D. G. (1976). Dynamic structure of wave-regenerated *Abies balsamea* forests in northeastern United States. *J. Ecol.* 64:889–911.

Sprugel, D. G. (1984). Density, biomass, productivity and nutrient cycling changes during stand development in wave-regenerated balsam fir forests. *Ecol. Monogr.* 54:165–186.

Stitt, M., and Quick, W. P. (1989). Photosynthetic carbon partitioning: Its regulation and possibilities for manipulation. *Physiol. Plant.* 77:633.

St. John, T. V. (1983). Response of tree roots to decomposing organic matter in two lowland Amazonian rain forests. *Can. J. For. Res.* 13:346–349.

St. John, T. V., Coleman, D. C., and Reid, C. P. P. (1983). Growth and spatial distribution of nutrient-absorbing organs: Selective exploitation of soil heterogeneity. *Plant Soil* 71:487–493.

Swain, T. (1977). Secondary compounds as protective agents. *Annu. Rev. Plant Physiol.* 8:479–501.

Taylor, G. E., Norby, R. J., McLaughlin, S. B., Johnson, A. H., and Turner, R. S. (1986). Carbon assimilation and growth of red spruce (*Picea rubens* Sarg.) seedlings in response to ozone, precipitation chemistry, and soil. *Oecologia* 70:163–171.

Teskey, R. O., and Whitehead, D. (1994). Photosynthesis and carbon gain by plants. *In* "A Comparative Analysis of Pine Forest Productivity" (H. L. Gholz, S. Linder, and R. E. McMurtrie, eds.). Swedish Natural Science Research Council, Ecol. Bull. In press.

Teskey, R. O., Grier, C. C., and Hinckley, T. M. (1984). Changes in photosynthesis and water relations with age and season in *Abies amabilis*. *Can. J. For. Res.* 14:77–84.

Thomas, R. B., and Strain, B. R. (1991). Root restriction as a factor in photosynthetic acclimation of cotton seedlings grown in elevated carbon dioxide. *Plant Physiol.* 96: 627–634.

Tingey, D. T., Wilhour, R. G., and Standley, C. (1976). The effect of chronic ozone exposures on the metabolite content of ponderosa pine seedlings. *For. Sci.* 22:234–241.

Tingey, D., Manning, M., Grothaus, L., and Burns, W. (1980). Influence of light and temperature on monoterpene emission from slash pine. *Plant Physiol.* 65:797–801.

Tjoelker, M. G., and Luxmoore, R. J. (1991). Soil nitrogen and chronic ozone stress influence physiology, growth and nutrient status of *Pinus taeda* L. and *Liriodendron tulipifera* L. seedlings. *New Phytol.* 119:69–81.

Tseng, E. C., Seiler, J. R., and Chevone, B. I. (1988). Effects of ozone and water stress on greenhouse-grown fraser fir seedling growth and physiology. *Environ. Exp. Bot.* 28: 37–41.

Turner, N. C. (1986). Adaptation to water deficits: A changing perspective. *Aust. J. Plant Physiol.* 13:175–190.

Turunen, M., and Huttunen, H. M. (1990). A review of the response of epicuticular wax of conifer needles to air pollution. *J. Environ. Qual.* 19:35–45.

Uren, N. C., and Reisenauer, H. M. (1988). The role of root exudates in nutrient acquisition. *In* "Advances in Plant Nutrition" (B. Tinker and A. Lauchli, eds.), 2nd Ed., pp. 79–114. Praeger, New York.

Ursino, D. J., and Paul, J. (1973). The long-term fate and distribution of ^{14}C photoassimilated by young white pines in late summer. *Can. J. Bot.* 51:683–687.

Van Cleve, K., Barney, R. J., and Schlentner, R. L. (1981). Evidence of temperature control on production and nutrient cycling in two interior Alaska black spruce ecosystems. *Can. J. For. Res.* 11:258–273.

Vogt, D. J. (1987). Douglas-fir ecosystems in western Washington: Biomass and production as related to site quality and stand age. Ph.D. Dissertation, Univ. of Washington, Seattle.

Vogt, K. A., Grier, C. C., Meier, C. E., and Edmonds, R. L. (1982). Mycorrhizal role in net primary production and nutrient cycling in *Abies amabilis* ecosystems in western Washington. *Ecology* 63:370–380.

Vogt, K. A., Grier, C. C., and Vogt, D. J. (1986). Production, turnover, and nutrient dynamics of above- and below-ground detritus of world forests. *Adv. Ecol. Res.* 15: 303–377.

Vogt, K. A., Vogt, D. J., Moore, E. E., Fatuga, B., Redlin, M. R., and Edmonds, R. L. (1987). Conifer and angiosperm fine-root biomass in relation to stand age and site productivity in Douglas-fir forests. *J. Ecol.* 75:857–870.

Vogt, K. A., Vogt, D. J. Gower, S. T., and Grier, C. C. (1990). Carbon and nitrogen interactions for forest ecosystems. *In* "Above- and Below-ground Interactions in Forest Trees in Acidified Soils." Air Pollution Report of the Environmental Research Programme of the Commission of the European Communities.

Vose, J. M., and Allen, H. L. (1988). Leaf area, stemwood growth, and nutrition relationships in loblolly pine. *For. Sci.* 34:547–563.

Walter, H. (1979). "Vegetation of the Earth and Ecological Systems of the Geo-Biosphere," 2nd Ed. Springer-Verlag, New York.

Waring, R. H. (1991). Response of evergreen trees to multiple stresses. *In* "Response of Plants to Multiple Stresses" (H. A. Mooney, W. E. Winner, and E. J. Pell, eds.), pp. 371–390. Academic Press, San Diego.

Waring, R. H., and Franklin, J. F. (1979). Evergreen coniferous forests of the Pacific Northwest. *Science* 204:1380–1385.

Waring, R. H., and Pitman, G. B. (1985). Modifying lodgepole pine stands to change susceptibility to mountain pine beetle attack. *Ecology* 66:889–897.

Waring, R. H., and Running, S. W. (1978). Sapwood water storage: Its contribution to transpiration and effect upon water conductance through stems of old-growth Douglas-fir. *Plant, Cell Environ.* 1:131–140.

Waring, R. H., Schroeder, P. E., and Oren, R. (1982). Application of the pipe model theory to predict canopy leaf area. *Can. J. For. Res.* 12:556–560.

Webb, W. L. (1977). Seasonal allocation of photoassimilated carbon in Douglas fir seedlings. *Plant Physiol.* 60:320–322.

Webb, W. L. (1991). Dynamics of photoassimilated ^{14}C and ^{12}C in *Pseudotsuga menziesii* seedlings: Photosynthesis, respiration, and biomass. *Photosynthesis* 25:323–331.

Wilkinson, T. G., and Barnes, R. L. (1973). Effects of ozone on $^{14}CO_2$ fixation patterns in pine. *Can. J. Bot.* 51:1573–1578.

Woodward, F. I., Thompson, G. B., and McKee, I. F. (1991). The effects of elevated concentration of carbon dioxide on individual plants, populations, communities and ecosystems. *Ann. Bot.* 67 (Suppl.):23–38.

Wyse, R. E. (1986). Sinks as determinants of assimilate partitioning: possible sites for regulation. *In* "Phloem Loading" (J. Cronshaw, W. J. Lucas, and R. T. Giaquinta, eds.), pp. 197–209. Alan R. Liss, New York.

Zahner, R. (1968). Water deficits and growth of trees. *In* "Water Deficits and Plant Growth" (T. T. Kozlowski, ed.), Vol. 2, pp. 191–254. Academic Press, New York.

8

Respiration from the Organ Level to the Stand

Douglas G. Sprugel, Michael G. Ryan, J. Renée Brooks, Kristiina A. Vogt, and Timothy A. Martin

I. Introduction

As several reviews have emphasized, respiration is a major factor in plant, stand, or ecosystem energy budgets, estimated to consume anywhere from 30 to 70% of total carbon fixed (Sprugel and Benecke, 1991; Hagihara and Hozumi, 1991; Ryan, 1991a). Respiration has been an area of particular interest and concern recently because of the possibility that CO_2-induced global warming might lead to substantial increases in respiration in temperate and boreal ecosystems, which could decrease net primary productivity (Ryan, 1991a). This concern, coupled with a general interest in scaling of physiological processes (e.g., Ehleringer and Field, 1993), has led to a resurgence of interest in respiration and in techniques for estimating respiration of stands and ecosystems.

Before we begin our discussion of respiration, a few definitions are in order. This chapter will deal with dark respiration (henceforth simply respiration), the process by which glucose is enzymatically combined with oxygen to liberate chemical energy and CO_2. It is generally assumed that most respiration in trees is through the "normal" cytochrome-mediated pathway, but there is an alternative; cyanide-resistant or salicylhydroxamic acid-sensitive respiration is a nonphosphorylating respiration pathway that generates only 40–50% as much chemical energy per glucose oxidized (Lambers, 1980). In most ecophysiological studies respiration is defined as CO_2 evolution, and if the results are the only thing being considered, it makes little difference which pathway is involved. However, when the respiration costs of producing plant tissue are estimated from tissue analyses (see Section V), it is implicitly assumed that any ATP required for synthesis is produced by cytochrome-mediated respiration. If alternate-path respiration is important, these calculations could be in error. Photorespiration is a by-product of photosynthesis in C_3 plants in which ribulose bisphosphate carboxylase/oxygenase (Rubisco) binds to O_2 instead of CO_2. It manifests as a reduction in the rate of photosynthesis and will not be considered further here.

Respiration is unusual among major physiological processes in that until recently (see Section V) there have been no techniques for estimating stand-level respiration that do not rely directly or indirectly on extrapolation from small-scale samples. Most physiologically based ecosystem processes can be quantified on a stand scale by measuring some whole-system property that represents the ecosystem-scale manifestation of the physiological process. Net primary productivity, for example, can be estimated by measuring litterfall, biomass increment, and grazing (with a potentially major error in estimating the belowground component). For transpiration there are more possibilities; stand- or watershed-level evapotranspiration can be estimated by using weighing lysimeters or whole-tree sap-flux gauges; through micrometeorological methods (e.g., eddy correlation, Bowen ratio, or boundary layer gradients); or by calculating the difference between precipitation inputs and streamwater outputs. For respiration, though, no such stand-level estimator has been available; until the development of eddy correlation methods for forests (see Section V), the only way to estimate stand-level respiration was to make physiological measurements on individual organs throughout the canopy and then sum or integrate to get a stand-level value.

The absence of such a large-scale proxy measurement of stand-level respiration has had several important consequences. First, because integrating and scaling up is much more laborious and time-consuming than whole-system measurements, there have simply not been very many attempts to estimate total autotrophic respiration for a stand or an ecosystem. Although there have been hundreds of estimates of whole-stand NPP for ecosystems around the world, the estimates of whole-stand respiration, or even total aboveground respiration, could be counted on the fingers of one or two hands. Moreover, the absence of a system-level estimator has made it difficult to check those estimates that have been made. For some processes it has been possible to achieve closure (Kaufmann and Landsberg, 1991) by measuring all components of a system and checking the total against an independent measurement. With respiration, this has not been possible (but see discussion in Section V of the possible use of eddy correlation to achieve this end). As a result, considerable effort has been put into developing techniques for scaling individual measurements up to stand level. This chapter will focus on these scaling techniques and how they deal with known sources of variation in respiration.

As in all scaling efforts, the crucial problem in extrapolating from small- to large-scale respiration estimates is finding an index that is easily measured on both large and small samples and is also closely tied to respiration of some important stand component. The search for such an index is made more difficult by the fact that respiration is an active process, highly responsive to environmental and physiological factors, and also highly variable within the plant. It is crucial to understand the nature and source of this variation before attempting to extrapolate from shoot-scale measurements up to the next higher level of organization (Dawson and Bliss, 1993). For example, past studies have often measured respiration per unit foliage biomass or per unit stem surface area for a few samples and then extrapolated these estimates up to the stand level simply by multiplying by the total foliage biomass or stem surface of the stand. However, Brooks *et al.* (1991) found that at a given date and time, respiration of mature *Abies amabilis* foliage varied by a factor of 7 within the canopy of a single stand; Sprugel (1990) found that in the same stand respiration per unit stem surface area varied by up to 20-fold. Given variation of this magnitude, uncritical use of these indices, particularly if combined with unlucky choice of sampling locations, could lead to disastrously bad estimates of total stand respiration. This large variation makes an understanding of within-stand variation critical both in choosing sample locations for respiration measurements and in selecting an index for extrapolating from those locations to the whole tree or stand.

A. Construction versus Maintenance Respiration

In attempting to scale respiration estimates up to stand level, or even simply to understand the factors controlling respiration, it is frequently useful to distinguish construction and maintenance components of respiration (Johansson, 1933a; Lambers *et al.*, 1983; Sprugel and Benecke, 1991; Ryan, 1991a). This distinction is often referred to as the "functional model" of respiration (McCree, 1970; Thornley, 1970) and is expressed by the following equation:

$$R = g(\Delta W/\Delta t) + mW, \tag{1}$$

where R is total respiration (grams C/day), $\Delta W/\Delta t$ is growth rate (grams C/day), g is the respiration cost of producing a unit of tissue (grams C/gram C), W is plant weight (grams C), and m is the cost of maintaining a unit of tissue (grams C/gram C/day). Thus the first term represents construction respiration (the cost of producing new tissue) and the second represents maintenance respiration (the cost of maintaining current tissue). Numerous alternative formulations of this model, and particularly of the "growth coefficient" g, are given in Lambers *et al.* (1983).

The functional model of respiration is not universally accepted because (1) biochemically, the process of glucose oxidation is the same regardless of what the energy is used for and (2) experimentally, it is often very difficult to separate the two components. Moreover, the model does not account for storage, which can be a major component of tree energy budgets at some times of the year [but see Thornley (1977) for an improved version]. Finally, the model as expressed in Eq. (1) is not quantitatively applicable to whole trees because different tissues have different construction and maintenance costs (Chung and Barnes, 1977), and even the same tissue may have different maintenance costs at different locations in the tree (Brooks *et al.*, 1991), so g and m are not well-defined on a whole-tree basis. However, from a conceptual standpoint, the functional model does provide an excellent logical framework within which to structure respiration studies, and in many cases this advantage outweighs some practical difficulties.

Construction respiration is defined more precisely as the carbon cost of constructing new tissue, exclusive of the materials that are actually incorporated into the tissue. The cost of producing a given amount of a particular type of tissue (grams of C respired/gram of C of new tissue produced) is determined solely by the chemical composition of the tissue and the oxidation state of the minerals used to produce it, and is unaffected by the temperature at which the synthesis occurs or any other environmental or genetic factors (Penning de Vries *et al.*, 1974; Penning de Vries, 1975a). (Temperature may indirectly affect construction respiration rates by changing the rate at which new tissue is produced, but the respiration per gram of tissue should be constant.) Because construction respiration is an inevitable consequence of producing new tissue and is linearly related to the amount of new tissue produced, it is often modeled simply as an "overhead" added to the cost of the tissue (which means, of course, that to model construction respiration accurately one must first be able to model growth accurately). Methods of calculating the total construction respiration involved in producing different types of tissues based on chemical or elemental composition are outlined in Section V.

Maintenance respiration, on the other hand, is the cost of maintaining established tissue once it has been formed (grams of C respired/gram of tissue per day). It includes the costs of protein synthesis and replacement, membrane repair, and the maintenance of ion gradients, and is very temperature sensitive (Penning de Vries, 1975b). The term "maintenance" is somewhat misleading, because it seems to imply a relatively passive process. In fact, the rate of maintenance respiration is highly variable even among leaves on the same tree (Brooks *et al.*, 1991), and is affected by many factors, including the overall level of metabolic activity and stress in the tissue. Accurate prediction of the maintenance respiration of a stand thus requires detailed understanding of its variation and the factors responsible for it.

Some of the most important classical and current questions about "respiration" become much more tractable when construction and maintenance respiration are clearly distinguished. For example, ecologists have wondered for years if net primary productivity decreases with stand age primarily because respiration uses an increasingly large fraction of total photosynthesis (Møller *et al.*, 1954; Kira and Shidei, 1967; Sprugel, 1984, 1985; Ryan and Waring, 1992). On examination it becomes clear that the only type of respiration that is likely to increase in older forests is maintenance respiration, because if net productivity goes down, construction respiration also decreases. So the question really is, does maintenance respiration increase enough with stand age to cause the observed decrease in NPP? To answer this, one needs to be able to estimate maintenance respiration directly rather than attempting to work with a blend of maintenance and construction respiration. Questions about the effect of climate change on respiration (Woodwell, 1987; Melillo *et al.*, 1990; McGuire *et al.*, 1992, 1993) are similar. No one is really very interested in how global warming will affect total respiration, because an increase in total respiration could result from increased maintenance respiration (which might lead to decreased biomass production), increased construction respiration (which would imply increased biomass production), or a combination of the two, and each would have vastly different consequences. The real question is whether maintenance respiration will increase so much that net productivity (and construction respiration) will decline as a result (Ryan, 1991a). In all of these cases distinguishing maintenance and construction respiration tends to clarify the issue and make it more amenable to logical and experimental analysis.

Numerous methods have been suggested for partitioning respiration into construction and maintenance components (e.g., Lambers *et al.*, 1983; Amthor, 1984; Sprugel and Benecke, 1991). Most of these are rather difficult to apply to trees in the field, but some are useful, and will be discussed in the following sections.

B. Temperature Effects on Respiration

It is well known that respiration is highly sensitive to temperature (Berry and Raison, 1981; Amthor, 1984, 1989; Ryan, 1991a). This sensitivity can be dealt with in several different ways. If the intent is simply to standardize measurements for comparison purposes, the simplest technique is to measure respiration at the same temperature for all samples using a temperature-controlled cuvette. However, in many studies temperature response curves are necessary, either (1) to allow the investigator to "remove" the effect of temperature from rates measured at ambient temperatures so that influences of other plant and environment variables can be ascertained, or (2) to predict respiration over a range of temperatures (as is often the case in scaling studies). In such cases there are two options. If a temperature-controlling device such as a growth chamber or a temperature-controlled cuvette is available, one can measure respiration over a predetermined range of temperatures to generate the desired response curve or surface. If measurements are to be made in the field without a temperature-controlled cuvette (or if temperature control is impractical due to the large thermal mass of the sample, as may be the case with large stems), respiration can be measured at ambient temperatures over a period of 1 or 2 days. However, temperature response curves derived from natural temperature variation must be used with caution, because respiration rates may also be altered by changes in plant water status and carbohydrate storage. Curves developed during periods of rapid developmental change (e.g., the time immediately after bud break) are even more tenuous and should be avoided.

The shape of the respiration–temperature curve has been a matter of extended and so far inconclusive debate. The usual presumption has been that respiration increases exponentially with an increase in temperature (Amthor, 1984). The response is usually expressed in terms of Q_{10}, the proportional change in respiration resulting from a 10°C increase in temperature:

$$R(t) = R_0(Q_{10})^{(T/10)}, \tag{2}$$

where R_0 is the respiration at 0°C. For most plants the Q_{10} of respiration is about 2.0 (i.e., respiration doubles when temperature increases by 10°C), but for woody plants the value may be higher (Ryan, 1991a).

Although it is convenient for modeling and has substantial empirical and theoretical support, the assumption that respiration increases exponentially with increasing temperature is probably accurate only over a fairly narrow temperature range. Over a broader temperature range, it is found that the exponential Q_{10} response applies only between about 10 and 25°C (for temperate species) (Larcher, 1983; Fitter and Hay,

1987). Below 10° C, the response to temperature is nearly linear; above 35° C, the response flattens out (and at very high temperatures becomes negative as thermal death approaches). In terms of Eq. (2), this is equivalent to saying that Q_{10} decreases at high temperatures and increases at low temperatures. Paembonan *et al.* (1991) found such a pattern in the respiration of a whole *Chamaecyparis obtusa* tree; Q_{10} varied from about 1.5 in August, when the mean temperature was 20–25° C, to 3.0–3.5 in January (mean temperature 5° C). Interpreting these data is difficult, though, because they could include seasonal as well as temperature-related changes in Q_{10}, and because the observations at various times of the year include both maintenance respiration and construction respiration (i.e., growth) in varying proportions. But other studies have also questioned the uncritical use of the Q_{10} relationship; for example, Brooks *et al.* (1991) found that over the range 5–35° C the instantaneous temperature response of *A. amabilis* foliage fitted a straight line better than an exponential curve. It may be that the use of Q_{10} values for describing temperature responses of respiration has been overemphasized, and that it would be better to focus more on comparing temperature response over a broad range of temperatures. In any case, the simplest solution is to measure the temperature response of respiration whenever it is needed to interpret the results of measurements.

In this paper, we review methods for obtaining stand-level estimates of autotrophic dark respiration in conifers. Because each type of tissue poses different problems and opportunities, we consider foliage, woody tissues, and fine roots separately. For each tissue type, we document methods for measuring respiration, examine the variability in respiration rates reported, and discuss whether existing models account for the observed variability. Then, we discuss extrapolating chamber measurements to the stand using our understanding of what controls variability. Finally, we discuss methods for obtaining independent stand-level estimates of autotrophic respiration, and system-level constraints on those rates.

II. Foliar Respiration

The confusion of construction and maintenance respiration that pervades measurements of other tissues is less of a problem in foliar respiration, because there are an abundance of measurements of respiration of mature leaves (Table I), which presumably represent maintenance respiration only. Construction respiration is less commonly measured; we know of no attempts to measure and isolate the construction costs of conifer foliage by regularly or continuously measuring CO_2 efflux from

Table I Maintenance Respiration Rates of Conifer Foliage[a]

Species	Location	Temperature (°C)	R_{PLA} (μmol/m²/sec)	R_{TLA} (μmol/m²/sec)	R_g (nmol/g/sec)	Reference
Abies amabilis	Pacific Northwest, United States	15	0.2–1.86	—	0.6–5.9	Brooks *et al.* (1991)
Abies veitchii	Japan	15	—	—	1–5	Matsumoto and Negisi (1982)
Chamaecyparis obtusa	Japan	14	0.1–1.3	—	—	Hagihara and Hozumi (1977)
Laris decidua × *leptolepis*	Germany	?	1[b]	0.45[c]	11	Matyssek and Schulze (1988)
Picea abies	Scotland	?	0.6–1.6	—	—	Barns *et al.* (1990)
Picea excelsa	Austria	15	—	—	1.9–3.2	Pisek and Winkler (1958)
Picea mariana	Canada	?	—	0.3–0.65	2.4–5.5[c]	Greenway *et al.* (1992)
Picea rubens	Southeastern United States	9.5	—	—	2.5–10.8	McLaughlin *et al.* (1991)
Picea sitchensis	Scotland	16	0.45–2.2	—	—	Leverenz and Jarvis (1979)
Pinus cembra	Austria	15	—	—	1.4–2.6	Pisek and Winkler (1958)
Pinus contorta	Scotland	13–22	0.5–1.5	—	—	Dick *et al.* (1991)
Pinus elliottii	Southeastern United States	15	0.16–0.6[b]	0.05–0.18[c]	0.5–1.6	Cropper and Gholz (1991)
Pinus radiata	New Zealand	15	0.4–1[b]	0.13–0.3	—	Benecke (1985)
Pinus sylvestris	Sweden	18	2.5	—	—	Leverenz (1988)
Pinus taeda	Southeastern United States	15	0.34[b]	0.11	1.3	R. O. Teskey, personal communication (1992)
Pseudotsuga menziesii	Pacific Northwest, United States	15	0.3–1.2	—	2–8.6	J. R. Brooks (unpublished data, 1989)

[a] Rates are for saplings or mature trees measured in the field. The rates included have been measured on mature, fully developed needles. Respiration rates are expressed on a projected leaf area basis (R_{PLA}), a total leaf area basis (R_{TLA}), and a unit weight basis (R_g).

[b] Converted from total leaf area by multiplying by 3.14 for *Pinus* spp. (R. O. Teskey, personal communication, 1992) and 2.2 for *Larix* spp.

[c] Calculated from author's specific leaf area measurements.

developing leaves. The only direct measurements of construction respiration in mature woody plants are those of Merino *et al.* (1984) for chaparral shrubs and Wullschleger *et al.* for *Quercus alba* and *Liriodendron tulipifera* (Wullschleger and Norby, 1992; Wullschleger *et al.*, 1992). (Several authors have also estimated construction costs of foliage using various chemical methods, as described in Section V.) There are also a substantial number of measurements of foliage respiration during leaf expansion that combine construction and maintenance respiration, but for scaling purposes these are of limited utility because it is impossible to partition out the components after the fact and the measurements are very sensitive to the exact stage of leaf expansion. In the following discussion, "respiration" of foliage means maintenance respiration unless specified otherwise.

A. Measurement Techniques

Infrared gas analyzers (IRGA) are the most common and most accurate way to measure CO_2 gas exchange; therefore, the techniques discussed here concern only the use of carbon dioxide exchange systems incorporating IRGAs. More details concerning gas exchange techniques in general are available in reviews by Leverenz and Hällgren (1991) and Field *et al.* (1991), and in an older classic by Šesták *et al.* (1971).

Equipment used to measure photosynthesis can generally also be used to measure respiration with only minor modification. The main difference is that, because the rate of foliage respiration is much lower than the rate of photosynthesis, systems designed for measuring photosynthesis may need to be modified to increase the accuracy of measuring CO_2. This is especially true if the study involves measuring respiration of old, shaded foliage at ambient temperatures, because on a cold day the CO_2 evolution of a cuvette-full of old needles can be very low indeed. If an open system is used, it may be helpful to lower the flow rates from that used for photosynthesis. In a closed system, it may be helpful to use a smaller cuvette to reduce the volume of the system so that the increase of CO_2 can be detected more accurately. For both open and closed systems it will be important to maximize the leaf area within the cuvette.

An obvious but critical decision for measurements of dark respiration is whether to measure respiration at night or during the day. The advantage of measuring respiration at night is that one need not control light and that within a few hours after darkness it can be assumed that foliage is completely dark acclimated (see below). The disadvantages of measuring respiration at night include physical difficulties in conducting the experiment, the fact that stomatal closure may restrict CO_2 evolution from the needles, and the fact that nighttime temperatures are generally low (especially in montane and boreal forest) and not very variable.

Under such conditions, if a temperature-controlled cuvette is not used, respiration can also be very low, resulting in small changes in CO_2 concentration that may be difficult to measure accurately. If a temperature-controlled cuvette is used when ambient temperatures are substantially lower than the standard temperature, there may be some concern about the effect of the rapid changes in temperature that occur when foliage is placed in the cuvette and the time required for complete acclimation to a new temperature level. Finally, some IRGAs (and many physiologists) do not function well below 5° C, adding mechanical difficulties to the other inconveniences of darkness.

The advantages of measuring respiration during the day include convenience and (usually) more satisfactory temperature conditions. The disadvantage is that foliage must be shaded during measurement and for a period of time prior to measuring so that the effect of recent illumination on respiration can be avoided. There is, however, substantial debate as to just how long this effect lasts. Using tomatoes, Ludwig *et al.* (1975) found that respiration rates are greatest directly after a period of illumination, and the brighter the illumination the greater the respiration response. The illumination effect lasted from 10 to 30 minutes, after which respiration reached a steady state. Brooks *et al.* (1991) also found that there were no significant differences between respiration rates measured in the field after 1 hour of shading and rates measured on the same samples in the laboratory after 12 hours of darkness. However, M. G. Ryan (unpublished, 1991) found that respiration of 1-year-old foliage of *Pinus strobus*, *Pinus ponderosa*, *Pinus resinosa*, and *Pinus elliottii* at a given temperature was 20–40% lower at night than during the day after 1 hour of shade. At a minimum, foliage should be dark acclimated for 30–60 minutes prior to measuring respiration during the day, but caution is still needed in interpreting the results of such measurements. Further research to clarify this situation would be very useful.

B. Rates

Conifers generally have lower respiration rates than broadleaf plants. Even within the conifer group, considerable variation exists both within and between species. Table I contains respiration rates of mature needles compiled from a number of studies to represent a broad range of coniferous species. Rates ranged between 0.1 and 2.5 μmol/m^2/sec and between 0.5 and 11 nmol/g/sec for temperatures ranging from 13 to 22° C. The observed variation in rates within a single species spans much of the ranges stated above. This wide variation makes comparisons on a species level very difficult, but in examining just the maximum rates reported for each species, some trends become evident. On a unit weight basis, the highest rates of respiration were observed in the deciduous

Larix, but on a per unit area basis, the rates are comparable to other species. This is due to the fact that *Larix* spp. have higher specific leaf area (based on total leaf area, 323–243 cm^2/g) than other coniferous species (Matyssek and Schulze, 1988; Gower and Richards, 1990). The lowest rates of respiration, on both a unit weight and area basis, were measured on *P. elliottii* (Cropper and Gholz, 1991).

Generally, baseline respiration rates increase with the elevation at which an individual or species originated (Billings *et al.*, 1971). In a classic study, Pisek and Winkler (1958) found that respiration was higher in *Picea excelsa* grown in the mountains compared to individuals grown in the lower elevation valley. This difference was detected even in the winter when respiration rates were much lower. Cropper and Gholz (1991) attributed the low rates of *P. elliottii* as an adaptation of the species to the warm subtropical climate of the southeastern United States. Rook (1969) found that seedlings of *Pinus radiata* had higher respiration rates when they were preconditioned under cold temperature conditions; however, respiration rates declined rapidly when seedlings were moved to warmer growing temperatures, indicating that this species acclimates rapidly to the temperature regime under which it grows. McLaughlin *et al.* (1991) observed higher respiration rates of *Picea rubens* growing at high-elevation sites compared to those at low-elevation sites. In contrast to the conclusions of Pisek and Winkler (1958), McLaughlin *et al.* (1991) concluded that nutrient deficiencies at their high-elevation sites (induced either naturally or by acid deposition) caused the increase in rates. On a per unit area basis *A. amabilis*, a high-elevation conifer growing at 1150 m, respired at twice the rate of *Pseudotsuga menziesii*, a lower elevation conifer growing at 160 m (1.86 vs. 0.76 $\mu mol/m^2/sec$ measured at 15°C on 1-year-old upper canopy sun foliage), although on a per unit weight basis, differences were much smaller (5.86 vs. 5.24 nmol/g/sec) (J. R. Brooks, unpublished data, 1989).

C. Variation within the Canopy

Foliage respiration has a distinct pattern of variation within the canopy. This pattern is similar to that of other foliage processes such as photosynthesis and transpiration. In general, foliage respiration is highest in regions near where growth is occurring (i.e., close to strong carbohydrate sinks) and where photosynthetic rates are high. These regions are usually in the upper exposed parts of the canopy (Hagihara and Hozumi, 1977; Matyssek and Schulze, 1988; Brooks *et al.*, 1991).

The degree to which respiration varies within a canopy depends on the species being examined. For example, Brooks *et al.* (1991) found that within a dense stand of *A. amabilis*, respiration rates were five times higher in the upper canopy than in the lower canopy. In a more open canopy of *P. elliottii*, Cropper and Gholz (1991) measured upper canopy rates that were less than twice that of the lower canopy. In a young stand of *P. menziesii* that had not achieved crown closure, no significant difference between upper and lower canopy positions were observed (J. R. Brooks, unpublished data, 1989).

The primary factors determining this pattern of variation appear to be light, growth, foliage age, and nitrogen content. These factors either directly or indirectly affect the photosynthetic rate and production of carbohydrates, which can in turn increase respiration rates by increasing enzyme turnover ion transport. Therefore, these factors can affect the rate of respiration, and may be useful indices in efforts to scale the variability of respiration within conifer canopies.

1. Ambient Light The most obvious factor that decreases with depth into the canopy is light. Hagihara and Hozumi (1977) found that foliage respiration rates were well correlated with light levels, and used light levels in the canopy as an index to estimate whole-canopy foliage respiration. Brooks (1993) also found a close correlation between light availability and foliage respiration rates in *A. amabilis*. The fact that respiration varied more in an *A. amabilis* stand than in a *P. menziesii* stand is also consistent with a strong control by light, because at the base of the *A. amabilis* canopy, light was less than 1% of the light at the top of the canopy, whereas at the base of the *Pseudotsuga* canopy, light was more than 20% of light at the top.

Light can affect foliage respiration in two ways: first, a short-term (≤1 hour) effect on metabolic activity and carbohydrate levels, and second, a long-term (days to years) effect on foliage morphology and photosynthetic capacity. As noted in Section II,A, the short-term effects of light on respiration have been noted repeatedly, but in most field studies sampling protocols are designed specifically to avoid the postillumination burst of respiration. Over a longer time period, respiration per unit area of foliage formed in sun is higher than that of foliage formed in the shade (Björkman, 1981; Larcher, 1983). The lower rates in shade foliage have been attributed to a lower demand for ATP and a reduced respiratory machinery, as well as a reduced leaf thickness (Björkman, 1981). Leverenz and Jarvis (1979) found that respiration was twice as high in sun foliage as in shade foliage on *Picea sitchensis*. In *P. radiata*, sun foliage respired at approximately three times the rate of shade foliage (Benecke, 1985). Although *Pinus* and *Picea* spp. do not form palisade mesophyll and thus cannot form true sun and shade foliage, they do have some ability to acclimate both morphologically and physiologically to the different light environments. *Tsuga* and *Abies* spp. do form palisade mesophyll cells and thus should have greater ability to acclimate to differences in light.

In evergreen conifers where foliage can live for numerous years, a canopy does not consist simply of sun foliage in the sun and shade foliage in the shade. Foliage that was formed in the sun is shaded as the tree continues to grow, so that a tremendous quantity of sun foliage eventually is in the shade. Although the morphological aspects of light acclimation are fixed after development is completed, mature foliage of *A. amabilis* can continue to acclimate physiologically to changes in light (Brooks, 1993). Brooks *et al.* (1991) noted that respiration decreased with depth in the canopy for any age class and that this decrease was not related to changes in foliage morphology as measured by specific leaf area. The ability to acclimate after leaf expansion is complete may be quite significant in conifers, because in some evergreen canopies (especially in shade-tolerant species) a large proportion of the canopy may undergo a dramatic light change in the years between development and senescence (Leverenz and Jarvis, 1979; Sprugel *et al.*, 1992).

2. Seasonal Growth Activity A large component of respiration is associated with the production of new tissue. As tissues expand, there is a burst of construction respiration in the developing tissue, and a strong local demand for carbohydrates that affects the maintenance respiration rate of nearby source foliage. As a result, "maintenance" respiration rates of mature foliage fluctuate seasonally and spatially with the growth of the tree (Hagihara and Hozumi, 1977; Matsumoto and Negisi, 1982; Brooks *et al.*, 1991). Teskey *et al.* (1984) observed that respiration rates of 1-year-old foliage were highest immediately following bud break and during shoot elongation for *A. amabilis*. However, not all studies have found seasonal differences in respiration rates. Cropper and Gholz (1991) observed no seasonal pattern of respiration in *P. elliottii*, but they point out that in northern Florida, *P. elliottii* is physiologically active year round. And in *P. radiata*, which also grows in a fairly equitable climate, Benecke (1985) found seasonal differences in current-year foliage only when he included the bud respiration in the sample; when the bud was excluded from the sample chamber there was no distinct seasonal pattern.

The degree of variation in respiration within the canopy may also be influenced by the seasonal pattern of growth activity. In *A. amabilis*, Brooks *et al.* (1991) found that the greatest variation in respiration within the canopy occurred during the early part of the growing season and that respiration in mature foliage was high in regions of intensive growth and low in regions of minimal growth. The respiration pattern within the canopy was far less distinct at the end of the growing season. The growth of reproductive structures also influences the pattern of foliage respiration. Dick *et al.* (1991) noted that the rate of foliage respiration for *Pinus contorta* was highest on branches that bore female cones.

They also found that foliage respiration fluctuated seasonally, but differences between branches were still detectable in autumn even though respiration rates were lower.

3. Foliage Age Foliage respiration rates change throughout the life span of a leaf. Respiration rates are at a maximum during growth and development, when construction respiration is occurring. Once a leaf matures the rate of respiration decreases dramatically. Respiration has also been noted to increase during the senescence of foliage as transport activity increases when nutrients and amino acids are rapidly exported from the leaf (Chabot and Hicks, 1982).

Conifer needle life spans range from 3 months to over 50 years (Reich *et al.*, 1994). This tremendous variation in leaf life span makes it difficult to generalize about the effects of age on any process. In *A. amabilis*, whose needles live 15 or more years, Brooks *et al.* (1991) found that even for mature needles, respiration decreased with age in the upper canopy, but in the lower canopy there was no detectable difference between age classes. They hypothesized that the decrease in respiration with age in the upper canopy was not simply a function of leaf age but was also related to the fact that as leaves age they become increasingly shaded as the tree grows around them, and are also more distant from active sinks. However, for *Abies veitchii*, Matsumoto and Negisi (1982) found a decrease in respiration with foliage age for seedlings grown in both the open and the shade. Greenway *et al.* (1992) did not observe any variation in foliage respiration with leaf age in *Picea mariana* growing on nutrient-poor sites; however, they also noted that the photosynthetic capacity of these needles did not differ until after the age of 6 or 8 years. How respiration changes with age has not been addressed in most studies involving conifers; yet, if respiration follows a trend similar to photosynthesis and leaf nitrogen content, then respiration should decrease with age in other conifer species as well as in those noted in the studies above. Hom and Oechel (1983) found both decreasing nitrogen content and photosynthetic capacity with needle age in *P. mariana*, and Linder and Troeng (1980) noted that photosynthetic capacity decreased with age in *Pinus sylvestris*. However, Nebel and Matile (1992) observed no photosynthetic decrease with age in *Pinus cembra*.

4. Nitrogen It has been suggested that nitrogen content of leaves should be a good indicator of foliage photosynthetic capacity because of the relationship between nitrogen content and the amount of reactants for the dark reaction of photosynthesis (Field and Mooney, 1986; Evans, 1989). If there is a correlation between respiration and photosynthetic capacity, then it would be expected that respiration and nitrogen content would also be correlated. Kawahara *et al.* (1976) showed a correlation between respiration and foliar N concentration within a *Pinus densiflora* canopy, and M. G. Ryan and R. M. Hubbard (unpublished, 1993) found correlations between respiration and N content in canopies of unfertilized *Tsuga heterophylla*, *P. ponderosa*, *P. resinosa*, and *P. elliottii*. Brooks (1993) found a correlation between respiration and N content in an *A. amabilis* canopy, although the correlation was considerably weaker than correlation between respiration and light ($R^2 = 0.28$ for N vs. $R^2 = 0.45$ for light; temperature is included in both correlations). Results of fertilization studies, however, tend to cloud this picture. Cropper and Gholz (1991) found no differences in respiration between fertilized and unfertilized trees of *P. elliottii*. R. O. Teskey (personal communication, 1992) also observed no differences in foliar respiration in *Pinus taeda* trees even though there were differences in foliar nitrogen content. These plants may have been experiencing luxury levels of nitrogen, in which case increases in N content would not affect the photosynthetic capacity or the respiration rate, although the nitrogen content of the fertilized plants was approximately 1.3%, which is not exceptionally high for conifers.

The lack of correlation between respiration and N content in fertilized trees may reflect the fact that in a N-rich environment, N in plant tissues may be present as amino acids rather than protein (van Dijk and Roelofs, 1988). In this case, foliar N estimated from total Kjeldahl nitrogen will be a poor estimator of foliar protein content. Because the relationship between foliar N and maintenance respiration depends on the N being in protein, if N is in amino acids or storage proteins, foliar N will likely not be a good predictor of maintenance respiration. For this reason, the relationship between respiration rates and N may not be valid for fertilized conifers. Similar problems have been noted for the relationship between photosynthesis and foliar N content (e.g., Reich and Schoettle, 1988). Further research on this question would be exceedingly useful.

D. Techniques for Extrapolating to Stand-Level and/or Long-Term Estimates

Scaling foliage respiration from measurements on a twig to an estimate of stand respiration requires knowledge of both the variation of respiration and the distribution of foliage within the canopy. As discussed previously, foliage respiration rates vary significantly both temporally and spatially. Ideally one would like to find an index closely related to respiration rates whose spatial distribution could be easily mapped, but few studies to date have taken this approach. A notable exception is Hagihara and Hozumi (1977), who related foliar respiration in *Chamaecyparis obtusa* to light levels and then used Beer's law of light extinction to estimate total canopy respiration. Most other studies, however, have used relatively complex methods of quantifying foliage distributions within canopies, and these cannot easily be transferred to other systems (Norman and Jarvis, 1974; Rook and Corson, 1978; Ågren *et al.*, 1980; Brooks, 1987; Hashimoto, 1990; Wang and Jarvis, 1990; Cropper and Gholz, 1990; Mori and Hagihara, 1991).

An additional factor that must be considered when examining the distribution of respiration within canopies is the within-canopy variation in temperature, because temperature has such a strong influence on the rate of respiration. Temperature varies both temporally and spatially within canopies (Fritschen *et al.*, 1973; Brooks *et al.*, 1991), with higher leaf temperatures in the upper exposed parts of the canopy and decreases in temperature with depth in the canopy. Brooks (1987) found that a fivefold difference in modeled respiration rates between the top and bottom of the canopy was increased to a sevenfold difference when temperature differences within the canopy were included in the calculations. This may represent an extreme case, because many conifer canopies are better coupled to the atmosphere than the dense *A. amabilis* canopy (McNaughton and Jarvis, 1991), but it does show that in at least some canopies temperature variation can play a significant role.

III. Woody-Tissue Respiration (Stems, Branches, and Coarse Roots)

Respiration of woody tissues (stems, branches, and coarse roots) has been measured much less often than respiration of foliage, but of those measurements that have been made, a relatively high proportion have been aimed at stand-level estimates (e.g., Hagihara and Hozumi, 1981; Linder and Troeng, 1981; Benecke, 1985; Sprugel, 1989; Ryan and Waring, 1992). Several methods have been used for scaling, but recent work (Ryan, 1990; Sprugel, 1989, 1990; Sprugel and Benecke, 1991; Ryan and Waring, 1992) has led to the development of relatively straightforward and general techniques that appear to hold great promise for the future.

A. Measurement Techniques

Techniques for measuring respiration of woody tissues have recently been reviewed in detail by Sprugel and Benecke (1991) and will be discussed only briefly here. Historically, two methods have received the most use. In the first, excised branch, stem, or root sections are removed from the tree and brought back to the laboratory for measurement of CO_2 evolution (or occasionally O_2 uptake) (Müller, 1924; see additional references in Sprugel and Benecke, 1991). This method requires little in the way of sophisticated equipment and permits the use of a large sample size, but it is difficult to be confident that respiration of an excised stem is not affected in some way by the harvesting process (e.g., through wounding or changes in water status, internal CO_2 concentration, carbohydrate content, or hormonal supplies). It is also impossible to make repeat measurements on the same sample to follow respiration trends through the year, and adequate sampling involves considerable destruction of trees in or near the study area.

The main alternative to excision methods has been *in situ* methods in which sections of living stems, branches or roots are enclosed in permanent cuvettes and CO_2 evolution through the bark is monitored continuously over prolonged periods of time (Johannson, 1933a; for others, see Sprugel and Benecke, 1991). This technique is much less intrusive than excision techniques, but it generally requires expensive and complex equipment to maintain airflow through the cuvettes, measure changes in CO_2 concentration, and record the data for future analysis. The main problem with this technique is that the extensive hardware requirements usually result in a very small sample size (one or two sections is typical), which makes it difficult or impossible to determine which characteristics of the sampled branch or stem are most important in determining respiration rates and thus to extrapolate to other parts of the tree.

A third technique that has been gaining in popularity involves the use of a moderate number of removable cuvettes that can be clamped onto a stem, branch, or root section while a measurement is being made, then removed until the next sampling date (Sprugel, 1990; Ryan, 1990; Sprugel and Benecke, 1991). This technique shares with the continuous monitoring technique the advantage that measurements are made *in situ* on living trees, but because CO_2 is not continually monitored, hardware requirements are much smaller, and a significantly larger sample size is possible. These larger sample sizes, coupled with the ability to make repeat measurements on the same site, have been crucial in recent efforts to separate construction and maintenance respiration in woody tissues.

A continuing problem with all measurements of woody-tissue respiration is the concern that CO_2 produced by respiration in the sapwood may be carried away by the transpiration stream instead of being released through the bark, so that measured CO_2 evolution would underestimate the actual respiration of the sample section (Boysen Jensen, 1933; Negisi, 1972, 1975, 1978, 1979; Martin, 1992; Martin *et al.*, 1994). Negisi (1975) reported that CO_2 evolution from a *P. densiflora* stem was reduced about 50% at midday compared with what would have been expected based on temperature alone. It is also possible that CO_2 produced by root or sapwood respiration might be carried upward into the stem or branches and released there, in which case measured CO_2 evolution higher in the tree would overestimate actual sapwood respiration.

However, no examples of excess CO_2 release during periods of high sap flow have been reported, with the possible exception of high rates of CO_2 release from *A. amabilis* branches in the spring that significantly exceeded what would have been expected from the sapwood volume and growth rate (Sprugel, 1990). After 60 years of controversy (cf. Boysen Jensen, 1933; Johannson, 1933b), the problem of CO_2 movement in the translocation stream remains a central concern in all measurements of woody-tissue respiration (Sprugel and Benecke, 1991).

An additional issue in measuring woody-tissue respiration is refixation of respiratory CO_2 by bark photosynthesis. Bark photosynthesis is reviewed in detail by Sprugel and Benecke (1991); here it suffices to say that for thin-barked trees, bark photosynthesis can refix a very substantial fraction of the CO_2 produced in stem respiration. Most studies of refixation have focused on angiosperms such as *Populus tremuloides* [where refixation rates of up to 90% have been reported (Schaedle, 1975)], but data in Sprugel and Benecke (1991) suggest that bark refixation reduced apparent respiration of a *P. radiata* branch by 35–40% over 24 hours. The possibility of significant refixation must thus be taken into account in any study of woody tissue respiration. If the intent is simply to measure respiration per se, then opaque cuvettes should be used to eliminate bark photosynthesis. If, on the other hand, the intention is to determine the contribution of the stem and branches to the tree's carbon budget, then it may be desirable to use clear cuvettes for at least some measurements and to minimize the shading by analysis equipment. (This, however, adds several additional complications, including the problem of heating of the enclosed stem if airflow rates are low and the difficulty of scaling bark photosynthesis over the entire surface of the tree).

B. Rates

There have been a substantial number of studies of woody-tissue respiration in conifers, including *Abies amabilis* (Sprugel, 1989, 1990), *Abies balsamea* (Lavigne, 1987, 1988), *Abies sachalinensis* (Yoda *et al.*, 1965), *Chamaecyparis obtusa* (Hagihara and Hozumi, 1981; Negisi, 1982), *Cryptomeria japonica* (Yoda *et al.*, 1965), *Larix decidua*, *Larix leptolepis*, and *Larix sibirica* (= *Larix russica*) (Johannson, 1933a), *Larix decidua* × *leptolepis* (Matyssek, 1985; Matyssek and Schulze, 1988), *Picea abies* (Johannson, 1933a), *Picea glehni* (Yoda *et al.*, 1965), *Picea engelmannii* (Ryan, 1990), *Picea jezoensis* (Yoda *et al.*, 1965), *Pinus cembra* (Havranek, 1981), *Pinus contorta* (Benecke and Nordmeyer, 1982; Ryan, 1990; Ryan and Waring, 1992), *Pinus densiflora* (Yoda *et al.*, 1965; Negisi, 1974, 1975, 1979, 1982), *Pinus radiata* (Benecke, 1985), *Pinus sylvestris* (Johannson, 1933a; Linder and Troeng, 1981; Zabuga *et al.*, 1983; Zabuga and Zabuga, 1985), *Pinus taeda* (Martin, 1992), and *Pinus thunbergii* (Yoda *et al.*, 1965). These studies contain a wide variety of information about respiration in conifer stems (and in a few cases, branches and roots), including seasonal variation, species comparisons, and relation of respiration to volume. Unfortunately, most of the actual respiration rates reported are given as respiration per unit surface area, which is relatively uninformative, because it varies tremendously for different individuals, and even for different locations on the same tree. For example, Sprugel (1990) found that on a single day in June, CO_2 efflux from stems of 28-year-old *A. amabilis* trees (corrected to 15°C) varied from 0.12 μmol/m^2/sec on a small, slow-growing tree to 2.67 μmol/m^2/sec on a large, fast-growing tree. Variation in respiration per unit area of branches was similarly dramatic, ranging from 0.10 to 1.61 μmol/m^2/sec. When respiration per unit area varies by more than 20-fold within a single species on a single site on a single date, trying to compare values among species or sites is relatively futile, and so we have refrained from tabulating such values.

The idea that woody-tissue respiration should be expressed on a per-unit-area basis was apparently based on the assumption that the largest component of woody-tissue respiration was maintenance respiration of the cambium and phloem, so that respiration per unit area might be expected to be relatively constant. The assumption in turn was apparently based on a misinterpretation of the seminal work of Goodwin and Goddard (1940), who measured O_2 uptake by cut stems and found that respiration was highest near the outer surface of the stem. Later authors, reading their work, apparently did not realize that what Goodwin and Goddard were measuring was almost entirely *construction* respiration resulting from the production and maturation of xylem and phloem cells (Havranek, 1981)—and hence highly variable from site to site and season to season, making respiration per unit surface area too variable to be a useful index for species comparisons or scaling. In fact, other studies (Ryan, 1990; Sprugel, 1989, 1990) have shown that when construction respiration is absent (i.e., once the growing season is over), phloem and cambium respiration are insignificant compared with sapwood maintenance respiration—and of course the amount of sapwood under a given area of surface, as well as the rate of new wood production, will vary drastically depending on the size and vigor of the tree. Thus there is no reason to expect respiration per unit surface area to be constant from tree to tree or even within a single tree, and measurements expressed on a surface area basis are of little use for comparisons.

If respiration per unit surface area is too variable to be a useful measure of woody-tissue respiratory activity, it is necessary to ask what units are better. As noted above, Ryan (1990) and Sprugel (1989, 1990) both found that when the growing season is over, stem respiration is strictly

Table II Maintenance Respiration Rates of Conifer Sapwood[a]

Species	Location	Respiration (nmol/g/sec)	Reference
Abies amabilis	Washington	0.05	Sprugel (1990) and D. G. Sprugel, unpublished (1986)
Pinus contorta	Colorado	0.032	Ryan (1990)
Pinus elliottii	Florida	0.020	Ryan *et al.* in review
Pinus ponderosa	Montana	0.020	Ryan *et al.* in review
Pinus resinosa	Wisconsin	0.022	Ryan *et al.* in review
Tsuga heterophylla	Oregon	0.015	Ryan *et al.* in review

[a] All values at 15°C.

proportional to the amount of sapwood present in the sample cuvette, suggesting that at a constant-temperature sapwood maintenance respiration per unit volume (or mass) is relatively constant throughout a stand. And as we have already noted, values of construction respiration per unit sapwood produced should be relatively constant as long as the chemical composition of the wood does not vary too much. Thus the most useful values for comparing woody-tissue respiration rates among species would be values for maintenance respiration per unit sapwood volume or dry weight, and construction respiration per unit of new wood and bark produced.

Unfortunately, only a few estimates of sapwood maintenance respiration are available for conifers (Table II). Those that are available suggest that respiration per gram of wood is relatively constant among conifers, and that there are no immediately obvious patterns of variation with climate. Estimates of the construction cost of conifer wood are somewhat more variable (Table III). Chemical methods (Penning de Vries, 1975a;

Table III Construction Respiration Rates of Conifer Sapwood

Species	Respiration (g C/g C)	Method	Reference
Generic wood	0.25	Chemical pathway analysis	Penning de Vries (1975a)
Abies amabilis	0.42	CO_2 efflux − maintenance respiration	Sprugel (1989) and D. G. Sprugel, unpublished (1986)
Larix decidua	0.87	See text	Havranek (1985)
Pinus elliottii	0.15	Chemical pathway analysis	Chung and Barnes (1977)
Pinus radiata	0.25	CO_2 efflux − maintenance respiration	U. Benecke (unpublished, 1984); cited in Sprugel and Benecke (1991)

Chung and Barnes, 1977; cf. Section V) suggest that the construction cost of wood should be about 0.25 g C/g C. U. Benecke (unpublished, 1984; cited in Sprugel and Benecke, 1991) monitored CO_2 efflux from two locations on a *P. radiata* stem throughout a summer and subtracted estimated maintenance respiration from the total CO_2 efflux, and estimated the construction cost of wood at 0.25 g C/g C, a perfect match to the theoretical value. However, Sprugel (1989 and D. G. Sprugel, unpublished, 1986), using essentially the same technique, measured the construction respiration cost of sapwood production in *A. amabilis* at 0.42 g C/g C. Havranek (1985), using a somewhat more indirect technique based on the differences between measured CO_2 efflux and wood increment in two consecutive years, estimated the respiration cost of wood production in *L. decidua* at 0.87 g C/g C. The fact that these last two estimates based on CO_2 efflux are substantially higher than those based on chemical pathway analysis may mean that there are substantial additional respiration costs involved with tissue production in addition to the direct chemical costs (e.g., the costs of loading and unloading carbohydrates from the phloem), or it could simply reflect experimental error or inaccurate assumptions involved in the calculations. However, the differences are quite striking and suggest that more study of the actual costs of wood production would be valuable.

C. Variation within the Canopy

Woody-tissue respiration varies seasonally, between trees and even between locations on the same tree. The limited data available so far suggest that most of the observed variation in stem respiration reflects variation in the rate of wood production at any given place and time and in the amount of respiring sapwood. Thus, the easiest way to explain intracanopy variation is to partition observed measurements of respiration into construction and maintenance components, and to relate these to the sapwood volume and growth rate.

Efforts to partition respiration into construction and maintenance components, and to use this partition to predict total respiration for whole stands, have proceeded farther in studies of stem respiration than for any other tissue (although still not very far). Techniques for partitioning stem respiration into growth and maintenance components are discussed in some detail by Sprugel and Benecke (1991), and are also reviewed in Martin (1992). The simplest technique for partitioning measurements of respiration into construction and maintenance components is to assume that respiration measured before or after the growing season is solely maintenance respiration, and that higher levels during the growing season represent construction plus maintenance. The summation of construction respiration over the whole growing season then represents the construction cost of the new wood that was produced that year. The key to this technique is the assumption that maintenance respiration at a constant temperature remains constant throughout the year, or at least for a month or two after growth stops (see discussion in Sprugel and Benecke, 1991). This assumption is necessary for most methods of partitioning respiration, but is virtually impossible to test, because maintenance respiration cannot be measured directly during the growing season. Sprugel (1990) attempted to check this assumption by regressing growing-season respiration against growth rate (with maintenance respiration left as the intercept when growth = 0), and although the results were rather noisy due to sampling variation, they were generally consistent with a constant rate of maintenance respiration.

Partitioning of branch respiration into maintenance and construction components has been less successful than for stem respiration. Sprugel (1990) found that measured branch respiration rates significantly exceeded the levels that would be expected for stems with similar growth rates and sapwood volumes, and that the observed respiration could not be readily partitioned into construction and maintenance components using the same techniques that worked for stems. Because the differences between predicted and measured values were greatest in the spring (when measured values exceeded predicted values by up to four or five times), and branch respiration rates were fairly well correlated with the amount of new growth on the branches, Sprugel suggested that some of the additional respiration might represent the cost of mobilizing carbohydrates stored in the parenchyma. However, little additional work has been done on this question and it basically remains a mystery.

Coarse root respiration has been systematically measured in only two studies (Linder and Troeng, 1981; Benecke, 1985), and in both cases only a single location was monitored. More studies with larger sample sizes will be needed before it is possible to determine the sources of variation in coarse root respiration and the possibility of using the construction : maintenance distinction to scale up to the stand level.

D. Scaling Up to the Stand Level

When stem respiration is partitioned into construction and maintenance components it becomes relatively easy to scale stem respiration measurements up to the stand level. Because maintenance respiration per unit of sapwood volume seems to be relatively constant throughout the main stem (at a given temperature), sapwood volume is a useful index for extrapolating small-scale measurements of stem maintenance respiration up to a stand level (Sprugel, 1989; Ryan and Waring, 1992). Although equations for estimating stem sapwood volume are available for only a few species (e.g., Ryan, 1989), there are no great technical difficulties in generating equations for estimating sapwood volume as a function of height and sapwood area, which is commonly estimated as an index of leaf area. Similarly, if the construction respiration cost of producing different kinds of tissues is estimated (either from pathway analysis or through measurement of CO_2 evolution), total annual construction respiration for the stems in a stand can be estimated easily from the annual production of stem wood, which is routinely estimated in studies of net primary productivity.

Techniques for scaling branch and coarse root respiration from samples to the stand are less well developed. If it turns out that biomass and growth rate are good indices of branch and coarse root respiration, then it should be possible to tap into a substantial literature on branch and wood biomasses in forests. For example, determination of coarse root biomass is relatively easy because good relationships exist between coarse root biomass and the tree diameter at breast height (Vogt and Persson, 1991). A sizable data base also exists on coarse root biomasses for forests (Cannell, 1982). However, until the nature of local and seasonal variation in branch and coarse root respiration is better understood, scaling these processes up to the stand level will remain difficult.

IV. Fine Root Respiration

In general, root respiration is far more difficult to measure unambiguously than foliage or woody-tissue respiration because of the inaccessibility of roots under natural conditions. All known techniques for measuring root respiration require either removing roots from the soil environment and measuring them under vastly different conditions in the laboratory or the field, or making very significant modifications of the soil, whose effects on root and heterotroph respiration cannot readily be predicted.

Most studies of belowground respiration have concentrated on determining the respiration rates of fine roots because fine roots contribute the bulk of the CO_2 from belowground autotrophic respiration. The following discussion generally considers only fine and small-diameter roots (i.e., roots less than 0.5 cm in diameter); methods used to measure and scale coarse roots are more akin to the methods used for stems and branches, and were considered in the previous section.

A. Measurement Techniques

Singh and Gupta (1977) provide a detailed discussion of the older methodologies used to estimate fine root respiration rates, and these will

Table IV Root Respiration Data for Coniferous Seedlings and Forests Using Indirect and Direct Methods

Species	Respiration[a]	Q_{10}	Temperature (°C)	Size class (cm)	Analysis approach	Reference
				Mass-based measurements		
Abies amabilis	0.6–8.7	—	5–14	0.03–0.1	30-year-old stand, same root sampled monthly (July–November)	K. A. Vogt, D. J. Vogt, and D. Sprugel, unpublished (1986)
	0.8–8.9	—	—	0.1–0.15		
	0.4–6.1	—	—	0.2		
	0.3–3.3	—	—	0.2		
	0.3–3.3	—	—	0.2–0.35		
	0.3–1.6	—	—	0.5		
	0.2–0.6	—	—	0.6		
Abies lasiocarpa	27.1–29.1[b]	1.9	12	Fine	Excised 1-cm root segments from small trees	Sowell and Spomer (1986)
Picea engelmannii	30.0–33.6[b]	2.0	12	Fine	Excised 1-cm root segments from small trees	Sowell and Spomer (1986)
Picea rubens	1.3–3.2	—	—	Fine	Excised roots	Keller, 1967 (in Singh and Gupta, 1977)
Pinus elliottii	2.5	1.94	12	Fine	Intact roots, 22-year-old stand, one-time field reading per root, forest floor only	Cropper and Gholz (1991)
Pinus radiata	3.0	2.8	10	Fine	Excised roots, seedlings	D. Hollinger, unpublished (1991)
	1.5	2.3	10	Fine	Excised soil blocks, mature forest	
Pinus sylvestris	1.1	—	?	<0.2	Excised roots	Ågren *et al.* (1980)
Pinus taeda	5.7–8.0[a]	2.0	17	Fine	Excised roots, field-planted seedlings	Boyer *et al.* (1971)
Pseudotsuga menziesii	2.8	2.7	10	Fine	Excised roots, seedlings	D. Hollinger, unpublished (1991)
	1.2	3.0	10	Fine	Excised soil blocks, mature forest	
				Ground area-based measurements		
Pinus elliottii	0.43	—	—	Fine + coarse	9-year-old stand	Ewel *et al.* (1987)
	0.81	—	—		29-year-old stand; compartmentalized budget	
Pinus taeda	0.06	—	—	Fine + coarse	Excised roots	Kinerson (1975)

[a] Units for mass-based measurements are nmoles/gram/second; units for ground area-based measurements are kilograms carbon/m^2/year.
[b] [illegible]

not be covered here. Interestingly, several of the earlier approaches to measuring fine root respiration are still utilized with slight modifications (Table IV).

Two major approaches have been used to estimate root respiration, which Singh and Gupta (1977) classified as direct and indirect approaches. The direct approach usually involves measuring CO_2 evolution or O_2 uptake of fine roots *in situ* or after they have been removed from the soil. This is by far the most common approach and is discussed in more detail below. The indirect approach typically consists of measuring total CO_2 evolution or O_2 uptake from intact soil surfaces in the field and either comparing different sites with varying root biomass or comparing nonmanipulated sites with sites that have been manipulated by removing all or part of the plants. In this approach the root contribution to total soil respiration is calculated by comparing total soil respiration rates from sites with known variation in root biomass. No studies of this type have been reported for conifer forests.

Only two studies have measured root respiration directly on intact roots in the field (Cropper and Gholz, 1991; K. A. Vogt, D. J. Vogt, and D. G. Sprugel, unpublished, 1986). Although studying the same coarse root throughout a year is very feasible, and the approach can mimic those described for branches (see previous section), respiration of intact fine roots has rarely been measured in the field because it is difficult to isolate fine roots from the soil without disturbing the system or causing the senescence of roots when they are exposed to the atmosphere (Vogt *et al.*, 1989). Cropper and Gholz (1991) measured the respiration rate of intact fine roots of *P. elliottii* in the forest floor using a one-time placement of roots in a chamber followed by harvesting of the roots for dry weight determinations. K. A. Vogt, D. J. Vogt, and D. G. Sprugel, (unpublished, 1986) used the removable chambers described in Vogt *et al.* (1989) to measure short-term respiration rates of intact *A. amabilis* roots monthly over a 5-month period, reburying the roots after each measurement (Table IV) (Vogt *et al.*, 1989). In this cool, damp subalpine stand there appeared to be little loss of vitality when the same fine roots (ranging in diameter from <1 to 6 mm) were measured repeatedly over 5 months.

Most other "direct" measurements of root respiration have been made under laboratory conditions. Two basic methods have been used. In the first, excised roots are collected from plants growing in pots or from the soil of a mature stand, and then CO_2 evolution is measured in the laboratory under controlled conditions. In the second method, CO_2 evolution is measured from intact soil cores collected in the field or the root systems of plants growing in pots in growth chambers, and the measured respiration is then compartmentalized into root (autotrophic) and soil (heterotrophic) components by separately analyzing the respiration rates of the individual components (Coleman, 1973; Edwards and Sollins, 1973).

All of the laboratory methods mentioned above have considerable problems that may skew the respiration rates measured for roots or for intact soil cores. Excised roots are commonly used to measure respiration rates, but no study has been conducted to compare intact and excised roots in conifers, or to determine whether the root injury that occurs when soil cores are collected alters the respiration rate of the roots in the cores. As noted in Section III,A, there has long been concern that harvesting branches or stem sections could cause significant changes in their respiration rates. These problems are magnified in dealing with roots because the roots are not only wounded and separated from their parent tree, but also in most cases are removed from the dense, CO_2-rich soil environment and placed into a vastly different physical environment (Evans, 1972). Redmann and Abouguendia (1978) (in grasslands) and Medina *et al.* (1980) (in rainforests) both found that when small cores were brought back to the laboratory, there was an immediate increase in CO_2 evolution, although it is not clear how long this response persists. Moreover, Billings *et al.* (1977), found that in the Arctic graminoids *Carex aquatilis* and *Dupontia fischeri* the respiration rate was higher in excised roots as compared with intact roots, and the Q_{10} was also greater. However, Holthausen and Caldwell (1980) found little effect of excision in the desert shrub *Atriplex confertifolia*. Clearly considerably more study is needed before measurements from excised roots or intact soil cores can be confidently assumed to be equivalent to respiration of roots in their natural environment.

When respiration from intact soil cores of whole soils is compartmentalized into root and soil components, there are problems in separating respiration due to roots from the respiration of microbes that are in close association with roots. A significant portion of the respiration in the rhizosphere may also be from microbes utilizing the carbohydrates exuded into the rhizosphere (Wanner, 1970; Hendrickson and Robinson, 1984). Furthermore, the presence of mycorrhizal fungi on fine root tips will modify respiration rates, but by how much is not known. Data of Barnard and Jorgensen (1977) and Harley and Smith (1983) suggest that they will increase respiration rates. It is also not known whether different species of mycorrhizal fungi colonizing root tips will have different effects on root respiration. Any study that separates roots from the surrounding soil may introduce unknown biases by disrupting these rhizosphere relationships and mycorrhizal associations.

Many early studies attempted to compartmentalize soil respiration into root and heterotrophic components either by excluding living roots from the sites in the field (using small minitrench plots) or by bringing soil cores back to the laboratory and comparing the respiration rates of the total core to the respiration of the components (Wiant, 1967; Schlesinger, 1977; Singh and Shekhar, 1986). Unfortunately, most studies using the small minitrench plot approach never quantified how much living root biomass and decaying root tissue was under the plot where vegetation was excluded. Many studies also used static chemical absorption systems to measure CO_2 evolution from the soil cores or plots; this method has been shown to underestimate CO_2 evolution rates, and so results from these studies must be regarded as questionable (Reiners, 1968).

Several recent studies have utilized new methods based on total carbon budgets (Ågren *et al.*, 1980; Ewel *et al.*, 1987). Ewel *et al.* (1987) subtracted CO_2 attributable to litter-layer roots and soil organic matter decomposition from total soil CO_2 evolution to obtain root respiration rates. Edwards and Harris (1977) used standing crops of litter, small and large roots, and their decay rates to partition C transfers. These carbon-budget-based approaches are useful for estimating maximum root respiration rates, but this approach is confounded by the usual problems of estimating ecosystem components by subtraction (Kaufmann and Landsberg, 1991); errors in estimating the other C components of an ecosystem accumulate and can result in an error in estimating root respiration that vastly exceeds the proportional error of the other components.

B. Rates

Published (and some unpublished) estimates of conifer root respiration are summarized in Table IV, along with the diameter sizes measured and the methods used. Those studies reporting only the percentage of total soil respiration attributed to roots (i.e., without reporting actual rates) were not included in Table IV, but can be found in Singh and Gupta (1977). Unfortunately, it is difficult to draw any significant conclusions by comparing values from different studies because (1) there are few data on how the different methods used to obtain these estimates affect the values observed, (2) only a few species have been analyzed, and (3) basic information on patterns of variation in root respiration have not been established (see Section IV,C).

C. Variation within the Rooting Zone

Root respiration may vary with temperature, soil moisture, season, root diameter, presence of mycorrhizal fungi on root tips, and possibly tree age and distance from the parent tree. The broad outlines of some of these patterns of variation are known, but few quantitative data are available.

As noted in Section I,B, temperature is well known to have an important effect on respiration in all tissues. However, studies in rainforest trees and in grasslands have shown that day and night respiration rates may not vary in a consistent direction (Medina *et al.*, 1980; Grahammer *et al.*, 1991). If soil cores are being used, then soil moisture variation must also be accounted for, because soil water content also influences the rate of soil respiration (Grahammer *et al.*, 1991). It has been suggested that under water stress conditions the fraction of soluble carbohydrates allocated to root growth increases as roots become a greater sink for carbohydrates compared to shoots, resulting in higher root respiration rates (Grahammer *et al.*, 1991). In addition, root respiration is apparently directly dependent on current photosynthate, which suggests that anything affecting shoot photosynthesis activity will feed back on root respiration levels. For example, Szaniawski and Adams (1974) found that root respiration rates were highest for *Tsuga canadensis* seedlings when the shoots were photosynthesizing, suggesting a dependence on current photosynthesis.

The few studies that have examined changes in fine root respiration rates by diameter show definite increases in respiration with decreasing diameters. For example, in *A. amabilis* (see Table IV), roots <2 mm in diameter consistently had higher respiration rates compared to larger roots. Few other data exist comparing roots of different sizes in conifers, but Cox (1975) found that smaller diameter excised roots from yellow poplar seedlings had higher respiration rates than did larger diameter roots (see Table IV) when measured at 20°C using a differential respirometer. Using similar methods, Chapman (1979) found that root respiration increased significantly with decreasing diameter of roots from coarse (5–7 mm) to very fine (<0.5 mm) in lowland *Calluna* heathland. Chapman (1979) recorded the highest respiration rates in roots <1 mm in diameter.

The stage of plant development may also affect the respiration rates of root tissues. Root respiration has been determined mainly on seedlings, but as stands mature, the carbon allocation patterns will change and should be reflected in the respiratory demands of tissues (Sprugel and Benecke, 1991). D. Hollinger (unpublished, 1991) found that in both *P. radiata* and *P. menziesii*, root respiration rates of seedlings were consistently higher than respiration rates for mature trees of the same species (Table IV). This suggests that great caution must be used in extrapolating from seedling measurements to mature stands to estimate fine root respiration (D. Hollinger, unpublished, 1991).

It is not clear if root respiration rates might change with increasing distance from a tree bole. Boyer *et al.* (1971) measured no differences in fine root respiration rates of root tips compared to root segments located several centimeters from the tips. In an *A. amabilis* stand (Table IV) in Washington, roots of similar diameters (~5 mm) at different distances from the bole did not have different respiration rates. Thus current evidence seems to suggest that distance from tree bole is not an important factor to consider, but further studies are needed.

D. Stand-Level Conversions

Converting fine root respiration data to a stand level is more difficult than for many other types of tissues because fine root biomass data are scarce for forest ecosystems (Vogt, 1991) and would have to be specifically collected on the site being studied. Moreover, so little is known quantitatively about the sources of variation in fine root respiration that it is not even obvious what kinds of data should be collected in many cases. For example, fine root biomass data have been collected using a great variety of root diameter size classes, depending on the question being asked (Vogt and Persson, 1991). For scaling purposes, fine roots should probably be separated into functional diameter classes that reflect their respiration rates. It may be necessary to collect fine root biomass data in at least three different diameter classes (e.g., <2, 2–5, and 5–10 mm) and measure respiration rates specifically by these diameter class groupings. This type of sorting is very time consuming, but it may be essential if we are to have accurate estimates of root respiration at a stand level. Even with careful sorting, though, it is often difficult to tell the difference between roots that are senescing and those that are not, especially when mycorrhizal fungi are present (Vogt and Bloomfield, 1991). Thus at the present time, both a lack of information on factors affecting root respiration and great difficulties in collecting appropriate stand-level data on root characteristics present critical barriers to scaling respiration measurements realistically from the sample level to the stand.

V. Proxy Methods for Estimating Respiration without Tissue-Specific Measurements

We study autotrophic respiration (R_a), among other reasons, because (1) carbon used for respiration may vary with the environment, type of vegetation, and successional stage; (2) the balance between photosynthesis and autotrophic respiration determines dry matter production; and (3) fluxes from respiration are important in determining the global carbon budget. Instantaneous fluxes of CO_2 from plant respiration indicate current physiological status, but are difficult to obtain and are not generally useful for determining the role or importance of respiration in ecosystem energetics. For this reason estimates of autotrophic respiration need to be placed in the context of the entire carbon budget, and examined over one to several years.

Fuller understanding of the role of autotrophic respiration in regulating ecosystem productivity will come only when we have assembled carbon budgets from many different ecosystems and different successional stages. Toward this goal, for estimating respiration we outline methods that can be applied without detailed physiological measurements. In this section, we describe methods for estimating instantaneous fluxes and annual budgets of CO_2 from respiration for whole ecosystems. First, we examine some of the disadvantages of the use of tissue-specific rates, and present evidence that element concentrations (primarily N and C) of tissues can provide estimates of respiration. Next, we discuss a method for estimating belowground autotrophic respiration, and its limitations. Finally, we discuss eddy correlation as a technique useful for estimating CO_2 fluxes, and derive equations for estimating instantaneous respiration for forests. The methods described may be useful where tissue-specific rates or annual budgets are lacking, and also serve as an independent estimate where detailed measurements have been made.

A. Estimating R_m from Tissue N Content

Several physiologically based models of forest production use the functional model of R_a outlined in Section I and tissue-specific rates to estimate R_a for forest ecosystems [for example, Forest-BGC (Running and Coughlin, 1988; Running and Gower, 1991), HYBRID (Friend, 1991; Friend *et al.*, 1993), BIOMASS (McMurtrie *et al.*, 1990; McMurtrie, 1991, and GEM (Rastetter *et al.*, 1991)]. Typically, these models estimate net canopy photosynthesis in the light (equivalent to a cuvette measurement), sum photosynthesis for a period, subtract maintenance respiration, and determine dry matter production from the remainder. However, the use of tissue-specific rates to estimate respiration may lead to uncertain estimates because of the high variability in respiration rates per unit biomass or surface area (see previous discussions). Also, using tissue-specific respiration rates precludes feedback with other ecosystem processes. Maintenance respiration and tissue nitrogen content are strongly correlated within a species [see, for example, Jones *et al.* (1978) and Irving and Silsbury (1987) for crop plants, and Kawahara *et al.* (1976) for conifers]. This relationship exists because most of the organic nitrogen in plants is in protein and roughly 60% of maintenance respiration supports protein repair and replacement (Penning de Vries, 1975b). Ryan (1991a) recently derived a relationship between maintenance respiration and tissue N content using a variety of crop, shrub, and tree species:

$$R_m = 2.95 \cdot 10^{-6} N, \tag{3}$$

where R_m (moles C/sec at 20° C) is maintenance respiration and N is Kjeldahl N content of plant tissue (moles). The relationship is linear over the range of tissue nitrogen contents normally encountered (about 0.04–6% of dry weight). In conifers, Kawahara *et al.* (1976) showed that tissue N content could account for much of the variability in respiration rates among tissues in a *P. densiflora* forest. This relationship estimated annual maintenance respiration costs for a *P. elliottii* stand within 10% of the value reported in Cropper and Gholz (1991) that was estimated from detailed chamber measurements.

To the extent that tissue N is correlated with respiration, using N to estimate maintenance respiration offers several advantages over the use of tissue-specific rates. First, photosynthesis has been strongly linked with foliar N content (Field and Mooney, 1986; Evans, 1989, 1993a,b), and the use of N to estimate respiration would provide a strong link between the two processes. However, this relationship is strongest in fast-growing and somewhat weaker in slow-growing evergreen sclerophylls. Second, a close relationship between N and respiration would make it easier to predict how respiration could respond to the nutrient status of ecosystems, changing as climate, atmospheric deposition of N, or succession changed tissue nutrient contents. Third, N content, photosynthetic capacity, and respiration decrease from the top to the bottom of the canopy (Brooks *et al.*, 1991), and N might be a more conservative estimator of canopy processes than an average canopy rate. Finally, modeling respiration in different ecosystems might be simpler, because in many cases tissue N contents are simpler to measure than respiration rates.

Although the relationship between respiration and nitrogen has been shown to exist in *P. densiflora* (Kawahara *et al.*, 1976), we lack evidence for this relationship for other conifers. In crop plants, R_m per unit of plant nitrogen varies with growth rate (McCree, 1982; McCree and Amthor, 1982), species (McCree, 1983), and tissue type (Szaniawski and Kielkiewicz, 1982), perhaps because protein content and protein turnover are not strictly related. However, variability in maintenance rates per unit tissue nitrogen is small when compared to the range of respiration rates expressed on a dry weight basis. M. G. Ryan and R. M. Hubbard, (unpublished, 1993) found a positive correlation between dark respiration rates and foliar N content in *P. ponderosa*, *P. radiata*, *P. resinosa*, and *Tsuga heterophylla*.

As noted in Section II, in a N-rich environment such as a fertilized forest, N in plant tissues may be present as amino acids rather than protein (van Dijk and Roelofs, 1988). Because the relationship between foliar N and maintenance respiration depends on the N being in protein, if N is in amino acids or storage proteins, tissue N will likely not be a good predictor of maintenance respiration. Similar problems have been noted for the relationship between photosynthesis and foliar N content (e.g., Reich and Schoettle, 1988).

B. Estimating R_c from Elemental Composition

Synthesis of organic compounds requires both carbon skeletons and energy, and the total carbon cost of synthesis is the sum of these two components. Construction respiration (R_c) is the energetic cost of biosynthesis. The value of R_c varies among molecules because of their complexity, the energy contained within chemical bonds, and the energy used in synthesis. For example, construction of 1 g of lignin requires 2.5 g glucose (R_c = 0.49 g C respired/g C in product), whereas 1 g of cellulose requires only 1.2 g glucose (R_c = 0.07 g C/g C) (Williams *et al.*, 1987).

Penning de Vries *et al.* (1974) pioneered the use of anabolic biochemical pathways to obtain quantitative estimates of R_c. Because this approach requires very detailed knowledge of the molecular composition of plant tissue, it is impractical for most models or growth studies. However, other researchers developed practical empirical relationships for estimating R_c (McDermitt and Loomis, 1981; Vertregt and Penning de Vries, 1987; Williams *et al.*, 1987) that correlate well with values derived from pathway analysis. These methods assume a value for conversion efficiency of about 0.88 g glucose into structure/g glucose substrate, and use elemental analysis (McDermitt and Loomis, 1981), heat of combustion and ash and organic nitrogen (Williams *et al.*, 1987), or carbon content and ash content (Vertregt and Penning de Vries, 1987) to estimate R_c.

For the fruits of various crop plants with a wide variety of chemical compositions, Vertregt and Penning de Vries (1987) found that R_c (g C/kg dry matter synthesized) was

$$R_c = (12/44)[4.24C_{dm} + 1.17\ \mathrm{Ash} - 1744], \tag{4}$$

where C_{dm} and Ash are the carbon and ash contents of tissue (dry matter) in grams/kilogram. For tissue with a C content equal to 50% of dry weight and 0.5% ash, Eq. (4) calculates R_c as 0.21 g C/g C in dry matter. For comparison, Chung and Barnes (1977) used biochemical pathway analysis to estimate R_c for slash pine as 0.26 g C/g C for needles and 0.15 g C/g C for stems.

The method of Williams *et al.* (1987) includes the energy to reduce nitrogen, which may be important if NO_3^- is the N source:

$$C' = \{(0.06968 \cdot \Delta H_c - 0.065)(1 - A) + (kN/14.0067)(180.15/24)\}(1/E_G), \tag{5}$$

where C' is construction cost in grams glucose/gram dry weight, ΔH_c is the ash-free heat of combustion in kilojoules/gram, A is the ash fraction, k is the oxidation state of nitrogenous substrate (-3 for ammonium, $+5$ for nitrate), N is the Kjeldahl organic nitrogen fraction, and E_G is the conversion efficiency (0.89). If dry matter (DM) is assumed to be 50% carbon, R_c (in g C/g C dry matter synthesized) is

$$R_c = [C'(72 \text{ g C}/180.15 \text{ g glucose})(2 \text{ g DM}/1 \text{ g C})] - 1. \quad (5a)$$

With this method, estimates of R_c ranged from 0.23 to 0.40 g C/g C for chaparral leaves (Williams *et al.*, 1987). If nitrate can be reduced inside leaves (Smirnoff *et al.*, 1984; Smirnoff and Stewart, 1985) from reductant generated directly in photosynthesis (Abrol *et al.*, 1983; Wallsgrove *et al.*, 1983), nitrogen would not affect R_c. However, gymnosperms show lower nitrate reductase activities in leaves than do other woody plants, and may reduce nitrate in roots using ATP or energy derived from respiration (Smirnoff and Stewart, 1985).

All of these methods, incidentally, assume that glucose is broken down through the cytochrome-mediated pathway; if the energy required by biosynthesis were derived from cyanide-resistant respiration, all estimates would have to be increased substantially.

To use the methods described above requires knowing dry matter production and chemical composition. For aboveground production in a study plot, this information can be obtained with routine sampling. For belowground production, or for estimating costs over large areas, obtaining the information is more problematic.

C. Estimating Belowground Respiration Costs

Belowground respiration costs are difficult to estimate, either directly or indirectly. To use tissue N to estimate R_m and fine root production to estimate R_c requires knowing fine root biomass and turnover—estimates of which are uncertain and difficult to obtain. Recently, Raich and Nadelhoffer (1989) described a method for indirectly estimating the total amount of carbon allocated belowground for root production and maintenance. This method can be adapted to estimate respiration costs, if the ratio of R_a to production is assumed constant.

Raich and Nadelhoffer (1989) found that total belowground allocation (estimated as the difference between soil CO_2 efflux and aboveground litter input) was strongly correlated with aboveground litter production over a range of litter production (L_f) of 70–500 g C/m^2/year. (However, the strength of the correlation was derived in part from the wide range of the data used; at any given point, and especially in the less productive sites, there was a severalfold range in the values of CO_2 efflux associated with a given litterfall level.) Belowground allocation (B_{p+c+m}, g C/m²/year) is given by

$$B_{p+c+m} = 130 + 1.92L_f. \qquad (6)$$

If fine root production is known, R_c for roots can be estimated with Eq. (4) or (5), and R_m can be estimated by difference. If fine root biomass and N content are known, R_m can be estimated with Eq. (3), and production of fine roots and R_c can be estimated by difference. If neither production or biomass is known, R_a can be assumed to be 50% of B_{p+c+m} (Amthor, 1989). Examples of applying this method to conifer forests are given in Ryan (1991b).

D. Estimating Whole-Stand Respiration Fluxes Using Eddy Correlation

Extrapolating fluxes measured in small enclosures to ecosystems is difficult, particularly for forests. Conditions in chambers are different than those outside. Estimated fluxes vary among samples because of differences in tissue chemistry, metabolic activity, and sampling technique. Gradients of light, temperature, humidity, and wind speed exist within canopies; these gradients and their effect on fluxes can be modeled, but estimates are uncertain. To overcome some of these problems, direct measurement of CO_2 exchange between a forest and the atmosphere by eddy correlation can provide an independent, system-level estimate of fluxes (e.g., Valentini *et al.*, 1991).

Using micrometeorological theory, it is possible to measure the flux of CO_2 by averaging the product of fluctuations of horizontal wind, vertical wind, and atmospheric [CO_2]. In practice, the method requires placing fast-response sensors above the canopy in such a way that the sensors do not disturb airflow. Campbell (1977) and Baldocchi *et al.* (1988) provide a good introduction to the theory and practice. Sensors placed on towers integrate fluxes over tens to hundreds of meters, whereas aircraft-based sensors can measure flux over tens to hundreds of kilometers (Desjardins, 1992). Reliable fast-response sensors are available but costly, and the technique has not yet been widely applied (Desjardins, 1992).

The advantages of eddy correlation are that fluxes are integrated over a large area, total flux is measured, and vegetation is not disturbed by measurement. Eddy correlation thus provides the most direct estimate available of net ecosystem productivity (assimilation minus autotrophic and heterotrophic respiration). However, technical problems often limit applicability of eddy correlation (Baldocchi *et al.*, 1988). For example, flux estimates are very dependent on measurement height, sensor response time, and averaging period and are currently limited to gently sloping, even terrain (Baldocchi *et al.*, 1988). Also, large variations in CO_2 over the averaging period yield unreliable flux estimates (Baldocchi *et al.*, 1988). Direct estimates of ecosystem respiration can therefore be problematic because large fluctuations in CO_2 concentration occur only at night. Finally, fluxes are measured with eddy correlation for hours or days or at most weeks or months, but annual budgets are needed to increase ecological understanding.

Despite limitations, eddy correlation can increase our understanding of ecosystem processes by providing realistic estimates of total flux (Valentini *et al.*, 1991). These estimates can be used to set bounds on predictions from current models, and perhaps to derive new empirical models. Additionally, diurnal and seasonal variations in CO_2 flux can be linked with variations in radiation, temperature, soil moisture, and humidity to "decompose" the flux trace and validate component models for photosynthesis and respiration.

E. Estimating Instantaneous Fluxes of CO_2 from R_a

Fluxes estimated with eddy correlation sum assimilation, autotrophic respiration, and heterotrophic respiration. To interpret these fluxes, models for estimating instantaneous fluxes for the separate components are required. Below, we outline theory for estimating instantaneous fluxes of CO_2 from R_a, using Eqs. (3)–(5).

Instantaneous CO_2 flux from R_a can be estimated as the sum of maintenance and construction respiration. Instantaneous R_m can be estimated from tissue nitrogen [Eq. (3)] and tissue temperature using the common exponential model of R_a response to temperature (Amthor 1989):

$$R_m = (7.4 \cdot 10^{-7})N \exp(\beta T), \tag{7}$$

where R_m is in units of moles C/second, N is moles of Kjeldahl nitrogen in living biomass, and β is 0.07 for R_a that doubles with a 10°C increase in temperature.

Instantaneous R_c can be estimated by assuming (1) current growth will be the same as in the previous growing season and (2) that growth will vary exponentially with temperature over the growing season. (This last assumption might be questionable, especially in areas where drought limits growth during a substantial part of the warm season.) For a growing season of n days, construction respiration is approximately αD, where D = annual dry matter production in grams of C and α is calculated from Eq. (4) or (5):

$$\alpha D = \sum^{n \text{ days}} R_0 \exp(\beta T) I_0(\beta A), \tag{8}$$

where T is average daily temperature in °C, A is daily temperature amplitude [(max − min)/2], $I_0(0.07A)$ corrects for amplitude effects on the integration of what is approximately exp(sin x) (Ågren and Axelsson, 1980), and R_0 is construction respiration at 0°C. Solving for R_0 and using a simple exponential model for temperature response, instantaneous R_c (for the growing season only) is

$$R_c = \{\alpha D/[86400 \cdot 12 \sum^{n \text{ days}} R_0 \exp(\beta T) I_0(\beta A)]\} \exp \beta T, \tag{9}$$

where R_c is in units of moles C/second. For a daily temperature amplitude of 5°C, $I_0(0.07A)$ is approximately 1.05. For a mature lodgepole pine forest, maintenance respiration summed for the entire year was about 70% of total CO_2 flux from R_a (Ryan and Waring, 1992).

The approach outlined estimates R_m from the standing crop of N, but requires knowledge of past growth to estimate R_c. However, if a model of canopy assimilation is validated for conifer ecosystems, R_c can be estimated from photosynthesis and R_m (e.g., Running and Coughlin, 1988; Bonan, 1991).

VI. Conclusions

The status of efforts to estimate respiration of conifers varies sharply from one tissue to another. There have been numerous measurements of foliage respiration in conifers, but relatively few measurements of within-stand variation in reference to parameters that might be used for scaling. However, a number of logical models for scaling have been proposed (e.g., light, age, or N) and general directions for future research seem well established. There are far fewer measurements of woody-tissue respiration that might be useful for scaling, but some consensus seems to have developed that the use of sapwood biomass and growth rates as indices may provide the key to scaling woody-tissue respiration up to the stand level. Root respiration is still bogged down by a plurality of methods, each of which seems to have some serious disadvantages, so that even the nature of within-stand variation is poorly known. Successful and believable scaling of root respiration from tissue-specific measurements to the stand level seems to be far in the future. Finally, proxy measurements such as litterfall and N concentration can and have been used to estimate respiration for whole stands without measuring tissue-specific rates at all, but all of these techniques require assumptions that need further testing before they will be generally accepted.

References

Abrol, Y. P., Sawhney, S. K., and Naik, M. S. (1983). Light and dark assimilation of nitrate in plants. *Plant, Cell Environ.* 6:595–599.

Ågren, G. I., and Axelsson, B. (1980). Population respiration: A theoretical approach. *Ecol. Modell.* 11:39–54.

Ågren, G. I., Axelsson, B., Flower-Ellis, J. G. K., Linder, S., Persson, H., Staaf, H., and Troeng, E. (1980). Annual carbon budget for a young Scots pine. *In* "Structure and Function of Northern Coniferous Forests: An Ecosystem Study" (T. Persson, ed.), *Ecol. Bull.* (Stockholm) 32:307–313.

Amthor, J. S. (1984). The role of maintenance respiration in plant growth. *Plant, Cell Environ.* 7:561–569.

Amthor, J. S. (1989). "Respiration and Crop Productivity." Springer-Verlag, New York.

Baldocchi, D. D., Hicks, B. B., and Meyers, T. P. (1988). Measuring biosphere–atmosphere exchanges of biologically related gases with micrometeorological methods. *Ecology* 69: 1331–1340.

Barnard, E. L., and Jorgensen, J. R. (1977). Respiration of field-grown loblolly pine roots as influenced by temperature and root type. *Can. J. Bot.* 55:740–743.

Barns, J. D., Eamus, D., and Brown, K. A. (1990). The influence of ozone, acid mist and soil nutrient status on Norway spruce [*Picea abies* (L.) Karst.]. *New Phytol.* 115:149–156.

Benecke, U. (1985). Tree respiration in steepland stands of *Nothofagus truncata* and *Pinus radiata*, Nelson, New Zealand. *In* "Establishment and Tending of Subalpine Forest: Research and Management" (H. Turner and W. Tranquillini, eds.), *Eidg. Anst. Forstl. Versuchswes.* 270:61–70.

Benecke, U., and Nordmeyer, A. H. (1982). Carbon uptake and allocation by *Nothofagus solandri* var. *cliffortioides* (Hook. f.) Poole and *Pinus contorta* Douglas ex Loudon ssp. *contorta* at montane and subalpine altitudes. *In* "Carbon Uptake and Allocation in Subalpine Ecosystems as a Key to Management: Proceedings of an IUFRO Workshop" (R. H. Waring, ed.), pp. 9–21. Forest Research Laboratory, Oregon State Univ., Corvallis, Oregon.

Berry, J. A., and Raison, J. K. (1981). Responses of macrophytes to temperature. *In* "Encyclopedia of Plant Physiology, Volume 12A: Physiological Plant Ecology I. Responses to the Physical Environment" (O. L. Lange, P. S. Nobel, C. B. Osmond, and H. Ziegler, eds.), pp. 277–338. Springer-Verlag, Berlin.

Billings, W. D., Godfrey, P. J., Chabot, B. C., and Bourque, D. P. (1971). Metabolic acclimation to temperature in arctic and alpine ecotypes of *Oxyria digyna*. *Arct. Alp. Res.* 3: 277–289.

Billings, W. D., Peterson, K. M., Shaver, G. R., and Trent, A. W. (1977). Root growth, respiration, and carbon dioxide evolution in an arctic tundra soil. *Arct. Alp. Res.* 9: 129–137.

Björkman, O. (1981). Responses to different quantum flux densities. *In* "Encyclopedia of Plant Physiology, Volume 12A: Physiological Plant Ecology I. Responses to the Physical Environment" (O. L. Lange, P. S. Nobel, C. B. Osmond, and H. Ziegler, eds.), pp. 57–107. Springer-Verlag, Berlin.

Bonan, G. B. (1991). Atmosphere–biosphere exchange of carbon dioxide in boreal forests. *J. Geophys. Res.* 96:7301–7312.

Boyer, W. D., Romancier, R. M., and Ralston, C. W. (1971). Root respiration rates of four tree species grown in the field. *For. Sci.* 17:492–493.

Boysen Jensen, P. (1933). Respiration i stamme og grene af traer. *Sven. Skogsvaardsfoeren. Tidskr. (Skogsvaardsfoeringengs Tidskrift)* 18:239–241.

Brooks, J. R. (1987). Foliage respiration of *Abies amabilis*. M.S. Thesis, University of Washington, Seattle.

Brooks, J. R. (1993). Light acclimation in *Abies amabilis* canopies. Ph.D. Dissertation, University of Washington, Seattle.

Brooks, J. R., Hinckley, T. M., Ford, E. D., and Sprugel, D. G. (1991). Foliage dark respiration in *Abies amabilis* (Dougl.) Forbes: Variation within the canopy. *Tree Physiol.* 9: 325–338.

Campbell, G. S. (1977). "An Introduction to Environmental Biophysics." Springer-Verlag, New York.

Cannell, M. G. R. (1982). "World Forest Biomass and Primary Production Data." Academic Press, London.

Chabot, B. F., and Hicks, D. J. (1982). The ecology of leaf life spans. *Annu. Rev. Ecol. Syst.* 13:229–259.

Chapman, S. B. (1979). Some interrelationships between soil and root respiration in lowland *Calluna* heathland in southern England. *J. Ecol.* 67:1–20.

Chung, H., and Barnes, R. L. (1977). Photosynthate allocation in *Pinus taeda*. I. Substrate requirements for synthesis of shoot biomass. *Can. J. For. Res.* 7:106–111.

Coleman, D. C. (1973). Compartmental analysis of "total soil respiration": An exploratory study. *OIKOS* 24:361–366.

Cox, T. L. (1975). Seasonal respiration rates of yellow-poplar roots by diameter classes. *For. Sci.* 21:185–188.

Cropper, W. P., and Gholz, H. L. (1990). Modelling the labile carbon dynamics of a Florida slash pine plantation. *Silva Carelica* 15:121–130.

Cropper, W. P., and Gholz, H. L. (1991). *In situ* needle and fine root respiration in mature slash pine (*Pinus elliottii*) trees. *Can. J. For. Res.* 21:1589–1595.

Dawson, T. E., and Bliss, L. C. (1993). Plants as mosaics: Leaf-, ramet-, and gender-level variation in the physiology of the dwarf willow, *Salix arctica*. *Funct. Ecol.* 7:293–304.

Desjardins, R. L. (1992). Review of techniques to measure CO_2 flux densities from surface and airborne sensors. *In* "Advances in Bioclimatology" (G. Stanhill, ed.), pp. 1–23. Springer-Verlag, Berlin.

Dick, J. M., Jarvis, P. G., and Leakey, R. R. B. (1991). Influence of male and female cones on needle CO_2 exchange rates of field-grown *Pinus contorta* Doug. trees. *Funct. Ecol.* 5:422–432.

Edwards, N. T., and Harris, W. F. (1977). Carbon cycling in a mixed deciduous forest floor. *Ecology* 58:431–437.

Edwards, N. T., and Sollins, P. (1973). Continuous measurements of carbon dioxide evolution from partitioned forest floor components. *Ecology* 54:406–412.

Ehleringer, J. R., and Field, C. B. (eds.) (1993). "Scaling Physiological Processes: Leaf to Globe." Academic Press, New York.

Evans, G. C. (1972). "The Quantitative Analysis of Plant Growth," Studies in Ecology, Vol. 1. Univ. of California Press, Berkeley and Los Angeles.

Evans, J. R. (1989). Photosynthesis and nitrogen relationships in leaves of C_3 plants. *Oecologia* 78:1–19.

Evans, J. R. (1993a). Photosynthetic acclimation and nitrogen partitioning within a Lucerne canopy. I. Canopy characteristics. *Aust. J. Plant Physiol.* 20:55–67.

Evans, J. R. (1993b). Photosynthetic acclimation and nitrogen partitioning within a Lucerne canopy. II. Stability through time and comparison with a theoretical optimum. *Aust. J. Plant Physiol.* 20:69–82.

Ewel, K. C., Cropper, W. P., Jr., and Gholz, H. L. (1987). Soil CO_2 evolution in Florida slash pine plantations. II. Importance of root respiration. *Can. J. For. Res.* 17:330–333.

Field, C. B., and Mooney, H. A. (1986). The photosynthesis–nitrogen relationship in wild plants. *In* "On the Economy of Plant Form and Function" (T. J. Givnish, ed.), pp. 25–55. Cambridge Univ. Press, Cambridge.

Field, C. B., Ball, J. T., and Berry, J. A. (1991). Photosynthesis: Principles and field techniques. *In* "Plant Physiological Ecology: Field Methods and Instrumentation" (R. W. Pearcy, J. R. Ehleringer, H. A. Mooney, and P. W. Rundel, eds.), pp. 209–253. Chapman & Hall, London.

Fitter, A. H., and Hay, R. K. M. (1987). "Environmental Physiology of Plants," 2nd Ed., Academic Press, London.

Friend, A. D. (1991). Use of a model of photosynthesis and leaf microenvironment to predict optimal stomatal conductance and leaf nitrogen partitioning. *Plant, Cell Environ.* 14:895–905.

Friend, A. D., Shugart, H. H., and Running, S. W. (1993). A physiology-based model of forest dynamics. *Ecology* 74:792–797.

Fritschen, L. J., Cox. L., and Kinerson, R. (1973). A 28-meter Douglas-fir in a weighing lysimeter. *For. Sci.* 19:256–261.

Goodwin R. H., and Goddard, D. R. (1940). The oxygen consumption of isolated trees. *Am. J. Bot.* 27:234–237.

Gower, S. T., and Richards, J. H. (1990). Larches: Deciduous conifers in an evergreen world. *BioScience* 40:818–826.

Grahammer, K., Jawson, M. D., and Skopp, J. (1991). Day and night soil respiration from a grassland. *Soil Biol. Biochem.* 23:77–81.

Greenway, K. J., Macdonald, S. E., and Lieffers, V. J. (1992). Is long-lived foliage in *Picea mariana* an adaptation to nutrient-poor conditions? *Oecologia* 91:184–191.

Hagihara, A., and Hozumi, K. (1977). Estimation of canopy respiration and its seasonal change in a *Chamaecyparis obtusa* plantation. *J. Jpn. For. Soc.* 59:405–413.

Hagihara, A., and Hozumi, K. (1981). Respiration consumption by woody organs in a *Chamaecyparis obtusa* plantation. *J. Jpn. For. Soc.* 63:156–164.

Hagihara, A., and Hozumi, K. (1991). Respiration. *In* "Physiology of Trees" (A. S. Raghavendra, ed.), pp. 87–110. Wiley, New York.

Harley, J. L., and Smith, S. E. (1983). "Mycorrhizal Symbiosis." Academic Press, London.

Hashimoto, R. (1990). Analysis of the morphology and structure of crowns in a young sugi (*Cryptomeria japonica*) stand. *Tree Physiol.* 6:119–134.

Havranek, W. M. (1981). Stammatmung, Dickenwachstum, und Photosynthese einer Zirbe (*Pinus cembra* L.) an der Waldgrenze. [Stem respiration, radial growth and photosynthesis of a cembran pine tree (*Pinus cembra* L.) at the timberline.] *Mitt. Forstl. Bundesversuchsanst. Wien,* 142. 443–467.

Havranek, W. M. (1985). Gas exchange and dry matter allocation in larch at the alpine timberline on Mount Patscherkofel. *In* "Establishment and Tending of Subalpine Forest: Research and Management" (H. Turner and W. Tranquillini, eds.), *Eidg. Anst. Forstl. Versuchswes.* 270:135–142.

Hendrickson, O. Q., and Robinson, J. B. (1984). Effects of roots and litter on mineralization processes in forest soil. *Plant Soil* 80:391–405.

Holthausen, R. S., and Caldwell, M. M. (1980). Seasonal dynamics of root system respiration in *Atriplex confertifolia. Plant Soil* 55:307–317.

Hom, J. L., and Oechel, W. C. (1983). The photosynthetic capacity, nutrient content, and nutrient use efficiency of different needle age-classes of black spruce (*Picea mariana*) found in interior Alaska. *Can. J. For. Res.* 13:834–839.

Irving, D. E., and Silsbury, J. H. (1987). A comparison of the rate of maintenance respiration in some crop legumes and tobacco determined by three methods. *Ann. Bot.* 59:257–264.

Johansson, N. (1933a). Sambandet mellan vedstammens andning och dess tillväxt. (The relation between the tree-stem's respiration and growth.) *Sven. Skogsvaardsfoeren. Tidskr. (Skogsvaardsfoeringengs Tidskrift)* 31:53–134. (in Swedish w/English summary).

Johansson, N. (1933b). Om förvedade stammars andning, dess fastställande och betydelse. *Sven. Skogsvaardsfoeren. Tidskr. (Skogsvaardsfoeringengs Tidskrift)* 18:242–249.

Jones, M. B., Leafe, E. L., Stiles, W., and Collett, B. (1978). Pattern of respiration of a perennial ryegrass crop in the field. *Ann. Bot.* 42:693–703.

Kaufmann, M. R., and Landsberg, J. J. (eds.) (1991). Advancing toward closed models of forest ecosystems. *Tree Physiol.* 9:1–324.

Kawahara, T., Hatiya, K., Takeuti, I., and Sato, A. (1976). Relationship between respiration rate and nitrogen concentration of trees. *Jpn. J. Ecol.* 26:165–170.

Keller, T. (1967). Beitrag zur kenntnis der wurzelatmung von koniferenjungpflanzen. *Proc. 14th Internat. Union of Forestry Research Organizations Congress,* Munich 4:329–340.

Kinerson, R. S. (1975). Relationship between plant surface area and respiration in pine. *J. Appl. Ecol.* 12:965–971.

Kira, T., and Shidei, T. (1967). Primary production and turnover of organic matter in different forest ecosystems of the western Pacific. *Jpn. J. Ecol.* 17:70–87.

Lambers, H. (1980). The physiological significance of cyanide-resistant respiration in higher plants. *Plant, Cell Environ.* 2:293–302.

Lambers, H., Szaniawski, R. K., and de Visser, R. (1983). Respiration for growth, maintenance and ion uptake. An evaluation of concepts, methods, values and their significance. *Physiol. Plant.* 58:556–563.

Larcher, W. (1983). "Physiological Plant Ecology," 2nd Ed., Springer-Verlag, Berlin.

Lavigne, M. B. (1987). Differences in stem respiration responses to temperature between balsam fir trees in thinned and unthinned stands. *Tree Physiol.* 3:225–233.

Lavigne, M. B. (1988). Stem growth and respiration of young balsam fir trees in thinned and unthinned stands. *Can. J. For. Res.* 18:483–489.

Leverenz, J. W. (1988). The effects of illumination sequence, CO_2 concentration, temperature and acclimation on the convexity of the photosynthetic light response curve. *Physiol. Plant.* 74:332–341.

Leverenz, J. W., and Hällgren, J.-E. (1991). Measuring photosynthesis and respiration of foliage. *In* "Techniques and Approaches in Forest Tree Ecophysiology" (J. P. Lassoie and T. M. Hinckley, eds.), pp. 303–328. CRC Press, Boca Raton, Florida.

Leverenz, J. W., and Jarvis, P. G. (1979). Photosynthesis in Sitka spruce. VIII. The effect of light flux density and direction on the rate of net photosynthesis and the stomatal conductance of needles. *J. Appl. Ecol.* 16:919–932.

Linder, S., and Troeng, E. (1980). Photosynthesis and transpiration of a 20-year-old Scots pine. *In* "Structure and Function of Northern Coniferous Forests: An Ecosystem Study" (T. Persson, ed.), *Ecol. Bull.* (Stockholm) 32:165–181.

Linder, S., and Troeng, E. (1981). The seasonal variation in stem and coarse root respiration of a 20-year-old Scots pine (*Pinus sylvestris* L.). *In* "Radial Growth in Trees" (W. Tranquillini, ed.), *Mitt. Forstl. Bundesversuchsanst. Wien* 142:125–140.

Ludwig, L. J., Charles-Edwards, D. A., and Withers, A. C. (1975). Tomato leaf photosynthesis and respiration in various light and carbon dioxide environments. *In* "Environmental and Biological Control of Photosynthesis" (R. Marcelle, ed.), pp. 17–28. Dr. W. Junk Publ., The Hague.

McCree, K. J. (1970). An equation for the rate of dark respiration of white clover plants grown under controlled conditions. *In* "Prediction and Measurement of Photosynthetic Productivity" (I. Setlik, ed.), pp. 221–229. Centre for Agricultural Publishing and Documentation, Wageningen, The Netherlands.

McCree, K. J. (1982). Maintenance requirements of white clover at high and low growth rates. *Crop Sci.* 22:345–351.

McCree, K. J. (1983). Carbon balance as a function of plant size in sorghum plants. *Crop Sci.* 23:1173–1177.

McCree, K. J., and Amthor, M. E. (1982). Effects of diurnal variation in temperature on the carbon balances of white clover plants. *Crop Sci.* 22:822–827.

McDermitt, D. K., and Loomis, R. S. (1981). Elemental composition of biomass and its relation to energy content, growth efficiency, and growth yield. *Ann. Bot.* 48:275–290.

McGuire, A. D., Melillo, J. M., Joyce, L. A., Kicklighter, D. W., Grace, A. L., Moore III, B., and Vorosmärty, C. J. (1992). Interactions between carbon and nitrogen dynamics in estimating net primary productivity for potential vegetation in North America. *Global Biogeochem. Cycles* 6:101–124.

McGuire, A. D., Joyce, L. A., Kicklighter, D. W., Melillo, J. M., Esser, G., and Vorosmärty, C. J. (1993). Productivity response of climax temperate forests to elevated temperature and carbon dioxide: A North American comparison between two global models. *Clim. Change* 24:287–310.

McLaughlin, S. B., Andersen, C. P., Hanson, P. J., Tjoelker, M. G., and Roy, W. K. (1991). Increased dark respiration and calcium deficiency of red spruce in relation to acidic deposition at high-elevation southern Appalachian Mountian sites. *Can. J. For. Res.* 21:1234–1244.

McMurtrie, R. E. (1991). Relationship of forest productivity to nutrient and carbon supply—A modeling analysis. *Tree Physiol.* 9:87–99.

McMurtrie, R. E., Rook, D. A., and Kelliher, F. M. (1990). Modelling the yield of *Pinus radiata* on a site limited by water and nitrogen. *For. Ecol. Manage.* 39:381–413.

McNaughton, K. G., and Jarvis, P. G. (1991). Effects of spatial scale on stomatal control of transpiration. *Agric. For. Meteorol.* 54:279–301.

Martin, T. A. (1992). Causes of temperature-independent diurnal fluctuations in CO_2 efflux from loblolly pine (*Pinus taeda* L.) stems. M.S. Thesis, University of Georgia, Athens.

Martin, T. A., Teskey, R. O., and Dougherty, P. M. (1994). Movement of respiratory CO_2 in stems of loblolly pine (*Pinus taeda* L.) seedlings. *Tree Physiol.* 14:481–495.

Matsumoto, Y., and Negisi, K. (1982). Photosynthesis and respiration in advanced growth *Abies veitchii* growing under a deep canopy and in the open. *J. Jpn. For. Soc.* 64: 165–176.

Matyssek, R. (1985). The carbon balance of three deciduous larch species and an evergreen spruce species near Bayreuth (W. Germany). *In* "Establishment and Tending of Subalpine Forest: Research and Management" (H. Turner and W. Tranquillini, eds.). *Eidg. Anst. Forstl. Versuchswes.* 270:123–133.

Matyssek, R., and Schulze, E.-D. (1988). Carbon uptake and respiration in above-ground parts of a *Larix decidua* × *leptolepis* tree. *Trees* 2:233–241.

Medina, E., Klinge, H., Jordan, C., and Herrera, R. (1980). Soil respiration in Amazonian rain forests in the Rio Negro Basin. *Flora (Jena)* 170:240–250.

Melillo, J. M., Callaghan, T. V., Woodward, F. I., Salati, E., and Sinha, S. K. (1990). Effects on Ecosystems. *In* "Climate Change: The IPCC Scientific Assessment" (J. T. Houghton, G. J. Jenkins, and J. J. Ephraums, eds.), pp. 284–310. Cambridge Univ. Press, Cambridge.

Merino, J., Field, C., and Mooney, H. A. (1984). Construction and maintenance costs of mediterranean-climate evergreen and deciduous leaves. *Acta Oecol. Oecol. Plant.* 5: 211–229.

Møller, C. M., Müller, D., and Nielsen, J. (1954). Respiration in stem and branches of beech. *Det. forstliche Forsøgsvaesen* 21:273–301.

Mori, S., and Hagihara, A. (1991). Crown profile of foliage area characterized with the Weibull distribution in a hinoki (*Chamaecyparis obtusa*) stand. *Trees* 5:149–152.

Müller, D. (1924). Studies on traumatic stimulus and loss of dry matter by respiration in branches from Danish forest trees. *Dan. Bot. Arch.* 4(6):1–33.

Nebel, A., and Matile, P. (1992). Longevity and senescence of needles in *Pinus cembra* L. *Trees* 6:156–161.

Negisi, K. (1972). Diurnal fluctuation of CO_2 release from the bark of a standing *Magnolia obovata* tree. *J. Jpn. For. Soc.* 54:257–263.

Negisi, K. (1974). Respiration rates in relation to diameter and age in stem or branch sections of young *Pinus densiflora* trees. *Bull. Tokyo Univ. For. (Tokyo Daigaku Nogakubu Enshurin Hokoku)* 66:209–223.

Negisi, K. (1975). Diurnal fluctuation of CO_2 release from the bark of standing young *Pinus densiflora* trees. *J. Jpn. For. Soc.* 57:375–383.

Negisi, K. (1978). Daytime depression in bark respiration and radial shrinkage in stem of a standing young *Pinus densiflora* tree. *J. Jpn. For. Soc.* 60:380–382.

Negisi, K. (1979). Bark respiration rate in stem segments detached from young *Pinus densiflora* trees in relation to velocity of artificial sap flow. *J. Jpn. For. Soc.* 61:88–93.

Negisi, K. (1982). Diurnal fluctuation of the stem bark respiration in relationship to the wood temperature in standing young *Pinus densiflora*, *Chamaecyparis obtusa*, and *Quercus myrsinaefolia* trees. *J. Jpn. For. Soc.* 64:315–319.

Norman, J. M., and Jarvis, P. G. (1974). Photosynthesis in Sitka spruce (*Picea sitchensis* (Bong.) Carr.) III. Measurements of canopy structure and interception of radiation. *J. Appl. Ecol.* 11:375–398.

Paembonan, S. A., Hagihara, A., and Hozumi, K. (1991). Long-term measurement of CO_2 release from the aboveground parts of a hinoki forest tree in relation to air temperature. *Tree Physiol.* 8:399–405.

Penning de Vries, F. W. T. (1975a). Use of assimilates in higher plants. *In* "Photosynthesis and Productivity in Different Environments" (J. P. Cooper, ed.), pp. 459–480. Cambridge Univ. Press, London.

Penning de Vries, F. W. T. (1975b). The cost of maintenance processes in plant cells. *Ann. Bot.* 39:77–92.

Penning de Vries, F. W. T., Brunsting, A. H. M., and van Laar, H. H. (1974). Products, requirements and efficiency of biosynthesis: A quantitative approach. *J. Theor. Biol.* 45:339–377.

Pisek, A., and Winkler, E. (1958). Assimilationsvermögen und Respiration der Fichte (*Picea excelsa* Link) in verschiedener Höhenlage und der Zirbe (*Pinus cembra* L.) an der alpinen Waldgrenze. *Planta* 51:518–543.

Raich, J. W., and Nadelhoffer, K. J. (1989). Belowground carbon allocation in forest ecosystems: Global trends. *Ecology* 70:1346–1354.

Rastetter, E. B., Ryan, M. G., Shaver, G. R., Melillo, J. M., Nadelhoffer, K. J., Hobbie, J. E., and Aber, J. D. (1991). A general biogeochemical model describing the responses of the C and N cycles in terrestrial ecosystems to changes in CO_2, climate, and N deposition. *Tree Physiol.* 9:101–126.

Redmann, R. E., and Abouguendia, Z. M. (1978). Partitioning of respiration from soil, litter and plants in a mixed-grassland ecosystem. *Oecologia* 36:69–79.

Reich, P. B., and Schoettle, A. W. (1988). Role of phosphorus and nitrogen in photosynthetic and whole plant carbon gain and nutrient use efficiency in eastern white pine. *Oecologia* 77:25–33.

Reich, P. B., Koike, T., Gower, S. T., and Schoettle, A. W. (1994). Causes and consequences of variation in conifer leaf life span. *In* "Ecophysiology of Coniferous Forests" (W. K. Smith and T. M. Hinckley, eds.), pp. 225–254. Academic Press, San Diego.

Reiners, W. A. (1968). Carbon dioxide evolution from the floor of three Minnesota forests. *Ecology* 49:471–483.

Rook, D. A. (1969). The influence of growing temperature on photosynthesis and respiration of *Pinus radiata* seedlings. *N. Z. J. Bot.* 7:43–55.

Rook, D. A., and Corson, M. J. (1978). Temperature and irradiance and the total daily photosynthetic production of a crown of a *Pinus radiata* tree. *Oecologia* 36:371–382.

Running, S. W., and Coughlin, J. C. (1988). A general model of forest ecosystem processes for regional applications. I. Hydrologic balance, canopy gas exchange and primary production processes. *Ecol. Modell.* 42:125–154.

Running, S. W., and Gower, S. T. (1991). Forest-BGC, a general model of forest ecosystem processes for regional applications II. Dynamic carbon allocation and nitrogen budgets. *Tree Physiol.* 9:147–160.

Ryan, M. G. (1989). Sapwood volume for three subalpine conifers: Predictive equations and ecological implications. *Can. J. For. Res.* 19:1397–1401.

Ryan, M. G. (1990). Growth and maintenance respiration in stems of *Pinus contorta* and *Picea engelmannii*. *Can. J. For. Res.* 20:48–57.

Ryan, M. G. (1991a). The effect of climate change on plant respiration. *Ecol. Appl.* 1: 157–167.

Ryan, M. G. (1991b). A simple method for estimating gross carbon budgets for vegetation in forest ecosystems. *Tree Physiol.* 9:255–266.

Ryan, M. G., and Waring, R. H. (1992). Maintenance respiration and stand development in a subalpine lodgepole pine forest. *Ecology* 73:2100–2108.

Ryan, M. G., Gower, S. T., Hubbard, R. M., Waring, R. H., Gholz, H. L., Cropper, W. P., and Running, S. W. (in review). Stem maintenance respiration of four conifers in contrasting climates. *Oecologia* (Submitted).

Schaedle, M. (1975). Tree photosynthesis. *Annu. Rev. Plant Physiol.* 26:101–115.

Schlesinger, W. H. (1977). Carbon balance in terrestrial detritus. *Annu. Rev. Ecol. Syst.* 8:51–81.

Šesták, Z., Catsky, J., and Jarvis, P. G. (eds.) (1971). "Plant Photosynthetic Production: Manual of Methods." Dr. W. Junk, Publ., The Hague.

Singh, J. S., and Gupta, S. R. (1977). Plant decomposition and soil respiration in terrestrial ecosystems. *Bot. Rev.* 43:450–528.

Singh, K. P., and Shekhar, C. (1986). Seasonal patterns of total soil respiration, its fractionation and soil carbon balance in a wheat–maize rotation cropland at Varanasi. *Pedobiologia* 29:305–318.

Smirnoff, N., and Stewart, G. R. (1985). Nitrate assimilation and translocation by higher plants: Comparative physiology and ecological consequences. *Physiol. Plant.* 64: 133–140.

Smirnoff, N., Todd, P., and Stewart, G. R. (1984). The occurrence of nitrate reduction in the leaves of woody plants. *Ann. Bot.* 54:363–374.

Sowell, J. B., and Spomer, G. G. (1986). Ecotypic variation in root respiration rate among elevational populations of *Abies lasiocarpa* and *Picea engelmannii*. *Oecologia* 68:375–379.

Sprugel, D. G. (1984). Density, biomass, productivity, and nutrient cycling changes during stand development in wave-regenerated balsam fir forests. *Ecol. Monogr.* 54:165–186.

Sprugel, D. G. (1985). Natural disturbance and ecosystem energetics. *In* "The Ecology of Natural Disturbance and Patch Dynamics" (S. T. A. Pickett and P. S. White, eds.), pp. 335–352. Academic Press, Orlando, Florida.

Sprugel, D. G. (1989). Components of woody-tissue respiration in *Abies amabilis*. *Ann. Sci. For.* 46(Suppl.):532s–535s.

Sprugel, D. G. (1990). Components of woody-tissue respiration in young *Abies amabilis* (Dougl.) Forbes trees. *Trees* 4:88–98.

Sprugel, D. G., and Benecke, U. (1991). Measuring woody-tissue respiration and photosynthesis. *In* "Techniques and Approaches in Forest Tree Ecophysiology" (J. P. Lassoie and T. M. Hinckley, eds.), pp. 329–355. CRC Press, Boca Raton, Florida.

Sprugel, D. G., Brooks, J. R., and Hinckley, T. M. (1992). Distribution of foliar sun/shade morphology and age within an *Abies amabilis* canopy. *Bull. Ecol. Soc. Am.* 73(2):352–353.

Szaniawski, R. K., and Adams, M. S. (1974). Root respiration of *Tsuga canadensis* seedlings as influenced by intensity of net photosynthesis and dark respiration of shoots. *Am. Midl. Nat.* 91:464–468.

Szaniawski, R. K., and Kielkiewicz, M. (1982). Maintenance and growth respiration in shoots and roots of sunflower plants grown at different root temperatures. *Physiol. Plant.* 54:500–504.

Teskey, R. O., Hinckley, T. M., and Grier, C. C. (1984). Temperature-induced change in the water relations of *Abies amabilis* (Dougl.) Forbes. *Plant Physiol.* 74:77–80.

Thornley, J. H. M. (1970). Respiration, growth, and maintenance in plants. *Nature (London)* 227:304–305.

Thornley, J. H. M. (1977). Growth, maintenance, and respiration: A re-interpretation. *Ann. Bot.* 41:1191–1203.

Valentini, R., Scarascia-Mugnozza, G. E., de Angelis, P., and Bimbi, R. (1991). An experimental test of the eddy correlation technique over a Mediterranean macchia canopy. *Plant, Cell Environ.* 14:987–994.

van Dijk, H. F. G., and Roelofs, J. G. M. (1988). Effects of excessive ammonium deposition on the nutritional status and condition of pine needles. *Physiol. Plant.* 73:494–501.

Vertregt, N., and Penning de Vries, F. W. T. (1987). A rapid method for determining the efficiency of biosynthesis of plant biomass. *J. Theor. Biol.* 128:109–119.

Vogt, K. A. (1991). Carbon budgets of temperate forest ecosystems. *Tree Physiol.* 9:69–86.

Vogt, K. A., and Bloomfield, J. (1991). Root turnover and senescence. *In* "Plant Roots: The Hidden Half" (Y. Waisel, A. Eshel, and U. Kafkafi, eds.), pp. 287–308. Dekker, New York.

Vogt, K. A., and Persson, H. (1991). Measuring growth and development of roots. *In* "Techniques and Approaches in Forest Tree Ecophysiology" (J. P. Lassoie and T. M. Hinckley, eds.), pp. 477–502. CRC Press. Boca Raton, Florida.

Vogt, K. A., Vogt, D. J., Moore, E. E., and Sprugel, D. G. (1989). Methodological considerations in measuring biomass, production, respiration and nutrient resorption for tree roots in natural ecosystems. *In* "Applications of Continuous and Steady-State Methods to Root Biology" (J. G. Torrey and L. J. Winship, eds.), pp. 217–232. Kluwer Academic Publ., Dordrecht, The Netherlands.

Wallsgrove, R. M., Keyes, A. J., Lea, P. J., and Miflin, B. J. (1983). Photosynthesis, photorespiration and nitrogen metabolism. *Plant, Cell Environ.* 6:301–309.

Wang, Y. P., and Jarvis, P. G. (1990). Description and validation of an array model—MAESTRO. *Agric. For. Meteorol.* 51:257–280.

Wanner, H. (1970). Soil respiration, litter fall and productivity of tropical rain forest. *J. Ecol.* 58:543–547.

Wiant, H. V., Jr. (1967). Contribution of roots to forest "soil respiration." *Adv. Front. Plant Sci.* 18:163–167.

Williams, K., Percival, F., Merino, J., and Mooney, H. A. (1987). Estimation of tissue construction cost from heat of combustion and organic nitrogen content. *Plant, Cell Environ.* 10:725–734.

Woodwell, G. M. (1987). Forests and climate: Surprises in store. *Oceanus* 29:71–75.

Wullschleger, S. D., and Norby, R. J. (1992). Respiratory cost of leaf growth and maintenance in white oak saplings exposed to atmospheric CO_2 enrichment. *Can. J. For. Res.* 22:1717–1721.

Wullschleger, S. D., Norby, R. J., and Gunderson, C. A. (1992). Growth and maintenance respiration in expanding leaves of *Liriodendron tulipifera* L. saplings exposed to long-term carbon dioxide enrichment under field conditions. *New Phytol.* 121:515–524.

Yoda, K., Shinozaki, K., Ogawa, H., Hozumi, K., and Kira, T. (1965). Estimation of the total amount of respiration in woody organs of trees and forest communities. *J. Biol. Osaka City Univ.* 16:15–26.

Zabuga, V. F., and Zabuga, G. A. (1985). Interrelationships between respiration and radial growth of the trunk in Scotch pine. *Sov. Plant Physiol. (Engl. Transl.)* 32:718–723.

Zabuga, V. F., Zabuga, G. A., and Zinov'eva, V. P. (1983). Ecological characteristics of respiratory CO_2 exchange in the growing trunk of Scotch pine. *Sov. Plant Physiol. (Engl. Transl.)* 29:934–939.

9

Water Transport Dynamics in Trees and Stands

S. G. Pallardy, J. Čermák, F. W. Ewers,
M. R. Kaufmann, W. C. Parker, and J. S. Sperry

I. Water Absorption

A. Inputs and Pathways

1. The Soil–Plant–Atmosphere Continuum By far, most of the water lost by trees in transpiration is absorbed by roots. The pathway of liquid water movement, often termed the soil–plant–atmosphere continuum, or SPAC, is a complex network of individual pathways that may be divided into segments (Phillip, 1966). By the widely accepted cohesion–tension theory of sap ascent (reviewed by Zimmermann, 1983; Boyer, 1985), water is normally transported under negative pressure, with the drops in Ψ in going from the soil, through the plant, and into the atmosphere being the driving force for water movement. In some plants positive root pressures can develop when transpiration is least (e.g., at night or during rainstorms). However, such root pressures are not characteristically found in conifers. Volume flow rate, or flux, of water in the soil–plant–atmosphere system is analogous to the flow of electricity in a conducting system, and it therefore can be described by an analog of Ohm's Law, where

$$\text{Flux} = F = \text{difference in } \Psi_W/\text{resistance}. \quad (1)$$

This concept can be applied to steady-state flux through a plant as follows:

$$F = \frac{\Psi_{\text{soil}} - \Psi_{\text{root surface}}}{r_{\text{soil}\rightarrow\text{root}}} = \frac{\Psi_{\text{root surface}} - \Psi_{\text{xylem}}}{r_{\text{root}}} = \frac{\Psi_{\text{xylem}} - \Psi_{\text{leaf cells}}}{r_{\text{xylem}\rightarrow\text{leaf cells}}} = \frac{C_{\text{leaf}} - C_{\text{air}}}{r_{\text{leaf}} + r_{\text{air}}}, \tag{2}$$

where C corresponds to concentration of water vapor.

Analogous reasoning can be used to derive an equation relating flux and driving forces to the hydraulic conductance (k), which is similar in concept and application to the reciprocal of the flow resistance described above:

$$F = k\,\Delta\Psi. \tag{3}$$

The hydraulic conductivity (k_h) is equal to k multiplied by the length of the pathway. The continuum concept provides a useful, unifying theory in which water movement through soil, roots, stems, and leaves, and its evaporation into the air, can be studied in terms of the driving forces and resistances operating in each segment. Liquid water flow within the system is proportional to the driving force ($\Delta\Psi$) for water movement and inversely proportional to the resistance to water flow within a portion. Although we will accept the model as valid within the context of this chapter, several simplifying assumptions are made in its application, and the reader is referred to Richter (1973), Jarvis (1975), Hinckley *et al.* (1978), Sinclair and Ludlow (1985), Boyer (1985), Kaufmann and Fiscus (1985), Kramer (1988), Passioura (1988a), Schulze *et al.* (1988), Boyer (1989), and Pallardy *et al.* (1991) for a historical perspective of the model and its adherents and critics.

2. Soil Water Uptake Conceptually, the root system of a conifer or any other plant can be viewed as a complex absorptive surface. The amount of water moving through this surface depends on the abundance of the water in the soil, soil physical properties relating to water retention and transport, and the physical and physiological determinants of water uptake by roots. The most important aspects of these topics will be discussed below.

a. Pathways Soil texture has a substantial influence on the relationship between flow resistance and Ψ. Soils of coarse texture show much sharper increases in flow resistance with loss of water than do fine-textured soils. Presumably this occurs because larger particles promote internal discontinuities within the soil and between soil and roots (Sands and Theodorou, 1978). Peat soils may present an especially large bar-

Table I Effects of Removal of Unsuberized Roots on Surface Area and Water Uptake of *Pinus taeda* Seedlings[a]

Description	Total surface area (cm^2)[c]	Water absorption: Absorption ($cm^3/sec \times 10^4$)	Water absorption: Rate [$cm^3/(cm^2\ sec) \times 10^6$]
Unpruned root systems[b]	147.3 ± 28.8	6.73 ± 0.85	4.69 ± 0.67
Approximately half of unsuberized root surface removed	112.0 ± 23.8 (76%)	4.72 ± 0.55 (70%)	4.28 ± 0.61 (91%)
All unsuberized roots removed	86.0 ± 16.3 (58%)	3.08 ± 0.46 (46%)	3.61 ± 0.42 (77%)

[a] Seedlings were 1 year old and were grown in nutrient solution. Reprinted from Chung and Kramer (1975), with permission.

[b] Values for unpruned root systems are averages of eight, and those for pruned root systems are averages of four root systems ± S.D.

[c] Numbers in parentheses are values expressed as percentages of values for unpruned root systems.

rier to water movement to roots, particularly when dry (Orlander and Due, 1986). Uptake of frozen soil water is negligible. Therefore, winter transpiration needs of evergreen conifers must be met by absorption of unfrozen water from deeper in the soil or by redistribution of internal water during periods of thawing.

Anatomical features of roots (the radial perspective) also influence the flow resistance. Unsuberized roots of conifer species typically present lower resistance to water flow than do suberized roots (Sands *et al.*, 1982). However, results typified by those of Chung and Kramer (1975) emphasize that water uptake through suberized root surfaces of conifers (in this case *Pinus taeda*) may be significant (Table I) (Büsgen, 1901; Addoms, 1946; Hoffman, 1974; Haussling *et al.*, 1988; van Rees and Comerford, 1990). MacFall *et al.* (1990), using magnetic resonance imaging, provided some evidence showing that depletion of soil water in proximity to woody taproots, lateral roots, and mycorrhizae of *P. taeda* occurred first around the taproot. Suberization of roots of conifers depends on environmental factors such as drought and low temperature. These factors promote increased suberization and an increased ratio of suberized to unsuberized surface area (Bartsch, 1987; Kozlowski *et al.*, 1991). Given the year-round potential for transpiration of largely evergreen conifer species, water absorption during the winter must necessarily occur through suberized roots (Kozlowski *et al.*, 1991).

Currently, debate continues on the nature of the radial pathway (apoplastic versus symplastic) from root surface to root xylem in young roots. It is unclear to what extent water moves in cell walls (the apoplast) versus the connected protoplasm (symplast) of the root cortex external to the vascular cylinder (Boyer, 1985; Passioura, 1988b). However, presence of a Casparian strip at the endodermis of young roots indicates that water must enter the symplast at *some* point from soil to xylem. Hanson *et al.* (1985) demonstrated that apoplastic flow into the root xylem was less than 1% of total flow in *Pinus resinosa* seedlings unless root systems were subjected to hypoxic conditions. After disruption of the Casparian strip associated with initiation of lateral root formation and initiation of secondary growth, the proportion of apoplastic flow must increase.

Root contraction in transpiring plants has been suspected for a long time, and measurements of roots of transpiring plants growing at the faces of rhizotron observation windows indicate that they may contract up to 40% in diameter over a 24-hour period (Huck *et al.*, 1970; Cole and Alston, 1974; Tinker, 1976). The presence of an air gap at the root–soil interface would greatly alter the radial resistance to water flow. However, the artificial environment required to measure root diameter (one must be able to see the root, hence isolating the observed portion from soil particles with which it would otherwise be in intimate contact) led Taylor and Willatt (1983) to conduct *in situ* experiments employing neutron radiography. The results suggested diurnal root shrinkage under mild conditions was minimal, but it might be significant if plants were subjected to high evaporative demands. Plants produce mucilaginous substances (mucigel) that may prevent detachment from adjacent particles, at least in younger roots (Coutts, 1982; Bowen and Nambiar, 1984). Coutts (1982) observed tenacious adherence of roots of *Picea sitchensis* to soil particles when plants were grown in dry, but not in wet, soil.

Mycorrhizal associations of conifers have been the subject of much research, motivated by the sometimes spectacular effects they have on growth and survival. Most attention has been directed toward nutritional aspects of mycorrhizae, but anecdotal reports and oblique assertions in the literature that mycorrhizae can improve "drought tolerance" of plants have motivated study of the impacts of the association on water relations and water uptake properties of root systems. The results of these experiments using conifer species have been variable. Sometimes no effect or negative impacts on Ψ of mycorrhizal associations of conifer plants are observed (e.g., Sands and Theodorou, 1978; Sands *et al.*, 1982; Lehto, 1989; Coleman *et al.*, 1990). It must be emphasized that the efficacy of mycorrhizae in ameliorating the water relations of plants cannot be evaluated solely by interpretation of trends in Ψ. This is because many potentially beneficial effects of mycorrhizae may cause apparent *increases* in water stress. For example, stimulation of transpiration (and photosynthesis) may cause lower leaf Ψ, but this will be beneficial as long as Ψ does not decline excessively. Additionally, the normally larger mycorrhizal plants nearly always deplete soil moisture sooner than do nonmycorrhizal plants in pot studies. Lower Ψ is an unambiguous indicator of water stress *only* under conditions wherein the influences of plant size, flow, and other effects have been eliminated or taken into account. Further, even in these "compensated" cases, lower Ψ in mycorrhizal plants may not indicate disadvantage per se. Any deleterious response to modest increases in water stress may be outweighed by other benefits (e.g., net photosynthesis or even mineral nutrition).

In some cases the presence of specific mycorrhizal associations has resulted in more favorable water status. For example, Walker *et al.* (1989) demonstrated that *P. taeda* seedlings that had been inoculated with *Pisolithus tinctorius* and had been growing on a reclaimed surface mine for 7 years exhibited midday Ψ_l (leaf water potential) values that were more than 0.2 MPa higher than control seedlings (Table II). Earlier research provided similar results for *Pinus virginiana* (Walker *et al.*, 1982). The most likely mechanisms by which mycorrhizae may increase water absorption by roots involve increased total root surface area by modification of root morphology and enhanced access to soil water provided by extramatrical hyphae (Mudge *et al.*, 1987; Duddridge *et al.*, 1980). Boyd *et al.* (1986) observed that transpiration declined substantially within 10 minutes after hyphal contact to moist soil was severed in *Pinus sylvestris–Suillus bovinues* associations. Lamhamedi *et al.* (1992) observed that the beneficial effects of inoculation of *Pinus pinaster* seedlings with *Piso-*

Table II Xylem Water Potential of *Pinus taeda* Plants[a]

Treatment	Height (cm)	Root collar diameter (mm)	Soil water potential (Ψ_s): dawn (MPa)	Soil water potential (Ψ_s): midday (MPa)	Xylem pressure potential (Ψ_p): dawn (MPa)	Xylem pressure potential (Ψ_p): midday (MPa)
Pisolithus tinctorius	52.4a	9.2a	−0.060a	−0.060a	−0.87a	−1.59a
Control (nonfertilized)	43.0a	6.7a	−0.058a	−0.059a	−1.05b	−1.86b

[a] Plants were either inoculated with *Pisolithus tinctorius* ectomycorrhizae or left uninoculated at planting on a reclaimed coal surface mine in Tennessee. For a given measurement, means sharing a common letter do not differ significantly between treatments at $\alpha = 0.05$. Reprinted from the *Forest Science* (vol. 35, no. 2), after Walker *et al.* (1989), published by the Society of American Foresters, 5400 Grosvenor Lane, Bethesda, MD 20814–2198. Not for further reproduction.

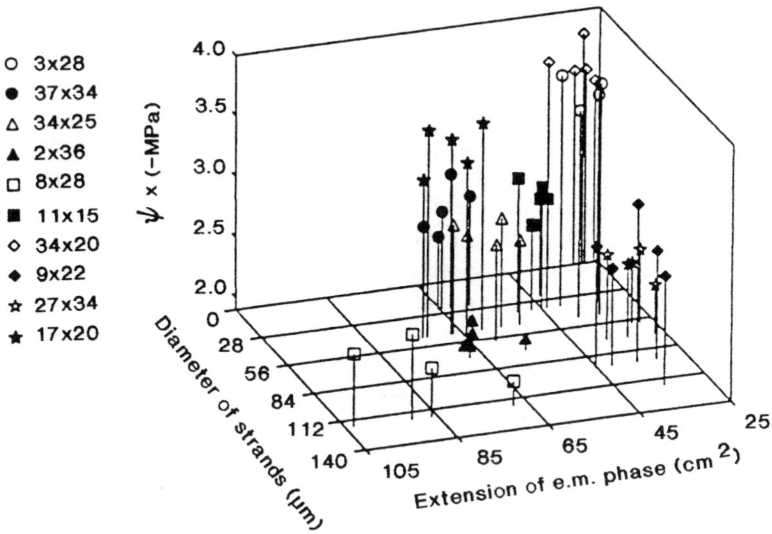

Figure 1 Influence of the extension of extramatrical hyphae (cm^2) and diameter of mycelia (μm) of *Pisolithus tinctorius* on xylem water potential (Ψ_x). Symbols indicate particular strain of *P. tinctorius*. Reprinted from Lamhamedi *et al.* (1992), with permission.

lithus tinctorius on Ψ_l and total root system flow resistance during drought were significantly correlated with the degree of extension of the extramatrical phase and hyphal strand diameter characteristics of the genetic strains with which seedlings were inoculated (Fig. 1).

b. Emergent Properties of Pathways—Hydraulic Resistances of Soil, Soil–Root Interface, and Roots The root systems of conifer species provide substantial resistance to water uptake. Several experiments have been conducted in which entire aboveground portions of trees were severed from the roots and connected to water reservoirs, effectively removing the soil and root resistance portions of the SPAC [see Eq. (2)]. Soil-plus-root resistance is thus derived by subtracting shoot resistance from total resistance. Roberts (1977) found that the root resistance of 16-m-tall *Pinus sylvestris* trees so treated accounted for 53% of total SPAC liquid flow resistance. Running (1980a) reported on similar experiments with *Pinus contorta*. In the latter study, more rapid recovery in Ψ_l from diurnal depressions was clearly apparent in severed trees. Although these estimates of "root" resistance often included any resistances associated with movement of water in the bulk soil and across the soil–root interface, the existence of substantial resistances even in moist soil (in which soil resistances are presumably very low) strongly suggests a substantial root resistance. Among conifers, there appear to be species differences in the relative importance of root resistance relative to that of stems, branches, and leaves (Whitehead and Jarvis, 1981). For example, Roberts (1978) noted that recovery of Ψ_l was less when stems of *Picea abies* were placed in water compared to that observed in *P. sylvestris*.

It must be emphasized that estimates of SPAC resistances must be interpreted carefully because they are sensitive to the manner in which they are measured (e.g., Richter, 1973) and to a wide array of influences, including environmental factors such as soil moisture and temperature (see below), root–shoot allocation patterns (which may vary with age and size), transpiration rates, and the physiological state of the plants under consideration (Jarvis, 1975; Cannell and Willett, 1976; Kaufmann and Fiscus, 1985; Pallardy *et al.*, 1991). Grossnickle and Russell (1990) provided a good example of the possible impact of physiological state in a study of water flow through *Chamaecyparis nootkatensis* seedlings removed from cold storage (Fig. 2). Whole-plant flow resistance progressively decreased as seedlings emerged from dormancy and began to grow new roots.

3. Water Uptake from the Atmosphere Water uptake from the atmosphere (from dew or fog) by plants likely occurs to some extent. Stone and Fowells (1955) noted that *Pinus ponderosa* seedlings survived longer during drought if needles were sprayed with a fine mist of water. However, given the effective barrier to water loss provided by the cuticle, it

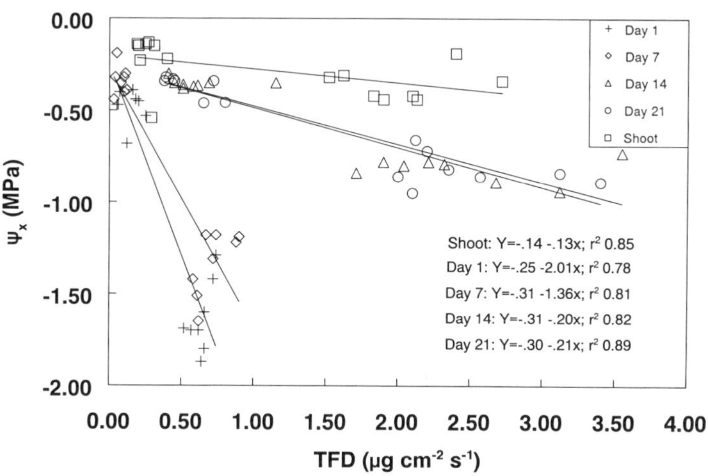

Figure 2 Relationship between shoot xylem water potential (Ψ) and transpirational flux density (TFD) for *Chamaecyparis nootkatensis* seedlings over a period of 21 days after removal from cold storage. Decreasing negative slope represents a reduction in liquid flow resistance in the soil–plant system. Each datum represents a measurement on an individual plant. Reprinted from Grossnickle and Russell (1990), with permission.

is unlikely that water uptake via this pathway is substantial, although it may be ecologically important in certain circumstances. Losses of water intercepted by the crowns of conifers may be considerable (McNaughton and Jarvis, 1983). Hence wetting of crowns may exert greater ecological influence by moving the evaporating water surface *outside* the plant body, substituting evaporation of external water for internal water loss by transpiration. Stands of evergreens appear to experience more interception losses than most broad-leafed deciduous forests (McNaughton and Jarvis, 1983) (see Section IV,B).

B. Adaptations of Root Systems to Access Soil Water Resource

1. Architectural Features Root systems of conifer species are diverse and defy simple categorization. Root system distribution patterns of conifers generally indicate concentration of roots at shallow depth, particularly at depths above 30 cm below the soil surface (Table III) (Moir and Bachelard, 1969; Roberts, 1976a,b; U.S.D.A. Forest Service, 1980; Strong and La Roi, 1983; Tryon and Chapin, 1983; Smith, 1985; Gholz *et al.*, 1986; Farrish, 1991; Nambiar and Sands, 1992). The shallow rooting in some conifer forests may be related to unusual forcing conditions of the soil. Soils characterized by (1) a high water table and its attendant effects on aeration (Nisbet and Mullins, 1986; Lieffers and Rothwell, 1986; Fisher and Stone, 1990a,b; Yamamoto *et al.*, 1987), (2) low temperatures (in particular, permafrost) (Tryon and Chapin, 1983; Oechel

Table III Total Root Length of *Pinus sylvestris* per Unit Soil Volume (Lv)[a]

Horizon (cm)		Lv (cm/cm^3)
1	0–15	5.26
2	15–30	1.25
3	30–45	0.61
4	45–61	0.34
5	61–76	0.19
6	76–91	0.18
7	91–106	0.08 (0.16)
8	106–122	0.09 (0.20)
9	122–137	0.06 (0.13)
10	138–153	0.06 (0.18)
11	153–165	0.03 (0.09)
12	168–183	0.04 (0.18)

[a] Data from different soil depths were obtained from a 40-year-old plantation in the United Kingdom. Values in parentheses represent the means of only those cores containing roots. Reprinted from Roberts (1976a), by permission of Kluwer Academic Publishers.

and Lawrence, 1985), (3) presence of restrictive layers of various kinds within the soil profile (Zisa *et al.*, 1980; Heilman, 1981; Froehlich *et al.*, 1986; Nambiar and Sands, 1992), or (4) nutrient distribution (Grier *et al.*, 1981) may truncate root system development.

Stone and Kalisz (1991) surveyed the literature for the maximum extent of root systems of a wide variety of species, including gymnosperms and angiosperms. General correlations of maximum depth of rooting with climate are apparent, based on the finding that conifer species native to semiarid regions (e.g., certain *Pinus* and *Juniperus* species) have very deep roots (>20 m) whereas other species may possess very superficial root systems. Similarly, conifer taxa with a wide ecological distribution across climatic regions (e.g., *Pinus*) show a far greater range in maximum root depth than do taxa that are for the most part restricted to humid regions and habitats (e.g., *Abies, Larix, Picea, Tsuga*). The importance of deep roots, even if sparse, to water relations will be discussed below. Strong and La Roi (1983) noted that a vertical distribution of roots in *Pinus banksiana, Larix laricina, Picea mariana, Picea glauca*, and *Abies balsamea* responded more to soil conditions, particularly soil texture, than did a horizontal root system development. Lateral root system extent thus appears less closely related to climatic and habitat factors, but the data emphasized the potential for conifers to exploit soil resources at great distance, substantially beyond the influence of the crown. The extent of root development also will depend on both genetic and stand factors, the latter including competing species and stand density. Shainsky *et al.* (1992) showed that root biomass of 5-year-old Douglas fir trees was greatly reduced by increasing red alder density in mixed plantings of the two species (Fig. 3). In contrast, root biomass of red alder was far less sensitive to increasing density of Douglas fir. Similar impacts of *Eucalyptus obliqua* invasion on rooting density of fine roots of *Pinus radiata* also have been reported (Bi *et al.*, 1992).

2. Influences of Life History Stage Water absorption characteristics of conifer root systems reflect the spatial and temporal attributes of both resources and plant. As such, it is not surprising that access to an expanding soil moisture resource base develops during the lifetime of a tree. Seeds of most conifers are quite small and possess limited reserves for heterotrophic growth after germination. Root growth capacity of germinating conifer seeds is relatively weak compared with many angiosperm trees (e.g., heavy-fruited species such as *Quercus, Carya* and *Juglans*) and thus conifer seeds require a mineral seed bed so that root growth potential is not exhausted within a thick, dry litter layer. The few exceptions to this rule (e.g., seedlings of *Tsuga* spp. and *Larix* spp., which often survive on organic substrates) benefit from unusual environmental

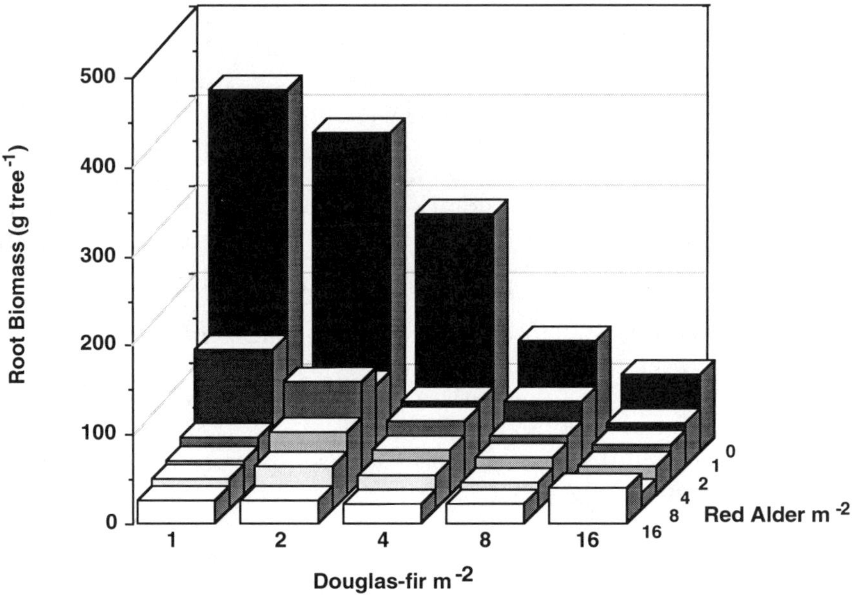

Figure 3 Root biomass of 5-year-old *Pseudotsuga menziesii* trees as a function of planting density and density of competing *Alnus rubra* plants. Adapted from Shainsky *et al.* (1992), with permission.

conditions such as exceptionally moist or wet habitats (U.S.D.A. Forest Service, 1990a,b).

Because they are small-seeded, conifer species also tend to produce small and initially slow-growing seedlings, which are poor competitors compared to other species that colonize the sunny habitats that are frequently required for successful regeneration. Numerous studies have shown that survival and early growth of conifer seedlings are substantially a function of seedling size (Mullin and Bowdery, 1978; van den Driessche, 1984). This combination of attributes has made improved seedling physiology (often focused on water relations), rapid early root and shoot growth, and effective vegetation management in conifer plantations important objectives of forest managers, especially for shade-intolerant species, for which dominance in a stand is a prerequisite for long-term survival (Cannell *et al.*, 1978; Ritchie and Dunlap, 1980; Ritchie, 1982, 1984; Sucoff *et al.*, 1985; Grossnickle and Blake, 1987a; Carlson and Miller, 1990; Brissette and Barnett, 1988; Boyle and Hellenbrand, 1991; Koppenaal *et al.*, 1991; Shainsky *et al.*, 1992).

3. Angiosperm–Conifer Comparisons With regard to water absorption by root systems, the biggest potential difference between angiosperms and conifers in temperate regions is the need to support year-round transpiration in predominantly evergreen conifers. Interestingly, there have been few comparisons of the two groups with regard to seasonal patterns of root system function and its integration into whole-plant water transport and balance (Pallardy, 1989), and further study in this area should be attractive to physiological ecologists. Lyr and Hoffman (1967) found no systematic differences in the persistence of root growth into the winter in a group of three conifer (*Pinus sylvestris*, *Picea abies*, and *Larix leptolepis*) and four angiosperm (*Quercus rubra*, *Populus trichocarpa*, *Robinia pseudoacacia*, and *Betula verrucosa*) tree species. What little information is available suggests that the root–shoot ratios in conifer and angiosperm species of temperate forests do not differ substantially (Kozlowski *et al.*, 1991), nor is there an ecologically significant difference in inherent capacity for deep rooting (Stone and Kalisz, 1991).

4. Environmental Effects on Root System Development and Water Absorption

a. Plasticity of Root–Shoot Allocation under Resource Stresses Scarcity of water and nutrient resources belowground induces preferential allocation of carbon to belowground growth. This was demonstrated by Keyes and Grier (1981), who found substantially lower belowground net primary production for *Pseudotsuga menziesii* growing on a highly productive versus a poor site. Similar results have been reported in other conifer systems and plantations (Santantonio and Hermann, 1985; Linder, 1987; Gower *et al.*, 1992) (see also Sections V,B,4 and V,B,5).

b. Influences of Other Soil Properties (Aeration, Bulk Density, pH) Many studies have shown that persistent flooding and chronically high water tables limit root system development of conifer species to well-aerated portions of the soil. The best documented instance of such limitation is that for *Picea sitchensis* plantations in the United Kingdom (Fraser and Gardiner, 1967; Armstrong *et al.*, 1976; Sanderson and Armstrong, 1978; Coutts and Philipson, 1978a,b; Philipson and Coutts, 1978, 1980). There is enough evidence for other conifers to suggest that root system truncation in hypoxic soil is a general phenomenon within this group (e.g., Lorio *et al.*, 1972; Sena Gomes and Kozlowski, 1980). The sensitivity of root growth to flooding has been found to differ among conifer species and among families of the same species (Zinkan *et al.*, 1974; Coutts and Philipson, 1978a,b; Topa and McLeod, 1986a). Root initiation and elongation in a number of conifer species are reduced or halted under flooded conditions, with mortality and progressive degeneration of the root system occurring if flooding stress is prolonged (Coutts and

Philipson, 1978a,b; Tsukahara and Kozlowski, 1984; Topa and McLeod, 1986a).

One noteworthy exception to the trend noted in the previous paragraph appears to be *Pinus elliottii*, a species that commonly occupies sites with high water tables in the southeastern United States and exhibits significant rooting below the high water table mark (van Rees and Comerford, 1986). *Pinus elliottii* trees retain the capacity for root growth and water and nutrient uptake in an hypoxic rhizosphere, presumably through O_2 transport from shoots (particularly through the development of hypertrophied lenticels) or well-aerated roots (Fisher and Stone, 1990a,b). Other conifers also apparently possess some O_2 transport capacity (Armstrong and Read, 1972; Hook and Brown, 1972; Philipson and Coutts, 1978, 1980; Topa and McLeod, 1986b).

As soil bulk density increases, soil strength presents a hindrance to root extension (Sands *et al.*, 1979; Masle and Passioura, 1987). The presence of restrictive layers in soil has been reported in some conifer forests. For example, *Pinus palustris* forests cover extensive areas that possess soil layers that render these sites very wet in spring but exceedingly deficient in total soil moisture during the summer. Other North American conifer species reported as occupying soils with a compacted layer include *Abies balsamea*, *Juniperus occidentalis*, and *Pinus contorta* (U.S.D.A. Forest Service, 1990a). However, as a group, conifers appear to favor coarse soils with good internal drainage and no restrictive layer.

Root penetration through channels left by dead roots appears to be a potentially important mechanism by which conifer roots may grow through high bulk density regions. Nambiar and Sands (1992) have recently demonstrated a striking capacity of *Pinus radiata* roots to grow through simulated root channels in a compacted layer of soil. Dense root mats formed within channels and penetrated to soil below the compacted zone (Fig. 4). Although the root length below the compacted layer remained a small fraction of the total root system (8.5% below the compacted layer perforated soil versus 2.2% for roots growing in an intact compacted soil), needle Ψ was higher for seedlings in the perforated soil through most of the growing season.

There has been little study of the impact of soil pH on root growth of conifers. As a group, they tend to grow in soils that are more acidic than those generally occupied by angiosperms (U.S.D.A. Forest Service, 1990a,b). This arises in part from the acidity of litter decomposition products in conifer forests, and so is self-induced. However, root growth inhibition of *Pinus sylvestris* and *Picea abies* at low pH has been reported (Bartsch, 1987). This topic deserves more study because of long-term impacts of anthropogenically enhanced acidic precipitation on soil reaction.

Figure 4 Vertical profile of the pattern of root penetration through perforation in a compacted subsoil. Reprinted from Nambiar and Sands (1992), with permission.

c. Selected Environmental Influences on Soil Water Absorption

i. Temperature Soil temperature has a dramatic impact on water absorption properties of root systems. First, soil temperature has an overriding influence on root growth, which will greatly control access to soil water (Rook and Hobbs, 1976; Nambiar *et al.*, 1979; Tryon and Chapin, 1983; Lopushinsky and Kaufmann, 1984; Andersen *et al.*, 1986). Further, temperatures influence the water uptake properties of existing roots. Kramer (1942) showed that root water absorption was reduced at low temperature for *Pinus*. Kaufmann (1975) demonstrated that soil–plant liquid flow resistance of *Picea engelmannii* seedlings nearly doubled as root zone temperature fell from 12.5 to 5° C, and he attributed most of the difference to changes in root permeability to water. Other factors being equal, greater flow resistances will be reflected in decreased hydration of shoots. Increased flow resistance, decreased water uptake, or increased shoot water stress with decreasing root zone temperature has been reported in numerous conifers, including *P. elliottii* (Shoulders and Ralston, 1975), *P. radiata* (Kaufmann, 1977a), *P. banksiana* (Grossnickle, 1988), *P. sylvestris* (Smit-Spinks *et al.*, 1984; DeLucia *et al.*, 1991), *P. contorta* (Running and Reid, 1980), *P. taeda* (Day *et al.*, 1991), *Picea glauca* (Grossnickle, 1988) and *Pseudotsuga menziesii* (Lopushinsky and Kaufmann, 1984).

Interestingly, there appear to be ecologically significant species differences in soil temperature responses. Shoulders and Ralston (1975) observed a 66% reduction in water uptake in roots of the subtropical *P. elliottii* with a decline in root zone temperature from 28 to only 16°C. Kaufmann (1975) showed that the subalpine *P. engelmannii* showed much less increase in liquid flow resistance from 20 to 5°C than did subtropical *Citrus*. Within the conifers, Day *et al.* (1991) noted that low root system temperatures had a larger impact on shoot Ψ of *P. taeda*, a species native to the warm-temperate southeastern United States, compared to *P. contorta*, which is native to cooler montane regions (DeLucia *et al.*, 1991). Similarly, Teskey *et al.* (1984) found less response of soil-to-needle conductance to low temperatures in *Abies amabilis* than in other conifer species native to the region but found at lower altitudes. The capacity for *some* water absorption in cold soil would appear to provide the necessary capacity for winter transpiration needs of many conifers (Lassoie *et al.*, 1983) and to prevent severe water stress from developing during the early part of the growing season in cold regions, when soil temperatures lag behind air temperatures because of high soil heat capacity and the presence of residual snow packs. However, it also should be noted that the presence of a snowpack in winter can keep soil water from freezing, thereby rendering it available for uptake.

ii. Water Deficits Effects of soil moisture on water absorption are diverse. First, water movement to and water uptake by roots varies with soil moisture content and attendant Ψ gradients and flow resistances in relevant portions of the SPAC [see Eq. (2)]. The impact of soil moisture on root system growth and uptake properties is quite complex, particularly with regard to carbon allocation to various plant parts. Increases in relative allocation to belowground growth have been reported in naturally growing individual conifers on poor sites (Keyes and Grier, 1981) or in plants subjected to artificial drought. An example of the latter response was provided by Bongarten and Teskey (1987), who reported greater root growth relative to stem growth in several seed sources of *Pinus taeda*. Axelsson and Axelsson (1986) reported reduced relative investment in the coarse root fraction (>2 mm) of irrigated 20-year-old *P. sylvestris* trees compared to control trees, but there was little reduction in the fine root fraction (<2 mm) with irrigation. Nguyen and Lamant (1989) demonstrated that root–shoot ratios increased in hydroponically grown seedlings of *Pinus pinaster* as substrate Ψ declined from −0.03 to −0.8 MPa.

Other investigators have reported contrasting results. For example, Seiler and Johnson (1988) found that root–shoot ratios of *P. taeda* decreased as the severity of water stress imposed on potted seedlings increased. Frequently, when drought increases the root–shoot ratio, it apparently arises through greater inhibition of shoot compared to root growth (i.e., root growth *is* inhibited to some extent). Such was the case in studies of *P. pinaster* (Nguyen and Lamant, 1989), *P. sylvestris* (Axelsson and Axelsson, 1986), and *P. taeda* (Bongarten and Teskey, 1987).

Additionally, reports of inhibition of root growth and fine root mortality under reduced soil moisture, in the absence of specific measures of root–shoot allocation, are quite common in conifers, e.g., *Picea sitchensis* (Nisbet and Mullins, 1986), *Pseudotsuga menziesii* (Marshall, 1986); *Pinus radiata* (Squire *et al.*, 1987); *Pinus taeda* (Hallgren *et al.*, 1991). In conifers, response of allocation to drought may be quite sensitive to experimental conditions, including the genetic materials used (Bongarten and Teskey, 1987; Joly *et al.*, 1989) and the degree of water stress imposed. Squire *et al.* (1987) suggested that a pattern of reduced elongation of individual roots extending to dormancy induction in root apical meristems was the response of *Pinus* to persistent drought. Severe drought may induce root system "quiescence" in conifers, delaying or aborting subsequent root system responses to rewetting of the soil (Kaufmann, 1968; Squire *et al.*, 1987; Seiler and Johnson, 1988).

C. Nutrient and Water Acquisition: Comparisons and Contrasts

As noted in Section I,B,1, root systems of most trees are concentrated in the upper portion of the soil. This distribution is advantageous from the standpoint of both water and nutrient absorption if the soil is reliably moist. In cold temperate and subarctic regions, where conifer forests frequently dominate, soil moisture deficiency is rare except when the soil is frozen (e.g., Oechel and Lawrence, 1985). Further, frequent observations of root concentrations near the surface are consistent with the prevalence of nutrient limitations in conifer forests and the scarcity of conifer-dominated forests in arid zones. However, these generalizations are not without significant exceptions, because there are several taxa (e.g., *Pinus* and *Juniperus*) that show very deep rooting capacities (Stone and Kalisz, 1991) (see also Section I,B,1).

II. Internal Transport and Storage

A. Tracheids as Conducting Units

1. Structure

a. Pitting of Tracheids Long-distance water flow in vascular plants occurs through tracheary elements (tracheids and vessels), which are nonliving at functional maturity. The cell walls of tracheary elements are impregnated with lignin, which impedes lateral water flow and makes the cell wall rigid. Flow between adjacent tracheary elements occurs through special areas of the wall known as pits. Pits are thin, porous, nonlignified areas in the wall where secondary wall material has not been deposited (Esau, 1965). Conifer xylem is simpler than that of angiosperms in that it contains tracheids but no fibers or vessel elements. Tracheids are nonspecialized tracheary elements in that they function both for water transport and mechanical support. However, the conifer tracheid appears to be quite resistant to permanent dysfunction as caused by embolism, as will be described below.

There is variation among conifer species in the structure of the pit membrane (Bauch *et al.*, 1972). In most conifers pit membranes are differentiated into a central, thickened, nonporous region called the torus, and a peripheral, porous area called the margo. When the conifer tracheid is air filled, the pits of that tracheid become "aspirated," meaning the torus becomes appressed against the edges of the pit border, in effect closing off air entry into adjacent tracheids (Bailey, 1958). A few conifers are classified as lacking a torus (e.g., *Thuja plicata*) because the central region of the pit membrane is not visibly thickened under the light microscope. However, in these species the central region of the pit membrane is nonporous as seen with an electron microscope (Panshin and de Zeeuw, 1980), and thus it may be functionally equivalent to a torus.

b. Variation in Tracheid Dimensions Tracheid dimensions are relevant to water flow, because, by Poiseuille's law, conductance is proportional to tracheid diameter to the fourth power. Modifications of Poiseuille's law (Calkin *et al.*, 1986) take into account limitations to water flow set by the pits of tracheids.

i. Among Conifer Species The dimensions of conifer tracheids vary considerably among species. For instance, conifer tracheids range in length from a mean of 1.2 mm in *Juniperus osteosperma* up to a mean of 7.4 mm in *Sequoia sempervirens*. They range in width from a mean of about 17 μm in *Taxus brevifolia* to a mean of 57 μm in *S. sempervirens* (Panshin and de Zeeuw, 1980). Maximum tracheid diameters range from 25 μm in wood of *Taxus brevifolia* to 80 μm in wood of *S. sempervirens*.

ii. Within a Plant Within a growth ring, the first-formed earlywood will have wider tracheids than the latewood. Overall, there is a tendency for tracheid diameters within a tree to increase basipetally from twigs to major branches to the trunk (Panshin and de Zeeuw, 1980; Zimmermann, 1983). Within a plant there is a good correlation between stem diameter and maximum tracheid diameter (Ewers and Zimmermann, 1984a,b). However, the widest tracheids occur in the woody roots (Reidl, 1937).

iii. With Age of the Plant With increasing stem age, the length and width of the tracheids that are produced increase, often plateauing only after many years (Sanio, 1872). At the extreme, some long-lived trees such as *S. sempervirens*, which can live more than 1000 years, do not reach their maximum tracheid dimensions until the tree is 200 to 300 years old (Panshin and de Zeeuw, 1980).

c. Axial and Ray Tracheids The wood of most species of the Pinaceae as well as many other conifer taxa contains both axial (or longitudinal) (Panshin and de Zeeuw, 1980) and ray tracheids (Esau, 1965). Axial tracheids, which are the most abundant type, are oriented with their long axis parallel to the plant axis and parallel to the major flow of water. Ray tracheids are oriented with their long axis in the radial plane and are shorter and usually have smaller pits than axial tracheids in the same wood. Ray tracheids permit radial flow of water in the wood, allowing movement between growth rings of different ages and hence circumvention of embolized or mechanically damaged axial tracheids. In contrast, ray tracheids and ray vessels in dicotyledonous wood are extremely rare.

2. Sapwood and Heartwood The long-distance transport of water occurs in the sapwood, which is the portion of the xylem containing living parenchyma cells. The inner, older portions of woody stems and roots of conifers and dicotyledons become heartwood. With heartwood formation the parenchyma cells often secrete gums or resins, and "extractives" encrust the pit membranes of the tracheary elements (Panshin and de Zeeuw, 1980; Wheeler, 1981). Death of the parenchyma cells soon follows. By the time heartwood formation is complete, tracheary elements are permanently nonconductive.

The ratios of sapwood area to projected leaf area (Huber value, or HV) vary considerably among conifer species (e.g., 1.3×10^{-4} m^2/m^2 for trunks of *Abies lasiocarpa* to 7.4×10^{-4} m^2/m^2 for *P. sylvestris* (Waring *et al.*, 1982). Intermediate values occur for trunks of *A. balsamea* (Fig. 5) (Ewers and Zimmermann, 1984a).

The number of growth rings of sapwood and heartwood varies with position in the plant and with growth rates (Ewers and Cruiziat, 1991). Conifers have many growth rings of sapwood in their xylem, with *P. palustris* having up to 200 growth rings of sapwood near the base of the trunk (Panshin and de Zeeuw, 1980). In contrast, dicotyledonous ring-porous trees have a narrow band of sapwood, with most of the water transport occurring in the current year's wood (Rogers and Hinckley, 1979; Ellmore and Ewers, 1986; Cochard and Tyree, 1990).

The wide sapwood areas of conifer stems are significant both for water storage and water transport. In conifers, the entire sapwood area is in-

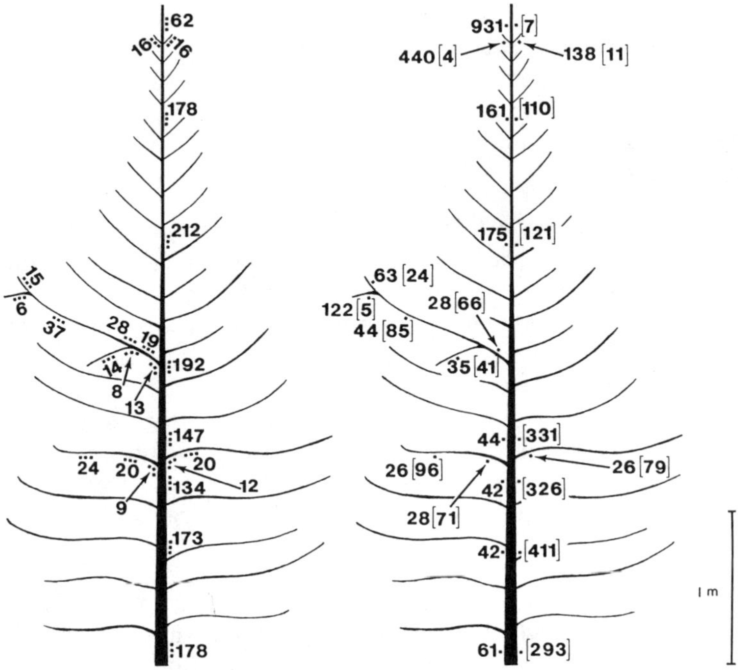

Figure 5 Hydraulic architecture of *Abies balsamea*. Left diagram shows leaf specific conductivities (k_l) in 10^{-5} kg/sec/m/MPa. Right diagram shows Huber values in 10^{-5} m^2 sapwood area per m^2 distal leaf area, and (in brackets) specific conductivity (k_s) values in 10^{-2} (kg)/(sec m MPa). Note the large k_l and Huber values in the leader and the reduced k_l values at junctions (hydraulic constrictions). Adapted from Ewers and Zimmermann (1984a).

volved with axial transport of water (Swanson, 1966; Lassoie *et al.*, 1977). The relatively low velocities of water flow and the often high water use by leaves of conifers make the water storage in the sapwood especially important. Water enters the xylem transport system first in the primary xylem and current year's wood. The presence of ray tracheids in the sapwood of many conifer species (Panshin and de Zeeuw, 1980) presumably enhances radial flow of water between growth rings, thus increasing the potential role of inner sapwood in storage and transport of water.

3. Vulnerability to Xylem Cavitation The transport of water under negative pressure is vulnerable to disruption by cavitation. Cavitation is the rapid vaporization of water initiated at a nucleation site, and it results in a vapor- and air-filled (embolized) xylem conduit that is no longer functional in water conduction. Cavitation in plants can occur as a result of water stress, freeze–thaw events, and mechanical damage to the cell wall. The origin of the nucleating site is quite different for these three stresses and, as a result, the relationship between xylem structure and vulnerability to cavitation also is very different. Cavitation appears to be an important limiting factor in the response of trees to environmental stresses; and adaptation for its avoidance and/or reversal is an important component of the physiological ecology of woody plants.

The exclusive occurrence of tracheids in conifer xylem has important consequences for how water transport in conifers is maintained through seasons of drought and freezing stress. To appreciate the unique aspects of a tracheid-based vascular system it is useful to compare it with the vessel system of angiosperms. Although the tracheid is generally less efficient in water conduction than the vessel because of its limited length and diameter, it possesses distinct advantages over the vessel when cavitation is considered.

a. Cavitation Caused by Water Stress Conifers and angiosperm trees alike appear to have a small margin of safety against the occurrence of cavitation by water stress. In both types of trees, minimum xylem pressures regularly approach or fall within the range causing cavitation (Sperry and Sullivan, 1992). Modeling studies have suggested that maximum transpiration rates of both conifers and angiosperms approach a critical value above which complete cavitation and die-back would be induced (Tyree and Sperry, 1988). This suggests that in all trees, cavitation represents a real limitation to transpiration and thus stomatal conductance. This limit could be particularly critical because cavitation is a purely physical process; once the xylem is mature there does not seem to be a mechanism whereby sensitivity to cavitation could be altered except by rupture or plugging of pit membranes. In this sense, vulnerability to cavitation could set limits to more readily adjustable biological responses to drought such as turgor maintenance and stomatal closure. The extent of cavitation during drought also may impair the recovery of stomatal conductance when drought is over.

Although conifer tracheids and vessels differ in many structural details, they do not differ in any consistent way in their vulnerability to cavitation by water stress. The relationship between xylem pressure and cavitation can be determined by measuring the loss of hydraulic conductivity resulting from embolism as xylem pressures become increasingly negative in water-stressed stems. This relationship varies considerably among species, with more drought-resistant species being more resistant to cavitation. There is no correlation with conduit type: conifer tracheids and angiosperm vessels show comparable variation in the xylem pressures required to cause cavitation (Fig. 6A) (Tyree and Sperry, 1989).

The fact that the smaller size of the tracheid confers no advantage for

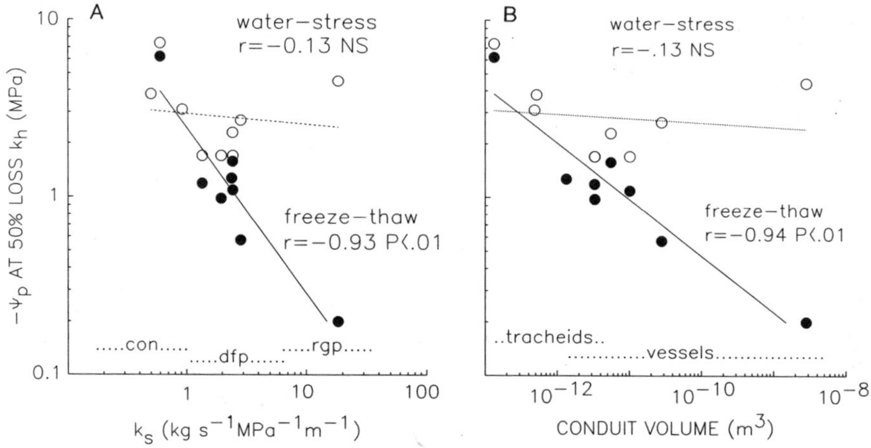

Figure 6 Xylem pressure required to cause 50% embolism as a function of (A) specific conductivity (k_s) and (B) conduit volume. In both cases there was no significant correlation for the water stress treatment (open circles, dashed line) but there was for freeze- and thaw-treated segments (solid circles, solid line). (A) Ranges are shown for conifers (con), diffuse-porous dicotyledons (dfp), and ring-porous dicotyledons (rgp). (B) Ranges are shown for tracheids and vessels. From Sperry *et al.* (1994).

withstanding cavitation by water stress is not surprising when the cavitation mechanism is considered. Considerable experimental evidence supports the air-seeding hypothesis (Zimmermann, 1983), with air entry occurring by failure of the interconduit pit membrane to confine embolism within single conduits (Crombie *et al.*, 1985; Sperry and Tyree, 1990). The nucleation of cavitation is caused by the air bubble being pulled into the functional conduit. Evidence suggests species that cavitate at more negative xylem pressures have pit membranes that can withstand a higher pressure difference before the seal is broken and air passes from embolized into functional conduits. In the strict sense, this mechanism is independent of conduit size. However, within a stem, the smaller latewood tracheids have pits that are less permeable to an air–water interface (Sperry and Tyree, 1990).

Although tracheids and vessels can be equally resistant to cavitation, they do appear to differ in how the air bubble crosses the pit membrane to nucleate cavitation. Intervessel pits normally lack a torus and the sealing action is by capillary forces at pores in the pit membrane. The bigger these pores are, the less negative the requisite xylem pressure that pulls air through them (Sperry and Tyree, 1988). Intertracheid pits in conifers have a torus, and although capillary forces at relatively large margo pores are initially involved in displacing the membrane, the ultimate sealing results from the torus pressing against the pit aperture. Evidence suggests that when cavitation occurs, the torus becomes displaced from the aperture and air passes around it through the large margo pores. Conifers that are vulnerable to cavitation (e.g., *Abies lasiocarpa*) (Sperry and Sullivan, 1992) have weak pit membranes that cannot hold the torus in place. In contrast, embolism-resistant species have strong pit membranes (e.g., *Juniperus virginiana*) (Sperry and Tyree, 1990).

The fact that conifers and other trees operate so closely to the limits set by cavitation suggests there is a competitive disadvantage to having an excessively large safety margin. The most obvious trade-off associated with increased cavitation resistance may be a loss of water-conducting efficiency. This would be expected because a pit membrane that is more resistant to air entry is probably also more resistant to water flow. However, conifers as a rule have lower specific conductivities (k_s, or hydraulic conductivity divided by sapwood transverse area) than angiosperms (Fig. 7), and yet conifers are not necessarily more resistant to cavitation caused by water stress (Fig. 6A). Apparently, any effect of pit membrane structure on k_s is overwhelmed by the influence of conduit size (Sperry and Sullivan, 1992).

Even within conifers, where conduit size is less variable than in angiosperms, there is not necessarily a trend for more cavitation-resistant species to have lower k_s values (Cochard, 1992). Again, this likely is attributable to the importance of conduit size rather than pit membrane structure for determining conducting capability. Modeling studies have indicated that the influence of pit membranes on k_h in fern tracheids increases with increasing diameter (Calkin *et al.*, 1986), suggesting the trade-off between cavitation resistance and k_s would be most obvious in conifers with large tracheids.

b. Cavitation Caused by Freeze–Thaw Events Although tracheids and vessels do not differ in their ability to withstand cavitation by water stress, tracheids are much superior to the average vessel in minimizing cavitation caused by freezing and thawing. When seasonal embolism patterns are monitored in temperate areas, peak values in both conifers and angiosperms generally occur in winter but conifers develop much lower maximum values than vessel-bearing species (Sperry, 1993; Sperry and Sullivan, 1992). In the laboratory, conifers also show much greater resistance to cavitation caused by freeze–thaw cycles than do most angiosperms (Sperry and Sullivan, 1992). The observed correlation between conduit volume and the incidence of cavitation by freeze–thaw cycles for both conifers and angiosperms suggests that it is the small volume of conifer tracheids that is the important factor contributing to their resistance to cavitation by freezing (Fig. 6B). There is no evidence for a role of the torus–margo pit membrane structure in minimizing freezing-induced cavitation in conifers, as has been suggested (Hammel, 1967).

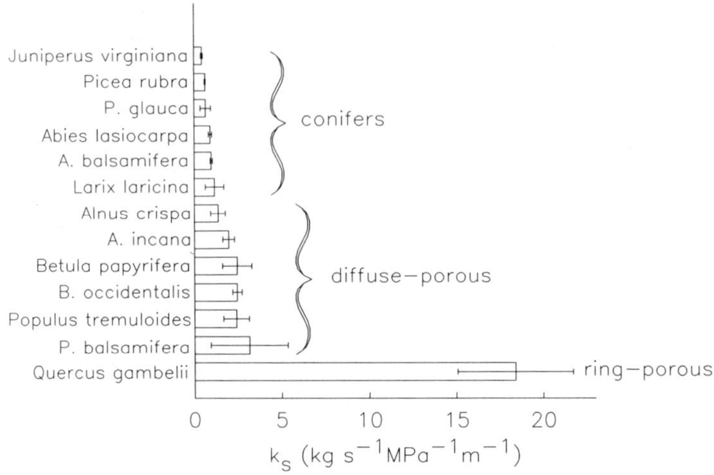

Figure 7 Specific conductivity (k_s) in stems of six conifers (lacking vessels), six diffuse-porous (narrow-vesseled) and one ring-porous (wide-vesseled) trees. From Sperry and Tyree (1990) and Sperry *et al.* (1994).

The greater vulnerability of larger conduits to cavitation by freezing and thawing is consistent with the way the nucleating site arises as a result of a freeze–thaw event. When xylem water freezes, dissolved gases freeze out and can nucleate cavitation when water is placed under negative pressure after thawing. The larger the conduit, the larger the gas bubbles, and the less negative xylem Ψ_p must be to cause cavitation (Fig. 6B) (Ewers, 1985; Sperry and Sullivan, 1992).

The importance of conduit size (particularly diameter) for determining hydraulic conductivity results in a strong trade-off between increasing k_s and decreasing resistance to cavitation caused by freeze–thaw events (Fig. 6 vs. Fig. 7) (Sperry and Sullivan, 1992). This relationship holds across both conifers and angiosperms.

c. Reversal of Embolism Conifers exhibit refilling of embolized tracheids in the absence of positive xylem Ψ_p. Several angiosperm species exhibit refilling, but it is generally associated with either root or stem pressures (Sperry *et al.*, 1987, 1988b). Recovery in conifers occurs in the spring after embolism has peaked during the winter (Sperry, 1993; Sperry and Sullivan, 1992), but in no case has this recovery been associated with positive root pressures. Recovery of hydraulic conductivity also is not necessarily associated with new xylem production and is therefore apparently due to refilling of embolized tracheids. Large changes in stem relative water content during the growing season also suggest that cavitation and at least partial refilling occur (Waring and Running, 1978).

Laboratory and field experiments indicate refilling occurs while xylem Ψ_p is negative, and that living cells are not required (Borghetti *et al.*, 1991). Capillary forces within embolized tracheids could in theory cause refilling when xylem pressures are as low as −46 kPa (Yang and Tyree, 1992; Sobrado *et al.*, 1992), but in some experiments refilling was observed at lower xylem pressures (Borghetti *et al.*, 1991). If this finding is substantiated, some mechanism other than capillarity must be responsible.

Although the mechanism of refilling in conifers requires more clarification, the fact that it occurs under natural conditions has important consequences for conifer water relations. It allows for hydraulic conductivity to recover following cavitation. This may be especially important for the reversal of embolism caused by water stress during late winter and early spring. Refilling also allows the xylem to contribute reversibly to the water storage capacity of the plant, yielding water during water stress and recharging when stress is low. As discussed below, water withdrawn from storage in the xylem contributes significantly to transpiration and mitigates to some extent the low hydraulic conductance of the conifer vascular system.

B. Capacitance

Trees undergo seasonal and diurnal fluctuations in water content, which can be viewed as water going into and out of storage. Such water storage occurs in roots, trunks, branches, and leaves. The relative importance of capacitance is related in part to the maximum transpiration rates that a plant experiences. At the extreme, succulent plants with low transpiration rates are quite dependent on water storage for survival (Nobel, 1976; Gibson and Nobel, 1986).

The Ohm's law analogy [Eqs. (1) and (2)] can be modified to allow for consideration of hydraulic capacitance, where the flux of water into a region (F_A) does not equal the flux of water going out (F_B). Each plant part can be assigned a discrete water storage capacity value, C_{AB} (in kg/MPa). Water moving into or out of storage is thus a function of the Ψ at a particular point in time. The water from storage is thus added to the F_A value to derive F_B. The capacitance is the storage water capacity (Q) expressed as a function of the water mass (w) and tissue volume (V) for stems or roots, or, for leaves, as a function of unit leaf area (A_l). For instance,

$$Q_{stem} = (\Delta w/\Delta\Psi)(1/V) \quad (4a)$$

$$Q_{leaf} = (\Delta w/\Delta\Psi)(1/A_l). \quad (4b)$$

In conifers, water storage in the trunk appears to be particularly significant (Ewers and Cruiziat, 1991). For instance, in *Pseudotsuga menziesii*, the percentage of the total water storage capacity was reported as follows: 74.8% in trunks, 19.8% in roots, 5.2% in branches, and just 0.1% in leaves. Water coming out of storage in the sapwood represented about 17% of the total daily transpiration stream during one 5-day period (Waring and Running, 1978). In *Pinus sylvestris*, 30–50% of the daily transpiration came from water stored in the sapwood (Waring *et al.*, 1979). However, Tyree and Ewers (1991) reported threefold greater diurnal water storage in leaves than in the trunk of *Thuja occidentalis*. They suggested that trunk water storage might be most important during long-term water stress, where it could account for the equivalent of 1.7 days of transpiration. In contrast, trunks of the temperate diffuse-porous tree *Acer saccharum* provide only enough water storage for 0.2 days of transpiration. Water storage is more important in conifers than in angiosperms, probably providing some water for transpiration when soil water is frozen and in part compensating for the high leaf areas and the stronger coupling of water flow to atmospheric conditions in conifer forests.

Changes in stored water can involve shrinking and swelling of living tissues, storage by capillarity in intercellular spaces and in embolized conduits, and storage by cavitation and refilling of nonliving cells (fibers and tracheary elements). The high water storage capacity of the trunks of conifer trees is related to their large sapwood volumes and short narrow conduits (tracheids) that can readily refill following embolism. Water storage in conifer leaves may be accentuated by the presence of transfusion tracheids surrounding the veins in the leaves and by the large amount of mesophyll tissue relative to the leaf surface area. Whole-tree water transport and storage are discussed further in Section IV,A.

C. Conducting Efficiency

The conducting efficiency of a plant part can be expressed in many different ways. For stems and woody roots, hydraulic conductivity [k_h; (kg m)/(sec MPa)] can be used to predict, for a particular pressure gradient, the axial volume flow rate of water through the organ. Within a plant, k_h will be proportional to stem diameter. Increases in k_h with increases in stem diameter could be due to greater sapwood transverse area, greater conduit diameters, or decreases in embolism. By itself, k_h does not indicate how well leaves will be supplied with water, because large-diameter trees may have to supply more leaves with water compared to small trees. These factors can be dealt with as described below.

1. Role of Embolism The impact of embolism on k_h is best detected by the conductivity method (Sperry *et al.*, 1988a). The initial k_h is measured as the flow rate through the isolated plant segment divided by the pressure gradient driving the flow. The segments are then perfused with degased fluid to remove all air emboli. The maximum (or final) k_h is then measured. The difference between the initial and maximum k_h is attributed to embolism.

2. Role of Sapwood Area The Huber value (HV) is equal to the sapwood area of the plant segment divided by the leaf area distal to that segment. This reflects the investment of sapwood area supplying the leaves. Huber values are much greater in conifers than in most dicotyledonous trees (Zimmermann, 1983; Ewers, 1985).

3. Specific Conductivity When k_h is divided by the sapwood transverse area, the result is the specific conductivity [k_s; (kg)/(sec m MPa)], a measure of the porosity of the sapwood. Values of k_s in conifers range from just 2.5% up to 50% of those values found in diffuse-porous angiosperm trees (Fig. 7) (Tyree *et al.*, 1991). The low k_s values in conifers are due to relatively narrow tracheids as opposed to wider vessels in angiosperms.

4. Role of Leaf Area The leaf specific conductivity [k_l; (kg)/(sec m MPa)] is equal to k_h divided by the distal leaf area. The k_l notation is used here instead of the more familiar LSC (*sensu* Tyree and Ewers, 1991) to make the terminology consistent with the other expressions of the ability to transport water, k, k_h, and k_s. The k_l is a measure of the conductive sufficiency of an axis to supply water to its leaves. If the transpiration rates [E; (kg)/(m^2 sec)] are presented on a leaf area basis, ignoring capacitance, the pressure gradient in the plant segment will be equal to E/k_l. Values of k_l generally are lower in gymnosperm than in angiosperm trees (Ewers, 1985; Tyree and Ewers, 1991). However, the pressure gradients in gymnosperm trees are not necessarily greater than in angiosperms because the transpiration rates per leaf area are generally lower in gymnosperms (Ginter-Whitehouse *et al.*, 1983). In fact, the low k_l values in conifers apparently are one of the factors that limit their stomatal opening and, hence, gas exchange.

The relationship among HV, k_l, and k_s is as follows:

$$k_l = \mathrm{HV}(k_s). \tag{5}$$

The relatively low k_l values in conifers are due to their extremely low k_s values, which are only partly compensated for by high HV. The low k_l values are attributable to narrow conduits, which offer low theoretical maximum conductance. It is interesting to note that one of the world's tallest conifers, *Sequoia sempervirens*, has the widest and longest tracheids of the group, features that presumably allow for high k_s values and low pressure gradients up the trunk. However, the long and wide conduits might also make these trees the most vulnerable of conifers to freezing-induced embolism, thus helping to limit their distribution to the moderate climate of the coastal fog belts of western North America.

5. Hydraulic Architecture of Roots The hydraulic architecture of conifer root systems is poorly understood. It is known that the roots of conifers contain larger tracheids than do the stems (Reidl, 1937; Bannan, 1941; Fayle, 1968). It is thus not surprising that in *Pinus resinosa* the k_s values are much greater in woody roots than in stems (Stone and Stone, 1975). The woody roots of conifers are apparently much less limiting to water flow than are the woody stems.

6. Shoot Hydraulic Architecture and Apical Control The pressure drop from the trunk to a leaf transpiring at a particular rate can be predicted by the hydraulic architecture. The hydraulic architecture of trees can allow the highest leaves on a tree to compete with the lower leaves on the same tree for water (Zimmermann, 1978; Tyree *et al.*, 1983; Tyree, 1988; Tyree and Ewers, 1991). This is attributable to the fact that the larger trunks and branches have greater k_l values than do smaller branches and twigs; the larger axes represent the favored pathways (Fig. 5). Larger stems in conifers have wider tracheids, which results in greater k_s and k_l values (Fig. 5) (Ewers and Zimmermann, 1984a,b). In addition, at junctions from a stem into a lateral branch there are hydraulic constrictions, which represent areas with reduced k_l due to narrower tracheid diameters (Ewers and Zimmermann, 1984a,b). Last, in certain conifers such as *Abies balsamea*, the main trunk has a much greater HV than do the lateral branches (Fig. 5) (Ewers and Zimmermann, 1984a). Consequently, the trunk has "more wood per leaf." This adds further to the k_l advantage of the main trunk over lateral branches. Species with strong apical control may thus have particularly strong "hydraulic dominance" of the main stem over laterals.

7. Implications for Seedlings Seedling establishment often is a weak link in the life cycle of trees, but the xylem transport system in seedlings remains largely uncharacterized. Before roots are well established, conifer seedlings might often be subject to drought-induced, "runaway" embolism. However, young conifer trees tend to have particularly short and narrow tracheids (Zimmermann, 1983), which should allow for faster recovery from drought-induced embolism if water supply to the roots is restored. With increasing age of the tree, the root system becomes better established, and the water storage capacity increases, both of which should make the trees less vulnerable to drought effects. The increased tracheid size with advancing age might compensate for the greater transport distances in large trees. Other life history considerations are given in Section I,B,1.

III. Water Stress Problems Unique to Conifers

The lower k_s of conifer xylem compared to that of angiosperms means that for the same thickness of branch and the same range of Ψ causing cavitation, less water can be moved without inducing cavitation. This is consistent with the basic observation that conifers tend to have lower transpiration rates per unit leaf area than broadleaf trees, whereas total leaf area may be similar or even greater for conifers (Groom, 1910; Ginter-Whitehouse *et al.*, 1983). In this way the presence of tracheids places constraints on the gas exchange rate of conifers relative to angiosperms. However, the "transpiration season" can be twice as long in conifers as in deciduous angiosperms (Table IV). Lassoie *et al.* (1983) showed that *Juniperus virginiana* maintains significant levels of carbon assimilation in the spring and fall when most angiosperms are leafless. The remarkable ability of conifer tracheids to withstand freezing-induced embolism and to recover seasonally from water stress-induced embolism may help allow for a longer growing season in temperate regions, and may make conifers particularly well suited for subalpine, subarctic, and other environments where assimilation is inherently limited.

Table IV Characteristics Affecting Annual Transpiration in Four Subalpine Species in the Central Rocky Mountains[a]

Factor	*Picea engelmannii*	*Abies lasiocarpa*	*Pinus contorta*	*Populus tremuloides*
Midday leaf conductance (cm/sec)				
Full sunlight	0.06	0.04	0.12	0.20
10% full sunlight	0.02	0.01	0.04	0.07
Leaf–air temperature difference (°C)	0	0	0	−1 to −5
Leaf area index (m^2/m^2)	15[b]	5[b]	6[c]	5[d]
Length of transpiration season (days)	210–245	210–245	210–245	105–115[e]

[a] From Kaufmann (1985).
[b] Well-stocked stands containing both *A. lasiocarpa* and *P. engelmannii*, with the latter constituting one-fourth the leaf area.
[c] For pure stands.
[d] Including both upper and lower leaf surface, with only the lower leaf surface transpiring.
[e] From the beginning of bud break until after all leaves have dried and turned brown.

The evergreen habit of most conifers poses unique problems for avoiding severe water stress during the winter. Evergreen leaves lose some water in the winter even when stomata remain closed. This is especially critical if the cuticle becomes degraded (Herrick and Friedland, 1991). Even small transpiration rates can result in low Ψ if k of the system or water absorption is decreased by frozen xylem or frozen soil, or inactive and cold roots (Sakai and Larcher, 1987). Erosion of the cuticle by blowing snow, leading to desiccation, is one of the important factors limiting growth of conifers in windy alpine and boreal environments (Hadley and Smith, 1986). Severe needle tissue damage resulting from winter desiccation is not reversible, and browning and abscission can occur even before the growing season begins. Protection from winter desiccation obviously is enhanced by stomatal closure under conditions where k is decreased by cold or frozen boles and soil, and by a cuticle with low conductance to water vapor.

Enhanced winter desiccation resistance of buds in conifers is associated with greater numbers of bud scales, the presence of resins on the surfaces of the bud scales, and a lower resistance to internal water movement between the xylem and the primordium (Sakai, 1983). Variation in winter desiccation resistance among conifer species may be an important ecophysiological determinant of the upper limit of the tree line and species distribution in subalpine forests (Tranquilini, 1976; Hadley and Smith, 1990).

Gymnosperms as a group are less flood tolerant than woody angiosperms (Kozlowski, 1982a). Flooded plants typically exhibit a rapid (1–3 days) reduction in stomatal aperture and E, followed in some species by a degree of stomatal reopening even when flooding stress persists (Pereira and Kozlowski, 1977a). Reduced gas exchange is sometimes (Zaerr, 1983) but not always preceded by development of shoot water deficits (Pereira and Kozlowski, 1977a; Lewty, 1990). These findings suggest hormonal, nutritional, and/or hydraulic rather than dehydration effects on stomatal behavior of some species during flooding (Pereira and Kozlowski, 1977a; Sojka and Stolzy, 1980; Bradford and Hsiao, 1982; Lewty, 1990).

Variation in flooding tolerance among and within conifer species has been linked to root and stem aeration capacity, the potential for stress-induced production of aerenchyma in existing or newly formed roots, and the formation of adventitious roots (Philipson and Coutts, 1978; Sanderson and Armstrong, 1980; Topa and McLeod, 1986b; Pezeshki, 1991). The close association of adventitious root formation and stomatal reopening in some species suggests that these aerenchymatous roots may play a key role in the recovery of gas exchange from flooding stress (Tang and Kozlowski, 1984; Topa and Cheeseman, 1992) (Section I,B,4). This response may be related to improved uptake of water and nutrients or restoration of a normal balance of plant growth regulators in the shoot.

IV. Transpiration: Trees to Stands

Transpiration and movement of water in the transpiration stream are driven by evaporation of water from foliage, and transpiration depends on those conditions that affect the gradient for vapor loss, conductance of the foliar evaporating surfaces, soil water availability, and internal resistances of trees. Estimating transpiration is difficult for large trees because there are no convenient techniques for measuring water transfer from foliage to the atmosphere in tall canopies. Furthermore, in stands of trees the total water loss includes not only transpiration, but also evaporation of intercepted water and of water in soil and litter. Methods for estimating transpiration and evapotranspiration were reviewed recently by Kaufmann and Kelliher (1991).

Regardless of the difficulty of estimating transpiration or evapotranspiration, the processes are important enough that some understanding of the transport characteristics affecting evaporative losses from trees and stands is worthwhile. Water movement is a process lending itself more readily than carbon or nutrient cycling to study and interpretation at very different hierarchical scales, from individual leaves or needles and branches to trees, stands, and watersheds, and for time periods of minutes to days to years. This following discussion focuses largely on the transport characteristics of the tree bole and transpiration for contrasting species, and on integration of evaporative losses at the spatial scale of the stand.

A. Transpiration of Individual Trees

1. Transpiration and Sap Flow Response to Environmental Conditions

Morning sunlight results in rapid stomatal opening and rapid increases in transpiration, although low vapor pressure deficits often initially keep transpiration low. Maximum transpiration rates for well-watered trees generally occur between several hours after sunrise and midday. In almost all cases transpiration declines gradually through the rest of the light period as a result of stomatal closure in response to increasing humidity deficits and finally to declining light. Under conditions of limiting soil water supply, stomatal conductance may decline rapidly, particularly as vapor pressure deficits increase through the day.

Rapid changes in environmental conditions may affect conifer transpiration. Clouds may decrease transpiration, but the response depends

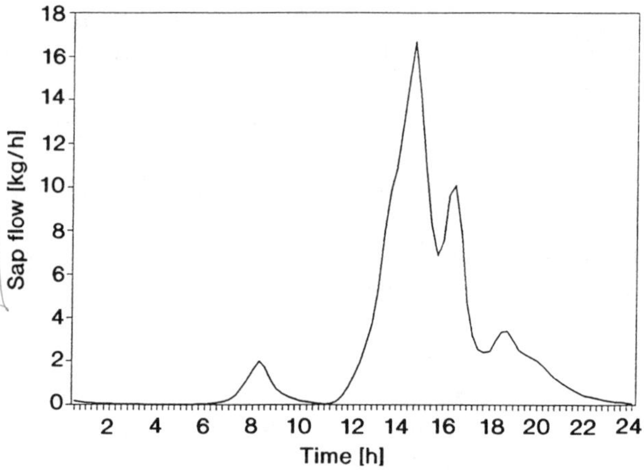

Figure 8 Diurnal course of the sap flow rate in Douglas fir under rapidly changing weather (Kootwijk, The Netherlands, August 21, 1992).

on cloud cover intensity and duration. A small cloud passing quickly (within seconds or minutes) may have almost no effect on stomatal control of transpiration because conifer stomata respond only slowly to midday changes in light intensity, requiring periods of tens of minutes or longer, and because leaf temperature changes may be small. But such a cloud has pronounced influence on transpiration and even the sap flow rate of rapidly responding species, such as oaks (Fig. 8). Overcast skies generally reduce transpiration both because of decreased stomatal opening and often decreased vapor pressure gradients. There is some evidence from spectral and coherency analysis, however, that even short-term fluctuations in bole sap flux in large trees are related to fluctuations in transpiration caused by larger turbulence eddies in the canopy (Hollinger *et al.*, 1992). These turbulence effects probably occur through reductions in boundary layer thickness in the canopy. Undoubtedly the most pronounced changes in transpiration occur as precipitation begins and the foliage is wetted. In moments, the transpiration rate may decrease from very high values to near zero, although even during rainfall events transpiration may occur at modest rates if the air does not become saturated or if leaf heating occurs.

Foliar water loss establishes a hydraulic gradient for water movement, and that gradient is propagated down through the transport pathway. The steepness of the gradient in any portion of the pathway depends on the flow rate and the hydraulic conductance of that pathway segment. Thus changes in conditions affecting transpiration from the foliage are reflected in changes in flow rates along the pathway, except when the sap flow is partially driven by osmotic factors (i.e., root pressure in some broad-leafed species during the spring) or if the conductive system is damaged. Kucera *et al.* (1977) and Cermak and Kucera (1991) described extremely fast changes of sap flow rate in trees caused by atmospheric, soil, and mechanical factors. The fastest change in flow caused by cutting the stem of ring-porous species was about 60% per minute. Rain caused the sap flow to decrease up to 30% per minute, while the fastest change in the sap flow caused by an abrupt decrease in soil water availability (in oak trees growing on sandy soil during severe drought) was around 10% per minute.

Usually soil factors cause much slower changes in the sap flow rate and transpiration, requiring hours or days for significant effects. Drought clearly affects diurnal courses of water flow in trees. The maximum rate of flow decreases and the diurnal curve reaches a plateau in the middle of the day. When soil water decreases to the wilting point, stomata open only partially and only in morning hours. During severe drought, transpiration may occur only through the cuticle and lenticels (and through stomata in cases where they are opened permanently) until the tree is able to absorb soil water (Cermak, 1977, 1986). Transpiration of trees also may decrease substantially under conditions of excessive soil moisture, when lack of air (hypoxia) becomes a limiting factor for absorption of water (Tripepi and Mitchell, 1984; Kozlowski, 1986; Smith *et al.*, 1990; see Section I). After heavy rains (150 mm/week) in midsummer, the transpiration in an old spruce stand decreased by 60%, and even under high evaporative demand remained low until the end of the growing season (Fig. 9) (Cermak and Kucera, 1990a).

2. Variability of Transpiration and the Sap Flow Rate within Trees Because transpiration is dependent on local micrometeorological factors, it varies both among and within trees with the position of the foliage in the canopy and with time. Higher portions of the canopy and exposed crown edges may have higher vapor pressure deficits and often have more light and therefore greater stomatal opening than the lower canopy or internal parts of crowns. In spruce, the upper one-quarter of needles transpired the same amount of water daily as the rest of the crown below (Cermak and Kucera, 1990b) (Fig. 10). Transpiration of a large tree growing in a dense canopy decreased by about 20% when the shadow of a neighbor tree passed over its crown for about half an hour (Cermak *et al.*, 1993).

The variability of xylem sap flow around the bole depends on root and foliage distribution and on soil water distribution. Variability may increase dramatically when the soil starts drying, and it is low in moist or very dry soil. Typically the tangential variation in flow velocity is only twofold or less, but in unusual cases of eccentric root distribution the

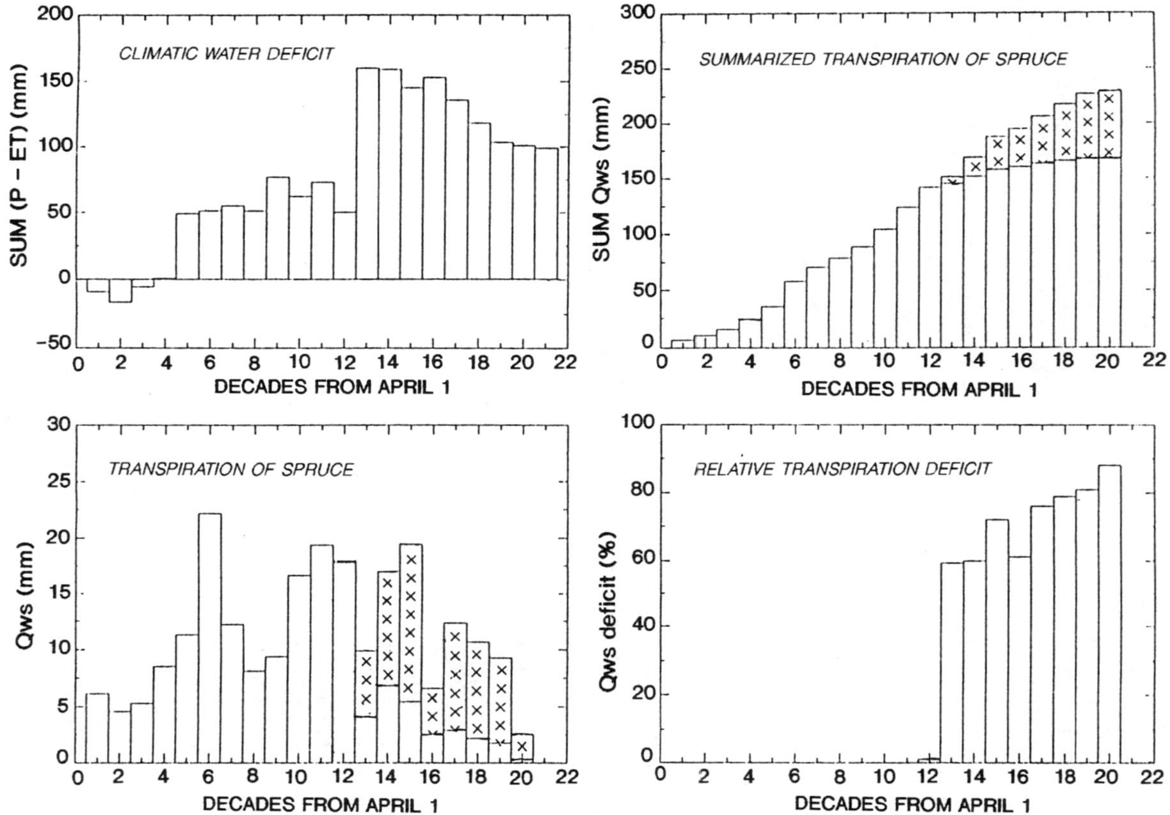

CLIMATIC WATER DEFICIT
SUM (P – ET) (mm)
DECADES FROM APRIL 1
SUMMARIZED TRANSPIRATION OF SPRUCE
SUM Qws (mm)
DECADES FROM APRIL 1
TRANSPIRATION OF SPRUCE
Qws (mm)
DECADES FROM APRIL 1
RELATIVE TRANSPIRATION DEFICIT
Qws deficit (%)
DECADES FROM APRIL 1

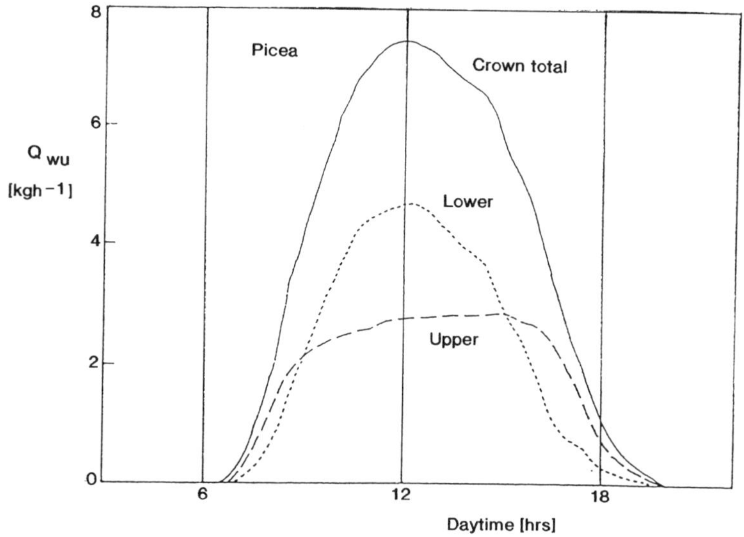

Figure 10 Diurnal courses of sap flow rate in upper and lower crown of spruce. Upper crown represents one-fourth of needle dry mass, lower crown represents three-fourths of needle dry mass. From Cermak and Kucera (1990b).

variation may be considerably higher. In a large spruce (*Picea abies*) tree in drying soil, flow rates on opposite sides of the tree were quite different, varying by a factor of about 3 at breast height (Fig. 11). This was caused by drought occurring unequally around the tree. The variation in flow was significantly lower (a factor of 1.5 or less) when the soil was moist or very dry (Cermak and Kucera, 1990c). Cohen *et al.* (1985) observed large tangential variability in one of four Douglas fir trees, and the pattern shifted through the diurnal period. Variability of sap flow rate is typically highest near the tree base and progressively decreases upward in trees having symmetrical crowns, particularly in conifers where lateral transport across tracheids is easier than across vessels in broad-

---

Figure 9 Seasonal course of transpiration rate in large spruce [*Picea abies* (L.) Karst] trees and climatic conditions under abrupt imposition of decreased soil aeration. Upper left: climatic water deficit (or simplified water balance) summarized from April 1; this value increased dramatically during the first week of August. Lower left: seasonal course of transpiration (Q_{ws}). Actual data are given in white columns and the differences between actual and theoretical data in the second part of the growing season were estimated using the dependence of transpiration on potential evapotranspiration throughout the whole season (×-marked columns). Upper right: actual decrease in seasonal transpiration total if compared with the theoretical value. Lower right: relative transpiration deficit, which reached 60–80% under hypoxic conditions. From Cermak and Kucera (1990a).

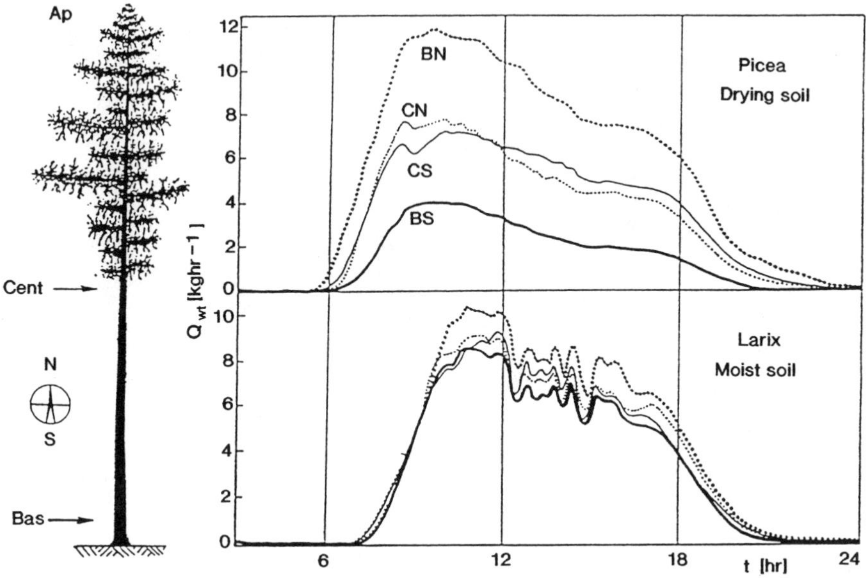

Figure 11 Diurnal pattern of xylem water flow (Q_{wt}) measured at the basal (Bas, 1.3 m) and central (Cent, base of crown) portions of the bole, on the south (BS, CS) and north (BN, CN) sides of a spruce and larch tree. From Cermak and Kucera (1990c).

leafed species. Even under higher variation of flow rate near the trunk base in spruce in dry soil, the variation diminishes substantially with tree height; differences in flow on opposite sides were almost negligible at midtrunk (Cermak and Kucera, 1990c). This is caused by conductive paths having spiral, zig-zag, and combined patterns in most species (an exception is *Thuja*) (Rudinsky and Svihra, 1970). The functional advantage of such systems is that if one root is damaged, no part of the crown suffers substantially because water is distributed equally into the whole crown (Rudinsky, 1959) (Fig. 12). In contrast, the variation of sap flow remains similar at different heights in broad-leafed species, especially those that are ring-porous, with a straight pattern of conduction. In these species the damage of one root usually is reflected by the decline of specific branches above the injured root.

The velocity of sap flow varies radially in the xylem sapwood. Heat pulse velocity (HPV) and staining methods have been used to evaluate specific features of the conductive pathways for conifer and broadleafed species. In conifer species, the flow velocity often increases from the youngest (outermost) annual rings of sapwood inward, reaches a maximum, and then decreases toward the heartwood. The radial velocity profile resembles a Gaussian distribution (Swanson, 1967; Cohen

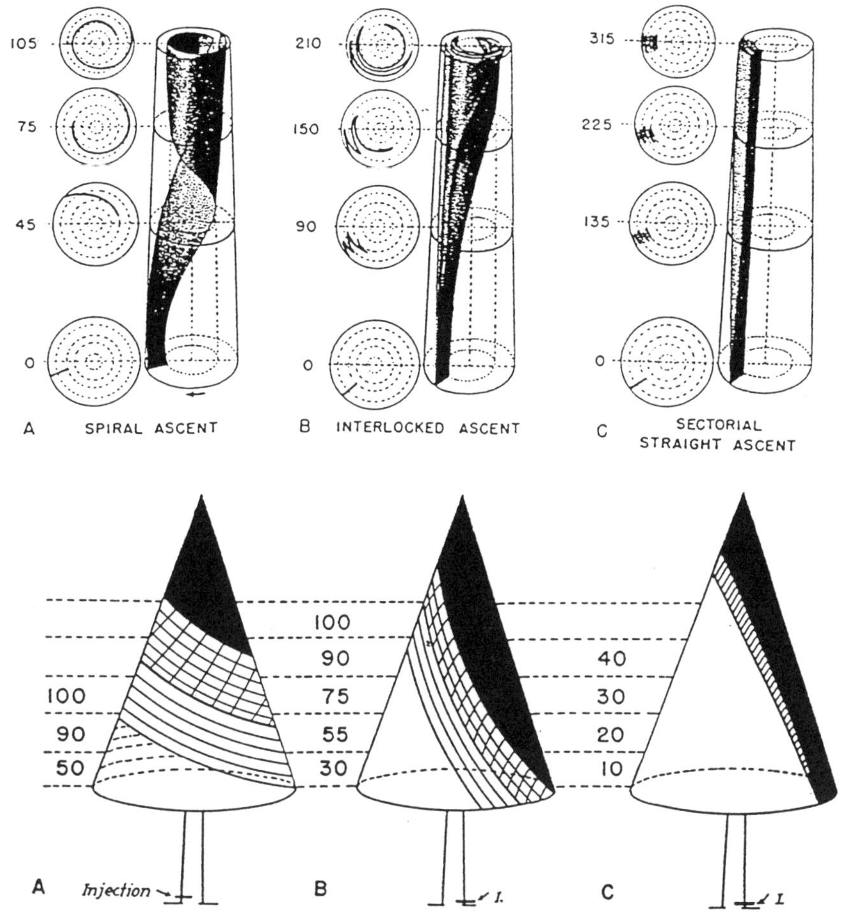

Figure 12 Representative types of water-conducting systems in various conifers (upper row) and the distribution of injected material in the crown (lower row). Spiral ascent was observed in *Abies*, *Picea*, *Larix*, and *Pinus*. Interlocked ascent was observed in *Sequoia*, *Libocedrus*, and *Juniperus*. Sectorial straight ascent was observed in *Thuja* and *Chamaecyparis*. The figures give the percentages of branches supplied by the material in various parts of the crown. Branches supplied with material by one growth ring are represented by a simple lined field, those connected by two rings are shown by a cross-lined field, and branches connected by several rings are shown as a dark field. From Rudinsky (1959).

et al., 1985; Cermak *et al.*, 1992) (Fig. 13). This pattern results from a gradual differentiation of bordered pits (erosion of matrix material from the pit membrane) in tracheid cells as sapwood rings age; in the oldest sapwood rings flow velocities decrease as pits become plugged (Mark and Crews, 1973). Sipchanov and Baurenska (1965) found maximum dye transport in only one or two annual rings in old spruce trees,

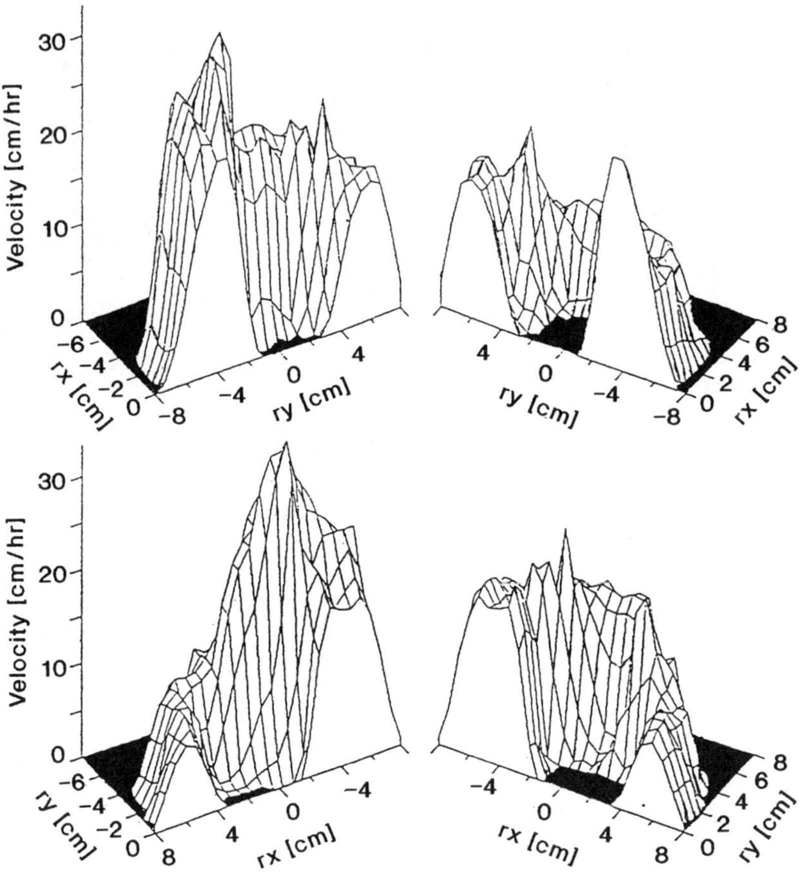

Figure 13 Acropetal velocity profile of a water or dye solution in a cross-section of the basal part of a Norway spruce trunk (dbh = 11 cm, diameter at the base = 16 cm, area of conducting xylem = 111 cm^2). The velocity profile was derived from the stained areas visible on the disks cut from the tree trunk after exposure. The path length and time of exposure were computed from a simplified matrix of 5 × 5 mm only, neglecting the slight spiral pattern. Cross-sections through the radial profile in the west–east–west (upper part of the figure) and north–south–north directions are given. From Cermak *et al.* (1992).

and Cohen *et al.* (1985) observed that the outer one-fourth of the sapwood in Douglas fir accounted for all flow early in the day. Nonuniform flow also was observed in diffuse-porous species such as willows (Cermak *et al.*, 1984a). In ring-porous broad-leafed species, the flow velocity is highest in the youngest growth ring and decreases almost exponentially inward because of embolisms or blockages in the conducting vessels (Miller *et al.*, 1980; Cermak *et al.*, 1992) (see also Section II,A,3) (Fig. 14). An example of high sap flow variability around boles is shown in Fig. 15.

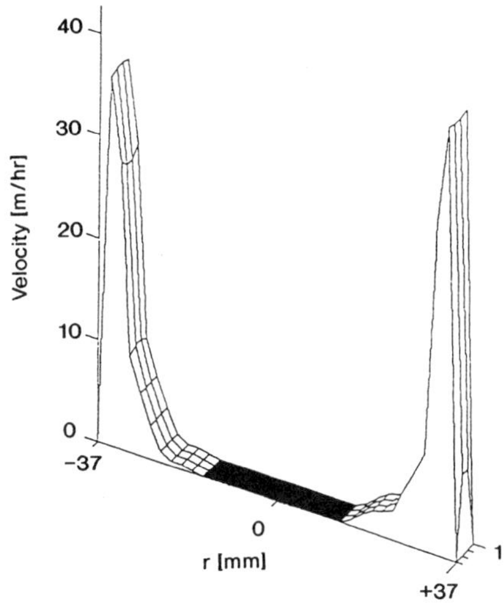

Figure 14 Acropetal velocity profile of a water or dye solution in a cross-section (north–south) of the basal part of a young oak trunk (dbh = 8 cm) as determined on disks obtained from the resampled tree trunk that had been stained by a dye-injection technique. From Cermak *et al.* (1992).

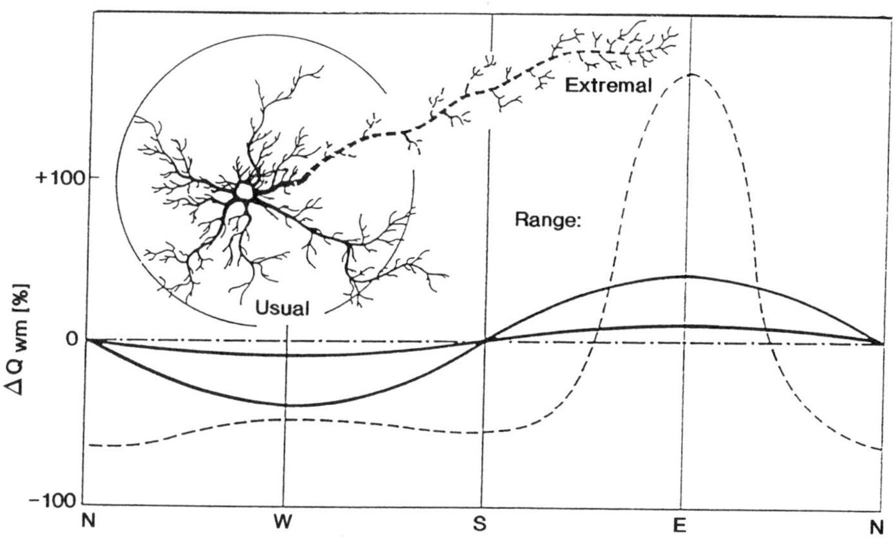

Figure 15 Transpiration flow rate variability around tree trunks (at 1.3 m) under usual and extreme conditions. The dashed-dot line corresponds to ideally symmetrical flow. Solid lines represent usual ranges of variation of transpiration flow, expressed as the deviation in flow (Q_{wm}, as a percentage of the mean) corresponding to ordinary distribution of roots around the tree (cardinal points are given for orientation only). The dashed line represents variation found in a tree growing on rocks when only one long root supplied the tree with water from deep soil layers, while other small roots surrounding the tree in shallow soil suffered drought. From Cermak and Kucera (1990b).

3. Sap Flow and Capacitance of Trees Because of capacitance in leaf and stem tissues, there is a time lag between a change in transpiration and sap flow response in lower bole sections. This time lag is very short during rainfall events, especially in species with large vessels (Kucera *et al.*, 1977), but is much longer during diurnal periods under other conditions. Asynchrony between transpiration and water absorption and transport is related to changes in stored water along the transport pathway. Transpiration is higher than the sap flow rate in the bole near ground during morning hours (or at periods when transpiration is increasing), almost equal around noon, and lower in the afternoon when transpiration is decreasing and the empty reservoirs are gradually refilled (Cermak *et al.*, 1982, 1984a; Schulze *et al.*, 1985) (Fig. 16). Rychnovska *et al.* (1980) observed that the sap flow rate followed transpiration with time shifts of only several minutes in small stems. Cermak *et al.* (1982) and Schulze *et al.* (1985) found that transpiration in tall *Quercus* trees and large *Larix* and *Picea* trees began 2 to 3 hours earlier than water flow in the bole and decreased at midday before the maximum xylem sap flow had occurred; transpiration decreased 2 to 3 hours earlier in the evening than xylem sap flow. Milne *et al.* (1983) observed that transpiration rate and shoot water potential in *Picea sitchensis* changed in phase with each other, but changes in stem radius at the base of the canopy, caused by sap flow out of the bole, lagged by up to 3 hours.

The magnitude of the delay depends on the amount of water made available from upper parts of the transport pathway as foliar water loss occurs (a capacitance, Q_{stem}, effect). This, in turn, is in part a function of k_h. Thus changes in transpiration may affect sap flow in the main stem almost immediately in small trees having high k_h and low Q_{stem}, or they may be detected quite slowly and with only modest sap velocity changes in large trees having lower k_h and large (Q_{stem}). When soil water is adequate, the net daily change in tissue water content along this pathway is near zero. Tyree (1988) concluded that water stored in the bole in *Thuja occidentalis* L. dramatically reduced the rate at which water was drawn from the soil during times of rapidly increasing transpiration. He also concluded that storage in the bole did not significantly affect the range of variation in needle water potential.

The capacitance effect depends on the size of individual trees and the degree to which soil water supplies to trees change over time. In larger trees, internally stored water within oak (*Quercus robur*) tissues were found to supply around 0.4 mm (20%) of daily transpiration (Cermak *et al.*, 1982). In other species, especially conifers, it may be equivalent to as much as 1.5 or 1.7 mm/day (Waring and Running, 1978; Waring *et al.*, 1979), with two-thirds to three-fourths of this water coming from bole sapwood. Leaves and shoots supply significant amounts of water for transpiration during morning hours but the bole sapwood may supply

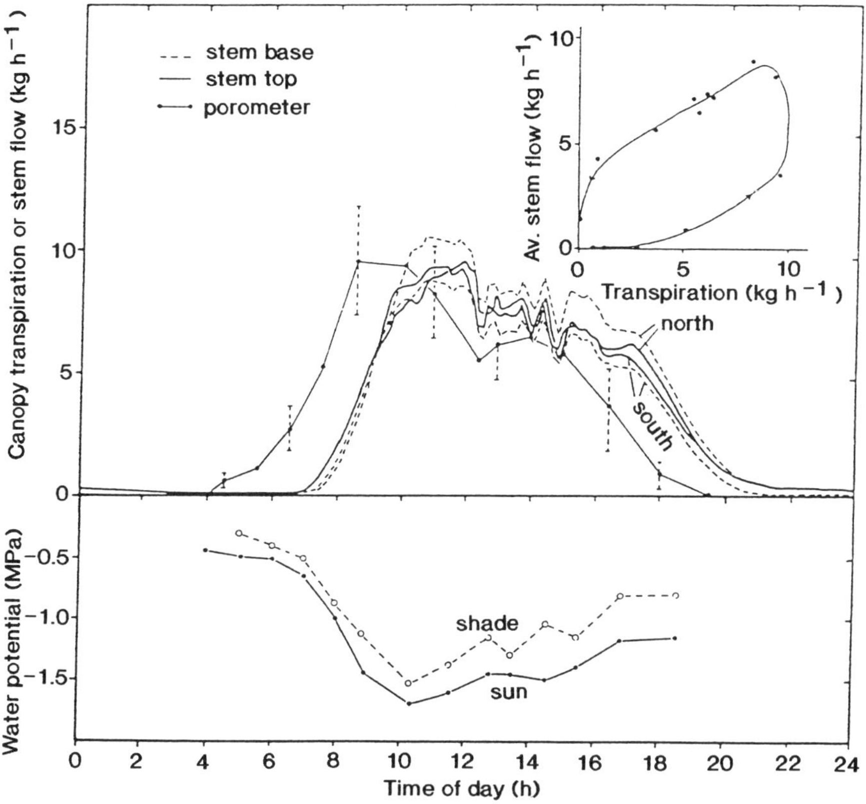

Figure 16 Top: Daily course of canopy transpiration (extrapolated on a leaf area basis from porometer measurements) and xylem water flow through the stem base (1.3 m) and stem top (12 m) on the north and south side of a larch tree. Stem flow rates are expressed as if the sapwood around the whole tree circumference would carry water equivalent to the measured rates at the north or south exposures. Total water flow was calculated as the mean of north and south sides for each height. Insert: The relationship between average xylem flow at the stem top and canopy transpiration. Bottom: Daily course of leaf water potentials in sun and shade needles. From Schulze *et al.* (1985).

more water over longer periods of time. With transpiration rates usually only around 2–4 mm/day, stored water may supply a large percentage of transpired water during part of the day. In smaller trees, however, stored water may account for less than one hour's worth of transpiration when transpiration rates are high (Running, 1980b). Because the water used from the internal storage is mostly refilled from the soil later during the same day, this amount is not very important hydrologically (Roberts, 1976b), although it may be critical for tree survival. Depletion of soil water results in decreased transpiration, sap flow, and stored water in the bole. When soil water supplies change appreciably, the water content of the tree may decrease or increase enough that transpiration and water absorption may not be equal for a given 24-hour period. These imbalances between bole transport (which correspond more to absorption) and transpiration should be considered especially when estimating the diurnal course of tree water use. The sap flow rate data obtained with HPV and THB methods can be used for estimating transpiration, but only for periods longer than 1 day or around midday. During diurnal periods sap flow is useful only if the time shift of sap flow behind transpiration is considered. Transpiration can be practically considered as equal to the sap flow rate for long-term studies, but not for short portions of soil drying cycles.

4. Sap Flow and Specific Conductivity of Xylem Tissue As noted earlier, the hydraulic properties of xylem tissue vary both among species and across the sapwood from the youngest to the oldest annual rings. Pothier and Margolis (1989) observed that saturated sapwood permeability at the base of the live crown increased with age in *Pinus banksiana*. Hydraulic properties also vary along the xylem pathway, with branches often having significantly lower k_h than the bole (Tyree, 1988), and second-order branches with primary nodes having lower k_l than first-order branches without nodes (Ewers and Zimmermann, 1984a,b; Tyree *et al.*, 1983) (see Section II,A). These features are important along with other measures of the transport system in determining the amount of water deficit that develops in the foliage.

Data summarized by Cermak (1992) in Table V compare representative specific conductivity properties and sap flow for three hardwood and two conifer species. These data indicate that sapwood flow rates are considerably higher in the hardwood species. The value of k_s, calculated for the entire bole basal area, the conductive (sapwood) portion of the basal area, and the effective lumen area of the xylem conductive elements, was found to differ among species, but the differences depended on the basis for comparison. Considering the entire basal area, specific conductivity was lower in both *Picea* and *Larix* than in the three hardwood species. On a conducting area basis, k_s was lowest in *Picea* whereas values for *Betula*, *Salix*, and *Larix* were intermediate and k_s was highest in *Quercus*. When only the last two growth rings of *Quercus* were used, k_s was nearly double that calculated for the entire sapwood conducting area. Specific conductivities calculated for measured lumen areas for the species indicated that *Quercus* conductivities were considerably higher than the rest, and *Betula* was the lowest.

5. Variability among Individual Trees Transpiration of individual trees is particularly sensitive to tree size, the environmental conditions to which trees are exposed, and the unique characteristics of individual

Table V Xylem Water Flow Rate (F_w), Flow Velocity (v_w), and Specific Conductivity (k_s) of Xylem in Different Species[a]

Species	F_w (kg/hour)	v_w (m/hour)		k_s [m²/(Pa sec)]			
						A_{lum}	
		Max	Mean	A_{bas}	A_{cx}	Max	Mean
Picea abies	5.5	1.2	0.6	4.5×10^{-10}	8.9×10^{-10}	5.8×10^{-9}	2.9×10^{-9}
Picea abies	10.0	1.2	0.6	3.4×10^{-10}	7.0×10^{-10}	4.0×10^{-9}	2.0×10^{-9}
Larix leptolepis × *decidua*	9.0	2.1	1.0	5.4×10^{-10}	1.2×10^{-9}	6.9×10^{-9}	3.4×10^{-9}
Betula alba	13.0	3	1.5	1.8×10^{-9}	2.6×10^{-9}	8.4×10^{-9}	8.4×10^{-9}
Quercus petraea	28.8	40	20	8.0×10^{-10}	1.6×10^{-8}	1.3×10^{-7}	1.3×10^{-7}
Salix fragilis	16.9	3	1.6	9.6×10^{-10}	1.5×10^{-9}	7.0×10^{-9}	2.6×10^{-9}

[a] Calculated for the whole basal area (A_{bas}), conductive part of the area (A_{cx}), and the effective lumen area of conductive elements in the xylem (A_{lum}). The value of k_s for oak, calculated for the last two growth rings only, is 2.8×10^{-8} m²/Pa/sec. From Cermak (1992).

species which affect their transpiration. In conifers, foliage temperature is relatively unaffected by irradiance because of air mixing within the canopy, and vapor pressure deficits remain relatively similar throughout the canopy of a stand (Landsberg, 1986). However, transpiration of individuals within a species differs with leaf area and with exposure of the foliage to incident radiation. Just as photosynthesis (A) depends on intercepted or absorbed photosynthetically active radiation, transpiration also is affected by the radiation environment.

Dominant individuals having a large amount of foliage in abundant light have higher transpiration than do suppressed or overtopped individuals having less foliage and growing in the shadows of their larger neighbors (Granier, 1987; Cermak and Kucera, 1990b). In forest stands of normal densities, the largest one-third of the total number of trees in the stand accounted for about two-thirds of total stand water consumption in several softwood and hardwood species (Cermak *et al.*, 1987; Cermak, 1989).

6. Variability among Species The maximum flow velocity reaches 1–2 m/hour in conifers and more (2–6 m/hour) in diffuse-porous hardwood species, whereas in ring-porous hardwoods the velocities may be as high as several tens (30–40) of meters per hour (Zimmermann and Brown, 1971). According to the Hagen–Poiseuille law, the highest velocities occur in the largest conductive elements of xylem. Because sapwood elements of various sizes occur in hardwood species and there is an abundance of elements of small sizes, the average velocity of sap flow within the conductive xylem is substantially lower.

Differences in transpiration also exist among species. In Table IV, some key characteristics affecting transpiration of individual trees in four subalpine species are given (Kaufmann, 1985). These characteristics were determined simultaneously for trees growing next to each other in a single stand. Leaf conductance in full sunlight varied threefold among three conifers and fivefold from *Abies lasiocarpa* to *Populus tremuloides*, with conductances of the latter nearly twice as high as those of any conifer. In conifers, the leaf-air temperature differences were very low. In *Populus*, however, leaves were 1–5°C cooler than the air. This cooling reduced the vapor gradient for transpiration, thereby offsetting some effects of higher conductance. The transpiration season, determined by daily temperature patterns, was well over 200 days for the conifers compared with slightly over 100 days for *Populus*. The consequences of these differences in stand transpiration are given below.

Within a stand, large differences in the dynamics of transpiration may occur between deep- and shallow-rooted species. Under conditions of severe drought on soils with higher hydraulic conductivity, transpiration usually decreases first in species with shallow root systems (*Picea*, *Betula*).

Deeply rooted species (*Pinus*, *Quercus*) may transpire over much longer periods during the growing season (Cermak *et al.*, 1980, 1983; Cermak, 1986). However, if shallow-rooted species survive, transpiration may resume after rains much more quickly than in deeply rooted species that require wetting of lower soil horizons (Cermak *et al.*, 1984b). Under certain site conditions, especially on heavy soils, transpiration of some tree species or tree forms can change abruptly after significant changes in stand density. For example, in old *Fagus* trees it was observed that transpiration decreased dramatically and trees started to decline after heavy thinning. Trees growing in an originally dense stand had root systems concentrated around their trunks. Transpiration did not change in specimens growing in open parts of canopy, or in other hardwood and conifer species that have extensive root systems. The cause of decline was the local drought around those trees in soils of low hydraulic conductivity (Cermak *et al.*, 1993).

B. Transpiration of Forest Stands

1. The Components of Evapotranspiration Stand evapotranspiration (ET) is composed of transpiration of plants, interception, and evaporation. Stand transpiration is the total evaporative loss occurring from stands by transpiration from individual trees. Stand transpiration also may include understory transpiration, but generally this component is small and is included in evaporative loss from the forest floor. Transpiration by the overstory often is the major component of evapotranspiration (excluding interception), accounting for 75–90% or more of ET in many situations (Rutter, 1966; Roberts *et al.*, 1980). On a seasonal or annual basis, the proportion of ET occurring as transpiration depends on leaf area index (LAI) and on interception of precipitation (see below). When LAI is high, little irradiance reaches the forest floor, and the foliage is dry; most ET occurs as transpiration by the overstory. Leaf area index is generally at its highest when young stands reach crown closure, and it thereafter declines slowly as stands mature. This decline in LAI may be related to various ecophysiological processes and to physical damage associated with sway of large trees.

Modeling of ET has shown that transpiration from forest canopies is influenced both by stomata and by radiation input [see Kaufmann and Kelliher (1991) for a recent review of measurement and modeling techniques]. The Penman–Monteith equation incorporates the portions of transpiration regulated by stomata and vapor pressure deficit and the complementary portion regulated by energy input. The proportioning of transpiration into these two components differs widely among types of canopies. For example, in conifer forests, aerodynamic mixing in the canopy generally is large, and most of the control of transpiration resides in stomata. In contrast, air mixing is limited in very dense tropical rain forest canopies or the canopies of field crops such as maize or alfalfa, and transpiration largely is influenced by radiation input rather than by stomatal control (McNaughton and Jarvis, 1983).

The distinction between these two forms of control of transpiration was examined by McNaughton and Jarvis (1983), who described the omega factor, Ω, as the dimensionless proportioning coefficient determining how much transpiration was controlled by stomata and how much by energy input. In dense canopies having low aerodynamic conductance, Ω is large and transpiration is said to be decoupled, meaning that stomatal control of transpiration is small (large variation in stomatal conductance may have little influence on transpiration). In aerodynamically rough canopies, Ω is small, and transpiration is said to be strongly coupled to stomatal behavior. Although Ω may be above 0.6 for tropical or broad-leafed tree canopies with large leaves, it may be around 0.2 for broad-leafed canopies with smaller leaves. The Ω factor is generally 0.1 or less for conifers. Thus transpiration by conifers is largely controlled by stomatal opening and only slightly by energy input (note, however, that energy in visible wavelengths is necessary for stomatal opening). Consequently, model estimates of transpiration in conifer forests are most successful if based on measurements or estimates of stomatal conductance.

Interception is the portion of precipitation that wets the foliage and bark of trees and evaporates rather than reaches the forest floor. The quantity of water intercepted by trees and temporarily stored in the canopy depends primarily on the surface area of foliage and bark. It is larger in dense canopies and progressively less as leaf area declines. Typically, interception amounts to 0.5–2 mm of precipitation (Rutter, 1975), and it is higher in conifers than in broad-leafed forests. The balance of precipitation reaches the forest floor as either throughfall or stemflow. Stemflow is normally only 1–2% of total precipitation in conifer forests, but in young stands (or in species such as *Fagus*, with erect branches and smooth bark) it may be considerably higher (Ford and Deans, 1978).

The percentage of total rainfall intercepted depends primarily on the duration, intensity, and direction of the precipitation event, the vapor pressure deficit, and the canopy turbulence during and at the end of storms. For example, wind motion may cause some intercepted water to fall as rain ceases, thereby reducing the quantity of water that might have evaporated. Landsberg (1986) and McNaughton and Jarvis (1983) reviewed interception studies and reported that as little as 5–10% or as much as 55–60% of total precipitation may be intercepted and evaporated back to the atmosphere. Evaporation of intercepted water may begin immediately after the foliage is wetted if the air does not become saturated, or it may begin when precipitation ceases and the vapor pressure gradient increases. The net effect of interception on stand water balance depends on the amount of interception loss and the amount of transpiration reduction that occurs during precipitation. Evaporation of intercepted water may occur at several times the rate of transpiration had the foliage not been wetted, but the net effect is the difference between the evaporation of intercepted water and the amount of water that would have been transpired.

While all forms of water loss in ET are evaporative, the evaporation term often is taken to mean the amount of water lost by evaporation from the forest floor. This usually includes both evaporation from the litter and mineral soil and also from understory vegetation, down logs, and woody debris. Evaporation from the forest floor may account for less than 10 or even 5% of the total ET in stands having dense overstories (high LAI). However, forest floor evapotranspiration may exceed overstory transpiration in heavily thinned stands, and in clearcuts the forest floor evaporation may amount to two-thirds of the total ET found in adjacent forested areas (Kaufmann *et al.*, 1987).

2. Scaling from Leaves or Trees to Stands Physiologists, ecologists, and hydrologists often require estimates of transpiration or evapotranspiration of forest stands. However, the methods used to estimate transpiration of individual trees cannot be applied without corrections to estimate the evapotranspiration of forest stands. The problem of scaling up is that the processes and controlling factors affecting water loss from plant communities are different from those for individual plants in subtle but important ways. To illustrate this, consider an extreme example of estimating ET from large land areas using general circulation models. We know that transpiration from individual trees is a major part of the ET from forests. However, estimating ET for large land grids may be relatively unaffected by the presence or absence of forest ecosystems, much less the presence of individual species, stand density, tree size, or the details of stomatal control of water vapor loss on tree branches. Yet at the level of individual trees, leaf area and stomatal behavior are quite important. Thus the details of processes important at one spatial or temporal scale may have significance at the next level of scale above or below, but generally not across several levels.

In scaling of transpiration from individual trees to ET for stands, certain process details important in regulating transpiration for trees remain important for stands, but other factors (feedbacks) come into consideration. For individual trees and shorter periods, internal water transport characteristics, water absorption, and the stomatal control of water loss are critical determinants of the transpiration process. As trees are aggregated into stands and when longer intervals are considered, however, factors such as the pattern of water flow velocity within bole sapwood decrease in importance, whereas other factors, such as boundary layer conductance in the canopy and the vapor pressure deficit of bulk air above the canopy, become more significant.

The introduction of new processes and the reduction in importance of others place limitations on using empirical measurements or modeled estimates from one scale and simply summing or multiplying up such results to a larger scale. Transpiration measured with cuvettes on branches may be used to estimate transpiration for single trees, provided information about stomatal behavior throughout the canopy is available. However, extrapolation of branch measurements to entire canopies is more uncertain, though acceptable results may be obtained (Kaufmann, 1984). Estimates of stand transpiration from bole sap flow measurements of representative trees in the stand involve less extrapolation, and the sap flow of representative trees accounts for boundary layer effects in the canopy. Swanson (1970, 1975) and Cermak and Kucera (1987, 1990b) successfully estimated stand transpiration by measuring daily transpiration for selected sample trees that represent the range of tree crown sizes and positions in the canopy. Estimates of understory evapotranspiration must be added to estimate total stand ET.

Scaling of transpiration to groups of individual trees also is possible in combination with remote sensing data. Good relationships have been observed between actual values of transpiration of large trees of different species measured continually with the THB method and the reflectance characteristics of tree crowns estimated at the same time from infrared images taken from an airplane (Balek *et al.*, 1985; Cermak and Kucera, 1990b). Appropriate equations can be used to calibrate images in terms of transpiration for stands or watersheds where the remote sensing characteristics are available.

3. Stand Comparisons Daily and seasonal estimates of stand transpiration differ among species, sites, and environmental conditions. McNaughton and Jarvis (1983) summarized transpiration rates for some conifer and broad-leafed forests, showing typical summer midday transpiration rates of 0.2–0.4 mm/hour for conifers and 0.3 to 0.45 mm/hour for broad-leafed stands. As a general rule, conifer stands transpire between 1 and 6 mm/day in the course of summer months. Several estimates of seasonal or annual transpiration are available based upon xylem sap flow or stomatal behavior and environmental conditions. Cermak (1992) summarized data of many studies on seasonal transpiration examined for several species (mostly in central Europe) using the THB method with internal heating (Table VI). Annual transpiration rates for three subalpine conifers and one hardwood species in the central Rocky Mountains were estimated using a model based on stomatal behavior

Table VI Daily Maximum and Seasonal Total Transpiration Rates of Conifer Tree Species[a]

Species	Forest and soil	Daily maxima		Seasonal totals		
		Q_{wu} (kg/m)	Q_{ws}/E_t (%)	Q_{ws} (mm)	Q_{ws}/E_t (%)	Q_{ws}/P (%)
Picea abies (L.) Karst	Fageto-quercetum, brown oligotrof loam	78	90	—	—	—
	Quercetum, oligotrof sandy loam	69	—	101	24	19
	Quercetum, oligotrof sandy loam	33	—	90	18	30
	Querceto-carpinetum, sandy loam	69	—	190	42	54
	Fageto-quercetum, loamy soil	105	66	—	—	—
	Fagetum, brown podsol cambisol	51	74	—	—	—
	Querceto-fagetum, brown loamy sand	79	88	—	—	—
	Querceto-carpinetum, sandy loam	40	—	—	—	—
	Fagetum, loamy clayish soil	79	76	190	40	46
	Fagetum, loamy clayish soil	54	54	90	19	16
	Fageto-quercetum, brown oligotrof loam	—	—	185	40	38
	Fageto-quercetum, brown oligotrof	58	—	—	—	—
	Fageto-quercetum, brown oligotrof soil (dry)	—	33	136	31	25
	Fageto-quercetum, brown oligotrof soil (wet)	—	74	296	62	72
Laris decidua L. × *Laris leptolepis* (Gord.)	Querceto-fagetum, sandy loam soil	68	—	—	—	—
Pinus sylvestris L.	Querceto-pinetum, sandy podsol	114	43	—	—	—
	Pinetum, shallow sandy soil	58	57	52	12	11
	Pinetum, shallow sandy soil	47	39	56	11	15
	Pinetum, deep sandy soil	173	—	160	38	33

[a] Transpiration is calculated for trees of unit size (Q_{wu}) given by its bole circumference at 1 m, and for stands (Q_{ws}) expressed as the ratio to potential evapotranspiration (E_t) and precipitation (P). From Cermak (1992).

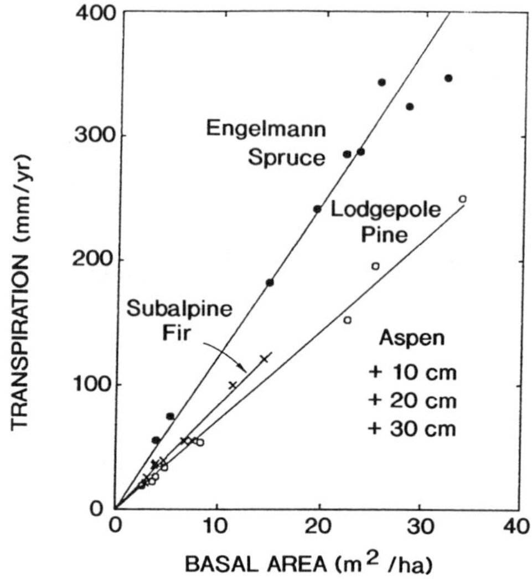

Figure 17 Annual transpiration of four subalpine tree species in the central Rocky Mountains estimated with a transpirational model for stands in two watersheds. From Kaufmann (1985).

(Kaufmann, 1985). These data (Fig. 17) indicated that for similar stand basal areas, wide differences were estimated for annual transpiration, with *Picea engelmannii* having the highest transpiration and *Populus tremuloides* the lowest. The differences in annual transpiration for these species are related to differences in stomatal behavior, leaf–air temperature difference, stand LAI, and length of transpiration season given in Table IV.

Data given in Table VI indicate wide variation in seasonal transpiration within species, depending in part on soil water conditions. For example, transpiration of *Pinus sylvestris* L. on deep sandy soil was 160 mm for the transpiration season compared with 50–60 mm on shallow sandy soil. Conifer transpiration rates were generally lower than those of broad-leafed species. The highest conifer seasonal transpiration observed was around 300 mm for *Picea abies,* whereas several hardwood species had higher values. The highest values observed were for several hardwoods on deep alluvial soils. Transpiration of some broad-leafed forest stands may be significantly enhanced by increasing their aerodynamic and optical roughness, when tree crowns of pronounced spatial structure (such as solitary growing trees) are illuminated from the sides during morning and evening hours (Cermak *et al.*, 1984b). In conifers, roughness is high even in fairly dense stands, and thinning may have less effect.

These and other studies indicate that transpiration is highly variable, depending on environmental conditions, canopy density and structure, and the unique structural and physiological features of each species affecting the transpiration process. However, the degree to which individual factors affect transpiration and the evapotranspiration of entire forest stands or watersheds depends on the importance of those factors when the water vapor transport process is scaled up to those spatial scales.

V. Physiological Response of Conifers to Water Deficits

The wide distribution of conifer species throughout the world results in their exposure to a variety of biotic and abiotic environmental stresses. Most conifers experience periods of chronic diurnal and seasonal plant water deficits associated with soil and/or atmospheric drought. Conifers also may experience excess soil moisture, for example, when plants are flooded or exposed to high water tables. The rapid decline in soil oxygen status in waterlogged soils results in alterations of plant water relations. As noted above, conifers also may experience water deficits associated with low soil temperatures and winter desiccation because of their evergreen leaf habit and distribution in subalpine and boreal forests.

Plants rarely encounter a single environmental stress, rather they are exposed to multiple stress events throughout their lives (Chapin *et al.*, 1987). For example, plants growing in drying soils experience a progressive reduction in availability and uptake of nutrients. The impact of additional stresses can have a significant influence on the ultimate physiological response of plants to water stress. Conversely, exposure to drought stress will alter the capacity of plants to cope with other natural or anthropogenic stresses (e.g., pathogens, atmospheric pollutants), resulting in additive or synergistic injury, or even protection from injury by these stresses (Mattson and Haack, 1987; Eamus *et al.*, 1989; Chevone *et al.*, 1990). The manner in which plant species respond at the cellular level to water deficits alone or in combination with other environmental stresses will affect processes such as carbon and nitrogen metabolism. At the stand and population levels, species responses to water stress will influence the structure, productivity, and succession of conifer and mixed conifer/angiosperm forests (Grier and Running, 1977; Gholz, 1982; Barton, 1993).

A. Metabolic Responses to Water Deficit

Most metabolic processes of plants are negatively affected by water stress (Hsiao, 1973; Hanson and Hitz, 1982). Water stress-induced alterations of metabolism can be viewed as lesions that in some respects are very similar to those exhibited by senescent tissues. Alternatively, metabolic responses may be adaptive and serve to improve capacity to survive and function through periodic drought (Hanson and Hitz, 1982). The literature concerning water stress impacts on metabolism in conifers is incomplete, and discussion of the subject must include some material from studies of agronomic species.

1. Photosynthesis Water stress limits photosynthesis by restricting diffusive uptake of CO_2 and through direct dehydration effects at the chloroplast level (Boyer, 1976; Kaiser, 1987; Chaves, 1991). Stomatal limitation of A typically occurs under conditions of moderate atmospheric and soil drought. Stomatal closure and gas-phase limitation of A during water stress may occur through negative feedback responses to declining leaf Ψ or RWC (Lopushinsky, 1969; Brix, 1979; Grossnickle and Blake, 1987b) or to decreased soil moisture (Blake and Ferrell, 1977; Sands and Mulligan, 1990), or from decreased mesophyll cell volume and the functional linkage between stomatal conductance (g_s) and the intrinsic photosynthetic capacity of the mesophyll (Wong *et al.*, 1979; Guehl and Aussenac, 1987). Under more severe, prolonged drought, when leaf relative water content (RWC) is decreased by 20–40%, nonstomatal components of A may be inhibited (Kaiser, 1982). Mesophyll limitations of A under dehydration stress may be associated with chloroplast shrinkage and the negative effects of increased solute concentration and ionic strength on enzymatic activity. Conformation of proteins and membranes also may change (Younis *et al.*, 1979; Kaiser, 1987; Santakumari and Berkowitz, 1990). The photosynthetic light reactions generally are less sensitive to water stress than are those of the dark reactions, unless dehydration is accompanied by elevated temperatures and high irradiance (Kaiser, 1987; Krause, 1988; Cornic *et al.*, 1989). Irreversible inhibition of A may occur when RWC is dramatically reduced (i.e., RWC $< 30\%$), the likely result of an inability to repair membranes during the rapid swelling of organelles on rehydration (Kaiser, 1987).

Whatever the precise mechanism by which plants respond to water stress, restricted entry of CO_2 into and fixation by the leaf mesophyll result in decreased A. Comparative differences among species in stomatal and photosynthetic responses to increased shoot water deficits have sometimes been linked to habitat preferences and drought tolerance (Puritch, 1973; Brix, 1979). Drought-tolerant conifers typically maintain positive rates of A to greater levels of shoot water deficits and exhibit more rapid recovery of A to prestress levels on reirrigation (Ormsbee *et al.*, 1976; Black and Bliss, 1980; Seiler and Cazell, 1990). The comparative limitations of A by stomata and mesophyll during the development of water stress also differ among species and change with plant water status (Beadle *et al.*, 1981; Melzack *et al.*, 1985). Stomatal limitation of photosynthesis was the dominant factor in dehydrating *Picea sitchensis* and *Picea rubens* shoots at Ψ of −1.4 MPa and increased further as shoot Ψ declined (Beadle *et al.*, 1981; Seiler and Cazell, 1990). Nonstomatal inhibition of *A* dominated in droughted *Pinus taeda* (Teskey *et al.*, 1986) and *Picea glauca* (Toivonen and Vidaver, 1988) seedlings. Available research indicates that conifer species apparently differ in the relative contribution of stomatal and nonstomatal components of photosynthetic inhibition during drought, but nonuniform stomatal closure and subsequent overestimation of intercellular CO_2 concentration may have influenced the results of some studies (Downton *et al.*, 1988; but see Gunasekera and Berkowitz, 1992).

The relationship between net photosynthetic carbon gain and unit transpirational water loss can be viewed as a relative measure of water use efficiency (WUE). Comparatively large expenditures of water per unit of carbon gain (lower WUE) generally are thought to be advantageous to plants growing under competition or exposed to short-term water stress. More frugal water use (greater WUE) may be adaptive for plants growing free of competing vegetation or during longer periods of water stress (Cohen, 1970; Cowan, 1982). Greater WUE and the manner in which *A* and *E* change with decreasing plant water status have sometimes (Bennett and Rook, 1978; Cui and Smith, 1991) but not always (Grieu *et al.*, 1988) been correlated with improved growth and survival of conifer species. The manner in which *A* and *E* change with dehydration differs among species and may influence plant growth and survival (Squire *et al.*, 1987; Conroy *et al.*, 1988; Gross, 1989). A larger proportional increase in WUE by *Pinus contorta* compared to *Abies lasiocarpa* and *Picea englemannii* saplings when growing in drier, more open habitats was suggested to favor the establishment of pine on drier, more exposed sites and perhaps explain the restriction of spruce and fir to more sheltered, mesic habitats (Carter and Smith, 1988). Intraspecific variation in WUE has also been reported for conifers, with greater WUE being characteristic of seedlings of more xeric geographic origin (Eickmeier *et al.*, 1975; Monson and Grant, 1989) and of more vigorous seedlings under herbaceous competition (Sands *et al.*, 1984).

Drought acclimation of gas exchange characteristics has been noted in conifer seedlings. Drought stress conditioning of *Pinus taeda* seedlings induced a capacity to maintain greater *A* through increased mesophyll conductance at a given needle Ψ, and lower *E* and increased daily WUE during subsequent drought stress (Seiler and Johnson, 1985, 1988). Exposure of *Picea mariana* seedlings to drought conditioning treatments increased stomatal sensitivity to water stress, a potentially beneficial conservation response, but did not affect the response of gas exchange to development of, or recovery from water stress of *Picea rubens* seedlings (Seiler and Cazell, 1990). The potential for photosynthetic acclimation to water stress apparently differs among conifers, but whether this phenotypic flexibility typifies species or families from drought-prone habitats is not clear (Seiler and Johnson, 1988). The development of mycorrhizal associations also is related to acclimation of gas exchange in some conifers (Parke *et al.*, 1983; Boyle and Hellenbrand, 1991; Lamhamedi *et al.*, 1992) (see Section I,A,2).

*2. **Respiration*** Dark respiration generally is much less sensitive to water stress than A (Boyer, 1976; Hanson and Hitz, 1982). For example, dark respiration of droughted *Pinus taeda, Pinus halepensis* (Melzack *et al.*, 1985), *Pseudotsuga menziesii, Tsuga heterophylla* (Zavitkovski and Ferrell, 1970; Brix, 1979), and four species of *Abies* (Puritch, 1973; Brix, 1979) decreased with Ψ and the stress sensitivity of dark respiration was much less than that of A.

Photorespiration is increased in water-stressed C_3 plants following stomatal closure because diffusion of O_2 through stomata and into the mesophyll is much less restricted by decreased g_s than is CO_2. Increased CO_2 compensation points of droughted *Picea sitchensis* and *P. halepensis* were interpreted as increased rates of photorespiration in these species (Beadle *et al.*, 1981; Melzack *et al.*, 1985). A potential photoprotective role for photorespiration in water-stressed plants exposed to high light levels has been suggested and is discussed elsewhere (Osmond *et al.*, 1980; Krause, 1988; Cornic *et al.*, 1989).

*3. **Nitrogen Metabolism*** Nitrogen metabolism of plants involves the uptake and assimilation of inorganic nitrogen into protein, nucleotides, and nucleic acids. Quantitative changes in free amino acids, decreased levels of soluble protein, and the formation of water stress proteins have been reported for water-stressed conifers (Hulbert *et al.*, 1988; Tschaplinski *et al.*, 1993). Inhibition of protein synthesis was exhibited in *Pinus ponderosa* var. *scopulorum* seedlings droughted to shoot predawn Ψ (Ψ_{pd}) as low as -4.5 MPa, with seedlings subjected to shading showing greater sensitivity to stress (Vance *et al.*, 1990). Synthesis of 11 novel polypeptides occurred in needles, but only in unshaded seedlings. The presence of these "stress proteins" was associated with comparatively greater survival and recovery of protein synthesis following alleviation of severe stress ($\Psi_{pd} = -4.5$ MPa), whereas their absence in shaded seedlings was correlated with reduced recovery from a much lower stress level ($\Psi_{pd} = -1.8$ MPa). Synthesis of novel water stress proteins was not observed in seedlings of three *P. taeda* families (Hulbert *et al.*, 1988).

The accumulation of specific amino acids such as proline during exposure to water stress may be adaptive, with these compounds playing a role in NH_3 detoxification, or serving as osmotica or compatible solutes (Hanson and Hitz, 1982; Rabe, 1990). Increased content of total free amino acids and amino acid profiles have been reported for shoots and roots of several conifer species under some form of water stress (Cyr *et al.*, 1990; Zwiazek and Blake, 1990; Meier *et al.*, 1992; Tan *et al.*, 1992b). Needles of droughted *P. ponderosa* seedlings that survived severe stress accumulated increased amounts of arginine, proline, and glutamine, which together represented 89% of the total free amino acid pool (Vance and Zaerr, 1990). It appears now that drought-induced shifts in free amino acid profiles and increased total free amino acids may not play a primary role in osmotic adjustment (Cyr *et al.*, 1990; Zwiazek and Blake, 1990), but instead may reflect internal refixation and sequestration of NH_3/NH_4^+ released during proteolysis (Rabe, 1990; Vance and Zaerr, 1990; Tschaplinski *et al.*, 1993).

*4. **Lipid Synthesis and Membrane Function*** Alterations in the molecular composition and structural integrity of plant membranes are induced during water stress cycles (Simon, 1974; Vieira da Silva, 1976) and can disrupt water and ion transport, disturb compartmentation, and decrease enzyme activity (Simon, 1974; Vieira da Silva, 1976; Kuiper, 1985).

The effects of water stress on membrane structure and function of conifers have received little attention. Early attempts to quantify "drought resistance" often involved the determination of injury to plant tissues exposed to various relative humidities or tissue water contents (Parker, 1956, 1968). Differences in lethal injury levels or viability indexes among conifers were reported and likely provided an indirect, qualitative measure of dehydration effects on membrane integrity (Pharis, 1966; Pharis and Ferrell, 1966; Parker, 1968). More recent research concerning water stress and membrane injury in conifer plants has used electrolyte leakage for quantifying dehydration effects, with electrical conductivity of tissue leachate being positively correlated with shoot water deficits (Keller, 1986; van den Dreissche, 1989; Zwiazek and Blake, 1991).

*5. **Plant Growth Regulations*** Plant growth regulators are chemical messengers that coordinate plant development and physiology. They also initiate stress responses to environmental signals through activation or repression of gene expression. Environmental regulation of the stress response is manifested through simultaneous changes in the balance and physiological activity of all growth regulators and not to changes in levels and activity of a single compound. In well-watered plants, high concentrations of auxin (IAA), gibberellin (GA), and cytokinin (CK), coupled with comparatively low concentrations of abscisic acid (ABA) and ethylene, are associated with normal, active plant growth and physiological processes. Under water stress, the comparative balance of these compounds is reversed, a situation that leads to accelerated senescence, abscission of plant parts, and stress adaptation (Morgan, 1990; Davies and Zhang, 1991).

Very little information exists concerning the response of plant growth regulators to water deficits in conifer species. Increased endogenous ABA or application of exogenous ABA decreases leaf growth and g_s and alters a variety of physiological processes (Davies and Zhang, 1991; Bray, 1991; Morgan, 1990). Only the effects of foliar ABA on g_s in water-stressed conifers have been examined. Elevated needle ABA in conifers was correlated with reduction in g_s (Johnson and Ferrell, 1982), and reduced E (Roberts and Dumbroff, 1986) was correlated with differences in stomatal sensitivity to ABA among species, perhaps associated with adaptation to xeric habitats (Roberts and Dumbroff, 1986). Other studies have shown an absence of a strong relationship between leaf ABA and g_s, and differences in this relationship between successive stress cycles (Newville and Ferrell, 1980; Squire *et al.*, 1988).

A significant result of early work on stress-induced ABA accumulation in conifers was the much stronger relationship of g_s and needle ABA with soil rather than with shoot water status (Blake and Ferrell, 1977) (Fig. 18). These results, coupled with observations that mechanical damage, drought, and waterlogging of root systems reduced E independent of shoot water status, caused speculation that chemical signals released to the xylem by stressed root systems initiated stomatal closure (Coutts 1980, 1981, 1982). A great deal of research addressing root–shoot communication and nonhydraulic signals in response to drying soils has since been conducted. Drought-induced increases in xylem ABA and reduced xylem CK concentrations, changes in sap pH and ionic composition, and perhaps other factors have been implicated in the sensing of soil water status by roots (Davies and Zhang, 1991; Schurr *et al.*, 1992). The prevalence of this feed-forward type of stomatal response to drought, its potential mode of action in seedlings versus large trees, and its ecophysiological role in drought adaptation of conifer species remain to be investigated (Hinckley and Ceulemans, 1989; Schulze, 1991).

Increased levels of endogenous ethylene often exhibited in water-stressed plants are associated with growth inhibition; epinasty; accelerated senescence; abscission of leaves, flowers, and fruits; and root initiation (Abeles, 1973; Wang *et al.*, 1991). Ethylene may induce stomatal closure in some but not all species, and could have a direct negative effect on A through stomatal and nonstomatal factors (Pallas and Kays,

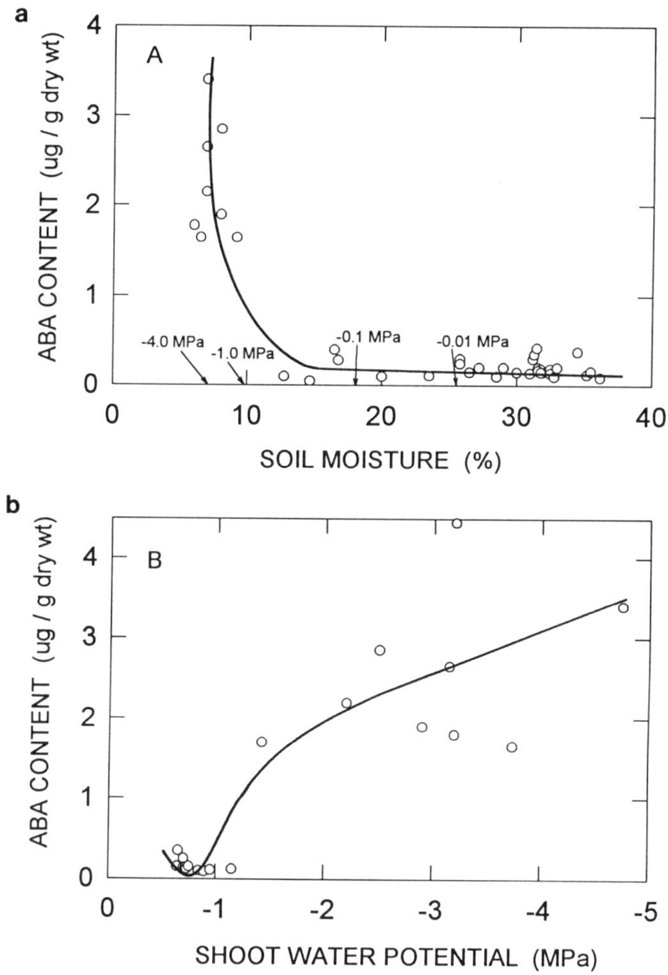

Figure 18 Association between needle abscisic acid content (ABA), (a) soil moisture content (percent oven dry wt), and (b) shoot water potential in progressively water-stressed *Pseudotsuga menziesii* seedlings. Adapted from Blake and Ferrell (1977).

1982; Gunderson and Taylor, 1988). Ethylene production in *P. taeda* seedlings of two seed sources differed between needles and roots and varied with needle Ψ during a drought stress cycle (Stumpff and Johnson, 1987). Differences in patterns of drought-induced ethylene production between a coastal plain and a more drought-tolerant seed source were not related in any obvious way to the stomatal response or relative drought tolerance of the two families. Recent evidence that enhanced ethylene production in droughted tissues is detected in detached tissues but not *in situ* calls into question much early work in this area (Morgan *et al.*, 1990).

B. Plant Growth and Development

An adequate water supply is essential to the normal growth and development of trees and stands. Moreover, the amount and seasonal distribution of precipitation have a significant impact on plant morphogenesis, with each ontogenetic event having a critical window of vulnerability. Physiological processes associated with distinct developmental events, such as seed germination and the formation of sexual reproductive structures, also are affected by the soil moisture regime. Water availability will affect plant growth at the cellular and whole-plant level through effects on cell division and expansion, metabolism, carbon allocation patterns, and plant growth regulators (Hsiao, 1973; Dale, 1988). At the stand level, net primary productivity is intimately related to the effects of seasonal water supply on leaf area dynamics and dry matter distribution.

1. Seed Germination, Seedling Emergence, and Initial Establishment Soil moisture effects on seed germination are primarily a function of soil Ψ, because changes in this parameter are accompanied by simultaneous changes in the hydraulic conductivity of the soil and seed–soil contact area (Bachelard, 1985; Evans and Etherington, 1990). Many dormant seeds are extremely dehydration tolerant and the transition from the dormant to the active germinating state is associated with a pronounced loss of tolerance (Bewley, 1979). Soil moisture availability during germination must therefore be sufficient in both amount and duration to enable germinants to quickly become established and self-supportive in terms of water, nutrient, and CO_2 uptake.

The water relations of seed germination are triphasic (Baron, 1978), with each species having a characteristic seed sorption isotherm and critical seed Ψ and RWC below which germination will not occur (Larson and Schubert, 1969; Hillel, 1972; Thomas and Wein, 1985a). Because emerging seedlings are most sensitive to soil moisture conditions during the first summer of development, germination and emergence characteristics are critical to successful establishment (Cui and Smith, 1991). More rapid seed germination, as found for populations from drier, upland sites, compared to more mesic, lowland populations of conifer species, may be associated with greater first-year survival due to the longer period available for soil penetration by roots (Habeck, 1958; Larson, 1963; Wang and MacDonald, 1992). Improved survival and growth of first-year conifer seedlings have been correlated with greater rates of radicle elongation and root penetration to deeper, wetter soil depths and greater dehydration tolerance of radicles (Noble and Alexander, 1977; Knapp and Smith, 1982; Thomas and Wein, 1985a,b; Selter *et al.*, 1986).

2. Shoot Growth Shoot and needle elongation of conifers are quite sensitive to even a slight decline in soil moisture (Kaufmann, 1968; Sands and Correll, 1976). Needle elongation of *Pinus sylvestris* seedlings was decreased by soil Ψ as high as -0.05 MPa (Sands and Rutter, 1959). Delayed flushing of shoots, decreased shoot and needle elongation, and earlier cessation of shoot and needle development during periods of soil water deficit have been reported for several conifer species (Stransky and Wilson, 1964; Coutts, 1982; Becker *et al.*, 1987; Hallgren and Helms, 1988). The comparative response of shoot and needle elongation to water deficits differs among species with determinate and indeterminate growth patterns (Kozlowski, 1982b). In species exhibiting recurrent flushing, the inhibitory effects of early season drought on elongation and development of spring shoots may be partially overcome by production of summer shoots if water is abundant later in the season (Hallgren and Helms, 1988, 1992). In species characterized by determinate shoot growth, greater comparative inhibition of total needle and terminal shoot elongation occurs during periods of severe, prolonged drought stress (Kaufmann, 1968, 1977b). However, compensatory shoot growth during periods of adequate soil moisture may result in little effect of brief, early season water stress on final shoot and needle length (Kaufmann, 1977b).

Water stress results in earlier initiation of bud set (Griffin and Ching, 1977; Young and Hanover, 1978) and decreases in the size and number of buds formed (Lotan and Zahner, 1963; Clements, 1970; Becker *et al.*, 1987), as well as decreases in the rate of bud and primordium development (Hallgren and Helms, 1992) and the number of leaf primordia initiated (Pollard and Logan, 1979; Macey and Arnott, 1986). This alteration in normal bud morphogenesis results in shoot and needle characteristics predetermined in large part by the environmental conditions experienced during bud development the previous growing season (Hallgren and Helms, 1988; Raison *et al.*, 1992). Needle length, the number of fascicles, and lengths of shoots originating from buds of *Pinus resinosa* seedlings formed during a previous year's drought were reduced regardless of water availability during the season of development (Clements, 1970). As a result, these new shoots were smaller, contained fewer, more closely spaced fascicles, and had shorter needles that emerged later in the season.

Because of the impact of soil moisture availability on the leaf area dynamics of conifer forests (Linder *et al.*, 1987; Raison and Myers,

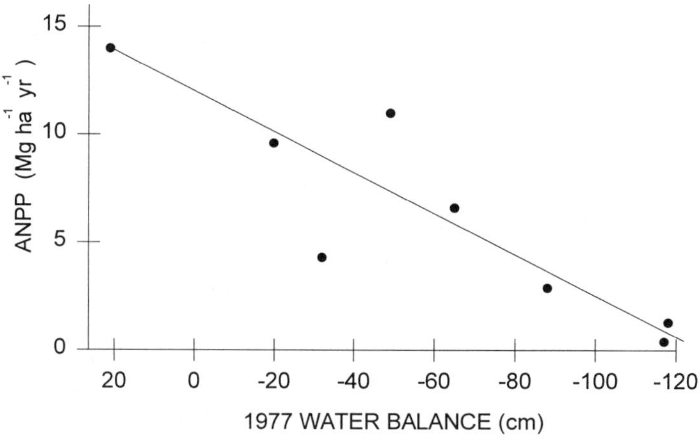

Figure 19 Relationship of aboveground net primary productivity (ANPP) to 1977 growing season water availability across sites representing a gradient in site water availability (water balance = precipitation − evaporation + soil water storage capacity in centimeters). Dominant canopy species across the gradient were *Tsuga heterophylla*, *Pseudotsuga menziesii*, *Abies grandis*, *Abies amabilis*, *Pinus ponderosa*, *Juniperus occidentalis*, and *Artemisia tridentata* var. *tridentata*. From Gholz (1982).

1992), and the relationship between leaf area and light interception, water is an important environmental determinant of the productivity of conifer forests (Teskey *et al.*, 1987; Hennessey *et al.*, 1992). Growing season site water balance indices were significantly correlated with leaf area index, total aboveground biomass, and aboveground net primary productivity (ANPP) for mature forests of several major vegetation zones from the Pacific coast to central Oregon (Fig. 19) (Gholz, 1982). The importance of other environmental factors such as temperature, light, and nutrients has been noted in this and other studies of this type, with soil moisture tending to increase in importance in habitats at lower elevations with warmer temperatures (Waring and Major, 1964; Zobel *et al.*, 1976; Gholz, 1982).

3. Cambial and Radial Stem Growth The seasonal duration of cambial growth, the number, size, and wall thickness of xylem cells, and the initiation, duration and proportion of latewood to total xylem produced respond to the amount and distribution of precipitation over a growing season (Kozlowski, 1971; Sheriff and Whitehead, 1984). As a result, radial stem growth and total stem wood production may be greatly reduced in comparatively dry years or during drought events, but resumption of growth can occur following periods of significant rainfall (Zahner, 1962; Bassett, 1964; Braekke *et al.*, 1978). Soil moisture content and various measures of cumulative seasonal plant water stress have been correlated with radial and basal area growth of tree stems (Myers, 1988; Newton and Preest, 1988; Pavlik and Barbour, 1991). The effects of soil water deficits on cambial and radial growth are likely expressed through the combined effects of plant water status on cell division and enlargement and the translocation of carbohydrates and plant growth regulators (Little, 1975; Braekke *et al.*, 1978; Aloni, 1987).

4. Root Growth Root growth of trees in temperate regions is periodic, with seasonal peaks in growth generally occurring in the spring before bud break and following the completion of shoot elongation and maturation (Reich *et al.*, 1980). This seasonal rhythm is genetically determined but may be greatly influenced by environmental factors that act directly on root water status, source–sink activity, carbon allocation patterns, and translocation of plant growth regulators. Soil moisture deficiency can have a significant impact on the growth, morphology, spatial distribution, and mycorrhizal formation of root systems of conifer species (Coutts, 1982; Becker *et al.*, 1987; Feil *et al.*, 1988). Often these impacts will be most severe in the seedling establishment phase. Droughted root systems of conifer seedlings exhibit decreased root elongation and numbers of active root tips, but may respond rapidly to increased soil moisture availability through activation of root apices, increased root elongation rates, and formation of mycorrhizal laterals (Kaufmann, 1968; Bartsch, 1987; Feil *et al.*, 1988).

Maintenance of a large standing crop of fine roots less than 1–2 mm in diameter and up to a few centimeters in length is more important for uptake of water and nutrients than is a large root biomass (Lyr and Hoffmann, 1967; Sands and Mulligan, 1990). Fine roots are comparatively short-lived and often are shed on exposure to dry soil, with subsequent production of a new crop of fine roots originating from live coarse roots on rewetting of the soil. Reduction in fine root growth and increased fine root mortality were reported in droughted conifer seedlings, with sensitivity differing among species (Marshall, 1986; Hallgren *et al.*, 1991). At the stand level, production and turnover of fine root biomass respond seasonally to site water balance, with longevity of these roots increasing at greater levels of soil moisture (Reynolds, 1970; Nisbet and Mullins, 1986; Cannell, 1989; Vogt and Bloomfield, 1991). The high energy cost of root turnover in terms of net primary production occurs at the expense of aboveground production, but may be a necessary consequence of habitation of poor sites (Keyes and Grier, 1981; Comeau and Kimmins, 1989).

5. Carbon Allocation Patterns The cyclic, coordinated growth activity of shoots and roots exhibited over a growing season is genetically determined but subject to alteration by environmental factors. Shoot and root growth typically are inversely related, with periods of high dry matter allocation in shoots coinciding with depressed allocation to roots (Ursino *et al.*, 1968; Reich *et al.*, 1980; Drew, 1982). A functional equilibrium between the shoot and root systems is attained in a particular habitat such that the shoot capacity for gas exchange is proportional to the potential for water and nutrient absorption by the roots (Thornley, 1972; Richards, 1977).

Under soil or atmospheric conditions conducive to development of plant water deficits, root growth may be stimulated at the expense of shoot growth, resulting in increased root–shoot ratio and improved adaptive balance between the potential for water uptake and transpiration (Schulze *et al.*, 1983; Bongarten and Teskey, 1987). The specific effects of water stress on carbon allocation patterns are expressed through its combined influence on source–sink relationships, translocation, plant growth regulator levels, and the comparative growth and metabolic activity of roots and shoots (Setter, 1990; Kuhns and Gjerstad, 1991). The adaptive advantage of comparatively high root–shoot ratios in water-limited habitats was reported for *P. taeda*, with families exhibiting better growth under mild water stress being characterized by relatively greater dry matter allocation to roots (Cannell *et al.*, 1978). The relative balance between aboveground and belowground biomass production in conifer forests also varies with site water relations, because a greater fraction of total plant biomass and ANPP is found in the root system on drought-prone sites (Keyes and Grier, 1981; Comeau and Kimmins, 1989; Gower *et al.*, 1992).

6. Reproductive Growth Reproductive growth generally is more sensitive to water stress than is vegetative growth (Chiariello and Gulmon, 1991). Environmental effects on reproductive growth may be direct, or indirect through their impact on vegetative structures that are the sources of water, nutrients, carbohydrates and plant growth regulators for reproductive tissues (Bonnet-Masimbert and Zaerr, 1987; Ross and Pharis, 1987). Drought stimulates cone development and maturation primarily through its influence on levels of less polar giberellic acids (GAs) (i.e., GA_4 and GA_7) during early, rapid phases of shoot elongation (Pharis and Kuo, 1977; Brix and Portlock, 1982; Ross, 1985). The interaction of drought and GA on cone induction may arise through decreased root and shoot sink activity, which results in greater availability of carbohydrates and GA for reproductive bud differentiation and development (Ross, 1985). Above-threshold levels of shoot GA may be required to induce reproductive growth, and drought may serve to decrease GA utilization by elongating shoots, thereby freeing this compound for stimulation of reproductive bud formation (Ross, 1985; Ross *et al.*, 1985).

Environmental control of sex expression has been reported for a number of herbaceous and woody species. Male plants in dioecious species or production of a greater proportion of male reproductive structures in monoecious species are favored on more xeric, wind-exposed, upland sites. These trends may be adaptive on water-limited sites because of the comparatively lower carbon costs of producing male structures (Freeman *et al.*, 1976; Freeman and McArthur, 1982) or because female plants may be more sensitive to drought (Freeman and McArthur, 1982; Dawson and Bliss, 1989; Houssard *et al.*, 1992). In monoecious *Juniperus osteosperma*, preferential production of male cones was reported on sites characterized by greater water stress (Freeman *et al.*, 1981). Induction of dioecy and production of predominantly male cones in the normally monoecious *Pinus edulis* was suggested to be a stress adaptation of this species in hot, arid environments (Floyd, 1983). The physiological basis for differential sex expression in response to soil moisture availability is unknown but may be linked to the balance of plant growth regulators synthesized in the roots of plants exposed to drying soils (Freeman *et al.*, 1981; Bonnet-Masimbert and Zaerr, 1987; Crisosto *et al.*, 1992).

C. Drought Tolerance Adaptations

The growth and distribution of vegetation is limited more by soil moisture deficits than any other environmental factor (Kramer, 1983). Drought tolerance may be defined as the capacity of plants for maintenance of physiological processes during brief or prolonged periods of climatic drought. The precise mechanisms by which drought tolerance is conferred is species specific and tissue specific and is dependent on the stage of plant development and life history (Hinckley and Ceulemans, 1989; Cui and Smith, 1991). Further, site factors such as light, temperature, and nutrient availability may interact with soil moisture to influence water relations and drought tolerance (Bradbury and Malcolm, 1977; Brix and Mitchell, 1986; Livingston and Black, 1987; Vance and Zaerr, 1990). Differential drought tolerance of conifer species is well documented and linked to species segregation across habitats varying in soil moisture availability (Heth and Kramer, 1975; Pereira and Kozlowski, 1977b; Brix, 1979; Livingston and Black, 1987). Intraspecific variation in drought tolerance also exists in conifers endemic to wide geographic areas or localized but environmentally diverse habitats (Pallardy, 1981; Teskey *et al.*, 1987). Drought tolerance adaptations are categorized on the basis of their relationship to their contribution to either avoidance or tolerance of cellular dehydration (Kramer, 1983). Dehydration avoidance adaptations act to delay the onset of potentially injurious water deficits whereas dehydration tolerance mechanisms confer an improved potential for sustained functioning in the face of plant water deficits. Drought tolerance is a result of the combined, integrated action of a suite of avoidance and tolerance attributes, with species often exhibiting adaptations of both types.

1. Dehydration Avoidance Dehydration avoidance adaptations promote high plant water status during climatic drought through regulation of the balance between soil moisture uptake and transpirational water loss and through maintenance of a functional xylem transport system (see Section II,A,3). Numerous adaptations can be identified (Pallardy, 1981), but they can be categorized in functional categories as adaptations for (1) water acquisition [e.g., rooting patterns and growth, mycorrhizal associations, root–shoot allocation changes (Section I)], (2) efficient transport from roots to shoots [i.e., xylem characteristics (Sections II and III)], and (3) conservation of water by shoots.

A well-developed potential for avoidance of foliar water deficits through enhanced stomatal sensitivity to soil and/or atmospheric drought has been invoked to explain the drought tolerance adaptation of *Juniperus occidentalis* (Miller and Schultz, 1987), *Pinus banksiana* (Roberts and Dumbroff, 1986), and more xeric ecotypes of *Pseudotsuga menziesii* (Zavitkovski and Ferrell, 1968, 1970) and *P. taeda* (Bilan *et al.*, 1977). Increased stomatal control of E also was associated with improved survival in conifer seedlings (Livingston and Black, 1988; Cui and Smith, 1991). An enhanced sensitivity of stomata to leaf ABA and a comparatively gradual decline in ABA (and increase in E) following stress alleviation were associated with dehydration avoidance capacity of *P. banksiana* (Roberts and Dumbroff, 1986). Prior exposure of some conifer species to water stress results in acclimation whereby stomatal response to subsequent stress is characterized by lower g_s and enhanced stomatal sensitivity to declining shoot Ψ (Seiler and Johnson, 1985; Zwiazek and Blake, 1989).

A direct stomatal response to diurnal changes in atmospheric water saturation vapor pressure deficit (VPD) in the absence of significant changes in bulk leaf water status has been demonstrated for a number of conifers (Sandford and Jarvis, 1986; Grossnickle and Russell, 1991). This response provides a more efficient and sensitive mechanism for conservation of water than stomatal response to bulk leaf water status. Stomatal sensitivity to diurnal changes in VPD differs among conifer species, from very sensitive in *Picea glauca* (Goldstein *et al.*, 1985) and *Chamaecyparis nootkatensis* (Grossnickle and Russell, 1991), to little if any response in *P. taeda* (Teskey *et al.*, 1986). The response of stomatal aperture to VPD also varies with environmental factors such as light (Kaufmann, 1976; Beadle *et al.*, 1985a; Goldstein *et al.*, 1985), air and soil temperature (Beadle *et al.*, 1985a; Miller and Schultz, 1987), and soil Ψ (Tan *et al.*, 1977; Running, 1980c; Graham and Running, 1984). Plant factors such as shoot Ψ (Beadle *et al.*, 1985b; Goldstein *et al.*, 1985) or previous exposure to high VPD or low soil moisture may alter this stomatal response (Teskey *et al.*, 1987; Mansfield and Atkinson, 1990). Generally, the lower the shoot or soil Ψ, the more sensitive the response or the lower the value of g_s at a given VPD. The diurnal sensitivity of stomata to VPD may decrease over time with aging of the foliage (Kaufmann, 1976; Sandford and Jarvis, 1986; Guehl *et al.*, 1991).

Efficient plant water use (high A/E ratios) could be an important adaptation of plants growing in water-limited environments (Cowan, 1977). From the model of Cowan and Farquhar (1977), stomatal "optimization" of gas exchange will occur when the relative sensitivity of A and E to changes in g_s, $\delta A/\delta E$, remains constant despite environmental fluctuations that affect g_s. This stomatal optimization hypothesis has been tested for a number of conifer species, with stomatal response conforming to this hypothesis in some species (Meinzer, 1982; Sandford and Jarvis, 1986; Guehl *et al.*, 1991) but not in others (Sandford and Jarvis, 1986; Guehl and Aussenac, 1987; Fites and Teskey, 1988; Grieu *et al.*, 1988). The variability in results suggests that optimized stomatal behavior may simply be one of many potentially adaptive traits.

Differences among and within conifer species in stomatal response to humidity have been invoked to define comparative drought tolerance and habitat segregation, but conflicting results have been reported. Comparisons among woody species suggest that species with high vigor and growth rate are generally less sensitive to humidity than are more slower growing, evergreen species (Marshall and Waring, 1984; Korner, 1985). Comparatively low sensitivity of stomata to VPD was offered as explanation for the confinement of some conifer species to more humid, mesophytic environments (Grossnickle and Blake, 1987a,b; Higgins *et al.*, 1987). Populations of conifer species that originate in distinct humidity regimes also exhibit different g_s–VPD relationships and stomatal adaptations to humidity. Seedlings of a dry land family of *Abies alba* were characterized by a stomatal response indicative of greater dehydration tolerance (Guehl and Aussenac, 1987). Conversely, dry site progeny of *P. ponderosa* had lower g_s values at a given VPD compared with seedlings of more mesic, coastal origin, a stomatal adaptation for enhanced dehydration avoidance (Monson and Grant, 1989).

Stomatal closure at low T_s in the absence of changes in shoot water status also may constitute a drought avoidance mechanism. The involvement of chemical signals transported from the roots to shoots in this stomatal response was investigated in split-root experiments (Day *et al.*, 1991). No evidence for a nonhydraulic root signal was found in *P. taeda* seedlings when half the root system was exposed to 24°C and half to either 7 or 1°C. Instead, feedback inhibition of A by carbohydrate accumulation due to reduced root metabolic activity and sink strength at low T_s was suggested to be involved in this response (Day *et al.*, 1991). In other species this response may be related to plant growth regulators produced by roots or close coupling between xylem flux, epidermal turgor, and g_s (Teskey *et al.*, 1983; DeLucia *et al.*, 1991). Although the effect of low T_s on gas exchange may not be a direct consequence of plant water deficits, stomatal closure in response to low T_s may effectively avoid foliar water deficits associated with impaired water uptake from cold soil.

Preferential carbon allocation to root systems in drought-prone environments results in a more favorable functional balance between water absorptive and water-expending tissues and improved potential for dehydration avoidance. Seedlings of conifer species or families native to more xeric environments are sometimes characterized by larger root–shoot ratios (Bongarten and Teskey, 1987; Joly *et al.*, 1989), but not always (Strauss and Ledig, 1985; Barton and Teeri, 1993). The abscission of older foliage coupled with fine root mortality during seasonal periods of decreased soil moisture availability results in a readjustment in root–shoot balance. Further, this reduction in leaf area coincident with rapid production of fine roots following replenishment of soil moisture will result in an increased root–shoot ratio and may enhance plant capacity for stress recovery (Bartsch, 1987; Chaves and Pereira, 1992).

2. Dehydration Tolerance Dehydration tolerance adaptations are not well developed in higher plants compared to the situation in many epiphytic algae, lichens, bryophytes, and a few lower vascular plants (Benzing, 1990). In vascular plants one of the most fundamental limits to drought stress is the Ψ at which irreversible, runaway xylem embolism occurs (Tyree and Sperry, 1988, 1989). Conifers vary considerably in xylem vulnerability to water stress-induced embolism (Cochard, 1992). For example, the xylem of *Juniperus scopulorum* is adapted to withstand much lower Ψ than that of *Abies lasiocarpa* (Sperry and Sullivan, 1992).

Drought-induced alterations in the osmotic and elastic properties of shoot tissues represent another dehydration tolerance mechanism, but the putative benefits of these changes are uncertain. Leaf osmotic adjustment of herbaceous crop species has been shown to sustain A and growth through turgor maintenance under declining water status (Morgan, 1984; Ludlow, 1987). Although shoot osmotic adjustment has been demonstrated in droughted conifers, rigorous examinations of the impact of this response on physiological processes are few (Jackson and

Spomer, 1979; Buxton *et al.*, 1985; Zwiazek, 1991; Meier *et al.*, 1992). However, in one study, seedlings of *Picea mariana* characterized by rapid growth under drought exhibited greater stress-induced leaf osmotic adjustment, sustained gas exchange, and more rapid stress recovery of gas exchange than slower-growing families showing less osmotic adjustment (Tan *et al.*, 1992a,b) (Fig. 20). Accumulation of soluble carbohydrates that served to maintain turgor (and hence g_{wv} and A) under osmotic stress also conferred some measure of dehydration tolerance that facilitated a rapid recovery in gas exchange following relief from stress.

Decreases in the shoot osmotic potential at which turgor loss occurs in some conifers results from adjustments in tissue elasticity or reduced symplastic water volume and not from a net increase in solute content (Miller and Schultz, 1987; Andersen and McLaughlin, 1991; Blake *et al.*, 1991). Stress-induced adjustments in tissue elasticity result in maintenance of cell volume within a range conducive to sustained photosynthetic capacity (Saliendra and Meinzer, 1991). Shifts in biochemical composition of cell walls in needles of conifer seedlings exposed to water stress may be related to altered cell wall elasticity and potential for turgor maintenance as needle RWC declines (Joly and Zaerr, 1987; Vance and Zaerr, 1991; Zwiazek, 1991).

Symplastic solute accumulation may be important to reduction of injury to cellular constituents and plant survival (Morgan, 1984; Flower *et al.*, 1990). In order for sustained gas exchange under increasing water stress to be advantageous, there must be a concomitant elevation in the capacity of the protoplasm to resist dehydration-induced damage. Solute accumulation may provide both of these benefits under water stress, with solute species differing in their role in turgor maintenance and protection from cellular injury. Inorganic ions and low-molecular-weight soluble carbohydrates are the primary osmotic constituents of the vacuole, whereas compatible solutes such as proline, polyols, and high-molecular-weight sugars in the cytoplasm may protect membranes and macromolecules from dehydration injury (Morgan, 1984; Crowe *et al.*, 1988; Leopold, 1990).

Osmotic adjustment of roots in response to drought stress also has been reported for conifer species (Kandiko *et al.*, 1980; Pallardy *et al.*, 1982; Nguyen and Lamant, 1989; Koppenaal *et al.*, 1991). Root osmotic adjustment and turgor maintenance may improve root growth during water stress, related in part to enhanced potential for root penetration in drying soils of increased mechanical impedance (Greacen and Oh, 1972; Yau and Sands, 1992). A greater comparative degree of osmotic adjustment of primary root tips of drought-tolerant versus drought-sensitive families of *Pinus pinaster* during exposure to osmotic stress was

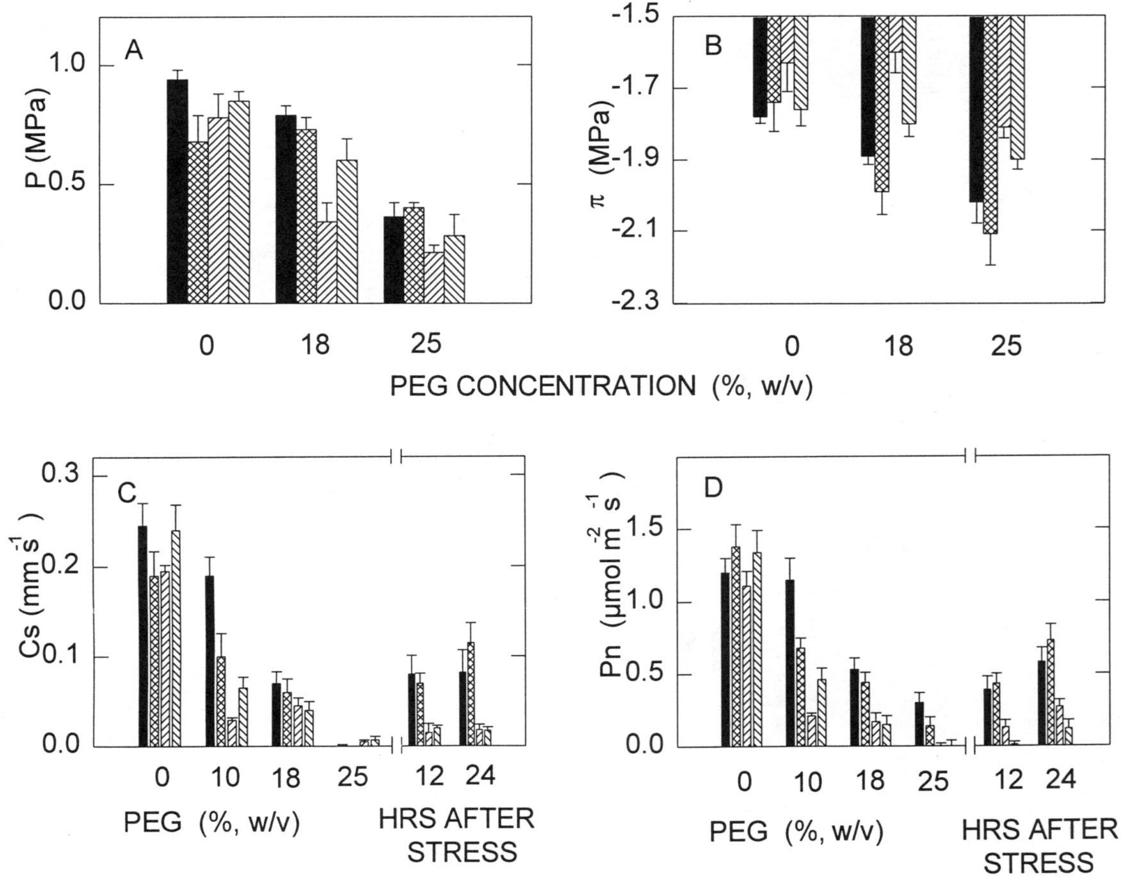
A
P (MPa)
1.0
0.5
0.0
0
18
25
B
π (MPa)
-1.5
-1.7
-1.9
-2.1
-2.3
0
18
25
PEG CONCENTRATION (%, w/v)
C
Cs (mm s-1)
0.3
0.2
0.1
0.0
0
10
18
25
12
24
PEG (%, w/v)
HRS AFTER STRESS
D
Pn (µmol m-2 s-1)
1.5
1.0
0.5
0.0
0
10
18
25
12
24
PEG (%, w/v)
HRS AFTER STRESS

associated with greater root elongation and increased root–shoot ratios (Nguyen and Lamant, 1989).

Altered gene expression and synthesis of stress proteins under drought may play an important role in the dehydration tolerance of conifers (Newton *et al.*, 1991). Substantial improvement in survival of a severe drought by *P. ponderosa* seedlings was correlated with the synthesis of several novel polypeptides (Vance *et al.*, 1990). Differential dehydration tolerance among conifers as judged on the basis of physiological recovery from drought stress and chlorophyll loss also has been reported, but the mechanistic basis for expression of this tolerance in conifer species remains unknown (Buxton *et al.*, 1985; Tan *et al.*, 1992a,b). A heightened potential for stress recovery and dehydration tolerance has been related to rapid reconstitution of cell membranes during rehydration (Bewley, 1979; Kaiser, 1982; Johnson-Flanagan and Singh, 1986).

VI. Conclusions

Water transport dynamics in trees and stands of conifers have certain features that are characteristic of this group and are at least rare among angiosperms. Among these features is the xylem transport system that is dependent on tracheids for long-distance water transport. Tracheid-containing xylem is relatively inefficient, a property that can reduce submaximum allowable rates of gas exchange, but tracheids also offer substantial capacity for water storage and high resistance to freezing-induced dysfunction. Thus, they are quite compatible with the typical evergreen habit and long transpiration season of conifers.

At the stand level, canopy transpiration in conifers is primarily controlled by stomatal conductance. In contrast, in dense canopies of angiosperms, particularly those of tropical forests with limited air mixing, stand transpiration is limited by radiation input rather than by stomatal control. Because of their evergreen habit a greater proportion of evapotranspiration in conifer forests is associated with evaporation of water intercepted by the tree crowns.

Other features of transport dynamics are characteristic of most coni-

---

Figure 20 Effects of osmotic stress on (A) pressure potential (p), (B) osmotic potential (π), (C) stomatal conductance (C_s), and (D) net photosynthetic rate (P_n) of black spruce progenies. Data obtained at 12 or 24 hours after relief from stress are included in C and D. Seedlings were 6 months of age. Vertical bars are SEM ($n = 6$). Progenies 22 (■) and 43 (⊠) were faster growing under drought; 25 (▨) and 46 (▧) were slower growing. From Tan *et al.* (1992a,b).

fers, but are not unique to this group. Among these features are typically shallow root systems that often must supply water in winter to replace transpiration needs of evergreen species, common occurrence of mycorrhizae that enhance mineral and water uptake, and drought tolerance adaptations that include elements of both dehydration avoidance (e.g., stomatal closure under water stress, shifts in allocation of dry matter to belowground sinks) and dehydration tolerance (e.g., capacity for acclimation of photosynthetic apparatus to drought, osmotic adjustment). Transpiration rates from conifer foliage often are lower than those of deciduous angiosperms, probably because of the lower maximum capacity of tracheid-bearing xylem to transport water.

References

Abeles, F. B. (1973). "Etylene in Plant Biology." Academic Press, New York.

Addoms, R. M. (1946). Entrance of water into suberized roots of trees. *Plant Physiol.* 21:109–111.

Aloni, R. (1987). Differentiation of vascular tissues. *Annu. Rev. Plant Physiol.* 38:179–204.

Andersen, C. P., and McLaughlin, S. B. (1991). Seasonal changes in shoot water relations of *Picea rubens* at two high elevation sites in the Smoky Mountains. *Tree Physiol.* 8: 11–21.

Andersen, C. P., Sucoff, E. I., and Dixon, R. K. (1986). Effect of root zone temperature on root initiation and elongation in red pine seedlings. *Can. J. For. Res.* 16:696–700.

Armstrong, W., and Read, D. J. (1972). Some observations on oxygen transport in conifer seedlings. *New Phytol.* 71:55–62.

Armstrong, W., Booth, T. C., Priestly, P., and Read, D. J. (1976). The relationship between soil aeration, stability, and growth of Sitka spruce (*Picea sitchensis* (Bong.) Carr.) on upland peaty gleys. *J. Appl. Ecol.* 13:585–591.

Axelsson, E., and Axelsson, B. (1986). Changes in carbon allocation patterns in spruce and pine trees following irrigation and fertilization. *Tree Physiol.* 2:189–204.

Bachelard, E. P. (1985). Effects of soil moisture stress on the growth of seedlings of three eucalypt species. I. Seed germination. *Aust. J. Plant Physiol.* 15:103–114.

Bailey, I. W. (1958). The structure of tracheids in relation to the movement of liquids, suspensions, and undissolved gasses. *In* "The Physiology of Forest Trees" (K. V. Thimann, ed.), pp. 71–82. Ronald, New York.

Balek, J., Cermak, J., Kucera, J., Prax, A., and Palous, M. (1985). Regional transpiration assessment by remote sensing. *In* "Hydrological Application of Space Technology," *Proceedings of the Cocoa Beach Workshop*, Florida, August 1985. IAHS Publ. No. 160, pp. 141–148. IAHS Press, Wallingford, UK.

Bannan, M. W. (1941). Variability in wood structure in roots of native Ontario conifers. *Bull. Torrey Bot. Club* 68:173–194.

Baron, F. J. (1978). Moisture and temperature in relation to seed structure and germination of sugar pine (*Pinus lambertiana* Dougl.). *Am. J. Bot.* 65:804–810.

Barton, A. M. (1993). Factors controlling plant distributions: Drought, competition, and fire in montane pines in Arizona. *Ecol. Monogr.* 63:367–397.

Barton, A. M., and Teeri, J. A. (1993). The ecology of elevational positions in plants:

Drought resistance in five montane pine species in southeastern Arizona. *Am. J. Bot.* 80:15–25.

Bartsch, N. (1987). Responses of root systems of young *Pinus sylvestris* and *Picea abies* plants to water deficits and soil acidity. *Can. J. For. Res.* 17:805–812.

Bassett, J. R. (1964). Tree growth as affected by soil moisture availability. *Soil Sci. Soc. Am. J.* 28:436–438.

Bauch, J., Liese, W., and Schultze, R. (1972). The morphological variability of the bordered pit membranes in gymnosperms. *Wood Sci. Technol.* 6:165–184.

Beadle, C. L., Neilsen, R. E., Jarvis, P. G., and Talbot, H. (1981). Photosynthesis as related to xylem water potential and carbon dioxide concentration in Sitka spruce. *Physiol. Plant.* 52:391–400.

Beadle, C. L., Neilson, R. E., Talbot, H., and Jarvis, P. G. (1985a). Stomatal conductance and photosynthesis in a mature Scots pine forest. I. Diurnal, seasonal and spatial variation in shoots. *J. Appl. Ecol.* 22:557–571.

Beadle, C. L., Jarvis, P. G., Talbot, H., and Neilson, R. E. (1985b). Stomatal conductance and photosynthesis in a mature Scots pine forest. II. Dependence on environmental variables of single shoots. *J. Appl. Ecol.* 22:573–586.

Becker, C. A., Mroz, G. D., and Fuller, L. G. (1987). The effects of plant moisture stress on red pine (*Pinus resinosa*) seedling growth and establishment. *Can. J. For. Res.* 17: 813–820.

Bennett, K. J., and Rook, D. A. (1978). Stomatal and mesophyll resistances in two clones of *Pinus radiata* D. Don. known to differ in transpiration and survival rate. *Aust. J. Plant Physiol.* 5:231–238.

Benzing, D. H. (1990). "Vascular Epiphytes." Cambridge Univ. Press, Cambridge.

Bewley, J. D. (1979). Physiological aspects of desiccation tolerance. *Annu. Rev. Plant Physiol.* 30:195–238.

Bi, H. Q., Turvey, N. D., and Heinrich, P. (1992). Rooting density and tree size of *Pinus radiata* (D. Don) in response to competition from *Eucalyptus obliqua* (Lherit). *For. Ecol. Manage.* 49:31–42.

Bilan, M. V., Hogan, C. T., and Carter, H. B. (1977). Stomatal opening, transpiration, and needle moisture in loblolly pine seedlings from two Texas seed sources. *For. Sci.* 23:457–462.

Black, R. A., and Bliss, L. C. (1980). Reproductive ecology of *Picea mariana* (Mill.) BSP. at tree line near Inuvik, Northwest Territories, Canada. *Ecol. Monogr.* 50:331–354.

Blake, J., and Ferrell, W. K. (1977). The association between soil and xylem water potential, leaf resistance, and abscisic acid content in droughted seedlings of Douglas-fir (*Pseudotsuga menziesii*). *Physiol. Plant.* 39:106–109.

Blake, T. J., Bevilacqua, E., and Zwiazek, J. J. (1991). Effects of repeated stress on turgor pressure and cell elasticity changes in black spruce seedlings. *Can. J. For. Res.* 21: 1329–1333.

Bongarten, B. C., and Teskey, R. O. (1987). Dry weight partitioning and its relationship to productivity in loblolly pine seedlings from seven sources. *For. Sci.* 33:255–267.

Bonnet-Masimbert, M., and Zaerr, J. B. (1987). The role of plant growth regulators in promotion of flowering. *Plant Growth Regul.* 6:13–35.

Borghetti, M., Edwards, W. R. N., Grace, J., Jarvis, P. G., and Raschi, A. (1991). The refilling of embolized xylem in *Pinus sylvestris* L. *Plant. Cell Environ.* 14:357–369.

Bowen, G. D., and Nambiar, E. K. S. (1984). "Nutrition of Plantation Forests." Academic Press, New York.

Boyd, R., Furbank, R. T., and Read, D. J. (1986). Ectomycorrhiza and the water relations of trees. Physiological and genetical aspects of mycorrhizae: Proceedings of the First European Symposium on Mycorrhizae, Dijon, July 1–5, 1985, pp. 689–693.

Boyer, J. S. (1976). Water deficits and photosynthesis. *In* "Water Deficits and Plant Growth" (T. T. Kozlowski, ed.), Vol. 4, pp. 153–190. Academic Press, New York.

Boyer, J. S. (1985). Water transport. *Annu. Rev. Plant Physiol.* 36:473–516.

Boyer, J. S. (1989). Water potential and plant metabolism: Comments on Dr. P. J. Kramer's article 'Changing concepts regarding plant water relations', volume 11, number 7, pp. 565–568, and Dr. J. B. Passioura's response, pp. 569–571. *Plant Cell Environ.* 12:213—

Boyle, C. D., and Hellenbrand, K. E. (1991). Assessment of the effect of mycorrhizal fungi on drought tolerance of conifer seedlings. *Can. J. Bot.* 69:1764–1771.

Bradbury, I. K., and Malcolm, D. C. (1977). The effect of phosphorus and potassium on transpiration, leaf diffusive resistance and water-use efficiency in Sitka spruce (*Picea sitchensis*) seedlings. *J. Appl. Ecol.* 14:631–641.

Bradford, K. J., and Hsiao, T. C. (1982). Stomatal behavior and water relations of waterlogged tomato plants. *Plant Physiol.* 70:1508–1513.

Braekke, F. H., Kozlowski, T. T., and Skroppa, T. (1978). Effects of environmental factors on estimated daily radial growth of *Pinus resinosa* and *Betula papyrifera*. *Plant Soil* 49:491–504.

Bray, E. A. (1991). Regulation of gene expression by endogenous ABA during drought stress. *In* "Abscisic Acid: Physiology and Biochemistry" (W. J. Davies and H. G. Jones, eds.), pp. 81–98. BIOS Scientific Publ., Oxford.

Brissette, J. C., and Barnett, J. P. (1988). Depth of planting and J-rooting affect loblolly pine seedlings under stress conditions. Proceedings of the Southern Silviculture Research Conference, pp. 169–175.

Brix, H. (1979). Effects of plant water stress on photosynthesis and survival of four conifers. *Can. J. For. Res.* 9:160–165.

Brix, H., and Mitchell, A. K. (1986). Thinning and nitrogen fertilization effects on soil and tree water stress in a Douglas-fir stand. *Can. J. For. Res.* 16:1334–1338.

Brix, H., and Portlock, F. T. (1982). Flowering response of western hemlock seedlings to gibberellin and water stress treatments. *Can. J. For. Res.* 12:76–82.

Büsgen, M. (1901). Einiges uber Gestalt und Wachstumsweise der Baumwurzeln. *Allg. Forst. Jagdztg.* 27:273–278.

Buxton, G. F., Cyr, D. R., Dumbroff, E. B., and Webb, D. P. (1985). Physiological responses of three northern conifers to rapid and slow induction of moisture stress. *Can. J. Bot.* 63:1171–1176.

Calkin, H. W., Gibson, A. C., and Nobel, P. S. (1986). Biophysical model of xylem conductance in tracheids of the fern *Pteris vittata*. *J. Exp. Bot.* 37:1054–1064.

Cannell, M. G. R. (1989). Physiological basis of wood production: A review. *Scand. J. For. Res.* 4:459–490.

Cannell, M. G. R., and Willett, S. C. (1976). Shoot growth phenology, dry matter distribution, and root:shoot ratios of provenances of *Populus trichocarpa*, *Picea sitchensis*, and *Pinus contorta* growing in Scotland. *Silvae Genet.* 25:49–59.

Cannell, M. G. R., Bridgwater, F. E., and Greenwood, M. S. (1978). Seedling growth rates, water stress responses and root–shoot relationships related to eight-year volumes among families of *Pinus taeda* L. *Silvae Genet.* 27:237–248.

Carlson, W. C., and Miller, D. E. (1990). Target seedling root system size, hydraulic conductivity, and water use during seedling establishment. *U.S.D.A. Forest Service, Rocky Mountain Forest Exp. Station, Gen. Tech. Rep.* RM-200, pp. 53–65.

Carter, G. A., and Smith, W. K. (1988). Microhabitat comparisons of transpiration and photosynthesis in three subalpine conifers. *Can. J. Bot.* 66:963–969.

Cermak, J. (1977). The course of transpiration flow rate in fully grown trees from viewpoint of different time periods (in Czech). *In* "Fotosynteza a Vodny Provoz Drevin"

(J. Huzulak and E. Masarovicova, eds.), Proc. Celostatny dendrofyziologicky seminar UEBE-SAV, pp. 47–54. Modra-Piesky, Czechoslovakia.

Cermak, J. (1986). Short- and long-term response of transpiration flow rate in full-grown trees to water stress. Proceedings of the 18th IUFRO World Congress Whole-Plant Physiology Working Party, pp. 187–193. Ljubljana, Yugoslavia.

Cermak, J. (1989). Solar equivalent leaf area as the efficient biometrical parameter of individual leaves, trees and stands. *Tree Physiol.* 5:269–289.

Cermak, J. (1992). Transpiration of trees and its importance in forest ecology (in Czech.). Assoc. Prof. Thesis, Agric. Univ. Brno, Czechoslovakia.

Cermak, J., and Kucera, J. (1987). Transpiration of fully grown trees and stands of spruce (*Picea abies* (L.) Karst) estimated by the tree-trunk heat balance method. Proceedings of the Forest Hydrology and Watershed Management Symposium, Vancouver, Canada. IAHS Publ. No. 167, pp. 311–317. IAHS Press, Wallingford, UK.

Cermak, J., and Kucera, J. (1990a). Water uptake in healthy and ill trees, under drought and hypoxia and non-invasive assessment of the effective size of root systems. *In* "Above and Belowground Interactions in Forest Trees in Acidified Soils." (H. Persson, ed.), Proc. COST 612 Workshop, pp. 185–195. Simlangsdalen, Sweden.

Cermak, J., and Kucera, J. (1990b). Changes in transpiration of healthy mature trees due to environmental conditions and of those with damaged water conductive system. *In* "Verification of Hypotheses and Possibilities of Recovery of Forest Ecosystems," (E. Klimo and J. Materna, eds.), Proc. Cs. MAB Natl. Comm./IUFRO International Workshop, Agr. Univ. Brno. pp. 275–286. Beskydy Mt., Czechoslovakia.

Cermak, J., and Kucera, J. (1990c). Scaling up transpiration data between trees, stands and watersheds. *Silva Carelica* 15:101–120.

Cermak, J., and Kucera, J. (1991). Extremely fast changes of xylem water flow rate in mature trees, caused by atmospheric, soil and mechanical factors. *In* "Methodologies to Assess the Impacts of Climatic Changes on Vegetation: Analysis of Water Transport in Plants and Cavitation of Xylem Transport in Plants and Cavitation of Xylem Conduits" (A. Raschi and M. Borghetti eds.), pp. 181–190, Proc. CEC International Workshop, May 29–31, 1991. Firenze, Italy. Cambridge Univ. Press, Cambridge, UK.

Cermak, J., Huzulak, J., and Penka, M. (1980). Water potential and sap flow rate in adult trees with moist and dry soil as used for assessment of the root system depth. *Biol. Plant.* 22:34–41.

Cermak, J., Ulehla, J., Kucera, J., and Penka, M. (1982). Sap flow rate and transpiration dynamics in the full-grown oak (*Quercus robur* L.) in floodplain forest exposed to seasonal floods as related to potential evapotranspiration and tree dimensions. *Biol. Plant.* 24:446–460.

Cermak, J., Kucera, J., Stepankova, M., and Bednarova, E. (1983). Measurement and evaluation of transpiration of trees at the water balance site in Nedamov, central Bohemia (in Czech.). *Inst. For. Ecol. (Brno) Res. Rep.*

Cermak, J., Jenik, J., Kucera, J., and Zidek, V. (1984a). Xylem water flow in a crack willow tree (*Salix fragilis* L.) in relation to diurnal changes of environment. *Oecologia* 64:145–151.

Cermak, J., Kucera, J., Bednarova, E., and Stepankova, M. (1984b). Measurement and evaluation of transpiration of trees at the water balance site in Nedamov (in Czech.). Res. Rep. Inst. of Forest Ecology, Brno (for the Construction Geology Enterprise, Prague).

Cermak, J., Kucera, J., and Bednarova, E. (1987). Estimation of stand water consumption on the basis of measurement of transpiration flow rate in selected fully grown spruce trees in relation to environmental conditions (in Czech.). *Res. Rep. Inst. For. Ecol., Brno* (for the Water Management Research Institute, Prague).

Cermak, J., Cienciala, E., Kucera, J., Lindroth, A., and Hallgren, J. E. (1992). Radial velocity profiles of water flow in stems of spruce and oak and response of spruce tree to severing. *Tree Physiol.* 10:367–380.

Cermak, J., Matyssek, R., and Kucera, J. (1993). Rapid response of large, drought stressed beech trees to irrigation. *Tree Physiol.* 12:281–290.

Chapin, F. S., Bloom, A. J., Field, C. B., and Waring, R. H. (1987). Plant responses to multiple environmental factors. *BioScience* 37:49–57.

Chaves, M. M. (1991). Effects of water deficits on carbon assimilation. *J. Exp. Bot.* 42:1–16.

Chaves, M. M., and Pereira, J. S. (1992). Water stress, CO_2 and climate change. *J. Exp. Bot.* 43:1131–1139.

Chevone, B. J., Seiler, J. R., Melkionian, J., and Amundson, R. G. (1990). Ozone–water stress interactions. *In* "Stress Responses in Plants: Adaptation and Acclimation Mechanisms" (R. G. Alscher and J. R. Cumming, eds.), pp. 311–328. Wiley, New York.

Chiariello, N. R., and Gulmon, S. L. (1991). Stress effects on plant reproduction. *In* "Response of Plants to Multiple Stresses" (H. A. Mooney, W. E. Winner, and E. J. Pell, eds.), pp. 161–188. Academic Press, New York.

Chung, H. H., and Kramer, P. J. (1975). Absorption of water and ^{32}P through suberized and unsuberized roots of loblolly pine. *Can. J. For. Res.*, 5:229–235.

Clements, J. R. (1970). Shoot responses of young red pine to watering applied over two seasons. *Can. J. Bot.* 48:75–80.

Cochard, H. J. (1992). Vulnerability of several conifers to air embolism. *Tree Physiol.* 11:73–83.

Cochard, H., and Tyree, M. T. (1990). Xylem dysfunction in *Quercus*: Vessel sizes, tyloses, cavitation and seasonal changes in embolism. *Tree Physiol.* 6:393–407.

Cohen, D. (1970). The expected efficiency of water utilization in plants under different competition and selection regimes. *Isr. J. Bot.* 19:50–54.

Cohen, Y., Kelliher, F. M., and Black, T. A. (1985). Determination of sap flow in Douglas-fir trees using the heat pulse technique. *Can. J. For. Res.* 15:422–428.

Cole, P. J., and Alston, A. M. (1974). Effect of transient dehydration on absorption of chloride by wheat roots. *Plant Soil* 40:243–247.

Coleman, M. D., Bledsoe, C. S., and Smit, B. A. (1990). Root hydraulic conductivity and xylem sap levels of zeatin riboside and abscisic acid in ectomycorrhizal Douglas fir seedlings. *New Phytol.* 115:275–284.

Comeau, P. G., and Kimmins, J. P. (1989). Above- and belowground biomass and production of lodgepole pine on sites with differing soil moisture regimes. *Can. J. For. Res.* 19:447–454.

Conroy, J. P., Kuppers, M., Kuppers, B., Virgona, J., and Barlow, E. W. R. (1988). The influence of CO_2 enrichment, phosphorus deficiency and water stress on the growth, conductance and water use of *Pinus radiata* D. Don. *Plant, Cell Environ.* 11:91–98.

Cornic, G., Le Goullec, J.-L., Briantais, J. M., and Hodges, M. (1989). Effect of dehydration and high light on photosynthesis of two C_3 plants (*Phaseolus vulgaris* L. and *Elatostema repens* (Lour.) Hall f.). *Planta* 177:84–90.

Coutts, M. P. (1980). Control of water loss by actively growing Sitka spruce seedlings after transplanting. *J. Exp. Bot.* 31:1587–1597.

Coutts, M. P. (1981). Leaf water potential and control of water loss in droughted Sitka spruce seedlings. *J. Exp. Bot.* 32:1193–1201.

Coutts, M. P. (1982). Growth of Sitka spruce seedlings with roots divided between soils of unequal matric potential. *New Phytol.* 92:49–61.

Coutts, M. P., and Philipson, J. J. (1978a). Tolerance of tree roots to waterlogging. I. Survival of Sitka spruce and lodgepole pine. *New Phytol.* 80:63–69.

Coutts, M. P., and Philipson, J. J. (1978b). Tolerance of tree roots to waterlogging.

II. Adaptation of Sitka spruce and lodgepole pine to waterlogged soil. *New Phytol.* 80:71–77.

Cowan, I. R. (1977). Stomatal behaviour and environment. *Adv. Bot. Res.* 4:117–228.

Cowan, I. R. (1982). Regulation of water use in relation to carbon gain. *In* "Encyclopedia of Plant Physiology, New Series, Volume 12B Physiological Plant Ecology II. Water Relations and Carbon Assimilation" (O. L. Lange, P. S. Nobel, C. B. Osmond, and H. Ziegler, eds.), pp. 589–613. Springer-Verlag, Berlin.

Cowan, I. R., and Farquhar, G. D. (1977). Stomatal function in relation to leaf metabolism. *Symp. Soc. Exp. Biol.* 31:471–505.

Crisosto, C. H., Grantz, D. A., and Meinzer, F. C. (1992). Effects of water deficit on flower opening in coffee (*Coffea arabica* L.). *Tree Physiol.* 10:127–139.

Crombie, D. S., Hipkins, M. F., and Milburn, J. A. (1985). Gas penetration of pit membranes in the xylem of *Rhododendron* as the cause of acoustically detectable sap cavitation. *Aust. J. Plant Physiol.* 12:445–453.

Crowe, J. H., Crowe, L. M., Carpenter, J. F., Rudolph, A. S., Aurell Wistrom, C., Sprago, B. J., and Anchordoguy, T. J. (1988). Interactions of sugars with membranes. *Biochim. Biophys. Acta* 947:367–384.

Cui, M., and Smith, W. K. (1991). Photosynthesis, water relations, and mortality in *Abies lasiocarpa* seedlings during natural establishment. *Tree Physiol.* 8:37–46.

Cyr, D. R., Buxton, G. F., Webb, D. P., and Dumbroff, E. B. (1990). Accumulation of free amino acids in the shoots and roots of three northern conifers during drought. *Tree Physiol.* 6:293–303.

Dale, J. E. (1988). The control of leaf expansion. *Annu. Rev. Plant Physiol. Plant Mol. Biol.* 39:267–295.

Davies, W. J., and Zhang, J. (1991). Root signals and the regulation of growth and development of plants in drying soil. *Annu. Rev. Plant Physiol. Plant Mol. Biol.* 42:55–76.

Dawson, T. E., and Bliss, L. C. (1989). Patterns of water use and the tissue water relations in the dioecious shrub, *Salix artica*: The physiological basis for habitat partitioning between the sexes. *Oecologia* 79:332–343.

Day, T. A., Heckathorn, S. A., and DeLucia, E. H. (1991). Limitations of photosynthesis in *Pinus taeda* L. (loblolly pine) at low soil temperatures. *Plant Physiol.* 96:1246–1254.

DeLucia, E. H., Day, T. A., and Oquist, G. (1991). The potential for photoinhibition of *Pinus sylvestris* L. seedlings exposed to high light and low soil temperature. *J. Exp. Bot.* 42:611–617.

Downton, W. J. S., Loveys, B. R., and Grant, W. J. R. (1988). Non-uniform stomatal closure induced by water stress causes putative non-stomatal inhibition of photosynthesis. *New Phytol.* 110:503–509.

Drew, A. P. (1982). Shoot-root plasticity and episodic growth in red pine seedlings. *Ann. Bot.* 49:347–357.

Duddridge, J. A., Malibari, A. M., and Read, D. J. (1980). Structure and function of mycorrhizal rhizomorphs with special reference to their role in water transport. *Nature (London)* 287:834–836.

Eamus, D., Leith, I., and Fowler, D. (1989). Water relations of red spruce seedlings treated with acid mist. *Tree Physiol.* 5:387–397.

Eickmeier, W., Adams, M., and Lester, D. (1975). Two physiological races of *Tsuga canadensis* in Wisconsin. *Can. J. Bot.* 53:940–951.

Ellmore, G. S., and Ewers, F. W. (1986). Fluid flow in the outermost xylem increment of a ring-porous tree, *Ulmus americana*. *Am. J. Bot.* 73:1771–1774.

Esau, K. (1965). "Plant Anatomy," 2nd Ed. Wiley, New York.

Evans, C. E., and Etherington, J. R. (1990). The effect of soil water potential on seed germination of some British plants. *New Phytol.* 115:539–548.

Ewers, F. W. (1985). Xylem structure and water conduction in conifer trees, dicot trees, and lianas. *Int. Assoc. Wood Anat. Bull. N. S.* 6:309–317.

Ewers, F. W., and Cruziat, P. (1991). Measuring water transport and storage. *In* "Techniques and Approaches in Forest Tree Ecophysiology" (J. P. Lassoie and T. M. Hinckley, eds.), pp. 91–115. CRC Press, Boca Raton, Florida.

Ewers, F. W., and Zimmermann, M. H. (1984a). The hydraulic architecture of balsam fir (*Abies balsamea*). *Physiol. Plant.* 60:453–458.

Ewers, F. W., and Zimmermann, M. H. (1984b). The hydraulic architecture of eastern hemlock (*Tsuga canadensis*). *Can. J. Bot.* 62:940–946.

Farrish, K. W. (1991). Spatial and temporal fine-root distribution in three Louisiana forest soils. *Soil Sci. Soc. Am. J.* 55:1752–1757.

Fayle, D. C. F. (1968). Radial growth in tree roots. *Univ. Toronto Tech. Rep.* 9.

Feil, W., Kottke, I., and Oberwinkler, O. (1988). The effect of drought on mycorrhizal production and very fine root system development of Norway spruce under natural and experimental conditions. *Plant Soil* 108:221–231.

Fisher, H. M., and Stone, E. L. (1990a). Air-conducting porosity in slash pine roots from saturated soils. *For. Sci.* 36:18–33.

Fisher, H. M., and Stone, E. L. (1990b). Active potassium uptake by slash pine roots from O_2-depleted solutions. *For. Sci.* 36:582–598.

Fites, J. A., and Teskey, R. O. (1988). CO_2 and water vapor exchange of *Pinus taeda* in relation to stomatal behavior: Test of an optimization hypothesis. *Can. J. For. Res.* 18: 150–157.

Flower, D. J., Usha Rani, A., and Peacock, J. M. (1990). Influence of osmotic adjustment on the growth, stomatal conductance and light interception of contrasting sorghum lines in a harsh environment. *Aust. J. Plant. Physiol.* 17:91–105.

Floyd, M. E. (1983). Dioecy in five *Pinus edulis* populations in the southwestern United States. *Am. Midl. Nat.* 110:405–411.

Ford, E. D., and Deans, J. D. (1978). The effects of canopy structure on stemflow, throughfall and interception loss in a young Sitka spruce plantation. *J. Appl. Ecol.* 15:905–917.

Fraser, A. I., and Gardiner, J. B. H. (1967). Rooting and stability in sitka spruce. Bulletin, Forestry Commission of the United Kingdom No. 40.

Freeman, D. C., and McArthur, E. D. (1982). A comparison of twig water stress between males and females of six species of desert shrubs. *For. Sci.* 28:304–308.

Freeman, D. C., Klikoff, L. G., and Harper, K. T. (1976). Differential resource utilization by the sexes of dioecious plants. *Science* 193:597–599.

Freeman, D. C., McArthur, E. D., Harper, K. T., and Blauer, A. C. (1981). Influence of environments on the floral sex ratio of monoecious plants. *Evolution* 35:194–197.

Froehlich, H. A., Miles, D. W. R., and Robbins, R. W. (1986). Growth of young *Pinus ponderosa* and *Pinus contorta* on compacted soil in central Washington. *For. Ecol. Manage.* 15:285–291.

Gholz, H. L. (1982). Environmental limits on aboveground net primary production, leaf area, and biomass in vegetation zones of the Pacific northwest. *Ecology* 63:469–481.

Gholz, H. L., Hendry, L. C., and Cropper, W. P., Jr. (1986). Organic matter dynamics of fine roots in plantations of slash pine (*Pinus elliottii*) in north Florida. *Can. J. For. Res.* 16:529–538.

Gibson, A. C., and Nobel, P. S. (1986). "The Cactus Primer." Harvard Univ. Press, Cambridge, Massachusetts.

Ginter-Whitehouse, D. L., Hinckley, T. M., and Pallardy, S. G. (1983). Spatial and temporal aspects of water relations of three tree species with different vascular anatomy. *For. Sci.* 29:317–329.

Goldstein, G. H., Brubaker, L. B., and Hinckley, T. M. (1985). Water relations of white spruce (*Picea glauca* (Moench) Voss) at tree line in north central Alaska. *Can. J. For. Res.* 15:1080–1087.

Gower, S. T., Vogt, K. A., and Grier, C. C. (1992). Carbon dynamics of Rocky Mountain Douglas-fir: Influence of water and nutrient availability. *Ecol. Monogr.* 62:43–65.

Graham, J. S., and Running, S. W. (1984). Relative control of air temperature and water status on seasonal transpiration of *Pinus contorta*. *Can. J. For. Res.* 14:833–838.

Granier, A. (1987). Evaluation of transpiration in a Douglas-fir stand by means of sap flow measurements. *Tree Physiol.* 3:309–320.

Greacen, E. L., and Oh, J. S. (1972). Physics of root growth. *Nature (London)* 235:24–25.

Grier, C. C., and Running, S. W. (1977). Leaf area of mature northwestern coniferous forests: relation to site water balance. *Ecology* 58:893–899.

Grier, C. C., Vogt, K. A., Keyes, M. R., and Edmonds, R. L. (1981). Biomass distribution and above- and belowground production in young and mature *Abies amabilis* zone ecosystems of the Washington Cascades. *Can. J. For. Res.* 11:155–167.

Grieu, P., Guehl, J. M., and Aussenac, G. (1988). The effects of soil and atmospheric drought on photosynthesis and stomatal control of gas exchange in three coniferous species. *Physiol. Plant.* 73:97–104.

Griffin, A. R., and Ching, K. K. (1977). Geographic variation in Douglas-fir from the coastal ranges of California. *Silvae Genet.* 26:149–157.

Groom, P. (1910). Remarks on the ecology of the Coniferae. *Ann. Bot.* 24:241–269.

Gross, K. (1989). Effects of long-term water stress on net photosynthesis, growth, and water-use efficiency of conifers in the field. *Ann. Sci. For.* 46:401s–415s.

Grossnickle, S. C. (1988). Planting stress in newly planted jack pine and white spruce. 1. Factors influencing water uptake. *Tree Physiol.* 4:71–83.

Grossnickle, S. C., and Blake, T. J. (1987a). Water relations and morphological development of bare-root jack pine and white spruce seedlings: Seedling establishment on a boreal cut-over site. *For. Ecol. Manage.* 18:299–318.

Grossnickle, S. C., and Blake, T. J. (1987b). Water relation patterns of bare-root and container jack pine and black spruce seedlings planted on boreal cut-over sites. *New For.* 1:101–116.

Grossnickle, S. C., and Russell, J. H. (1990). Water movement in yellow-cedar seedlings and rooted cuttings: Comparison of whole plant and root system pressurization methods. *Tree Physiol.* 6:57–68.

Grossnickle, S. C., and Russell, J. H. (1991). Gas exchange processes of yellow-cedar (*Chamaecyparis nootkatensis*) in response to environmental variables. *Can. J. Bot.* 69:2684–2691.

Guehl, J. M., and Aussenac, G. (1987). Photosynthesis decrease and stomatal control of gas exchange in *Abies alba* Mill. in response to vapor pressure difference. *Plant Physiol.* 83:316–322.

Guehl, J. M., Aussenac, G., Bouachrine, J., Zimmermann, R., Pennes, J. M., Ferhi, A., and Grieu, P. (1991). Sensitivity of leaf gas exchange to atmospheric drought, soil drought, and water-use efficiency in some Mediterranean *Abies* species. *Can. J. For. Res.* 21:1507–1515.

Gunasekera, D., and Berkowitz, G. A. (1992). Heterogenous stomatal closure in response to water deficits is not a universal phenomenon. *Plant Physiol.* 98:660–665.

Gunderson, C. A., and Taylor, G. E. (1988). Physiological site of ethylene effects on carbon dioxide assimilation in *Glycine max* (L.) Merr. *Plant Physiol.* 86:85–92.

Habeck, J. R. (1958). White cedar ecotypes in Wisconsin. *Ecology* 39:457–463.

Hadley, J. L., and Smith, W. K. (1986). Wind effects on needles of timberline conifers: Seasonal influence on mortality. *Ecology* 67:12–19.

Hadley, J. L., and Smith, W. K. (1990). Influence of leaf surface wax and leaf area to water content ratio on cuticular transpiration in western conifers, U.S.A. *Can. J. For. Res.* 20: 1306–1311.

Hallgren, S. W., and Helms, J. A. (1988). Control of height growth components in seedlings of California red fir and white fir by seed source and water stress. *Can. J. For. Res.* 18:521–529.

Hallgren, S. W., and Helms, J. A. (1992). The effects of summer shoot production on height growth components of seedlings of California red fir and white fir. *Can. J. For. Res.* 22:690–698.

Hallgren, S. W., Tauer, C. G., and Lock, J. E. (1991). Fine root carbohydrate dynamics of loblolly pine seedlings grown under contrasting levels of soil moisture. *For. Sci.* 37: 766–780.

Hammel, H. T. (1967). Freezing of xylem sap without cavitation. *Plant Physiol.* 42:55–66.

Hanson, A. D., and Hitz, W. D. (1982). Metabolic responses of mesophytes to plant water deficits. *Annu. Rev. Plant Physiol.* 33:163–203.

Hanson, P. J., Sucoff, E. I., and Markhart, A. H. (1985). Quantifying apoplastic flux through red pine root systems using trisodium, 3-hydroxy-5,8,10-pyrenetrisulfonate. *Plant Physiol.* 77:21–24.

Haussling, M., Jorns, C. A., Lehmbecker, G., Hecht-Bucholz, C., and Marschner, H. (1988). *J. Plant Physiol.* 133:486–491.

Heilman, P. (1981). Root penetration of Douglas-fir seedlings into compacted soil. *For. Sci.* 27:660–666.

Hennessey, T. C., Dougherty, P. M., Cregg, B. M., and Wittwer, R. F. (1992). Annual variation in needle fall of a loblolly pine stand in relation to climate and stand density. *For. Ecol. Manage.* 51:329–338.

Herrick, G. T., and Friedland, A. J. (1991). Winter desiccation and injury of subalpine red spruce. *Tree Physiol.* 8:23–36.

Heth, D., and Kramer, P. J. (1975). Drought tolerance of pine seedlings under various climatic conditions. *For. Sci.* 21:72–82.

Higgins, S. S., Black, R. A., Radamaker, R. A., and Bidlake, W. R. (1987). Gas exchange characteristics and water relations of *Larix occidentalis*. *Can. J. For. Res.* 17:1364–1370.

Hillel, D. (1972). Soil moisture and seed germination. *In* "Water Deficits and Plant Growth" (T. T. Kozlowski, ed.), Vol. 3, pp. 65–89. Academic Press, New York.

Hinckley, T. M., and Ceulemans, R. (1989). Current focuses in woody plant water relations and drought resistance. *Ann. Sci. For.* 46:317s–324s.

Hinckley, T. M., Lassoie, J. P., and Running, S. W. (1978). Temporal and spatial variations in the water status of forest trees. *For. Sci. Monogr.* 20.

Hoffman, G. (1974). Beziehungen zwischen Bodenfeuchte und dem Wurzelwachstum von Forstgeholzen. *Beitr. Forstwirtsch.* 8:94–96.

Hollinger, D. Y., Kelliher, F. M., and Schulze, E.-D. (1992). Turbulent coupling of tree transpiration to the atmospheric boundary layer. *Bull. Ecol. Soc. Am. Suppl.* 73:210.

Hook, D. D., and Brown, C. L. (1972). Permeability of the cambium to air in trees adapted to wet habitats. *Bot. Gaz.* 133:304–310.

Houssard, C., Escarre, J., and Vartanian, N. (1992). Water stress effects on successional populations of the dioecious herb, *Rumex acetosella* L. *New Phytol.* 120:551–559.

Hsiao, T. C. (1973). Plant response to water stress. *Annu. Rev. Plant Physiol.* 24:519–570.

Huck, M. G., Klepper, B., and Taylor, H. M. (1970). Diurnal variations in root diameter. *Plant Physiol.* 45:529–530.

Hulbert, C., Funkhouser, E. J., Soltes, E. J., and Newton, R. J. (1988). Inhibition of protein synthesis in loblolly pine hypocotyls by mannitol-induced water stress. *Tree Physiol.* 4:19–26.

Jackson, P. A., and Spomer, G. G. (1979). Biophysical adaptations of four western conifers to habitat water conditions. *Bot. Gaz.* 140:428–432.

Jarvis, P. G. (1975). Water transfer in plants. *In* "Heat and Mass Transfer in the Biosphere in the Environment of Vegetation" (D. A. De Vries and N. H. Afgan, eds.), pp. 369–394. Scripta, Washington, D.C.

Johnson, J. D., and Ferrell, W. K. (1982). The relationship of abscisic acid metabolism to stomatal conductance in Douglas-fir during water stress. *Physiol. Plant.* 55:431–437.

Johnson-Flanagan, A. M., and Singh, J. (1986). Membrane deletion during plasmolysis in hardened and non-hardened plant cells. *Plant, Cell Environ.* 9:299–305.

Joly, R. J., and Zaerr, J. B. (1987). Alteration of cell-wall water content and elasticity in Douglas-fir during periods of water deficit. *Plant Physiol.* 83:418–422.

Joly, R. J., Adams, W. T., and Stafford, S. G. (1989). Phenological and morphological responses of mesic and dry site sources of coastal Douglas-fir to water deficit. *For. Sci.* 35:987–1005.

Kaiser, W. M. (1982). Correlation between changes in photosynthetic activity and changes in total protoplast volume in leaf tissue form hygro-, meso-, and xerophytes. *Planta* 154:538–545.

Kaiser, W. M. (1987). Effects of water deficit on photosynthetic capacity. *Physiol. Plant.* 71:142–149.

Kandiko, R. A., Timmis, R., and Worrall, J. (1980). Pressure-volume curves of shoots and roots of normal and drought conditioned western hemlock seedlings. *Can. J. For. Res.* 10:10–16.

Kaufmann, M. R. (1968). Water relations of pine seedlings in relation to root and shoot growth. *Plant Physiol.* 43:281–288.

Kaufmann, M. R. (1975). Leaf water stress in Engelmann spruce. Influence of the root and shoot environments. *Plant Physiol.* 56:841–844.

Kaufmann, M. R. (1976). Stomatal response of Engelmann spruce to humidity, light, and water stress. *Plant Physiol.* 57:898–901.

Kaufmann, M. R. (1977a). Soil temperature and drying cycle effects on water relations of *Pinus radiata*. *Can. J. Bot.* 56:2413–2418.

Kaufmann, M. R. (1977b). Soil temperature and drought effects on growth of Monterey pine. *For. Sci.* 23:317–325.

Kaufmann, M. R. (1984). A canopy model (RM-CWU) for determining transpiration of subalpine forests. II. Consumptive water use in two watersheds. *Can. J. For. Res.* 14: 227–232.

Kaufmann, M. R. (1985). Annual transpiration in subalpine forests: Large differences among four tree species. *For. Ecol. Manage.* 13:235–246.

Kaufmann, M. R., and Fiscus, E. L. (1985). Water transport through plants—Internal integration of edaphic and atmospheric effects. *Acta Hortic.* 171:83–93.

Kaufmann, M. R., and Kelliher, F. M. (1991). Measuring transpiration rates. *In* "Techniques and Approaches in Forest Tree Ecophysiology" (J. P. Lassoie and T. M. Hinckley, eds.), pp. 117–140. CRC Press, Boca Raton, Florida.

Kaufmann, M. R., Troendle, C. A., Ryan, M. G., and Mowrer, H. T. (1987). Trees—The link between silviculture and hydrology. *In* "Management of Subalpine Forests: Building on 50 Years of Research" (C. A. Troendle, M. R. Kaufmann, R. H. Hamre, and R. P. Winokur, tech. coordinators), pp. 54–60, Gen. Tech. Rep. RM-149. U.S.D.A. Forest Service, Rocky Mtn. For. Range Expt. Sta., Fort Collins, CO.

Keller, T. (1986). The electrical conductivity of Norway spruce needle diffusate as affected by certain air pollutants. *Tree Physiol.* 1:85–94.

Keyes, M. R., and Grier, C. C. (1981). Above- and belowground net production in 40-year-old Douglas-fir stands on low and high productivity sites. *Can. J. For. Res.* 11:599–605.

Knapp, A. K., and Smith, W. K. (1982). Factors influencing understory seedling establishing of Engelmann spruce (*Picea engelmanii*) and subalpine fir (*Abies lasiocarpa*) in southeast Wyoming. *Can. J. Bot.* 60:2753–2761.

Koppenaal, R. S., Tschaplinski, T. J., and Colombo, S. J. (1991). Carbohydrate accumulation and turgor maintenance in seedling shoots and roots of two boreal conifers subjected to water stress. *Can. J. Bot.* 69:2522–2528.

Korner, Ch. (1985). Humidity responses in forest trees: Precautions in thermal scanning surveys. *Archives for Meteorology, Geophysics, and Bioclimatology, Series B: Climatology, Environmental Meteorology, Radiation Research* 36:83–98.

Kozlowski, T. T. (1971). "Growth and Development of Trees. Volume II. Cambial Growth, Root Growth, and Reproductive Growth." Academic Press, New York.

Kozlowski, T. T. (1982a). Water supply and tree growth. II. Flooding. *For. Abstr.* 43: 145–161.

Kozlowski, T. T. (1982b). Water supply and tree growth. I. Water deficits. *For. Abstr.* 43:57–95.

Kozlowski, T. T. (1986). Soil aeration and growth of forest trees. *Scand. J. For. Res.* 1:113–123.

Kozlowski, T. T., Kramer, P. J., and Pallardy, S. G. (1991). "The Physiological Ecology of Woody Plants." Academic Press, New York.

Kramer, P. J. (1942). Species differences with respect to water absorption at low soil temperature. *Am. J. Bot.* 29:828–832.

Kramer, P. J. (1983). "Water Relations of Plants." Academic Press, New York.

Kramer, P. J. (1988). Opinion: Changing concepts regarding plant water relations. *Plant, Cell Environ.* 11:565–569.

Krause, G. H. (1988). Photoinhibition of photosynthesis. An evaluation of damaging and protective mechanisms. *Physiol. Plant.* 74:566–574.

Kucera, J., Cermak, J., and Penka, M. (1977). Improved thermal method of continual recording the transpiration flow rate dynamics. *Biol. Plant.* 19:413–420.

Kuhns, M. R., and Gjerstad, D. H. (1991). Distribution of ^{14}C-labelled photosynthate in loblolly pine (*Pinus taeda*) seedlings as affected by season and time after exposure. *Tree Physiol.* 8:259–271.

Kuiper, P. J. C. (1985). Environmental changes and lipid metabolism of higher plants. *Physiol. Plant.* 64:118–122.

Lamhamedi, M. S., Bernier, P. Y., and Fortin, J. A. (1992). Hydraulic conductance and soil water potential at the soil root interface of *Pinus pinaster* seedlings inoculated with different dikaryons of *Pisolithus* sp. *Tree Physiol.* 10:231–244.

Landsberg, J. J. (1986). "Physiological Ecology of Forest Production." Academic Press, New York.

Larson, M. M. (1963). Initial root development of ponderosa pine seedlings as related to germination date and size of seed. *For. Sci.* 9:456–460.

Larson, M. M., and Schubert, G. H. (1969). Effect of osmotic water stress on germination and initial development of ponderosa pine seedlings. *For. Sci.* 15:30–36.

Lassoie, J. P., Scott, D. R. M., and Fritschen, L. J. (1977). Transpiration studies in Douglas-fir using the heat pulse technique. *For. Sci.* 23:377–390.

Lassoie, J. P., Dougherty, P. M., Reich, P. B., Hinckley, T. M., Metcalf, C. M., and Dina, S. J. (1983). Ecophysiological investigations of understory eastern red cedar in central Missouri. *Ecology* 64:1355–1366.

Lehto, T. H. (1989). Effects of mycorrhiza and drought on photosynthesis and water relations of Sitka spruce. *Agric. Ecosyst. Environ.* 28:299–303.

Leopold, A. C. (1990). Coping with desiccation. *In* "Stress Responses in Plants: Adaptation and Acclimation Mechanisms" (R. G. Alscher and J. R. Cummings, eds.), pp. 37–56. Wiley, New York.

Lewty, M. J. (1990). Effects of waterlogging on the growth and water relations of three *Pinus* taxa. *For. Ecol. Manage.* 30:189–201.

Lieffers, V. J., and Rothwell, R. L. (1986). Effects of depth of water table and substrate temperature on root and top growth of *Picea mariana* and *Larix laricina* seedlings. *Can. J. For. Res.* 16: 1201–1206.

Linder, S. (1987). Responses to water and nutrients in coniferous ecosystems. *Ecol. Stud.* 61: 180–202.

Linder, S., Benson, M. L., Myers, B. J., and Raison, R. J. (1987). Canopy dynamics and growth of *Pinus radiata*. I. Effects of irrigation and fertilization during drought. *Can. J. For. Res.* 17: 1157–1165.

Little, C. H. A. (1975). Inhibition of cambial activity in *Abies balsamea* by internal water stress: Role of abscisic acid. *Can. J. Bot.* 53: 3041–3050.

Livingston, N. J., and Black, T. A. (1987). Stomatal characteristics and transpiration of three species of conifer seedlings planted on a high elevation south-facing clearcut. *Can. J. For. Res.* 17: 1273–1282.

Livingston, N. J., and Black, T. A. (1988). The growth and water use of conifer seedlings planted on a high-elevation south-facing clearcut. *Can. J. For. Res.* 18: 1234–1242.

Lopushinsky, W. (1969). Stomatal closure in conifer seedlings in response to moisture stress. *Bot. Gaz.* 130: 258–263.

Lopushinsky, W., and Kaufmann, M. R. (1984). Effects of cold soil on water relations and spring growth of Douglas-fir seedlings. *For. Sci.* 30: 628–634.

Lorio, P. L., Howe, J. V., and Martin, C. N. (1972). Loblolly pine rooting varies with microrelief on wet sites. *Ecology* 53: 1134–1140.

Lotan, J. E., and Zahner, R. (1963). Shoot and needle responses of 20-year-old red pine to current soil moisture regimes. *For. Sci.* 9: 497–506.

Ludlow, M. M. (1987). Contribution of osmotic adjustment to the maintenance of photosynthesis during water stress. *In* "Progress in Photosynthesis Research" (J. Biggens, ed.), Vol. 4, pp. 161–168. Nijhoff, Dordrecht, The Netherlands.

Lyr, H., and Hoffman, G. (1967). Growth rates and growth periodicity of tree roots. *Int. Rev. For. Res.* 2: 181–236.

Macey, D. E., and Arnott, J. T. (1986). The effects of moderate moisture and nutrient stress on bud formation and growth of container-grown white spruce seedlings. *Can. J. For. Res.* 16: 949–954.

MacFall, J. S., Johnson, G. A., and Kramer, P. J. (1990). Observation of a water-depletion region surrounding loblolly pine roots by magnetic resonance imaging. *Proc. Natl. Acad. Sci. U.S.A.* 87: 1203–1207.

McNaughton, K. G., and Jarvis, P. G. (1983). Predicting effects of vegetation changes on transpiration and evaporation. *In* "Water Deficits and Plant Growth" (T. T. Kozlowski, ed.), Vol. 7, pp. 1–47. Academic Press, New York.

Mansfield, T. A., and Atkinson, C. J. (1990). Stomatal behavior in water stressed plants. *In* "Stress Response in Plants: Adaptation and Acclimation Mechanisms" (R. G. Alscher and J. R. Cumming, eds.), pp. 241–264. Wiley, New York.

Mark, W. R., and Crews, D. L. (1973). Heat pulse velocity and bordered pit condition in living Engelmann spruce and lodgepole pine trees. *For. Sci.* 19: 291–296.

Marshall, J. D. (1986). Drought and shade interact to cause fine-root mortality in Douglas-fir seedlings. *Plant Soil* 91: 51–60.

Marshall, J. D., and Waring, R. H. (1984). Conifers and broadleaf species: Stomatal sensitivity differs in western Oregon. *Can. J. For. Res.* 14: 905–908.

Masle, J., and Passioura, J. B. (1987). The effect of soil strength on the growth of young wheat plants. *Aust. J. Plant Physiol.* 14: 643–656.

Mattson, W. J., and Haack, R. A. (1987). The role of drought in outbreaks of plant-eating insects. *BioScience* 37: 110–118.

Meier, C. E., Newton, R. J., Puryear, J. D., and Sen, S. (1992). Physiological responses of loblolly pine (*Pinus taeda* L.) seedlings to drought stress: Osmotic adjustment and tissue elasticity. *J. Plant Physiol.* 140: 754–760.

Meinzer, F. C. (1982). The effect of vapor pressure deficit on stomatal control of gas exchange in Douglas-fir (*Pseudotsuga menziesii*) saplings. *Oecologia* 54:236–242.

Melzack, R. N., Bravdo, B., and Riov, J. (1985). The effect of water stress on photosynthesis and related parameters in *Pinus halepensis*. *Physiol. Plant.* 64:295–300.

Miller, D. R., Vavrina, C. A., and Christensen, T. W. (1980). Measurement of sap flow and transpiration in ring-porous oaks using a heat pulse velocity technique. *For. Sci.* 26: 485–494.

Miller, R. F., and Schultz, L. M. (1987). Water relations and leaf morphology of *Juniperus occidentalis* in the northern great basin. *For. Sci.* 33:690–706.

Milne, R., Ford, E. D., and Deans, J. D. (1983). Time lags in the water relations of Sitka spruce. *For. Ecol. Manage.* 5:1–25.

Moir, W. H., and Bachelard, E. P. (1969). Distribution of fine roots in three *Pinus radiata* plantations near Canberra, Australia. *Ecology* 50:658–662.

Monson, R. K., and Grant, M. C. (1989). Experimental studies of ponderosa pine. III. Differences in photosynthesis, stomatal conductance, and water-use efficiency between two genetic lines. *Am. J. Bot.* 76:1041–1047.

Morgan, J. M. (1984). Osmoregulation and water stress in higher plants. *Annu. Rev. Plant Physiol.* 35:299–319.

Morgan, P. W. (1990). Effects of abiotic stress on plant hormone systems. *In* "Stress Responses in Plants: Adaptation and Acclimation Mechanisms" (R. G. Alscher and J. R. Cumming, eds.), pp. 113–146. Wiley, New York.

Morgan, P. W., He, C.-J., and Powell, R. D. (1990). Does water deficit stress promote ethylene synthesis in intact plants? *Plant Physiol.* 94:1616–1624.

Mudge, K. W., Diebolt, K. S., and Whitlow, T. H. (1987). Ectomycorrhizal effect on host plant response to drought stress. *J. Environ. Hortic.* 5:183–187.

Mullin, R. E., and Bowdery, L. (1978). Effects of nursery seedbed density and topdressing fertilization on survival and growth of 3 + 0 red pine. *Can. J. For. Res.* 8:30–35.

Myers, B. J. (1988). Water stress integral—a link between short-term stress and long-term growth. *Tree Physiol.* 4:315–323.

Nambiar, E. K. S., and Sands, R. (1992). Effects of compaction and simulated root channels in the subsoil on root development, water uptake and growth of radiata pine. *Tree Physiol.* 10:297–306.

Nambiar, E. K. S., Bowen, G. D., and Sands, R. (1979). Root regeneration and plant water status of *Pinus radiata* D. Don seedlings transplanted to different soil temperature. *J. Exp. Bot.* 30:1119–1131.

Newton, M., and Preest, D. S. (1988). Growth and water relations of Douglas-fir (*Pseudotsuga menziesii*) seedlings under different weed control regimes. *Weed Sci.* 36:653–662.

Newton, R. J., Funkhouser, E. A., Fong, F., and Tauer, C. G. (1991). Molecular and physiological genetics of drought tolerance in forest species. *For. Ecol. Manage.* 43:225–250.

Newville, E. G., and Ferrell, W. K. (1980). Abscisic acid levels and stomatal behavior during drought and recovery in Douglas-fir (*Pseudotsuga menziesii*). *Can. J. Bot.* 58:1370–1375.

Nguyen, A., and Lamant, A. (1989). Variation in growth and osmotic regulation of roots of water-stressed maritime pine (*Pinus pinaster* Ait.) provenances. *Tree Physiol.* 5: 123–133.

Nisbet, T. R., and Mullins, C. E. (1986). A comparison of live and dead fine root weight in stands of Sitka spruce in contrasting soil water regimes. *Can. J. For. Res.* 16:394–397.

Nobel, P. S. (1976). Water relations and photosynthesis of a desert CAM plant, *Agave deserti*. *Plant Physiol.* 58:576–582.

Noble, D. L., and Alexander, R. R. (1977). Environmental factors affecting natural regeneration of Engelmann spruce in the central Rocky Mountains. *For. Sci.* 23:420–429.

Oechel, W. C., and Lawrence, W. T. (1985). Taiga. *In* "Physiological Ecology of North American Plant Communities" (B. F. Chabot and H. A. Mooney, eds.), Chap. 4, pp. 66–94. Chapman & Hall, New York.

Orlander, G., and Due, K. (1986). Location of hydraulic resistance in the soil–plant pathway in seedlings of *Pinus sylvestris* L. grown in peat. *Can. J. For. Res.* 16: 115–123.

Ormsbee, P., Bazzaz, F. A., and Boggess, W. R. (1976). Physiological ecology of *Juniperus virginiana* in old fields. *Oecologia* 23: 75–82.

Osmond, C. B., Winter, K., and Powles, S. B. (1980). Adaptive significance of carbon dioxided cycling during photosynthesis in water-stressed plants. *In* "Adaptations of Plants to Water and High Temperature Stress" (P. J. Kramer and N. C. Turner, eds.), pp. 139–154. Wiley, New York.

Pallardy, S. G. (1981). Closely related woody plants. *In* "Water Deficits and Plant Growth" (T. T. Kozlowski, ed.), Vol. 6, pp. 511–548. Academic Press, New York.

Pallardy, S. G. (1989). Hydraulic architecture and conductivity: An Overview. *In* "Structural and Functional Responses to Environmental Stresses" (H. G. Kreeb and T. M. Hinckley, eds.), pp. 3–19. SPB Publ., The Hague, The Netherlands.

Pallardy, S. G., Parker, W. C., Dixon, R. K., and Garrett, H. E. (1982). Tissue water relations of roots and shoots of droughted ectomycorrhizal shortleaf pine seedlings. *In* "Proceedings of the Seventh North American Forest Biology Workshop: Physiology and Genetics of Intensive Culture" (B. A. Thielges, ed.), pp. 368–373. Univ. of Kentucky, Lexington.

Pallardy, S. G., Pereira, J. S., and Parker, W. C. (1991). Measuring the state of water in tree systems. *In* "Techniques and Approaches in Forest Tree Ecophysiology" (J. P. Lassoie and T. M. Hinckley, eds.), pp. 28–76. CRC Press, Boca Raton, Florida.

Pallas, J. E., and Kays, S. J. (1982). Inhibition of photosynthesis by ethylene—A stomatal effect. *Plant Physiol.* 70: 598–601.

Panshin, A. J., and de Zeeuw, C. (1980). "Textbook of Wood Technology." 4th Ed. McGraw-Hill, New York.

Parke, J. L., Linderman, R. G., and Black, C. H. (1983). The role of ectomycorrhizas in drought tolerance of Douglas-fir seedlings. *New Phytol.* 95: 83–95.

Parker, J. (1956). Drought resistance in woody plants. *Bot. Rev.* 22: 241–289.

Parker, J. (1968). Drought resistance mechanisms. *In* "Water Deficits and Plant Growth" (T. T. Kozlowski, ed.), Vol. 1, pp. 195–234. Academic Press, New York.

Passioura, J. B. (1988a). Response to Dr. P. J. Kramer's article 'changing concepts regarding plant water relations', volume 11, number 7 pp. 565–568. *Plant, Cell Environ.* 11: 569–572.

Passioura, J. B. (1988b). Water transport in and to roots. *Annu. Rev. Plant Physiol. Plant Mol. Biol.* 39: 245–265.

Pavlik, B. M., and Barbour, M. G. (1991). Seasonal patterns of growth, water potential and gas exchange of red and white fir saplings across a montane ecotone. *Am. Midl. Nat.* 126: 14–29.

Pereira, J. S., and Kozlowski, T. T. (1977a). Variations among woody angiosperms in response to flooding. *Physiol. Plant.* 41: 184–192.

Pereira, J. S., and Kozlowski, T. T. (1977b). Water relations and drought resistance of young *Pinus banksiana* and *P. resinosa* plantation trees. *Can. J. For. Res.* 7: 132–137.

Pezeshki, S. R. (1991). Root responses of flood-tolerant and flood-sensitive tree species to soil redox conditions. *Trees* 5: 180–186.

Pharis, R. P. (1966). Comparative drought resistance of five conifers and foliage moisture content as a viability index. *Ecology* 47: 211–221.

Pharis, R. P., and Ferrell, W. K. (1966). Differences in drought resistance between coastal and inland sources of Douglas-fir. *Can. J. Bot.* 44: 1651–1659.

Pharis, R. P., and Kuo, C. G. (1977). The physiology of giberellins in conifers. *Can. J. For. Res.* 7: 299–325.

Philipson, J. J., and Coutts, M. P. (1978). The tolerance of tree roots to waterlogging. III. Oxygen transport in lodgepole pine and Sitka spruce roots of primary structure. *New Phytol.* 80: 341–349.

Philipson, J. J., and Coutts, M. P. (1980). Tolerance of tree roots to waterlogging. IV. Oxygen transport in woody roots of Sitka spruce and lodgepole pine. *New Phytol.* 80: 489–494.

Phillip, J. R. (1966). Plant water relations: Some physical aspects. *Annu. Rev. Plant Physiol.* 17:245–268.

Pollard, D. F. W., and Logan, K. T. (1979). The response of bud morphogenesis in black spruce and white spruce provenances to environmental variables. *Can. J. For. Res.* 9: 211–217.

Pothier, D., and Margolis, H. A. (1989). Patterns of change of saturated sapwood permeability and sapwood conductance with stand development. *Can. J. For. Res.* 19:432–439.

Puritch, G. S. (1973). Effect of water stress on photosynthesis, respiration, and transpiration of four *Abies* spp. *Can. J. For. Res.* 3:293–298.

Rabe, E. (1990). Stress physiology: The functional significance of the accumulation of nitrogen-containing compounds. *J. Hortic. Sci.* 65:231–243.

Raison, R. J., and Myers, B. J. (1992). The biology of forest growth experiment: linking water and nitrogen availability to the growth of *Pinus radiata*. *For. Ecol. Manage.* 52: 279–308.

Raison, R. J., Myers, B. J., and Benson, M. L. (1992). Dynamics of *Pinus radiata* foliage in relation to water and nitrogen stress: I. Needle production and properties. *For. Ecol. Manage.* 52:139–158.

Reich, P. B., Teskey, R. O., Johnson, P. S., and Hinckley, T. M. (1980). Periodic root and shoot growth in oak. *For. Sci.* 26:590–598.

Reidl, H. (1937). Bau und Leistungen des Wurzeholzes. *Jahrb. Wiss. Bot.* 85:1–75.

Reynolds, E. R. C. (1970). Root distribution and the cause of its spatial variability in *Pseudotsuga taxifolia* (Poir.) Britt. *Plant Soil* 32:501–517.

Richards, D. (1977). Root–shoot interactions: A functional equilibrium for water uptake in peach (*Prunus persica* (L.) Batsch.). *Ann. Bot.* 41:279–281.

Richter, H. (1973). Frictional potential losses and total water potential in plants: a reevaluation. *J. Exp. Bot.* 23:983–994.

Ritchie, G. A. (1982). Carbohydrate reserves and root growth potential in Douglas-fir seedlings before and after cold storage. *Can. J. For. Res.* 12:905–912.

Ritchie, G. A. (1984). Assessing seedling quality. *In* "Forest Nursery Manual: Production of Bareroot Seedlings" (M. L. Duryea and T. D. Landis, eds.), pp. 243–259. Martinus Nijhoff/W. Junk, The Hague.

Ritchie, G. A., and Dunlap, J. R. (1980). Root growth potential: Its development and expression in forest tree seedlings. *N. Z. J. For. Sci.* 10:218–248.

Roberts, D. R., and Dumbroff, E. B. (1986). Relationships among drought resistance, transpiration rates, and abscisic acid levels in three northern conifers. *Tree Physiol.* 1:161–167.

Roberts, J. (1976a). A study of root distribution and growth in a *Pinus sylvestris* L. plantation in East Anglia. *Plant Soil* 44:607–621.

Roberts, J. (1976b). An examination of the quantity of water stored in mature *Pinus sylvestris* trees. *J. Exp. Bot.* 27:473–479.

Roberts, J. (1977). The use of tree-cutting techniques in the study of water relations of mature *Pinus sylvestris* L. *J. Exp. Bot.* 28:751–767.

Roberts, J. (1978). The use of 'tree-cutting' techniques in the study of water relations of Norway spruce, *Picea abies* (L.) Karst. *J. Exp. Bot.* 29:465–471.

Roberts, J., Pymar, C. F., Wallace, J. S., and Pitman, R. M. (1980). Seasonal changes in leaf area, stomatal and canopy conductances and transpiration from bracken below a forest canopy. *J. Appl. Ecol.* 17:409–422.

Rogers, R., and Hinckley, T. M. (1979). Foliar weight and area related to current sapwood area in oak. *For. Sci.* 25:298–303.

Rook, D. A., and Hobbs, J. F. F. (1976). Soil temperatures and growth of rooted cuttings of radiata pine. *N. Z. J. For. Sci.* 5:296–305.

Ross, S. D. (1985). Promotion of flowering in potted *Picea engelmannii* (Perry) grafts: Effects of heat, drought, gibberellin $A_{4/7}$, and their timing. *Can. J. For. Res.* 15:618–624.

Ross, S. D., and Pharis, R. P. (1987). Control of sex expression in conifers. *Plant Growth Regul.* 6:37–60.

Ross, S. D., Webber, J. E., Pharis, R. P., and Owens, J. N. (1985). Interaction between gibberellin $A_{4/7}$ and root pruning on the reproductive and vegetative process in Douglas-fir. I. Effects on flowering. *Can. J. For. Res.* 15:341–347.

Rudinsky, J. A. (1959). Systemics in the control of forest insects. *J. For.* 57:284–286.

Rudinsky, J. A., and Svihra, P. (1970). Structures of transpiration flow in conifers and their ecological and phylogenetical relations (in Slovak.). *Lesnicky Casopis* 16:143–156.

Running, S. W. (1980a). Field estimates of root and xylem resistance in *Pinus contorta* using root excision. *J. Exp. Bot.* 31:555–569.

Running, S. W. (1980b). Relating plant capacitance to the water relations of *Pinus contorta*. *For. Ecol. Manage.* 2:237–252.

Running, S. W. (1980c). Environmental and physiological control of water flux through *Pinus contorta*. *Can. J. For. Res.* 10:82–91.

Running, S. W., and Reid, C. P. P. (1980). Soil temperature influences on root resistance of *Pinus contorta* seedlings. *Plant Physiol.* 65:635–640.

Rutter, A. J. (1966). Studies on the water relations of *Pinus sylvestris* in plantation conditions. IV. Direct observations on the rates of transpiration, evaporation of intercepted water, and evaporation from the soil surface. *J. Appl. Ecol.* 3:393–405.

Rutter, A. J. (1975). The hydrological cycle in vegetation. *In* "Vegetation and the Atmosphere" (J. L. Monteith, ed.), Vol. 1, pp. 111–154. Academic Press, New York.

Rychnovska, M., Cermak, J., and Smid, P. (1980). Water output in a stand of *Phragmites communis* Trin. A comparison of three methods. *Acta Sci. Nat. (Brno)* 14:1–27.

Sakai, A. (1983). Comparative study on freezing resistance of conifers with special reference to cold adaptation and its evolutive aspects. *Can. J. Bot.* 61:2323–2332.

Sakai, A., and Larcher, W. (1987). "Frost Survival of Plants." Springer-Verlag, Berlin.

Saliendra, N. Z., and Meinzer, F. C. (1991). Symplast volume, turgor, stomatal conductance and growth in relation to osmotic and elastic adjustment in droughted sugarcane. *J. Exp. Bot.* 42:1251–1259.

Sanderson, P. L., and Armstrong, W. (1978). Soil waterlogging, root rot and conifer windthrow: Oxygen deficiency or phytotoxicity? *Plant Soil* 49:185–190.

Sanderson, P. L., and Armstrong, W. (1980). The response of conifers to some of the adverse factors associated with waterlogged soils. *New Phytol.* 85:351–362.

Sandford, A. P., and Jarvis, P. G. (1986). Stomatal response to humidity in selected conifers. *Tree Physiol.* 2:89–103.

Sands, K., and Rutter, A. J. (1959). Studies on the growth of young plants of *Pinus sylvestris* L. II. The relation of growth to soil moisture tension. *Ann. Bot.* 23:269–284.

Sands, R., and Correll, R. L. (1976). Water potential and leaf elongation in radiata pine and wheat. *Physiol. Plant.* 37:293–297.

Sands, R., and Mulligan, D. R. (1990). Water and nutrient dynamics and tree growth. *For. Ecol. Manage.* 30:91–111.

Sands, R., and Theodorou, C. (1978). Water uptake by mycorrhizal roots of radiata pine seedlings. *Aust. J. Plant Physiol.* 5:301–309.

Sands, R., Greacen, E. L., and Gerard, C. J. (1979). Compaction of sandy soils in radiata pine forests. I. A penetrometer study. *Aust. J. Plant Physiol.* 17:101–113.

Sands, R., Fiscus, E. L., and Reid, C. P. P. (1982). Hydraulic properties of pine and bean roots with varing degrees of suberization, vascular differentiation, and mycorrhizal infection. *Aust. J. Plant Physiol.* 9:559–569.

Sands, R., Kreidemann, P. E., and Cotterill, P. P. (1984). Water relations and photosynthesis in three families of radiata pine known to differ in their response to weed control. *For. Ecol. Manage.* 9:173–184.

Sanio, K. (1872). Über die Größe der Holzzellen beider gemeinen Kiefer (*Pinus sylvestris*). *Jahrb. Wiss. Bot.* 6:401–420.

Santakumari, M., and Berkowitz, G. A. (1990). Correlation between the maintenance of photosynthesis and *in situ* protoplast volume at low water potentials in droughted wheat. *Plant Physiol.* 92:733–739.

Santantonio, D., and Hermann, R. K. (1985). Standing crop, production, and turnover of fine roots on dry, moderate, and wet sites of mature Douglas-fir in western Oregon. *Ann. Sci. For.* 42:113–142.

Schulze, E.-D. (1991). Water and nutrient interactions with plants water stress. *In* "Response of Plants to Multiple Stresses" (H. A. Mooney, W. E. Winner, and E. J. Pell, eds.), pp. 89–101. Academic Press, New York.

Schulze, E.-D., Schilling, K., and Nagarajah, S. (1983). Carbohydrate partitioning in relation to whole plant production and water use of *Vigna unguiculata* (L.) Walp. *Oecologia* 58:169–177.

Schulze, E.-D., Cermak, J., Matyssek, R., Penka, M., Zimmermann, R., Vasicek, F., Gries, W., and Kucera, J. (1985). Canopy transpiration and water fluxes in the xylem of the trunk of *Larix* and *Picea* trees—A comparison of xylem flow, porometer and cuvette measurements. *Oecologia* 66:475–483.

Schulze, E. D., Steudle, E., Gollan, T., and Schurr, U. (1988). Response to Dr. P. J. Kramer's article, 'changing concepts regarding plant water relations,' volume 11, number 7, pp. 565–568. *Plant, Cell Environ.* 11:573–576.

Schurr, U., Gollan, T., and Schulze, E.-D. (1992). Stomatal response to drying soil in relation to changes in the xylem sap composition of *Helianthus annuus*. II. Stomatal sensitivity to abscisic acid imported from the xylem sap. *Plant, Cell Environ.* 15:561–567.

Seiler, J. R., and Cazell, B. H. (1990). Influence of water stress on the physiology and growth of red spruce seedlings. *Tree Physiol.* 6:69–77.

Seiler, J. R., and Johnson, J. D. (1985). Photosynthesis and transpiration of loblolly pine seedlings as influenced by moisture-stress conditioning. *For. Sci.* 31:742–749.

Seiler, J. R., and Johnson, J. D. (1988). Physiological and morphological responses of three half-sib families of loblolly pine to water-stress conditioning. *For. Sci.* 34:487–495.

Selter, C. M., Pitts, W. D., and Barbour, M. G. (1986). Site microenvironment and seedling survival of shasta and red fir. *Am. Midl. Nat.* 115:288–300.

Sena Gomes, A. R., and Kozlowski, T. T. (1980). Responses of *Pinus halepensis* seedlings to flooding. *Can. J. For. Res.* 10:308–311.

Setter, T. L. (1990). Transport/harvest index: Photosynthate partitioning in stressed plants. *In* "Stress Responses in Plants: Adaptation and Acclimation Mechanisms" (R. G. Alscher and J. R. Cumming, eds.), pp. 17–36. Wiley, New York.

Shainsky, L. J., Newton, M., and Radosevich, S. R. (1992). Effects of intra-specific and inter-specific competition on root and shoot biomass of young Douglas-Fir and red alder. *Can. J. For. Res.* 22:101–110.

Sheriff, D. D., and Whitehead, D. (1984). Photosynthesis and wood structure in *Pinus radiata* D. Don during dehydration and immediately after rewatering. *Plant, Cell Environ.* 7:53–62.

Shoulders, E., and Ralston, C. W. (1975). Temperature, root aeration and light influence slash pine nutrient uptake rates. *For. Sci.* 21:401–410.

Simon, E. W. (1974). Phospholipids and plant membrane permeability. *New Phytol.* 73: 377–420.

Sinclair, T. R., and Ludlow, M. M. (1985). Who taught plants thermodynamics? The unfulfilled potential of plant water potential. *Aust. J. Plant Physiol.* 12:213–217.

Sipchanov, I., and Baurenska, N. (1965). Study of water flow in conifers (in Bulgarian). *Gorskostop. Nauka* 2:35–441.

Smit-Spinks, B., Swanson, B. T., and Markhart, A. H. I. (1984). Changes in water relations, water flux, and root exudate abscisic acid content with cold acclimation of *Pinus sylvestris* L. *Aust. J. Plant Physiol.* 11:431–441.

Smith, G. S., Miller, S. A., and Judd, M. J. (1990). Kiwifruit and waterlogging. *Wispas (DSIR)* 45:4.

Smith, W. K. (1985). Western montane forests. *In* "Physiological Ecology of North American Plant Communities" (B. F. Chabot and H. A. Mooney, eds.), Chap. 5. pp. 95–126. Chapman & Hall, New York.

Sobrado M. A., Grace J., and Jarvis, P. G. (1992). The limits to xylem embolism recovery in *Pinus sylvestris* L. *J. Exp. Bot.* 43:831–836.

Sojka, R. E., and Stolzy, L. H. (1980). Soil-oxygen effects on stomatal response. *Soil Sci.* 130:350–358.

Sperry, J. S. (1993). Water embolism and spring recovery in *Betula cordifolia, Fagus grandifolia, Abies balsamea,* and *Picea rubens. In* "Water Transport in Plants Under Climatic Stress" (A. Raschi, M. Borghetti, and J. Grace, eds.), pp. 86–98. Cambridge Univ. Press, Cambridge.

Sperry, J. S., and Sullivan, J. E. M. (1992). Xylem embolism in response to freeze–thaw cycles and water stress in ring-porous, diffuse-porous, and conifer species. *Plant Physiol.* 100:605–613.

Sperry, J. S., and Tyree, M. T. (1988). Mechanism of water stress-induced xylem embolism. *Plant Physiol.* 88:581–587.

Sperry, J. S., and Tyree, M. T. (1990). Water stress induced xylem embolism in three species of conifers. *Plant, Cell Environ.* 13:427–436.

Sperry, J. S., Holbrook, N. M., Zimmermann, M. H., and Tyree, M. T. (1987). Spring filling of xylem vessels in wild grapevine. *Plant Physiol.* 83:414–417.

Sperry, J. S., Donnelly, J. R., and Tyree, M. T. (1988a). A method for measuring hydraulic conductivity and embolism in xylem. *Plant, Cell Environ.* 11:35–40.

Sperry, J. S., Donnelly, J. R., and Tyree, M. T. (1988b). Seasonal occurrence of xylem embolism in sugar maple. *Am. J. Bot.* 75:1212–1218.

Sperry, J. S., Nichols, K. L., Sullivan, J. E. M., and Eastlack, S. E. (1994). Comparative studies of xylem embolism in ring-porous, diffuse-porous and coniferous trees of northern Utah and interior Alaska. *Ecology* 75:1736–1752.

Squire, R. O., Attwill, P. M., and Neales, T. F. (1987). Effects of changes of available water and nutrients on growth, root development, and water use in *Pinus radiata* seedlings. *Aust. For. Res.* 17:99–111.

Squire, R. O., Neales, T. F., Loveys, B. R., and Attiwill, P. M. (1988). The influence of water deficits on needle conductance, assimilation rate and abscisic acid concentration of seedlings of *Pinus radiata* D. Don. *Plant, Cell Environ.* 11:13–19.

Stone, E. C., and Fowells, H. A. (1955). Survival value of dew as determined under laboratory conditions with *Pinus ponderosa. For. Sci.* 1:183–188.

Stone, E. L., and Kalisz, P. J. (1991). On the maximum extent of tree roots. *For. Ecol. Manage.* 46:59–102.

Stone, J. E., and Stone, E. L. (1975). Water conduction in lateral roots of red pine. *For. Sci.* 21:53–60.

Stransky, J. J., and Wilson, D. R. (1964). Terminal elongation of loblolly and shortleaf pine seedlings under soil moisture stress. *Soil Sci. Soc. Am. J.* 28:439–440.

Strauss, S. H., and Ledig, F. T. (1985). Seedling architecture and life history evolution in pines. *Am. Nat.* 125:702–713.

Strong, W. L., and La Roi, G. H. (1983). Root-system morphology of common boreal forest trees in Alberta, Canada. *Can. J. For. Res.* 13:1164–1173.

Stumpff, N. J., and Johnson, J. D. (1987). Ethylene production of loblolly pine seedlings associated with water stress. *Physiol. Plant.* 69:167–172.

Sucoff, E., Buschena, C., and Tamte, P. (1985). Desiccation and water potentials in the roots, leaves, and shoots of bare-root red pine and white spruce. *Can. J. For. Res.* 15: 989–992.

Swanson, R. H. (1966). Seasonal course of transpiration of lodgepole pine and Engelmann spruce. *In* "International Symposium of Forest Hydrology Proceedings" (W. E. Sopper and H. W. Lull, eds.), pp. 417–433. Pergamon, Oxford.

Swanson, R. H. (1967). Improving tree transpiration estimates based on heat pulse velocity measurements. *IUFRO 14th Cong. Proc.* 1:252–263.

Swanson, R. H. (1970). Sampling for direct transpiration estimates. *J. Hydrol.* 9:72–77.

Swanson, R. H. (1975). Water use by mature lodgepole pine. *In* "Proceedings of the Management of Lodgepole Pine Ecosystems Symposium" (D. M. Baumgartner, ed.), pp. 264–277. Washington State Univ. Coop. Extension Service, Pullman, Washington.

Tan, C. S., Black, T. A., and Nnyamah, J. U. (1977). Characteristics of stomatal diffusion resistance in a Douglas-fir forest exposed to soil water deficits. *Can. J. For. Res.* 7: 595–604.

Tan, W., Blake, T. J., and Boyle, T. J. B. (1992a). Drought tolerance of faster- and slower-growing black spruce (*Picea mariana*) progenies: II. Osmotic adjustment and changes in soluble carbohydrates and amino acids under osmotic stress. *Physiol. Plant.* 85: 645–651.

Tan, W., Blake, T. J., and Boyle, T. J. B. (1992b). Drought tolerance in faster- and slower-growing black spruce (*Picea mariana*) progenies: I. Stomatal and gas exchange response to osmotic stress. *Physiol. Plant.* 85:639–644.

Tang, Z. C., and Kozlowski, T. T. (1984). Water relations, ethylene production, and morphological adaptation of *Fraxinus pennsylvanica* seedlings to flooding. *Plant Soil* 77: 183–192.

Taylor, H. M., and Willatt, S. T. (1983). Shrinkage of soybean roots. *Agron. J.* 75:818–820.

Teskey, R. O., Hinckley, T. M., and Grier, C. C. (1983). Effect of interruption of the flow path on stomatal conductance of *Abies amabilis*. *J. Exp. Bot.* 34:1251–1259.

Teskey, R. O., Hinckley, T. M., and Grier, C. C. (1984). Temperature-induced change in the water relations of *Abies amabilis* (Dougl.) Forbes. *Plant Physiol.* 74:77–80.

Teskey, R. O., Fites, J. A., Samuelson, L. J., and Bongarten, B. C. (1986). Stomatal and nonstomatal limitations to net photosynthesis in *Pinus taeda* L. under different environmental conditions. *Tree Physiol.* 2:131–142.

Teskey, R. O., Bongarten, B. C., Cregg, B. M., Dougherty, P. M., and Hennessey, T. C. (1987). Physiology and genetics of tree growth response to moisture and temperature stress: An examination of the characteristics of loblolly pine (*Pinus taeda* L.) *Tree Physiol.* 3:41–61.

Thomas, P. A., and Wein, R. W. (1985a). Water availability and the comparative emergence of four conifer species. *Can. J. Bot.* 63:1740–1746.

Thomas, P. A., and Wein, R. W. (1985b). The influence of shelter and the hypothetical effect of fire severity on the postfire establishment of conifers from seed. *Can. J. For. Res.* 15:148–155.

Thornley, J. H. M. (1972). A balanced quantitative model for root:shoot ratios in vegetative plants. *Ann. Bot.* 36:431–441.

Tinker, P. B. (1976). Roots and water. Transport of water to plant roots in soil. *Philos. Trans. R. Soc. London B* 273:445–461.

Toivonen, P., and Vidaver, W. (1988). Variable chlorophyll *a* fluorescence and CO_2 uptake in water-stressed white spruce seedlings. *Plant Physiol.* 86:744–748.

Topa, M. A., and Cheeseman, J. M. (1992). Effects of root hypoxia and a low P supply on relative growth, carbon dioxide exchange rates and carbon partitioning in *Pinus serotina* seedlings. *Physiol. Plant.* 86:136–144.

Topa, M. A., and McLeod, K. W. (1986a). Response of *Pinus clausa*, *Pinus serotina*, and *Pinus taeda* seedlings to anaerobic solution culture. I. Changes in growth and root morphology. *Physiol. Plant.* 68:532–539.

Topa, M. A., and McLeod, K. W. (1986b). Aerenchyma and lenticel formation in pine seedlings: A possible avoidance mechanism to anaerobic growth conditions. *Physiol. Plant.* 68:540–550.

Tranquilini, W. (1976). Water relations and alpine timberline. *In* "Water and Plant Life: Problems and Modern Approaches, Ecological Studies" (O. L. Lange, L. Kappen, and E.-D. Schulze, eds.), Vol. 19, pp. 473–491. Springer-Verlag, Berlin.

Tripepi, R. R., and Mitchell, C. A. (1984). Metabolic response of river birch and European birch roots to hypoxia. *Plant Physiol.* 76:31–35.

Tryon, P. R., and Chapin III, F. S. (1983). Temperature control over root growth and root biomass in taiga forest trees. *Can. J. For. Res.* 13:827–833.

Tschaplinski, T. J., Norby, R. J., and Wullschleger, S. D. (1993). Responses of loblolly pine seedlings to elevated CO_2 and fluctuating water supply. *Tree Physiol.* 13:283–296.

Tsukahara, H., and Kozlowski, T. T. (1984). Effect of flooding on *Larix leptolepis* seedlings. *J. Jpn For. Soc.* 66:333–336.

Tyree, M. T. (1988). A dynamic model for water flow in a single tree: Evidence that models must account for hydraulic architecture. *Tree Physiol.* 4:195–217.

Tyree, M. T., and Ewers, F. W. (1991). The hydraulic architecture of trees and other woody plants. *New Phytol.* 119:345–360.

Tyree, M. T., and Sperry, J. S. (1988). Do woody plants operate near the point of catastrophic xylem dysfunction caused by dynamic water stress? Answers from a model. *Plant Physiol.* 88:574–580.

Tyree, M. T., and Sperry, J. S. (1989). Vulnerability of xylem to cavitation and embolism. *Annu. Rev. Plant Physiol. Mol. Biol.* 40:19–38.

Tyree, M. T., Graham, M. E. D., Cooper, K. E., and Bazos, L. J. (1983). The hydraulic architecture of *Thuja occidentalis*. *Can. J. Bot.* 61:2105–2111.

Tyree, M. T., Snyderman, D. A., Wilmot, T. R., and Machado, J. L. (1991). Water relations and hydraulic architecture of a tropical tree (*Schefflera morototoni*)—Data, models, and a comparison with two temperate species (*Acer saccharum* and *Thuja occidentalis*). *Plant Physiol.* 96:1105–1113.

Ursino, D. J., Nelson, C. D., and Krotkov, G. (1968). Seasonal changes in the distribution of photo-assimilated ^{14}C in young pine roots. *Plant Physiol.* 43:845–852.

U.S.D.A. Forest Service (1980). Root characteristics of some important trees of eastern forests: A summary of the literature. U.S.D.A. Forest Service Eastern Region.

U.S.D.A. Forest Service (1990a). "Silvics of North America," Vol. 1, Conifers. U.S.D.A. Forest Service, Agric. Hdbk. No. 654. Washington, D.C.

U.S.D.A. Forest Service (1990b). "Silvics of North America," Vol. 2. Hardwoods. U.S.D.A. Forest Service, Agric. Hdbk. No. 654. Washington, D.C.

Vance, N. C., and Zaerr, J. B. (1990). Analysis by high-performance liquid chromatography of free amino acids extracted from needles of drought-stressed and shaded *Pinus ponderosa* seedlings. *Physiol. Plant.* 79:23–30.

Vance, N. C., and Zaerr, J. B. (1991). Influence of drought stress and low irradiance on plant water relations and structural constituents in needles of *Pinus ponderosa* seedlings. *Tree Physiol.* 8:175–184.

Vance, N. C., Copes, D. O., and Zaerr, J. B. (1990). Differences in proteins synthesized in needles of unshaded and shaded *Pinus ponderosa* var. *scopulorum* seedlings during prolonged drought. *Plant Physiol.* 92:1244–1248.

van den Driessche, R. (1984). Relationship between spacing and nitrogen fertilization of seedlings in the nursery. Seedling mineral nutrition and outplanting performance. *Can. J. For. Res.* 14:431–436.

van den Driessche, R. (1989). Changes in osmotic potential of Douglas-fir (*Pseudotsuga menziesii*) seedlings in relation to temperature and photoperiod. *Can. J. For. Res.* 19: 413–421.

van Rees, K. C., and Comerford, N. B. (1986). Vertical root distribution and strontium uptake of a slash pine stand on a Florida spodosol. *Soil Sci. Soc. Am. J.* 50:1042–1046.

van Rees, K. C., and Comerford, N. B. (1990). The role of woody roots of slash pine seedlings in water and potassium absorption. *Can. J. For. Res.* 20:1183–1191.

Vieira da Silva, J. (1976). Water stress, ultrastructure and enzymatic activity. *In* "Water and Plant Life: Problems and Modern Approaches, Ecological Studies" (O. L. Lange, L. Kappen, and E.-D. Schulze, eds.), Vol. 19, pp. 207–225. Springer-Verlag, Berlin.

Vogt, K. A., and Bloomfield, J. (1991). Tree root turnover and senescence. *In* "Plant Roots: The Hidden Half" (Y. Waisel, A. Eshel, and U. Kafkafi, eds.), pp. 287–306. Dekker, New York.

Walker, R. F., West, D. C., and McLaughlin, S. B. (1982). *Pisolithus tinctorius* ectomycorrhizae reduce moisture stress of Virginia pine on a southern Appalachian coal spoil. *In* "Proceedings of the Seventh North American Forest Biology Workshop" (B. A. Thielges, ed.), pp. 374–383. Univ. of Kentucky, Lexington.

Walker, R. F., West, D. C., McLaughlin, S. B., and Amundsen, C. C. (1989). Growth, xylem pressure potential, and nutrient absorption of loblolly pine on a reclaimed surface mine as affected by an induced *Pisolithus tinctorius* infection. *For. Sci.* 35(2):569–581.

Wang, S. Y., Wang, C. Y., and Welburn, A. R. (1991). Role of ethylene under stress conditions. *In* "Stress Responses in Plants: Adaptation and Acclimation Mechanisms" (R. G. Alscher and J. R. Cummings, eds.), pp. 147–173. Wiley, New York.

Wang, Z. M., and MacDonald, S. E. (1992). Peatland and upland black spruce populations in Alberta, Canada: Isozyme variation and seed germination ecology. *Silvae Genet.* 41: 117–122.

Waring, R. H., and Major, J. (1964). Some vegetation of the California coastal redwood region in relation to gradients of moisture, nutrients, light, and temperature. *Ecol. Monogr.* 34:167–215.

Waring, R. H., and Running, S. W. (1978). Sapwood water storage: Its contribution to transpiration and effect upon water conductance through the stems of old-growth Douglas-fir. *Plant, Cell Environ.* 1:131–140.

Waring, R. H., Whitehead, D., and Jarvis, P. G. (1979). The contribution of stored water to transpiration in Scots pine. *Plant, Cell Environ.* 2:309–317.

Waring, R. H., Schroeder, P. E., and Oren, R. (1982). Application of the pipe model theory to predict canopy leaf area. *Can. J. For. Res.* 12:556–560.

Wheeler, E. A. (1981). Intervascular pitting in *Fraxinus americana* L. *Int. Assoc. Wood Anat. Bull. N. S.* 2:169–174.

Whitehead, D., and Jarvis, P. G. (1981). Coniferous forest plantations. *In* "Water Deficits and Plant Growth" (T. T. Kozlowski, ed.), Vol. 6, pp. 49–152. Academic Press, New York.

Wong, S. C., Cowan, I. R., and Farquhar, G. D. (1979). Stomatal conductance correlates with photosynthetic capacity. *Nature (London)* 282:424–426.

Yamamoto, F., Kozlowski, T. T., and Wolter, K. E. (1987). Effect of flooding on growth, stem anatomy and ethylene production of *Pinus halepensis* seedlings. *Can. J. For. Res.* 17:69–79.

Yang, S., and Tyree, M. T. (1992). A theoretical model of hydraulic conductivity recovery from embolism with comparison to experimental data on *Acer saccharum*. *Plant, Cell Environ.* 15:633–643.

Yau, D. P., and Sands, R. (1992). Soil water content, soil strength and the growth of elm root suckers. *Tree Physiol.* 11:161–169.

Young, E., and Hanover, J. W. (1978). Effects of temperature, nutrient and moisture stresses on dormancy of blue spruce seedlings under continuous light. *For. Sci.* 24:458–467.

Younis, H. M., Boyer, J. S., and Govindjee (1979). Conformation and activity of chloroplast coupling factor exposed to low chemical potential of water in cells. *Biochim. Biophys. Acta* 548:328–340.

Zaerr, J. B. (1983). Short-term flooding and net photosynthesis in seedlings of three conifers. *For. Sci.* 29:71–78.

Zahner, R. (1962). Terminal growth and wood formation by juvenile loblolly pine under two soil moisture regimes. *For. Sci.* 8:345–352.

Zavitkovski, J., and Ferrell, W. K. (1968). Effect of drought upon rates of photosynthesis, respiration, and transpiration of seedlings of two ecotypes of Douglas-fir. *Bot. Gaz.* 129:346–350.

Zavitkovski, J., and Ferrell, W. K. (1970). Effect of drought upon rates of photosynthesis, respiration, and transpiration of seedlings of two ecotypes of Douglas-fir. II. Two-year-old seedlings. *Photosynthetica* 4:58–67.

Zimmermann, M. H. (1978). Hydraulic architecture of some diffuse-porous trees. *Can. J. Bot.* 56:2286–2295.

Zimmermann, M. H. (1983). "Xylem Structure and the Ascent of Sap." Springer-Verlag, Berlin.

Zimmermann, M. H., and Brown, C. L. (1971). "Trees: Structure and Function." Springer-Verlag, Berlin.

Zinkan, C. G., Jeglum, J. K., and Harvey, D. E. (1974). Oxygen in water culture influences growth and nutrient uptake of jack pine, black spruce and white spruce seedlings. *Can. J. Plant Sci.* 54:553–558.

Zisa, R. P., Halverson, H. G., and Stout, B. B. (1980). Establishment and early growth of conifers on compact soils in urban areas. U.S.D.A. Forest Service. Northeast Forest Exp. Sta., Res. Paper NE-451.

Zobel, D. B., McKee, A., Hawk, G. M., and Dyrness, C. T. (1976). Relationships of environment to composition, structure, and diversity of forest communities of the central western Cascades of Oregon. *Ecol. Monogr.* 46:135–156.

Zwiazek, J. J. (1991). Cell wall changes in white spruce (*Picea glauca*) needles subjected to repeated drought stress. *Physiol. Plant.* 82:513–518.

Zwiazek, J. J., and Blake, T. J. (1989). Effects of preconditioning on subsequent water relations, stomatal sensitivity, and photosynthesis in osmotically stressed black spruce. *Can. J. Bot.* 67:2240–2244.

Zwiazek, J. J., and Blake, T. J. (1990). Effects of preconditioning on carbohydrate and amino acid composition of osmotically stressed black spruce. *Can. J. Bot.* 67:2240–2244.

Zwiazek, J. J., and Blake, T. J. (1991). Early detection of membrane injury in black spruce (*Picea mariana*). *Can. J. For. Res.* 21:401–404.

Index

Abscisic acid, contents in water stressed seedlings, 354–355
Absorbed photosynthetically active radiation, *see also* Light
  canopy, 23
  effects on dry matter production, 29
  needles, 5–9
Acclimation
  photosynthetic, to shade, 12–13
Aging
  foliage respiration, 268
Allocation, *see also* Carbon allocation and accumulation, 217–218
Angiosperms versus conifers, water transport dynamics, 310–311
Architecture
  effects on water transport, 308–309
Aridity, *see* Water stress
Assimilation controls, 219–222
Assimilatory-transpiration efficiency (ATE), 157–159
Atmospheric pollution, effects on
  carbon allocation, 240–241
  photosynthesis, 111–112
Axial and ray cell dimensions, 316–317

Branch level
  allocation in woody tissue, 234
  measurement techniques, 277–281
  resource use efficiency, 161–163
  respiration rates, 281
  source–sink–storage of carbon, 197–198
  water transport, sapwood and heartwood, 317–318

Cambial and radial stem growth response to water deficits, 358–359
Canopy level
  as sources and sinks of nitrogen and sulfur, 59–60
  CO_2 exchange measurement, 128–130
  light environment, 24–25
  MAESTRO model of CO_2 exchange, 130–131
  photosynthetic light response, 28–29
  respiration of, 265–266
  sap flow variation, 331, 333
Carbon allocation and accumulation, 217–218
  allocation controls, 226
    defense compounds, 235–236
    fine roots, 229–234
    foliage, 227–229
    leaf leachates/root exudates, 237–238
    reproduction, 236–237
    respiration, 226–227
    storage, 234–235
    woody tissue, 231, 234
  allocation patterns, 222–223
  assimilation controls, 219–222
  future research, 242–243
  geographic distribution, 218–219
    genetic factors, 223
    mechanisms, 223
  multiple resource limitations on, 239–242
  source–sink–storage, 179–180
    developmental effects on
      leaf, 197
      life cycle stages, 197–200
      shoot, 197–198
    electrical analog, 189–191
    photosynthetic response to sinks, 192–194
    sink activity, storage, photosynthate, 194–197
    source–sink case studies, 180–189
    source–sink feedback, 191–192
  stand development, effects on, 238–239
Carbon dioxide, effects on photosynthesis, 115–117

atmospheric, 115–117
intercellular, 107–108
Carbon dioxide, fluxes, estimating respiration, 289–290
Carbon transport, *see* Carbon allocation and accumulation
Cavitation, due to, *see also* Embolisms
freeze–thaw episodes, 321–322
water stress, 319–321
Competition
and root biomass, 309
with understory plants, 64–66
Conifers versus angiosperms, water transport dynamics, 310–311
Convexity of the photosynthetic light response, 10–12, 109–111
Crown level, *see* Canopy level

Defense compounds, allocation, 235–236
Drought stress, *see* Water stress

Ecological scaling
forest-to-stand transpiration, 343
components, 343–345
leaves-to-stand, 345–346
organ/stand respiration, 255–257
resource use efficiency, 146–151
branch/plant, 161–163
community/ecosystem, 168
Ecosystem level
growth efficiency, 170–171
resource use efficiency, 146, 168–169
thinning effects on light/water use efficiency, 169–170
Embolisms, *see also* Cavitation
effects on
conducting efficiency, 325
xylem pressure potentials, 319–320
reversal of, 323–324
Environmental effects on
carbon allocation and accumulation, 239–242
carbon assimilation, 219–222
photosynthesis, *see also* Photosynthesis; Water stress
root development, 311
water absorption, 313–315
Exudates, root, 237

Fine roots, *see* Mycorrhiza; Roots
Flooding tolerance, 328–329
Fungi, *see* Mycorrhiza

Growth efficiency, community/ecosystem, 170–171

Huber values, 325
Hydraulic conductivity (k_h), 324–328

Leachates, leaf, 237
Leaf area index (LAI) and absorbed light, 220–221
Leaf level, *see also* Needle level
light absorption, 5–7
permeability, 42–43
source–sink feedback regulation, 191–192
UV penetration, 7–8
Leaf life span and nutrient retranslocation, 154–155
Life cycles and source–sink relationships, 200
Life history, effects on nutrient acquisition models, 82–83
Light
canopy level, 23–24
photosynthetic acclimation to, 12–13
photosynthetic response to
by canopy, 28–29
by needles, 5–9
respiration response to, 266–267
Light absorption
effects of needle structure, 9–12
influence on dry matter production, 29–30
Light sacrifice and water conservation, 169–170
Light use efficiency, 149
branch, 163
leaves, 151
plant, 163

Models
canopy CO_2 exchange, MAESTRO model, 130–131
carbon processes, 125–126
efficiency of mycorrhizal hyphae, 93–94
forest carbon budget, 219–220
limitations, nutrient acquisition efficiency, 82–84
maintenance respiration, 284–286
nutrient acquisition, fine roots, and mycorrhiza, 76–82
source–sink–storage, 189–191
water–nitrogen feedback, 205–208

Multiple resource limitations, 239–242
Mycorrhiza
  ectomycorrhizae, 81–82
  effects on
    nitrogen/phosphorus uptake, 97–98
    seedling vigor, 61–62
    water uptake, 304–306
  endomycorrhizae, 81
  nutrient acquisition models, 75, 81–82, 93–94
    fine-root efficiency model, 76–78
    nutrient availability effects, 55
    water effects, 53

Needle level
  cuticle and epicuticular waxes, 6–7
  permeability, 42–43
  sun/shade
    photosynthesis, 13–14
    water potential, 339
  UV penetration, 7–8
Net primary production and seasonal water availability, 358
Nitrogen
  effects on
    biomass partitioning, 198–199
    photosynthesis, 107, 109, 187
  uptake by roots, 52
Nitrogen metabolism, water stress effects, 352–353
Nutrient acquisition
  effects of
    other nutrients, 55–57
    water, 46–51, 53–54
Nutrient acquisition by
  foliage, 40–41
  mature stands, 63–64
  roots, 51–57
  seedlings, 60–63
Nutrient partitioning within photosynthetic pathways, 156–157
Nutrient retranslocation, 59, 63–64, 154–155
Nutrient stress
  effects on
    photosynthesis, 107, 121
    resource use efficiency, 203–204
Nutrient use efficiency
  branch and plant, 163
  leaves, 151–157
Nutrient–water acquisition, comparisons, contrasts, 315
Nutrition, effects on sources/sinks, 203–204

Osmotic adjustment, 365–367

Phosphorus uptake by roots, 52–53
Photoinhibition and freezing temperatures, 17–18
Photon flux density, photosynthetically active, *see* Light
Photosynthesis
  acclimation, 12–13
  canopy models, 128
  direct effects of
    atmospheric pollution, 117
    carbon dioxide, 115–117
    ecotypic variation in, 182–183
    internal factors, nutrients and carbon balance, 107–109
    light, 109–112
    temperature, 113–115
    vapor pressure deficit, 119–120
    water, 112–113
  effects on
    growth, 64–65
    photosynthesis, 107
  indirect effects of
    cone production, 193–194
    growth, 122–123
    light, 118–119
    nutrients, 121
    simultaneous control, 123
    temperature, 121–122
    water, 119–121
  integration over time, 124–125, 127–128
  light absorption
    canopy, 23
    dry matter production, 29
    needles, 5–9
  light environment
    canopy, 24
  light response
    canopy, 28–29
    convexity, 10–12, 109, 111
    environmental effects, 29
    needles, 9–11, 189
    needle-to-shoot, 18–22
  nutrient partitioning, 156–157
  quantum yield of, 16–17
  scaling, structural and geographic, 126–127, 130
  shade acclimation, 12–13

sun/shade needles, 13–14
UV effects on, 111–112
Pollution, *see* Atmospheric pollution

Quantum yield of photosynthesis, 16–17

Radiation, *see* Light
Relative growth rates (RGR), 164–167
Resource use efficiency (RUE)
absorbed light effects, 159–161
branch and plant level, 161–163
interactions between resource use efficiencies, 164
light use efficiency, 163
light/water use efficiency and thinning, 169–170
nutrient use efficiency, 163
water use efficiency, 163–164
whole-plant resource use efficiency and growth analysis, 164–168
environmental influences
nutrients, 203–204
temperature, 200
water, 200–203
forest community and ecosystem levels, 168
ecosystem efficiency, 168–169
growth efficiency, 170–171
tree, 198–199
leaf level
interaction between resource use efficiency, 150–151
light use efficiency, 151
nutrient use efficiency, 151–157
water use efficiency, 157–159
scaling issues, 146
source–sink relationships
Douglas fir (*Pseudotsuga menziesii*), 183–184
loblolly pine (*Pinus taeda*), 186–187
Norway spruce (*Picea abies*), 180–183
radiata pine (*Pinus radiata*), 184–186
slash pine (*Pinus elliottii*), 187–188
source–sink–storage relationships
electrical analog, 189–191
photosynthetic response to remote sink, 192–194
sink activity, storage, and photosynthate, 194–197
source–sink feedback, leaves, 191–192
spatial scales, 150–151
values of
light, 149
nutrient, 149
water, 149–150
Respiration
construction versus maintenance, 257–260
estimation of, using proxy methods, 283–284
below ground respiration costs, 287–288
CO_2 fluxes, 289–290
elemental composition, 286–287
rooting zone variation, 281–283
tissue nitrogen content, 284–286
whole-stand respiration, using eddy correlation, 288–289
foliar, 261–263
roots, 277–281
sapwood, 273–275
scaling from organ to stand, 255–257
stand level, 269–270
temperature effects on, 260–261
measurement techniques, 263–264
respiration rates, 264–265
variation in canopy effects of, 265–266
foliage age, 268
light, 266–267
nitrogen, 268–269
seasonal growth, 267–268
woody tissue
measurement techniques, 277–281
rates, 281
Retention of nutrients, 59, 63–64, 154–155
Root efficiency, optimal versus maximal, 83–84
Root longevity
effects on
water acquisition, 88–91
Roots
adaptations for
accessing soil water, 308–310
water absorption, 311–314
competition effects on biomass, 309
contraction, 304
life history complexity, 82–83
nitrogen uptake, 52
nutrient acquisition, 51–57
carbon expenditure, 76–78
root longevity effects, 87–88
phosphorus uptake, 52–53

response rates of, 277–281
strategies, 75–76
suberization of, 303
water acquisition, 46–51, 54–55
root longevity effects, 88–91

Sap flow
diurnal sources, 330–331
environmental effects, 329–331
specific conductivity of xylem, 340
tree capacitance, 338–340
variability of transpiration, 331–338
Sapwood respiration rates, 274–275
Scaling
CO_2 exchange, 126–130
geographic, 130
needle-to-shoot, photosynthetic light response, 18–23
resource use efficiency, 146, 150–151
structural, 126–127
Soil–plant–atmosphere continuum (SPAC), 301–302
Source–sink feedback regulation, 191–192
Specific conductivity (k_s), 319–322
Stomatal conductance and photosynthesis, 112–113
Structure, *see also* Canopy level
effects on
canopy light requirements, 23–28
light absorption by needles, 5–8
photosynthetic light response of needles, 9–12
shade acclimation of needles, 12–14
interaction with photosynthetic light response at different levels, 18–24
Suberization of roots, 303

Temperature
effects on
photoinhibition, 17–18
photosynthesis, 113–115, 121–122
water uptake, 313–314, 321–322, 327–329
Thinning, effects on light/water use efficiency, 169–170
Tracheids
as conducting units, 315
conducting efficiency, effects of
embolisms, 325
leaf area, 325
root and shoot hydraulic architecture, 326
sapwood area, 325
seedling structure, 326
specific conductivity, 325
dimensions
of conifers, 316
within plant, 316
with plant age, 317
pitting of, 315
structure, 315
vulnerability to cavitation, 318–321
Transpiration, *see* Canopy level; Water; Water stress
Transpiration and water loss in winter, 328
Transpiration and xylem flow, 338–339
Transpiration, factors affecting annual rates, 327
Transpiration of
stands, 343–346
components, 343
interstand comparisons, 346–349
leaf-to-stand scaling, 345–346
trees, 329–343
between-tree uniformity, 340–342
sap flow and capacitance, 338–340
sap flow and specific conductivity effects, 340
seasonal differences, 343–349
variability between species, 342–343
within-tree variability, 331–338

Water
foliage acquisition, 40–41
root acquisition, 40, 46–51
demand effects, 57
nutrient effects, 54
seedling uptake and use, 60–62
stand uptake and use, 63–64
Water conducting systems of conifers, 334–337
Water–nutrient uptake by bark, 43–54
Water stress
avoidance of, 362–364
effects on
growth, 64–65
lipid synthesis and membrane function, 353
nitrogen metabolism, 352–353
osmotic adjustment, 365–367
photosynthesis, 112–113, 119–121, 351–352

plant growth regulators, 353–356
respiration, 352
tolerance to, 364–365
Water transport dynamics, 301
absorption, 301
adaptive water access by roots, 308
angiosperms versus conifers, 310–311
architectural, 308–309
life history effects, 309–310
conducting efficiency, factors influencing
embolisms, 325
leaf area, 325–326
nutrient acquisition, 87–88
root hydraulic architecture, 326
sapwood area, 325
shoot hydraulic architecture and apical control, 326
specific conductivity, 325
embolism reversal, 323–324
environmental effects on root development, 311
root/shoot allocations, 311
soil properties, 311
environmental effects on water absorption
temperature, 313–314
water deficits, 314–315
forest-stand transpiration, 343
components, 343–345
scaling from leaves to stands, 345–346
stand comparisons, 346–349
huber values, 325
hydraulic resistance, soil, soil–root, roots, 306–307
inputs and pathways, the soil–plant–atmosphere continuum, 301–306
internal transport and storage, 315
nutrient/water acquisition, comparisons, 315
organ-to-stand scaling, 255–257
freeze–thaw, 321–322
water stress, 319–321
plant–branch scaling, resource use efficiency, 161–163
resource use efficiency and growth, 164–168
sapwood and heartwood, 317–318
seedling conducting efficiency, 326–327
specific conductivity (k_s), 319–322
storage capacitance, 323–324
thinning effects on light/water use efficiency, 169–170
tracheids as conducting elements, 315
transpiration and sapflow, 329–330
individual trees, 329–330
tree-to-stand transpiration, 329
axial and ray, 317
dimensions, 316–317
unique water stress problems in conifers, 327–329
environmental responses, 329–331
interspecific variability, 342–343
inter-tree variability, 340–342
specific conductivity, 340
within-tree variation, 331–338
water stress, responses of, 349
xylem cavitation, 318–319, due to
lipid synthesis, membrane function, 353
nitrogen metabolism, 352–353
photosynthesis, 351–352
plant growth regulators, 353–356
respiration, 352
Winter hardening and freeze resistance, 17–18

Xylem flow
effects on transpiration, 338–339
for different species, 342–343
inside tree bole, 333–334
Xylem pressure and
embolisms, 319–320
photosynthesis, 113

Physiological Ecology

A Series of Monographs, Texts, and Treatises

Continued from page ii

F. S. CHAPIN III, R. L. JEFFERIES, J. F. REYNOLDS, G. R. SHAVER, and J. SVOBODA (Eds.). Arctic Ecosystems in a Changing Climate: An Ecophysiological Perspective, 1991

T. D. SHARKEY, E. A. HOLLAND, and H. A. MOONEY (Eds.). Trace Gas Emissions by Plants, 1991

U. SEELIGER (Ed.). Coastal Plant Communities of Latin America, 1992

JAMES R. EHLERINGER and CHRISTOPHER B. FIELD (Eds.). Scaling Physiological Processes: Leaf to Globe, 1993

JAMES R. EHLERINGER, ANTHONY E. HALL, and GRAHAM D. FARQUHAR (Eds.). Stable Isotopes and Plant Carbon–Water Relations, 1993

E.-D. SCHULZE (Ed.). Flux Control in Biological Systems, 1993

MARTYN M. CALDWELL and ROBERT W. PEARCY (Eds.). Exploitation of Environmental Heterogeneity by Plants: Ecophysiological Processes Above- and Belowground, 1994

WILLIAM K. SMITH and THOMAS M. HINCKLEY (Eds.). Resource Physiology of Conifers: Acquisition, Allocation, and Utilization, 1994

WILLIAM K. SMITH and THOMAS M. HINCKLEY (Eds.). Ecophysiology of Coniferous Forests, 1994